Sustaining the Earth
An Integrated Approach
EIGHTH EDITION

G. TYLER MILLER, JR.

President, Earth Education and Research

CONTRIBUTING EDITOR

SCOTT SPOOLMAN

THOMSON

BROOKS/COLE

Australia • Canada • Mexico • Singapore • Spain • United Kingdom • United States

THOMSON

BROOKS/COLE

Publisher: *Jack Carey*
Contributing Editor: *Scott Spoolman*
Production Project Manager: *Andy Marinkovich*
Technology Project Manager: *Fiona Chong*
Assistant Editor: *Carol Benedict*
Editorial Assistant: *Kristina Razmara*
Permissions Editor/Photo Researcher: *Abigail Reip*
Marketing Manager: *Kara Kindstrom*
Marketing Assistant/Associate: *Catie Ronquillo*
Advertising Project Manager: *Bryan Vann*
Print/Media Buyer: *Karen Hunt*

Production Management, Copyediting, and Composition: *Thompson Steele, Inc.*
Interior Illustration: *Precision Graphics; Sarah Woodward; Darwin and Vally Hennings; Tasa Graphic Arts, Inc.; Alexander Teshin Associates; John and Judith Walker; Rachel Ciemma; Victor Royer, Electronic Publishing Services, Inc.; J/B Woosley Associates; Kerry Wong, ScEYEnce Studios.*
Cover Image: *Pacific coast tree frog (Hyla regilla)* © Imagestate
Text and Cover Printer: *Thomson West*

For more information about our products, contact us at:
Thomson Learning Academic Resource Center
1-800-423-0563
For permission to use material from this text, contact us by:
Phone: 1-800-730-2214
Fax: 1-800-730-2215
Web: http://www.thomsonrights.com

Library of Congress Control Number: 2006930475

ISBN: 0-495-01597-0

Thomson Higher Education
10 Davis Drive
Belmont, CA 94002-3098
USA

Asia (including India)
Thomson Learning
5 Shenton Way #01-01
UIC Building
Singapore 068808

Australia
Nelson Thomson Learning
102 Dodds Street
South Melbourne, Victoria 3205
Australia

Canada
Nelson Thomson Learning
1120 Birchmount Road
Toronto, Ontario M1K 5G4
Canada

Europe/Middle East/Africa
Thomson Learning
High Holborn House
50/51 Bedford Row
London WC1R 4LR
United Kingdom

Latin America
Thomson Learning
Seneca, 53
Colonia Polanco
11560 Mexico D.F.
Mexico

Spain
Paraninfo Thomson Learning
Calle/Magallanes, 25
28015 Madrid
Spain

BRIEF CONTENTS

Introduction: Learning Skills   1

**PART I: HUMANS
AND SUSTAINABILITY:
AN OVERVIEW**

1   Environmental Problems, Their Causes, and
Sustainability   5

**PART II: SCIENCE, ECOLOGICAL
PRINCIPLES, AND SUSTAINABILITY**

2   Science, Matter, Energy, and Ecosystems:
Connections in Nature   20
3   Evolution and Biodiversity   52
4   Community Ecology, Population Ecology,
and Sustainability   70
5   Applying Population Ecology: The Human
Population   84

**PART III: SUSTAINING
BIODIVERSITY**

6   Sustaining Biodiversity: The Ecosystem
Approach   108
7   Sustaining Biodiversity: The Species
Approach   129

**PART IV: SUSTAINING NATURAL
RESOURCES**

8   Food, Soil, and Pest Management   148
9   Water Resources and Water Pollution   171
10   Energy Resources   201

**PART V: SUSTAINING
ENVIRONMENTAL
QUALITY**

11   Environmental Hazards and Human
Health   237
12   Air Pollution, Climate Change, and Ozone
Depletion   252
13   Solid and Hazardous Waste   283

**PART VI: HUMAN TOOLS FOR
SUSTAINABILITY**

14   Economics, Politics, Worldviews, and
the Environment   300

Supplements   S1

Glossary   G1

Index   I1

DETAILED CONTENTS

Introduction: Learning Skills   1

**PART I: HUMANS AND
SUSTAINABILITY: AN OVERVIEW**

**1   Environmental Problems, Their
Causes, and Sustainability   5**

Living More Sustainably   5

Population Growth, Economic Growth, and
Economic Development   7

Resources   9

Pollution   11

Environmental Problems: Causes and
Connections   12

Cultural Changes and the Environment   15

Sustainability and Environmental
Worldviews   15

**PART II: SCIENCE, ECOLOGICAL
PRINCIPLES, AND SUSTAINABILITY**

**2   Science, Matter, Energy, and
Ecosystems: Connections
in Nature   20**

The Nature of Science   20

Matter and Energy   23

Earth's Life-Support Systems: From
Organisms to the Biosphere   28

**Case Study: What Species Rule the
World?   28**

**Case Study: Have You Thanked the Insects
Today?   30**

Energy Flow in Ecosystems   37

Soil: A Renewable Resource   41

Matter Cycling in Ecosystems   44

3   Evolution and Biodiversity   52

Evolution and Adaptation   52

Ecological Niches and Adaptation   55

*Spotlight: Cockroaches: Nature's Ultimate
Survivor*   55

Speciation, Extinction, and Biodiversity   56

The Future of Evolution   58

Biomes: Climate and Life on Land   60

Life in Water Environments   64

**4   Community Ecology, Population
Ecology, and Sustainability   70**

Types of Species   70

**Case Study: Why Are Amphibians
Vanishing?   70**

**Case Study: Why Should We Protect
Sharks?   72**

Species Interactions   73

Ecological Succession: Communities in
Transition   75

Population Dynamics and Carrying
Capacity   77

Human Impacts on Natural Systems:
Learning from Nature   80

**5   Applying Population Ecology: The
Human Population   84**

Human Population Growth   84

Factors Affecting Human Population Size   85

**Case Study: Fertility and Birth Rates in the
United States   86**

**Case Study: U.S. Immigration (Economics and
Politics)   88**

Population Age Structure   89

Solutions: Influencing Population Size   92

Slowing Population Growth   94

Case Study: India   94

Case Study: China   95

Population Distribution: Urban Growth and
Problems   96

**Case Study: Urbanization in the United States
(Economics)   97**

Urban Environmental and Resource
Problems   99

**Case Study: Mexico City (Science, Politics
and Poverty)   101**

Transportation and Urban Development   102

Case Study: Motor Vehicles in the United States (Economics and Politics) 102

Making Urban Areas More Sustainable and Desirable Places to Live 104

Case Study: The Ecocity Concept in Curitiba, Brazil 106

PART III: SUSTAINING BIODIVERSITY

6 Sustaining Biodiversity: The Ecosystem Approach 108

Human Impacts on Biodiversity 108

Managing and Sustaining Forests 109

Forest Resources and Management in the United States 113

Individuals Matter: Butterfly in a Redwood Tree 114

Tropical Deforestation 115

Individuals Matter: Kenya's Green Belt Movement 118

National Parks 118

Case Study: Stresses on U.S. National Parks 118

Case Study: Reintroducing Wolves to Yellowstone 119

Nature Reserves 120

Case Study: Costa Rica—A Global Conservation Leader (Science, Politics, and Economics) 121

Case Study: Wilderness Protection in the United States (Science and Politics) 123

Ecological Restoration 123

Case Study: Ecological Restoration of a Tropical Dry Forest in Costa Rica (Science and Stewardship) 124

Sustaining Aquatic Biodiversity 125

What Can We Do? 127

7 Sustaining Biodiversity: The Species Approach 129

Species Extinction 129

Case Study: The Passenger Pigeon—Gone Forever 129

Importance of Wild Species 132

Case Study: Why Should We Care about Bats? (Science and Economics) 133

Causes of Premature Extinction of Wild Species 134

Case Study: A Disturbing Message from the Birds (Science) 136

Protecting Wild Species: The Legal and Economic Approaches 141

Case Study: Accomplishments of the Endangered Species Act (Science, Economics, and Politics) 144

Protecting Wild Species: The Sanctuary Approach 144

Case Study: Using Reconciliation Ecology to Protect Bluebirds (Science and Stewardship) 146

PART IV: SUSTAINING NATURAL RESOURCES

8 Food, Soil, and Pest Management 148

Food Security and Nutrition 148

Food Production 149

Case Study: Industrial Food Production in the United States (Science and Economics) 150

Soil Erosion and Degradation 152

Case Study: Soil Erosion in the United States (Science and Economics) 153

Sustainable Agriculture through Soil Conservation 154

The Green Revolution and its Environmental Impact 156

The Gene Revolution 158

Producing More Meat 160

Catching and Raising More Fish and Shellfish 161

Solutions: Toward Global Food Security 163

Protecting Food Resources: Pest Management 164

Individuals Matter: Rachel Carson 165

Spotlight: What Goes Around Can Come Around (Science, Economics and Politics) 167

Case Study: Integrated Pest Management—A Component of Sustainable Agriculture (Science) 168

Solutions: Sustainable Agriculture 169

9 Water Resources and Water Pollution 171

Water's Importance, Availability, and Renewal 171

Case Study: Freshwater Resources in the United States 173

Water Resource Problems: Too Little Water and Too Much Water 174

Case Study: Water Conflicts in the Middle East—A Growing Problem   174

Case Study: Living on Floodplains in Bangladesh (Science and Poverty)   176

Supplying More Water   178

Case Study: The California Experience (Science, Economics, and Politics)   180

Case Study: The Aral Sea Disaster (Science and Economics)   181

Increasing Water Supplies by Reducing Water Waste   182

Water Pollution: Sources, Types and Effects   186

Pollution of Freshwater Streams, Lakes, and Aquifers   188

Case Study: Pollution in the Great Lakes (Science, Economics, and Politics)   190

Ocean Pollution   192

Case Study: The Chesapeake Bay (Science, Economics, and Politics)   193

Preventing and Reducing Surface Water Pollution   195

Drinking Water Quality   198

10   Energy   201

Evaluating Energy Resources   201

Nonrenewable Fossil Fuels   203

Case Study: U.S. Oil Supplies (Science, Economics, and Politics)   205

Nuclear Energy   210

Case Study: The Chernobyl Nuclear Power Plant Accident (Science and Human Error)   214

Case Study: High-Level Radioactive Wastes in the United States (Science, Economics, and Politics)   217

Reducing Energy Waste and Improving Energy Efficiency   219

Using Renewable Energy to Provide Heat and Electricity   224

Geothermal Energy   231

Hydrogen   231

A Sustainable Energy Strategy   233

PART V: SUSTAINING ENVIRONMENTAL QUALITY

11   Environmental Hazards and Human Health   237

Risks and Hazards   237

Case Study: Death from Smoking (Science)   237

Biological Hazards: Disease in Developed and Developing Countries   238

Case Study: The Growing Global Threat from Tuberculosis (Science)   240

Case Study: The Global HIV/AIDS Epidemic   241

Case Study: Malaria (Science)   241

Chemical Hazards   243

Toxicology: Assessing Chemical Hazards   244

Risk Analysis   247

12   Air Pollution, Climate Change, and Ozone Depletion   252

Atmosphere, Weather, and Climate   252

Air Pollution   255

Case Study: Radioactive Radon Gas (Science)   260

Harmful Effects of Air Pollution   262

Preventing and Reducing Air Pollution   262

Climate Change and Human Activities   266

Factors Affecting the Earth's Temperature   270

Effects of Global Warming   271

Dealing with Global Warming   274

Ozone Depletion in the Stratosphere   278

Case Study: Skin Cancer (Science)   280

Protecting the Ozone Layer   281

Individuals Matter: Ray Turner and His Refrigerator   281

13   Solid and Hazardous Waste   283

Wasting Resources   283

Integrated Waste Management   284

Reuse   285

Case Study: Using Refillable Containers (Science and Economics)   286

Recycling   287

Case Study: Recycling Plastics (Science and Economics)   288

Burning and Burying Solid Waste   289

Hazardous Waste   292

Case Study: The Threat from Lead (Science)   295

Achieving a Low-Waste Society   296

PART VI: HUMAN TOOLS FOR SUSTAINABILITY

14 Economics, Politics, Worldviews, and the Environment 300

Economic Systems and Sustainability 300

Using Economics to Improve Environmental Quality 302

Individuals Matter: Ray Anderson 306

Reducing Poverty to Improve Environmental Quality and Human Well-Being 306

Solutions: Microloans to the Poor 307

Politics and Environmental Policy 309

Case Study: Managing Public Lands in the United States—Politics in Action 312

Case Study: Environmental Action by Students in the United States 316

Case Study: Threats to the U.S. Environmental Legal and Regulatory Structure 316

Global Environmental Policy 317

Environmental Worldviews and Ethics: Clashing Values and Cultures 318

Living More Sustainably 320

Supplements

1   Measurement Units S1
2   Major Events in U.S. Environmental History S3
3   Brief History of the Age of Oil S8
4   How to Analyze a Scientific Article S9

Glossary G1

Index I1

PREFACE FOR INSTRUCTORS

Sustainability as the Central Theme with Five Major Subthemes

Sustainability is the central theme of this introductory environmental science textbook, as shown in the Brief Contents on p. iii. Five major subthemes—natural capital, natural capital degradation, solutions, trade-offs, and how individuals matter—guide the way to sustainability. Figure 1-1 (p. 5) is a key figure that shows a path to sustainability based on these subthemes.

- *Natural capital.* *Sustainability* requires a focus on preserving *natural capital*—the natural resources and natural services that support all life and economies (Figure 1-2, p. 6). Diagrams in the book illustrate this and other subthemes. Examples are Figures 3-9, (p. 63); 6-3 (p. 109); and 7-4 (p. 133).

- *Natural capital degradation.* Certain human activities lead to *natural capital degradation*—the second subtheme, also described in text and illustrated. Examples are Figures 3-10 (p. 64); 6-5 and 6-6 (p. 111); and 12-5 (p. 258).

- *Solutions.* The next step is a search for *solutions* to natural capital degradation and other environmental problems. Many chapter sections, subsections, and figures present proven and possible solutions. Proposed solutions are presented in a balanced manner, and I challenge students to use critical thinking to evaluate them. Examples are Figures 5-21 (p. 106); 9-13 (p. 180); and 12-10 and 12-11 (p. 264).

- *Trade-offs.* The search for solutions involves *trade-offs*—the fourth subtheme—because any solution involves weighing advantages against disadvantages. Many chapters contain *Trade-Offs* diagrams, which present advantages and disadvantages of various environmental technologies and solutions to environmental problems. Examples are Figures 5-17 through 5-19 (p. 104); 6-8 (p. 112); and 10-8 (p. 208).

- *Individuals matter* Boxes entitled *Individuals Matter* describe what some scientists and concerned citizens have done to help us achieve sustainability in dealing with environmental problems. (See pages 118, 281, and 306). Also, *What Can You Do?* boxes describe how readers can deal with various environmental problems. Examples are Figures 6-16 (p. 123); 9-20 (p. 186); and 10-41 (p. 235).

Science-Based, Interdisciplinary, and Global in Scope

The path to sustainability is supported throughout this text by *sound* or *consensus* science—concepts and explanations widely accepted by scientists. In addition, material from economics, political science, and environmental ethics is integrated throughout. Subsection headings are tagged with labels such as *Economics and Politics, Science and Economics,* and *Science and Ethics* (pages 88, 108, and 281) to show this interdisciplinary approach.

This book assumes a global perspective on two levels. First, ecological principles reveal how all the world's life is connected and sustained within the biosphere (Chapter 2). Second, the book integrates information and images from around the world into its presentation. To emphasize this feature I have added the label *Global Outlook* to many of the book's subsection headings (pages 115, 189, and 306) and Figures 5-6 (p. 90); 8-9 (p. 157); and 11-5 and 11-6 (p. 241).

Flexible, Current, and Concise

A *flexible format* allows instructors to use this book with almost any course outline. I suggest that instructors use Chapter 1, to provide an overview of environmental problems and solutions, and Chapters 2 through 5 to provide a base of scientific principles and concepts. The remaining chapters—6 through 14—can be used in virtually any order or omitted as desired. In addition, sections within chapters can be omitted or rearranged to meet instructor needs.

This book has 47 Case Studies (see topics in bold-faced type in the Detailed Contents, pp. v–viii), and 25 *Guest Essays* are available on the website for this book. These features also add flexibility by providing more depth on specific environmental problems and their possible solutions.

Lending even more flexibility is *ThomsonNOW*, a new feature of this edition that allows students to enhance their scientific understanding by viewing animations, many of them interactive, available on the website for this book. There are 86 *ThomsonNOW* items for this text, some of them related to figures (pp. 29, 37, and 92) and others to text (pp. 49, 85, and 257).

This book is *current*. My recent research for this and my other three books in the field included over 4,000 updates based on information and data published in 2003, 2004, 2005, and 2006. Dozens of new or expanded topics include impacts of new affluent consumers in China and India (pp. 94–96); ecological footprints (pp. 10, 86, and 125); the 2005 Millennium Ecosystem Report (pp. 15, 37, and 125); biodiversity (defined on p. 16 and covered extensively in Chapters 3 (pp. 52–69), 6, and 7 (pp. 108–147)); and many more.

And finally, this text is *concise*—with 14 chapters covered in 323 pages—and *economical*. We have lowered the book's cost by reducing its size, printing it in black and white, and using a soft cover. Instructors who want a book covering this material with a different emphasis and length can use one of my three other books written for various types of environmental science courses: *Living in the Environment*, fifteenth edition (784 pages, Brooks/Cole, 2007), *Environmental Science*, eleventh edition (436 pages, Brooks/Cole, 2007), and *Essentials of Ecology*, fourth edition (528 pages, Brooks/Cole, 2007).

In-Text Study Aids

Most chapter titles are accompanied by labels appearing within shaded triangles representing the natural resources and unshaded triangles representing the natural services that are discussed in the chapters. When a new term is introduced and defined, it is printed in boldface type. A glossary of all key terms is located at the end of the book.

Each subsection heading is followed by a single sentence that summarizes the key material in each subsection. This provides readers with a built-in running summary of the material. I believe this is a more effective learning tool than a chapter-ending summary, but students can certainly follow each chapter reading with a review of the running summary.

Critical thinking is emphasized throughout this book, starting with the introduction on *Learning Skills* (describing critical thinking skills on pp. 3–4). Critical thinking questions appear in many boxes and figure captions. The 58 *How Would You Vote?* exercises (see pp. 88, 210, and 263) also serve as critical thinking exercises. Students can register their votes on-line and compare their responses with those of other students.

This book makes great use of *visual learning*. Its 286 illustrations, updated and refined in each edition, are designed to present complex ideas in understandable ways and to relate learning to the real world.

Each chapter ends with a set of *Critical Thinking* questions. The website for the book also contains a set of *Review Questions* covering *all* of the material in the chapter and useful as a study guide for students. Some instructors download this off the website and give students a list of the particular questions they expect their students to answer.

Online Study Aids

The *website* for this book at **www.thomsonedu.com /biology/miller** contains study aids and many ideas for further reading and research. For each chapter, there is a summary, review questions, flash cards for key terms and concepts, a multiple-choice practice quiz, interesting Internet sites, references, information on green careers, guest essays, and a guide for accessing thousands of InfoTrac® College Edition articles.

Qualified adopters of this textbook also have access to *WebTutor Advantage* on *WebCT* at

http://webtutor.thomsonlearning.com

It provides access to a full array of study tools, including flashcards (with audio), practice quizzes, online tutorials, and web links.

Teachers and students using *new* copies of this textbook also have free and unlimited access to *InfoTrac*® *College Edition*. This fully searchable online library gives users access to complete environmental articles from several hundred periodicals dating back over the past 24 years.

Other student learning tools include:

- *Audio version for study and review*. Students can listen to readings of chapters in this book while walking, traveling, sitting in their rooms, or working out. New editions of the book can include pin code access to the audiobook, if requested by the instructor. Students can use the pin code enclosed in every new copy of this book to download book chapters free from the Web to any MP3 player.

- Nearly 100 animations, many of them interactive, based on figures from the text and correlated by chapter, are available for use by students at ThomsonNOW.

- *Essential Study Skills for Science Students* by Daniel D. Chiras. This book includes chapters on developing good study habits, sharpening memory, getting the most out of lectures, labs, and reading assignments, improving test-taking abilities, and becoming a critical thinker. Instructors can have this book bundled FREE with the textbook.

- *Laboratory Manual* (fourth edition), by C. Lee Rockett and Kenneth J. Van Dellen. This manual includes a variety of laboratory exercises, workbook exercises, and projects that require a minimum of sophisticated equipment.

Supplements for Instructors

■ *PowerLecture CD-ROM.* This CD-ROM—free to adopters—makes it easy to create custom lectures. Each chapter's Microsoft® PowerPoint presentation consolidates all relevant resources—illustrations, photographs, animations, InfoTrac® College Edition references—into one powerful class tool. This program's editing tools allow use of slides from other lectures, modification or removal of figure labels and leaders, insertion of your own slides, saving slides as JPEG images, and preparation of lectures for use on the Web. The CD-ROM also includes Microsoft® Word files of the Instructor's Manual and Test Bank.

■ *Transparency Masters and Acetates.* Includes 100 color acetates of line art from Miller's textbooks and over 400 black-and-white master sheets of key diagrams for making overhead transparencies. Free to adopters.

■ *CNN^{TM} Today Videos.* These informative news stories, available either on VHS tapes or CD-ROMs while supplies last, contain close to 40 two- to three-minute video clips of current news stories on environmental issues from around the world. Student Workbook Included.

■ *ABC News: Environmental Science in the Headlines* (2005, 2006) on DVD. Thomson Brooks/Cole has expanded its portfolio of videos with numerous current segments from ABC News. These two- to three-minute video clips are available on DVD. Student Workbook Included.

■ *Instructor's Manual with Test Items.* Free to adopters.

■ *ExamView.* Allows an instructor to easily create and customize tests, see them on the screen exactly as they will print, and print them out.

Help Me Improve This Book

Let me know how you think this book can be improved; if you find any errors, bias, or confusing explanations, please e-mail them to me at

mtg89@hotmail.com

Most errors can be corrected in subsequent printings of this edition, as well as in future editions.

Acknowledgments

I wish to thank the many students and teachers who responded so favorably to the seven previous editions of *Sustaining the Earth,* the fourteen editions of *Living in the Environment,* and the eleven editions of *Environmental Science,* and who corrected errors and offered many helpful suggestions for improvement. I am also deeply indebted to the more than 250 reviewers, who pointed out errors and suggested many important improvements in the various editions of these three books. Any remaining errors and deficiencies are mine.

I am particularly indebted to Scott Spoolman, who served as a contributing editor for this new edition and who made numerous suggestions for improving this book. The members of the talented production team, listed on the copyright page, have made vital contributions as well. My thanks also go to production editors Andy Marinkovich and Nicole Barone, copy editor Andrea Fincke, layout expert Bonnie Van Slyke, artist Patrick Lane, Brooks/Cole's hard-working sales staff, and Keli Amann, Fiona Chong, and the other members of the talented team who developed the multimedia, website, and advertising materials associated with my books.

I also thank C. Lee Rockett and Kenneth J. Van Dellen for developing the *Laboratory Manual* to accompany this book; Jane Heinze-Fry for her work on concept mapping, *Environmental Articles,* and *Critical Thinking and the Environment: A Beginner's Guide;* and Kelly West for her excellent work on the *Instructor's Manual.*

My deepest thanks go to Jack Carey, biology publisher at Brooks/Cole, for his encouragement, help, 40 years of friendship, and superb reviewing system. It helps immensely to work with the best and most experienced editor in college textbook publishing.

I dedicate this book to the earth that sustains us and to Kathleen Paul Miller, my wife and research associate.

G. Tyler Miller, Jr.

Guest Essayists

Guest Essays by the following authors are available online at the website for this book: **M. Kat Anderson,** ethnoecologist with the National Plant Center of the USDA's Natural Resource Conservation Service; **Lester R. Brown,** president, Earth Policy Institute; **Alberto Ruz Buenfil,** environmental activist, writer, and performer; **Robert D. Bullard,** professor of sociology and director of the Environmental Justice Resource Center at Clark Atlanta University; **Michael Cain,** ecologist and adjunct professor at Bowdoin College; **Herman E. Daly,** senior research scholar at the school of Public Affairs, University of Maryland; **Lois Marie Gibbs,** director, Center for Health, Environment, and Justice; **Garrett Hardin,** professor emeritus (now deceased) of human ecology,

University of California, Santa Barbara; **John Harte,** professor of energy and resources, University of California, Berkeley; **Paul G. Hawken,** environmental author and business leader; **Jane Heinze-Fry,** environmental educator; **Paul F. Kamitsuja,** infectious disease expert and physician; **Amory B. Lovins,** energy policy consultant and director of research, Rocky Mountain Institute; **Bobbi S. Low,** professor of resource ecology, University of Michigan; **Lester W. Milbrath,** director of the research program in environment and society, State University of New York, Buffalo; **Peter Montague,** director, Environmental Research Foundation; **Norman Myers,** tropical ecologist and consultant in environment and development; **David W. Orr,** professor of environmental studies, Oberlin College; **Noel Perrin,** adjunct professor of environmental studies, Dartmouth College; **John Pichtel,** Ball State University; **David Pimentel,** professor of insect ecology and agricultural sciences, Cornell University; **Andrew C. Revkin,** environmental author and environmental reporter for the *New York Times;* **Vandana Shiva,** physicist, educator, and environmental consultant; **Nancy Wicks,** ecopioneer and director of Round Mountain Organics; and **Donald Worster,** environmental historian and professor of American history, University of Kansas.

Introduction: Learning Skills

Students who can begin early in their lives to think of things as connected, even if they revise their views every year, have begun the life of learning.

MARK VAN DOREN

Why Is It Important to Study Environmental Science?

Environmental science may be the most important course you will ever take.

Welcome to **environmental science**—an *interdisciplinary* study of how the earth works, how we interact with the earth, and how we can deal with the environmental problems we face.

Environmental issues affect every part of your life. Thus, the concepts, information, and issues discussed in this book and the course you are taking should be useful to you now and throughout your life.

In 1966, I heard Dean Cowie, a physicist with the U.S. Geological Survey, give a lecture on the problems of population growth and pollution. Afterward, I went to him and said, "If even a fraction of what you have said is true, I will feel ethically obligated to give up my research on the corrosion of metals and devote the rest of my life to research and education on environmental problems and solutions. Frankly, I do not want to change my life, and I am going into the literature to try to show that your statements are either untrue or grossly distorted."

After six months of study I was convinced of the seriousness of these and other environmental problems. Since then, I have been studying, teaching, and writing about them. This book summarizes what I have learned in more than four decades of trying to understand environmental principles, problems, connections, and solutions.

Understandably, I am biased. But *I strongly believe that environmental science is the single most important course in your education.* What could be more important than learning how the earth works, how we are affecting its life support system, and how we can reduce our environmental impact?

We live in an incredibly challenging era. There is a growing awareness that during this century we need to make a new cultural transition in which we learn how to live more sustainably by not degrading our life support system. I hope this book will stimulate you to become involved in this change in the way we view and treat the earth that sustains us, other life, and all economies.

Improving Your Study and Learning Skills

Learning how to learn is life's most important skill.

Maximizing your ability to learn ought to be one of your most important lifetime educational goals. This involves continually trying to *improve your study and learning skills.* Here are some general *suggestions for starting that process.*

Get organized. Becoming more efficient at studying gives you more time for other interests.

Write daily to-do lists. Put items in order of importance, focus on the most important tasks, and assign a time to work on these items. Because life is full of uncertainties, you might not accomplish all the goals on your daily list. Shift your schedule as needed to handle the most important items.

Set up a study routine in a distraction-free environment. Develop a written daily study schedule and stick to it. Study in a quiet, well-lighted space. Work sitting at a desk or table—not lying down on a couch or bed. Take breaks every hour or so. During each break, take several deep breaths and move around to help you stay more alert and focused.

Avoid procrastination—putting work off until another time. Do not fall behind on your reading and other assignments. Accomplish this by setting aside a particular time for studying each day and making it a part of your daily routine.

Do not eat dessert first. Otherwise, you may never get to the main meal (studying). When you have accomplished your study goals then reward yourself with play (dessert).

Make hills out of mountains. It is psychologically difficult to climb a mountain such as reading an entire book, reading a chapter in a book, writing a paper, or cramming to study for a test. Instead, break such large tasks (mountains) down into a series of small tasks (hills). Each day read a few pages of the book or chapter, write a few paragraphs of the paper, and review what you have studied and learned. As American automobile designer and builder Henry Ford put it, "Nothing is particularly hard if you divide it into small jobs."

Look at the big picture first. Get an overview of an assigned reading by looking at the main headings or chapter outline.

Ask and answer questions as you read. For example, what is the main point of this section or paragraph? To help you do this I follow the heading of each subsection with a single sentence (in italics) that describes the main idea of the subsection. You can use these short descriptions as a way to review what you have learned. Putting them all together gives you a summary of the chapter's key ideas. I find this running summary to be a more useful learning device than a relatively dense summary at the end of each chapter.

Focus on key terms. Use the glossary in your textbook to look up the meaning of terms or words you do not understand. This book shows all key terms in **boldfaced** type and lesser, but still important, terms in *italicized* type. Flash cards for testing your mastery of key terms for each chapter are also available on the website for this book, or you can make your own by putting a term on one side of a piece of paper and its meaning on the other side.

Interact with what you read. I do this by marking key sentences and paragraphs with a highlighter or pen. I put an asterisk in the margin next to an idea I think is important and double asterisks next to an idea I think is especially important. I write comments in the margins, such as *Beautiful, Confusing, Misleading,* or *Wrong.* I fold down the top corner of pages with highlighted passages and the top and bottom corners of especially important pages. This way, I can flip through a chapter or book and quickly review the key ideas.

Use the audio version of this book for study and review. You can listen to readings of the chapters in this book while walking, traveling, sitting in your room, or working out. Use the pin code enclosed in any new copy of this book to download book chapters free from the Web to any MP3 player.

Know your learning style. If you're a *visual learner,* you learn best from reading and viewing illustrations and diagrams. If you are an *auditory learner,* you may learn best by reading aloud while studying, tape recording lectures, using the audio version of this book as a study supplement, and discussing what you read with others.

Review to reinforce learning. Before each class, review the material you learned in the previous class and read the assigned material.

Become a better note taker. Do not try to take down everything your instructor says. Instead, take down main points and key facts using your own shorthand system. Review, fill in, and organize your notes as soon as possible after each class.

Write out answers to questions to focus and reinforce learning. Answer all questions at the end of each chapter, or those assigned to you, and the review questions on the website for each chapter. Do this in writing as if you were turning it in for a grade. Save your answers for review and preparation for tests.

Use the buddy system. Study with a friend or become a member of a study group to compare notes, review material, and prepare for tests. Explaining something to someone else is a great way to focus your thoughts and reinforce your learning. Attend any review sessions that might be offered by instructors or teaching assistants.

Learn your instructor's test style. Does your instructor emphasize multiple choice, fill-in-the-blank, true-or-false, factual, thought, or essay questions? How much of the test will come from the textbook and how much from lecture material? Adapt your learning and studying methods to this style. You may disagree with this style and feel that it does not adequately reflect what you know. But the reality is that your instructor is in charge.

Become a better test taker. Avoid cramming. Eat well and get plenty of sleep before a test. Arrive on time or early. Calm yourself and increase your oxygen intake by taking several deep breaths. Do this about every 10–15 minutes. Look over the test and answer the questions you know well first. Then work on the harder ones. Use the process of elimination to narrow down the choices for multiple-choice questions. Getting it down to two choices gives you a 50% chance of guessing the right answer if necessary. For essay questions, organize your thoughts before you start writing. If you have no idea what a question means, make an educated guess. You might get some partial credit and avoid a zero. Another strategy for getting some credit is to show your knowledge and reasoning by writing: "If this question means so and so, then my answer is _____."

Develop an optimistic outlook. Try to be a "glass is half-full" rather than a "glass is half-empty" person. Pessimism, fear, anxiety, and excessive worrying (especially about things you have no control over) are destructive and lead to inaction. Try to keep your energizing feelings of optimism slightly ahead of any immobilizing feelings of pessimism. Then you will always be moving forward.

Take time to enjoy life. Every day, take time to laugh and enjoy nature, beauty, and friendship. Becoming an effective and efficient learner is the best way to do this without getting behind and living under a cloud of guilt and anxiety.

Improving Your Critical Thinking Skills: Detecting Baloney

Learning how to think critically is a skill you will need throughout your life.

Critical thinking involves developing skills to help you analyze and evaluate the validity of information and ideas you are exposed to and to make decisions. Critical thinking helps you distinguish between facts and opinions, evaluate evidence and arguments, take and defend informed positions on issues, integrate information and see relationships, and apply your knowledge to dealing with new and different problems. Here are some basic skills for learning how to think more critically.

Question everything and everybody. Be skeptical, as any good scientist is. Do not believe everything you hear or read, including the content of this textbook, without evaluating the information you receive. Seek other sources and opinions. As the famous physicist and philosopher Albert Einstein put it, "The important thing is not to stop questioning."

Identify and evaluate your personal biases and beliefs. Each of us has biases and beliefs taught to us by our parents, teachers, friends, role models, and experience. What are your basic beliefs and biases? Where did they come from? What assumptions are they based on? How sure are you that your beliefs and assumptions are right, and why? According to the American psychologist and philosopher William James, "A great many people think they are thinking when they are merely rearranging their prejudices."

Be open-minded and flexible. Be open to considering different points of view. Suspend judgment until you gather more evidence, and be capable of changing your mind. Recognize that there may be a number of useful and acceptable solutions to a problem and that very few issues are black or white. One way to evaluate divergent views is to get into another person's head. How do they see or view the world? What are their basic assumptions and beliefs? Is their position logically consistent with their assumptions and beliefs?

Be humble about what you know. Some people are so confident in what they know that they stop thinking and questioning. To paraphrase American writer Mark Twain, "It's not what we don't know that's so bad. It's what we know is true, but just ain't so, that hurts us." Or as philosopher Will Durant put it, "Education is a progressive discovery of our own ignorance."

Evaluate how the information related to an issue was obtained. Are the statements you heard or read based on firsthand knowledge or research or on hearsay? Are unnamed sources used? Is the information based on reproducible and widely accepted scientific studies (*sound* or *consensus science,* as discussed on p. 21) or preliminary scientific results that may be valid but need further testing (*frontier science,* p. 21)? Is the information based on a few isolated stories or experiences (*anecdotal information*) or on carefully controlled studies? Is it based on unsubstantiated and widely doubted scientific information or beliefs (*junk science* or *pseudoscience,* p. 21)?

Question the evidence and conclusions presented. What are the conclusions or claims? What evidence is presented to support them? Does the evidence support them? Is there a need to gather further evidence to test the conclusions? Are there other, more reasonable conclusions?

Try to uncover differences in basic beliefs and assumptions. On the surface, most arguments or disagreements involve differences in opinions about the validity or meaning of certain facts or conclusions. Scratch a little deeper and you will find that most disagreements are usually based on different (and often hidden) basic assumptions about how we look at and interpret the world around us. Uncovering these basic differences can allow the parties involved to understand where each is "coming from" and agree to disagree about their basic assumptions or principles.

Try to identify and assess the assumptions and beliefs of those presenting evidence and drawing conclusions. What is their expertise in this area? Do they have any unstated assumptions, beliefs, biases, or values? Do they have a personal agenda? Can they benefit financially or politically from acceptance of their evidence and conclusions? Would investigators with different basic assumptions or beliefs take the same data and come to different conclusions?

Expect and tolerate uncertainty. Recognize that science is an ever-changing adventure that provides only a degree of certainty. And the more complex the system or process being investigated, the greater the degree of uncertainty. Scientists can disprove things but they cannot establish absolute proof or certainty.

Do the arguments used involve common logical fallacies or debating tricks? Here are six examples. *First,* attack the presenter of an argument rather than the argument itself. *Second,* appeal to emotion rather than facts and logic. *Third,* claim that if one piece of evidence or conclusion is false, then all other pieces of evidence and conclusions are false. *Fourth,* say that a conclusion is false because it has not been scientifically proven. (Scientists never prove anything absolutely but they can establish degrees of reliability, as discussed on p. 22.) *Fifth,* inject irrelevant or misleading information to divert attention from important points. *Sixth,* present only either/or alternatives when there may be a number of alternatives.

Do not believe everything you read on the Internet. The Internet is a wonderful and easily accessible source of information. It is also a useful way to find alternative information and opinions on almost any subject or issue—much of it not available in the mainstream media and scholarly articles. Web logs, or blogs, have become a major source of information, even more important than standard news media for some people. However, because the Internet is so open, anyone can post anything they want to a blog or other website with no editorial control or *peer review*—the method in which scientists and other experts review and comment on an article before it is accepted for publication in a scholarly journal. As a result, evaluating information on the Internet is one of the best ways in which you can put into practice the principles of critical thinking discussed here. Use and enjoy the Internet, but think critically and proceed with caution.

Develop principles or rules for evaluating evidence. Develop a written list of principles, concepts, and rules to serve as guidelines for evaluating evidence and claims and for making decisions. Continually evaluate and modify this list on the basis of your experience.

Become a seeker of wisdom, not a vessel of information. Many people believe that the main goal of education is to learn as much as you can by concentrating on gathering more and more information—much of it useless or misleading. I believe that the primary goal of education is to learn how to sift through mountains of facts and ideas to find the few *nuggets of wisdom* that are the most useful for understanding the world and for making decisions. This book is full of facts and numbers, but they are useful only to the extent that they lead to an understanding of useful ideas, scientific laws, concepts, principles, and connections. A major goal of the study of environmental science is to find out how nature works and sustains itself (*environmental wisdom*) and to use *principles of environmental wisdom* to help make our societies and economies more sustainable, more just, and more beneficial and enjoyable for all. As writer Sandra Carey put it, "Never mistake knowledge for wisdom. One helps you make a living; the other helps you make a life."

Critical thinking involves trying to separate useful from useless information. You will find critical thinking questions throughout this book—at the end of each chapter and within figure captions—that prod you to evaluate various solutions to environmental problems. There are no right or wrong answers to many of these questions, although experience may show that some answers are better than others. A good way to improve your critical thinking skills is to compare your answers with those of classmates and to discuss how you arrived at your answers.

Trade-Offs

There are no simple answers to the environmental problems we face.

There are always *trade-offs* involved in making and implementing environmental decisions. My challenge is to give a fair and balanced presentation of different viewpoints, advantages and disadvantages of various technologies and proposed solutions to problems, and good and bad news about the environment without injecting personal bias.

Studying a subject as important as environmental science and ending up with no conclusions, opinions, and beliefs would mean that both the teacher and student had failed. However, conclusions we do form must be based on sound science and should be reached only through critical thinking to evaluate different ideas and to understand the trade-offs involved.

Help Me Improve This Book

I welcome your help in improving this book.

Researching and writing a book that covers and connects ideas in such a wide variety of disciplines is a challenging and exciting task. Almost every day I learn about some new connection in nature.

In a book this complex, there are bound to be some errors—some typographical mistakes that slip through and some statements that you might question based on your knowledge and research. My goal is to provide you with an interesting, accurate, balanced, and challenging book that furthers your understanding of this vital subject.

I invite you to contact me and point out any bias, errors, and ways to improve this book. Over decades of teaching, my students and readers of my textbooks have been some of my best teachers. Please e-mail your suggestions to me at **mtg8@hotmail.com**.

Now start your journey into this fascinating and important study of how the earth works and how we can leave the planet in at least as good a shape as we found it. Have fun.

Study nature, love nature, stay close to nature. It will never fail you.

FRANK LLOYD WRIGHT

1 ENVIRONMENTAL PROBLEMS, THEIR CAUSES, AND SUSTAINABILITY

Alone in space, alone in its life-supporting systems, powered by inconceivable energies, mediating them to us through the most delicate adjustments, wayward, unlikely, unpredictable, but nourishing, enlivening, and enriching in the largest degree—is this not a precious home for all of us? Is it not worth our love?

BARBARA WARD AND RENÉ DUBOS

LIVING MORE SUSTAINABLY

What Is Environmental Science?

Environmental science is a study of how the earth works, how we interact with the earth, and how to deal with environmental problems.

The **environment** is the sum total of all living and non-living things that affect any organism. We are part of, not apart from, the environment around us.

This textbook is an introduction to **environmental science,** an interdisciplinary study that uses information and ideas from the physical and social sciences to learn how nature works, how we interact with the environment, and how we can live sustainably without degrading it. To show how information and ideas from the physical and social sciences are integrated, many subheadings throughout this book are tagged with the terms *science, economics, politics,* and *ethics* or some combination of these words.

A basic tool used by environmental scientists is **ecology,** a biological science that studies the relationships between living organisms and their environment. Environmental science and ecology should not be confused with each other or with **environmentalism,** a social movement dedicated to protecting the earth's life support systems for us and other species. It is political in nature, involving activities such as working to pass and enforce environmental laws and protesting harmful environmental actions.

We need to understand how the environment works and how we interact with it for three reasons. *First,* natural systems in the environment provide clean air and water, soil for producing food, and everything else we need to survive. *Second,* we need to be aware of how our actions affect these systems. *Third,* understanding this is necessary for developing solutions to the environmental problems we face.

A Path to Sustainability: Themes of This Book

Sustainability, the central theme of this book, is built on the subthemes of natural capital, natural capital degradation, solutions, trade-offs, and how individuals matter.

Sustainability is the ability of earth's various systems, including human cultural systems and economies, to survive and adapt to changing environmental conditions indefinitely. Figure 1-1 shows steps along a path to sustainability.

The first step is to understand the components and importance of **natural capital**—the natural resources

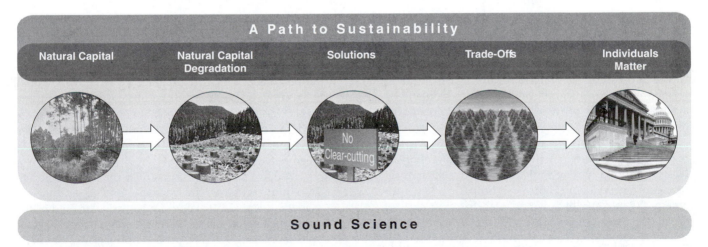

Figure 1-1 *A path to sustainability:* five subthemes are used throughout this book to illustrate how we can make the transition to more environmentally sustainable societies and economies, based on *sound science*—concepts widely accepted by scientists in various fields.

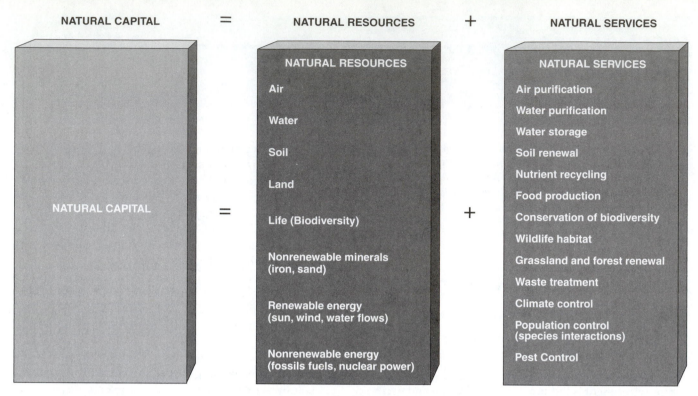

NATURAL CAPITAL = **NATURAL RESOURCES** + **NATURAL SERVICES**

NATURAL CAPITAL

NATURAL RESOURCES

Air

Water

Soil

Land

Life (Biodiversity)

Nonrenewable minerals
(iron, sand)

Renewable energy
(sun, wind, water flows)

Nonrenewable energy
(fossils fuels, nuclear power)

NATURAL SERVICES

Air purification

Water purification

Water storage

Soil renewal

Nutrient recycling

Food production

Conservation of biodiversity

Wildlife habitat

Grassland and forest renewal

Waste treatment

Climate control

Population control
(species interactions)

Pest Control

Figure 1-2 Natural capital: the natural resources (left) and natural services (right) that make up the earth's *natural capital* support and sustain the earth's life and economies. For example, *nutrients* or chemicals such as carbon and nitrogen, which plants and animals need as *resources*, are recycled through air, water, soil, and organisms by the natural process of *nutrient cycling.*

and natural services that keep us and other species alive and support our economies (Figure 1-2).

We can also think of energy from the sun as **solar capital** that warms the planet and supports photosynthesis, the process that plants use to provide food for themselves and for us and other animals. This direct input of solar energy also produces indirect forms of renewable solar energy such as wind and flowing water.

Natural capital is not fixed. It has changed over millions of years in response to environmental changes such as global warming and cooling and huge asteroids hitting the earth. Forests have grown and disappeared, as have grasslands, and deserts. Species have become extinct because of natural and human causes and new species have appeared. We have transformed many forests and grasslands into croplands—a more simplified form of natural capital created by humans.

The second step toward sustainability is to recognize that many human activities *degrade natural capital* by using normally renewable resources faster than nature can renew them (Figure 1-1, Step 2). A key variable is the *rate* at which we are transforming parts of the earth to meet our needs and wants. Most past natural environmental changes have taken place over many thousands of years. Humans are now making major changes in the earth's natural systems within

50 to 100 years. For example, we are clearing many forests much faster than nature can regrow them.

This leads us to search for *solutions* to these and other environmental problems (Figure 1-1, Step 3). For example, one solution might be to stop clear-cutting diverse mature forests.

The search for solutions often involves conflicts, and resolving these conflicts requires us to make *trade-offs,* or compromises (Figure.1-1, Step 4). To provide wood and paper, for example, we can promote the planting of tree plantations in areas that have already been cleared or degraded.

In the search for solutions, *individuals matter,* whether they are working alone or in groups. For example, one scientist might find a way to make paper by using crop residues instead of cutting down more trees. Or a group might work to get a law banning the clear-cutting of ancient forests while encouraging the planting of tree plantations in areas that have already been cleared or degraded.

The five steps to sustainability must be supported by **sound science**—the concepts and ideas that are widely accepted by experts in a particular field of the natural or social sciences. For example, sound science tells us that we need to protect and sustain the many natural services provided by diverse mature forests. It also guides us in the design and management of tree plantations and in finding ways to produce paper without using trees.

Environmentally Sustainable Societies: Protecting Natural Capital and Living off Its Income

An environmentally sustainable society meets the basic resource needs of its people in a just and equitable manner without degrading or depleting the natural capital that supplies these resources.

The ultimate human goal on a path to sustainability is an **environmentally sustainable society**—one that meets the current and future needs of its people for basic resources in a just and equitable manner without compromising the ability of future generations to meet their needs. *Living sustainably* means living off natural income replenished by soils, plants, air, and water and not depleting or degrading the earth's natural capital that supplies this income.

Imagine you win a million dollars in a lottery. Invest this capital at 10% interest per year, and you will have a sustainable annual income of $100,000 that you can spend without depleting your capital. If you spend $200,000 a year, your $1 million will be gone early in the seventh year. Even if you spend only $110,000 a year, you will be bankrupt early in the eighteenth year.

The lesson here is an old one: *Protect your capital and live off the income it provides.* Deplete, waste, or squander your capital, and you move from a sustainable to an unsustainable lifestyle.

The same lesson applies to the earth's natural capital. There is a general consensus among the world's environmental scientists that we are living unsustainably. They point out that we are using up finite supplies of nonrenewable resources, such as oil that has accumulated over millions of years, at an accelerating rate. We are also using renewable resources such as forests, soils, and freshwater faster than nature can renew them.

And in many areas we are adding various pollutants to the air and water faster than nature can dilute, absorb, and break them down. We are also increasing the earth's average temperature by adding certain gases, called greenhouse gases, to the atmosphere faster than nature can absorb them.

POPULATION GROWTH, ECONOMIC GROWTH, AND ECONOMIC DEVELOPMENT

Human Population Growth: Slowing but Still Rapid

The rate at which the world's population is growing has slowed, but the population is still increasing rapidly.

During the past 10,000 years, the human population has grown from several million to 6.7 billion (Figure 1-3). Rapid population has slowed but not ended. Between 1963 and 2006, the annual rate at which the world's population was growing decreased from 2.2% to 1.3%. This does not seem like a very high rate, but it added about 81 million people to the world's population in 2006, an average increase of 222,000 people a day, or 9,250 an hour. At this rate it takes only about 2.9 days to replace the 652,000 Americans killed in battle in all U.S. wars and only about 1.4 years to replace the 111 million soldiers and civilians killed in all wars fought in the twentieth century.

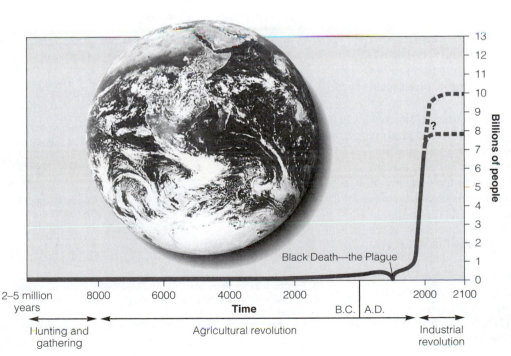

Figure 1-3 The *J*-shaped curve of past world population growth, with projections to 2100. The current world population of 6.7 billion people is projected to reach 8–10 billion people sometime this century. (This figure is not to scale.) (Data from the World Bank and United Nations; photo courtesy of NASA)

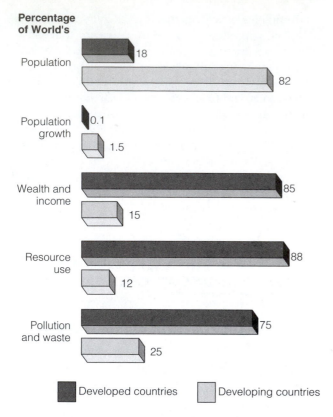

Percentage of World's

Population	18	82
Population growth	0.1	1.5
Wealth and income	85	15
Resource use	88	12
Pollution and waste	75	25

■ Developed countries   □ Developing countries

Figure 1-4 Global outlook: comparison of developed and developing countries, 2006. (Data from the United Nations and the World Bank)

Economic Growth and Economic Development

Economic growth provides people with more goods and services, and economic development uses economic growth to improve living standards.

Economic growth is an increase in the capacity of a country to provide people with goods and services. Accomplishing this increase requires population growth (more producers and consumers), more production and consumption per person, or both.

Economic growth is usually measured by the percentage change in a country's **gross domestic product (GDP):** the annual market value of all goods and services produced by all firms and organizations, foreign and domestic, operating within a country. Changes in a country's standard of living are measured by **per capita GDP:** the GDP divided by the total population at midyear. Between 1950 and 2006, there was more than a sevenfold increase in the world's economic growth. There is debate over how long the earth's natural capital can sustain such growth.

Economic development is the improvement of living standards by economic growth. The United Nations (U.N.) classifies the world's countries as economically

developed or developing based primarily on their degree of industrialization and their per capita GDP.

The **developed countries** (with 1.2 billion people) include the United States, Canada, Japan, Australia, New Zealand, and the countries of Europe. Most are highly industrialized and have a high average per capita GDP. All other nations (with 5.5 billion people) are classified as **developing countries,** most of them in Africa, Asia, and Latin America. Some are *middle-income, moderately developed countries* and others are *low-income countries.*

Figure 1-4 compares some key characteristics of developed and developing countries. About 97% of the projected increase in the world's population is expected to take place in developing countries.

Figure 1-5 summarizes some of the benefits and harm caused mostly by economic development. It

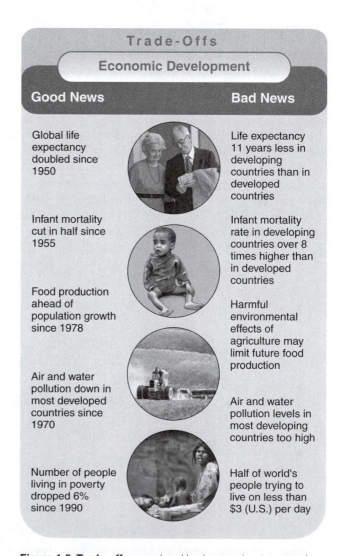

Figure 1-5 Trade-offs: good and bad news about economic development. QUESTION: *Which single advantage and which single disadvantage do you think are the most important?* (Data from the United Nations and World Health Organization)

shows effects of the wide and growing gap between the world's haves and have-nots.

During this century, many analysts call for us to put much greater emphasis on **environmentally sustainable economic development.** Its goals are to use political and economic systems to *encourage* environmentally beneficial and more sustainable forms of economic growth and to *discourage* environmentally harmful and unsustainable forms of economic growth.

RESOURCES

What Is a Resource? (Science)

We obtain resources from the environment to meet our needs and wants.

From a human standpoint, a **resource** is anything obtained from the environment to meet human needs and wants. Examples include food, water, shelter, and metals used to manufacture goods. On our short human time scale, we classify the material resources we get from the environment as *perpetual, renewable,* or *nonrenewable.*

Some resources, such as solar energy, fresh air, wind, fresh surface water, fertile soil, and wild edible plants, are directly available for use. Other resources, such as petroleum (oil), iron, groundwater (water found underground), and modern crops, are not directly available. They become useful to us only with some effort and technological ingenuity. For example, petroleum was a mysterious fluid until we learned how to find and extract it, and to refine it into gasoline, heating oil, and other products that we could sell at affordable prices.

Perpetual and Renewable Resources (Science)

Resources renewed by natural processes are sustainable if we do not use them faster than they are replenished.

Solar energy is called a **perpetual resource** because on a human time scale it is renewed continuously. It is expected to last at least 6 billion years as the sun completes its life cycle.

On a human time scale, a **renewable resource** can be replenished fairly rapidly (hours to several decades) through natural processes as long as it is not used up faster than it is replaced. Examples of renewable resources are forests, grasslands, wild animals, freshwater, fresh air, and fertile soil.

Renewable resources can be depleted or degraded. The highest rate at which a renewable resource can be used *indefinitely* without reducing its available supply is called **sustainable yield.**

When we exceed a renewable resource's natural replacement rate, the available supply begins to shrink, a process known as **environmental degradation.** Examples include urbanization of productive land, excessive topsoil erosion, pollution, clearing forests to grow crops, depleting groundwater, and reducing the earth's variety of wildlife (biodiversity) by eliminating habitats and species.

The Tragedy of the Commons (Economics and Politics)

Renewable resources that are freely available to everyone can be degraded.

One cause of environmental degradation of renewable resources is the overuse of **common-property** or **free-access resources.** Individuals do not own these resources, and they are available to all users at little or no charge. Examples include clean air, the open ocean and its fish, migratory birds, gases of the lower atmosphere, and space.

In 1968, biologist Garrett Hardin (1915–2003) called the degradation of renewable free-access resources the **tragedy of the commons.** It happens because each user reasons, "If I do not use this resource, someone else will. The little bit I use or pollute is not enough to matter, and such resources are renewable."

With only a few users, this logic works. Eventually, however, the cumulative effect of many people trying to exploit a free-access resource exhausts or ruins it. Then no one can benefit from it, and that is the tragedy.

One solution is to *use free-access resources at rates well below their estimated sustainable yields* by reducing population, regulating access to the resources, or both. Some communities have established rules and traditions to regulate and share their access to common-property resources such as ocean fisheries, grazing lands, and forests. Governments have also enacted laws and international treaties to regulate access to commonly owned resources such as forests, national parks, rangelands, and fisheries in coastal waters.

Another solution is to *convert free-access resources to private ownership.* The reasoning is that if you own something, you are more likely to protect your investment.

This sounds good, but private owners do not always protect natural resources they own when this goal conflicts with protecting their financial capital or increasing their profits. For example, some private forest owners can make more money by clear-cutting the timber, selling the degraded land, and investing their profits in other timberlands or businesses. Also, this approach is not practical for global common resources—such as the atmosphere, the open ocean, most wildlife species, and migratory birds—that cannot be divided up and converted to private property.

Our Ecological Footprints (Science)

Supplying each person with renewable resources and absorbing the wastes from such resource use creates a large ecological footprint or environmental impact.

The **per capita ecological footprint** is the amount of biologically productive land and water needed to supply individuals with resources and to absorb the wastes from such resource use (Figure 1-6, left). It is an estimate of the average environmental impact of individuals in a given country or area. The numbers shown in Figure 1-6 are only estimates, but they show relative differences in resource use and waste production by countries and geographical areas.

Humanity's ecological footprint per person exceeds by about 39% the earth's biological capacity to replenish renewable resources and absorb waste (Figure 1-6, right). If these estimates are correct, *it will take the resources of 1.39 planet earths to indefinitely support our current use of renewable resources.* You can estimate your ecological footprint by going to the website **www.redefiningprogress.org/**. Also, see the Guest Essay by Michael Cain on the website for this chapter.

According to the developers of the ecological footprint concept, William Rees and Mathis Wackernagel, with existing technology it would take the land area of about *four more planet earths* for the rest of the world to reach U.S. levels of consumption. See the Guest Essay on the website for this chapter by Norman Myers about the growing ecological footprint of China.

The *size* of our ecological footprint is not the only concern. A related concept is the *force* of our footprint—how heavily or lightly we walk on the earth through our lifestyles. The three things that have the greatest environmental impact are, in order, agriculture, transportation, and heating and cooling buildings.

Nonrenewable Resources (Science)

Nonrenewable resources can be economically depleted to the point where it costs too much to obtain what is left.

Nonrenewable resources exist in a fixed quantity or stock in the earth's crust. On a time scale of millions to billions of years, geological processes can renew such resources. But on the much shorter human time scale of hundreds to thousands of years, these resources can be depleted much faster than they are formed.

Such exhaustible resources include *energy resources* (such as coal, oil, and natural gas), *metallic mineral resources* (such as iron, copper, and aluminum), and *nonmetallic mineral resources* (such as salt, clay, and sand).

We never completely exhaust a nonrenewable mineral resource but it becomes *economically depleted* when the costs of extracting and using what is left exceed its economic value. At that point, we have five choices: try to find more, recycle or reuse existing supplies (except for nonrenewable energy resources, which cannot be recycled or reused), waste less, use less, or try to develop a substitute.

Some nonrenewable material resources, such as copper and aluminum, can be recycled or reused to extend supplies. **Recycling** involves collecting waste

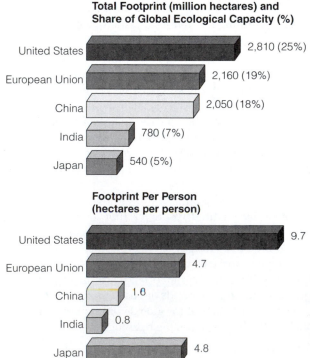

Total Footprint (million hectares) and Share of Global Ecological Capacity (%)

United States — 2,810 (25%)
European Union — 2,160 (19%)
China — 2,050 (18%)
India — 780 (7%)
Japan — 540 (5%)

Footprint Per Person (hectares per person)

United States — 9.7
European Union — 4.7
China — 1.6
India — 0.8
Japan — 4.8

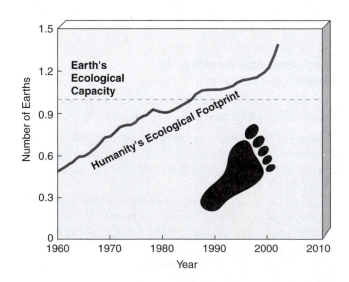

Figure 1-6 Natural capital use and degradation: relative *per capita and total ecological footprints* of the United States, the Netherlands, and India (left). By 2002, humanity's average ecological footprint was about 39% higher than the earth's ecological capacity (right). (Data from Worldwide Fund for Nature, UN Environment Programme, and Global Footprint Network)

materials, processing them into new materials, and selling these new products. For example, discarded aluminum cans can be crushed and melted to make new aluminum cans or other items. **Reuse** is using an item again in the same form. For example, glass bottles can be collected, washed, and refilled many times.

Recycling nonrenewable metallic resources takes much less energy, water, and other resources and produces much less pollution and environmental degradation than exploiting virgin metallic resources. Reusing such resources takes even less energy and other resources, and produces less pollution and environmental degradation than recycling. Most published estimates of the supply of a given nonrenewable resource refer to **reserves:** known deposits from which a usable mineral can be extracted profitably at current prices. Reserves can increase when new deposits are found or when higher prices or improved mining technology make it profitable to extract deposits that previously were too expensive to extract.

Depletion time is the time it takes to use up a certain proportion—usually 80%—of the reserves of a mineral at a given rate of use. Extracting the remaining 20% of a resource usually costs more than it is worth. Depletion times can vary according to how much a nonrenewable resource is recycled and reused, whether or not mining technologies are improved, and whether or not substitute resources are found.

Some resource experts believe that future supplies of widely used nonrenewable resources may be limited by the damage that extracting and processing them does to the environment in the form of land disturbance, soil erosion, and water and air pollution.

POLLUTION

Sources and Effects of Pollutants (Science)

Pollutants are chemicals found at high enough levels in the environment to cause harm to people or other organisms.

Pollution is the presence of chemicals at high enough levels in air, water, soil, or food to threaten the health, survival, or activities of humans or other organisms. Pollutants can enter the environment naturally such as from volcanic eruptions, or through human activities, such as burning coal. Most pollution from human activities occurs in or near urban and industrial areas, where pollutants are concentrated. Industrialized agriculture is also a major source of pollution. Some pollutants contaminate the areas where they are produced and some are carried by wind or flowing water to other areas.

The pollutants we produce come from two types of sources. **Point sources** of pollutants are single, iden-tifiable sources. Examples are the smokestack of a coal-burning power plant, the drainpipe of a factory, and the exhaust pipe of an automobile. **Nonpoint sources** of pollutants are dispersed and often difficult to identify. Examples are pesticides blown by the wind from the land into the atmosphere, and runoff of fertilizers and pesticides from farmlands, golf courses, and suburban lawns and gardens into streams and lakes. It is much easier and cheaper to identify and control pollution from point sources than from widely dispersed nonpoint sources.

Pollutants can have three types of unwanted effects. *First,* they can disrupt or degrade life-support systems for humans and other species. *Second,* they can damage wildlife, human health, and property. *Third,* they can be nuisances, as are noise and unpleasant smells, tastes, and sights.

Solutions: Prevention versus Cleanup (Science and Economics)

We can try to clean up pollutants in the environment or prevent them from entering the environment.

We use two basic approaches to deal with pollution. One is **pollution prevention,** or **input pollution control,** which reduces or eliminates the production of pollutants. The other is **pollution cleanup,** or **output pollution control,** which involves cleaning up or diluting pollutants after they have been produced.

Environmental scientists have identified three problems with relying primarily on pollution cleanup. *First,* it is only a temporary bandage as long as population and consumption levels grow without corresponding improvements in pollution control technology. For example, adding catalytic converters to car exhaust systems has reduced air pollution. But increases in the number of cars and in the distance each travels have reduced the effectiveness of this approach.

Second, cleanup often removes a pollutant from one part of the environment only to cause pollution in another. For example, we can collect garbage, but the garbage is then *burned* (perhaps causing air pollution and leaving toxic ash that must be put somewhere), *dumped* into streams, lakes, and oceans (perhaps causing water pollution), or *buried* (perhaps causing soil and groundwater pollution).

Third, once pollutants have entered and become dispersed into the environment at harmful levels, it usually costs too much or is impossible to reduce them to acceptable levels.

Both pollution prevention (front-of-the-pipe) and pollution cleanup (end-of-the-pipe) solutions are needed. Environmental scientists and some economists urge us to put more emphasis on prevention because it works better and because it is cheaper than cleanup.

ENVIRONMENTAL PROBLEMS: CAUSES AND CONNECTIONS

Key Environmental Problems and Their Basic Causes

The major causes of environmental problems are population growth, wasteful resource use, poverty, poor environmental accounting, and ecological ignorance.

We face a number of interconnected environmental and resource problems (Figure 1-7). As we run more and more of the earth's natural resources through the global economy, forests are shrinking, deserts are expanding, soils are eroding, the atmosphere is warming, water tables are falling, seas are rising, species are becoming extinct, resources are being wasted, and pollution is increasing. Most environmental problems are unintended results of activities designed to increase the quality of human life by providing goods and services.

The first step in dealing with these problems is to identify their underlying causes, listed in Figure 1-8.

We will discuss these causes and explore problems and potential solutions throughout this book.

Poverty and Environmental Problems (Economics and Politics)

Poverty is a major threat to human health and the environment, and the world's poorest people suffer the most from pollution and environmental degradation.

Poverty is the inability to meet one's basic economic needs. Many of the world's poor do not have access to the basic necessities for a healthy and productive life (Figure 1-9). Their daily lives are focused on getting enough food, water, and fuel (for cooking and heat) to survive. Desperate for land to grow enough food, many of the world's poor people deplete and degrade forests, soil, grasslands, and wildlife, for short-term survival. They do not have the luxury of worrying about long-term environmental quality or sustainability.

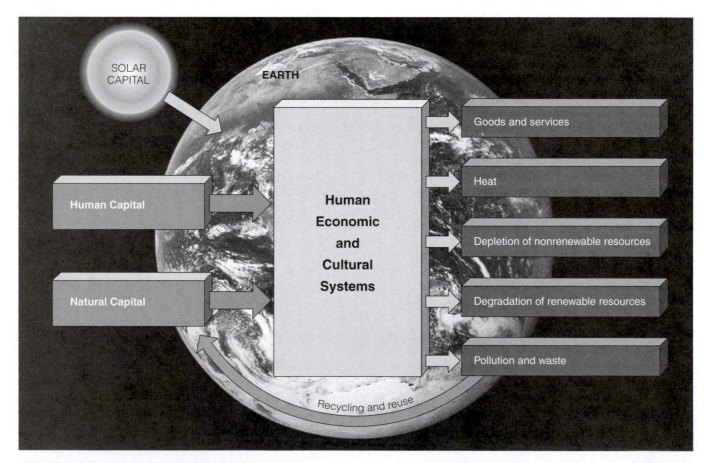

Figure 1-7 Natural capital use, depletion, and degradation: human and natural capital produce an amazing array of goods and services for most of the world's people. But the flow of material resources through the world's economic systems depletes nonrenewable resources, degrades renewable resources, and adds heat, pollution, and wastes to the environment.

Causes of Environmental Problems

Population growth

Unsustainable resource use

Poverty

Not including the environmental costs of economic goods and services in their market prices

Trying to manage and simplify nature with too little knowledge about how it works

Figure 1-8 Natural capital degradation: five basic causes of the environmental problems we face. QUESTION: *Can you think of any other basic causes?*

Poverty also affects population growth. Poor people often have many children as a form of economic security. Their children help them grow food, gather fuel (mostly wood and dung), haul drinking water, tend livestock, work, and beg in the streets. The children also help their parents survive into old age. Parents typically die in their 50s in the poorest countries.

Many of the world's desperately poor die from four preventable health problems. One is *malnutrition* from a lack of protein and other nutrients needed for good health. Another is increased susceptibility to nor-

mally nonfatal infectious diseases, such as diarrhea and measles, because of their weakened condition from malnutrition. A third is lack of access to clean drinking water. And a fourth is severe respiratory disease and premature death from inhaling indoor air pollutants produced by burning wood or coal for heat and cooking over open fires or on poorly vented stoves.

According to the World Health Organization, these four factors cause premature death for at least 7 million of the poor each year. *This premature death of about 19,200 people per day is equivalent to 48 fully loaded 400-passenger planes crashing every day with no survivors.* Two-thirds of those dying are children under age 5. This ongoing human tragedy is rarely covered in the daily news.

Resource Consumption and Environmental Problems (Economics and Ethics)

Many consumers in developed countries have become addicted to buying more and more stuff in their search for fulfillment and happiness.

Resource consumption is linked to both poverty and wealth. The poor *under consume* by not having enough food, water, and other resources to meet their basic needs. Many of the prosperous *over consume* by using and wasting far more resources than they need.

Affluenza (af-loo-en-zuh) is a term used to describe the unsustainable addiction to overconsumption and materialism exhibited in the lifestyles of affluent consumers in the United States and other developed countries. It is based on the assumption that buying more and more things will bring happiness. As humorist Will Rogers said, "Too many people spend money they haven't earned to buy things they don't want, to impress people they don't like."

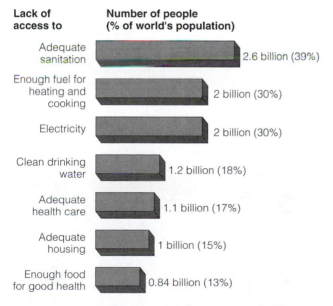

Lack of access to	Number of people (% of world's population)
Adequate sanitation	2.6 billion (39%)
Enough fuel for heating and cooking	2 billion (30%)
Electricity	2 billion (30%)
Clean drinking water	1.2 billion (18%)
Adequate health care	1.1 billion (17%)
Adequate housing	1 billion (15%)
Enough food for good health	0.84 billion (13%)

Figure 1-9 Natural capital degradation: some harmful results of poverty. QUESTION: *Which two of these effects do you think are the most harmful?* (Data from United Nations, World Bank, and World Health Organization)

Affluenza has an enormous environmental impact. It takes about 27 tractor-trailer loads of resources per year to support one American. This amounts to 8 billion truckloads a year to support the U.S. population. Stretched end-to-end, these trucks would reach beyond the sun—some 150 million kilometers (93 million miles away). And using these resources produces large amounts of pollution, environmental degradation, and wastes. Globalization and global advertising are now spreading the virus throughout much of the world. This problem and ways to deal with it are discussed in more depth in later chapters of this book.

Beneficial Effects of Affluence on Environmental Quality (Economics)

Affluent countries have more money for improving environmental quality.

Some analysts point out that affluence can lead people to become more concerned about environmental quality. And it provides money for developing technologies to reduce pollution, environmental degradation, and resource waste.

In the United States, the air is cleaner, drinking water is purer, most rivers and lakes are cleaner, and the food supply is more abundant and safer than in 1970.

Also, the country's total forested area is larger than it was in 1900 and most energy and material resources are used more efficiently. Similar advances have been made in most other affluent countries. Affluence financed these improvements in environmental quality.

A downside of wealth is that it allows the affluent to clean up the immediate environment of their homes, cities, and countries by transferring some of their wastes and pollution to more distant locations. It also allows them to obtain the resources they need from almost anywhere in the world without seeing the harmful impacts of their high-consumption life styles. However, environmentally educated affluent individuals also have the knowledge and means to make lifestyle choices to reduce such impacts.

Connections between Environmental Problems and Their Causes

Environmental quality is affected by interactions between population size, resource consumption, and technology.

Once we have identified environmental problems and their root causes, the next step is to understand how they are connected to one another. The three-factor model in Figure 1-10 is a starting point.

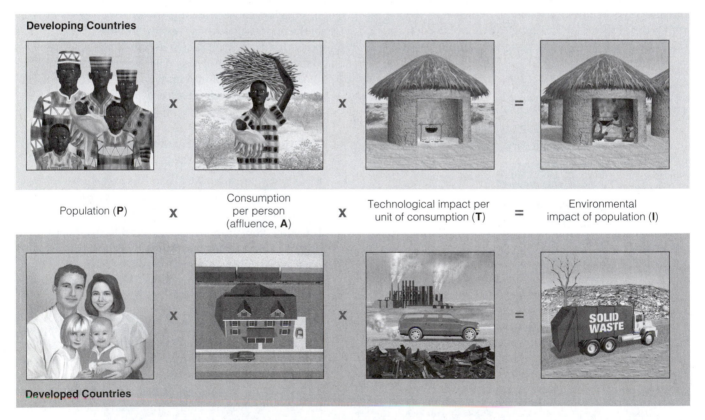

Figure 1-10 Connections: simplified model of how three factors—number of people, affluence, and technology—affect the environmental impact of the population in developing countries (top) and developed countries (bottom).

According to this simple model, the environmental impact (**I**) of population on a given area depends on three key factors: the number of people (**P**), average resource use per person (affluence, **A**), and the beneficial and harmful environmental effects of the technologies (**T**) used to provide and consume each unit of resource and to control or prevent the resulting pollution and environmental degradation.

In developing countries, population size and the resulting degradation of renewable resources as the poor struggle to stay alive tend to be the key factors in total environmental impact (Figure 1-10, top). In such countries, resource use per person is low.

In developed countries, high rates of resource use per person (affluenza) and the resulting high levels of pollution and environmental degradation per person usually are the key factors determining overall environmental impact (Figure 1-10, bottom) and a country's ecological footprint per person (Figure 1-6). For example, the average U.S. citizen consumes about 35 times as much as the average citizen of India and 100 times as much as the average person in the world's poorest countries. *Thus, poor parents in a developing country would need 70 to 200 children to reach the same lifetime resource consumption as 2 children in a typical U.S. family.*

Some forms of technology, such as polluting factories and motor vehicles and energy-wasting devices, increase environmental impact by raising the T factor in the equation. Other technologies, such as pollution control, solar cells, and energy-saving devices, lower environmental impact by decreasing the T factor. In other words, some forms of technology are *environmentally harmful* and some are *environmentally beneficial*.

CULTURAL CHANGES AND THE ENVIRONMENT

Human Cultural Changes

Since our hunter–gatherer days, three major cultural changes have increased our impact on the environment.

Evidence from fossils and studies of ancient cultures suggest that the current form of our species, *Homo sapiens sapiens*, has walked the earth for perhaps 90,000–195,000 years—less than an eye-blink in the earth's 3.7 billion years of life.

Until about 12,000 years ago, we were mostly hunter–gatherers who typically lived in small groups and moved as needed to find enough food for survival. Since then, three major cultural changes have occurred: the *agricultural revolution* (which began 10,000–12,000 years ago), the *industrial–medical revolution* (which began about 275 years ago), and the

information–globalization revolution (which began about 50 years ago).

These major cultural changes have greatly increased our impact on the environment. They have given us more energy and new technologies with which to alter and control more of the planet to meet our basic needs and increasing wants. They have allowed expansion of the human population, resource use, pollution, and environmental degradation as humans have cleared grasslands and forests for agriculture and built more and larger cities.

Supplement 2 at the end of this book provides a brief timeline overview of the environmental history of the United States as a series of interactions among environmental science, politics, and economics.

SUSTAINABILITY AND ENVIRONMENTAL WORLDVIEWS

Are Things Getting Better or Worse? A Millennium Assessment

There is good and bad environmental news.

Experts disagree about how serious our population and environmental problems are and what we should do about them. Some suggest that human ingenuity and technological advances will allow us to clean up pollution to acceptable levels, find substitutes for scarce resources, and keep expanding the earth's ability to support more humans. They accuse many environmental scientists and environmentalists of exaggerating the seriousness of the problems we face and failing to appreciate the progress we have made in improving quality of life and protecting the environment.

Many leading environmental scientists disagree with this view. In 2005, the Millennium Ecosystem Report was released. This four-year, $22-million study, prepared by nearly 1,400 experts from 95 countries, was sponsored by the United Nations, the World Bank, and various foundations. According to this study, human activities are degrading two-thirds of 24 important services provided by the earth's natural capital (Figure 1-2). In other words, we are living unsustainably. The report also tells us how we can change course. It sets out common-sense strategies for protecting species and habitats and sustaining the earth's natural capital.

The most useful answer to the question of whether things are getting better or worse is *both*—some things are getting better and some worse. Our challenge is not to get trapped into confusion and inaction by listening primarily to either of two groups of people. *Technological optimists* tend to overstate the solutions by telling us to be happy and not to worry, because technological innovations and conventional economic growth and development will lead to a wonderworld for everyone. In contrast, *environmental pessimists* over-state the problems to the point where our environmental situation seems hopeless. According to the noted conservationist Aldo Leopold, "I have no hope for a conservation based on fear."

✘ HOW WOULD YOU VOTE?* Is the society in which you live on an unsustainable path? Cast your vote online at www.thomsonedu.com/biology/miller.

Environmental Worldviews

The way we view the seriousness of environmental problems and how to solve them depends on our environmental worldview and our environmental ethics.

Differing views about the seriousness of our environmental problems and what we should do about them arise mostly out of differing environmental worldviews and environmental ethics. Your **environmental worldview** is a set of assumptions and values about how you think the world works and what you think your role in the world should be.

People with widely differing environmental worldviews can take the same data, be logically consistent, and arrive at quite different conclusions because they start with different assumptions and moral principles or values. Various environmental worldviews are discussed in detail in Chapter 14, but here is a brief introduction.

Some people in today's industrial consumer societies have a **planetary management worldview.** This view holds that we are separate from nature, that nature exists mainly to meet our needs and increasing wants, and that we can use our ingenuity and technology to manage the earth's life-support systems, mostly for our benefit. It assumes that economic growth is unlimited.

A second environmental worldview, known as the **stewardship worldview,** holds that we can manage the earth for our benefit but that we have an ethical responsibility to be caring and responsible managers, or *stewards,* of the earth. It says we should encourage environmentally beneficial forms of economic growth and discourage environmentally harmful forms.

A third environmental worldview is the **environmental wisdom worldview.** It holds that we are part of and totally dependent on nature and that nature exists for all species, not just for us. It also calls for encouraging earth-sustaining forms of economic growth and development and discouraging earth-degrading forms. According to this view, our success depends on

*To cast your vote, go the website for the book and then to the appropriate chapter (in this case, Chapter 1). In most cases, you will be able to compare your vote with those of others using this book throughout the United States and the rest of the world.

learning how the earth sustains itself and on integrating such environmental wisdom into the ways we think and act.

Four Principles of Sustainability: Copy Nature

We can develop more sustainable economies and societies by mimicking the four major ways that nature has adapted and sustained itself for billions of years.

How can we live more sustainably? According to ecologists, we can study how life on the earth has survived and adapted for 3.7 billion years and use what we learn as guidelines for our lives and economies.

The four basic components of the earth's *natural sustainability* are quite simple (Figure 1-11):

- **Reliance on Solar Energy:** the sun provides heat and supports photosynthesis used by plants to provide food for us and other animals.

- **Nutrient Recycling:** natural processes recycle all chemicals or nutrients that plants and animals need to stay alive and reproduce.

- **Biodiversity:** a great variety of genes, species, ecosystems, and ecological processes have provided many ways to adapt to changing environmental conditions throughout the 3.7-billion-year history of life on earth.

- **Population Control:** competition for limited resources among species places a limit on how much any one population can grow.

Figure 1-12 (p. 18) summarizes how we can live more sustainably by using these four amazingly simple fundamental lessons from nature (left side) in designing our societies, products, and economies (right side). *Figures 1-11 and 1-12 summarize the sustainability theme central to this book.*

Environmental scientists have used these principles of sustainability to formulate four guidelines for developing more sustainable societies and lifestyles:

1. *Our lives, lifestyles, and economies are totally dependent on the sun and the earth.* Everything we do, from breathing to building bridges, depends on natural capital. We are ultimately vulnerable to population control and extinction, like all species.

2. *Everything is connected to, and interdependent with, everything else.* Since we are part of it all, everything we do affects something else in nature and vice versa. We must seek to understand these connections.

3. *We can never do just one thing.* Any human intrusion into nature has unexpected and mostly unintended side effects. When we alter nature, we need to ask, "Now what will happen?"

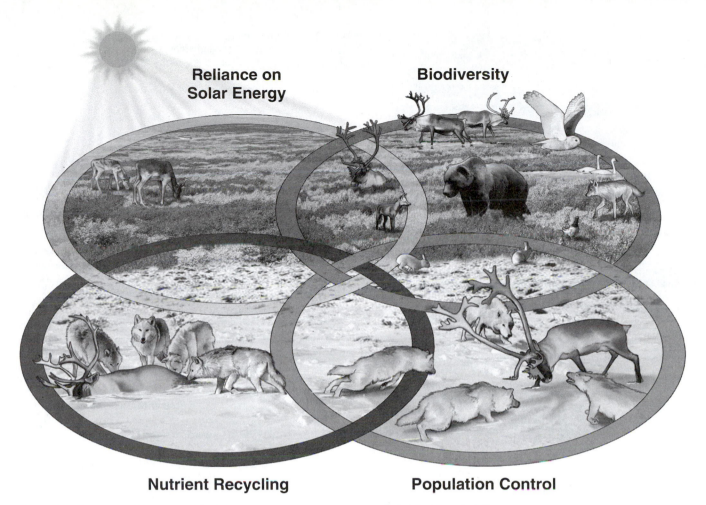

Reliance on Solar Energy

Biodiversity

Nutrient Recycling

Population Control

Figure 1-11 Four principles of sustainability: these four interconnected principles of sustainability are derived from learning how nature has sustained a variety of life on the earth for about 3.7 billion years.

4. *Only societies that live off the biological income provided by the earth's natural capital can be indefinitely sustained.* As environmental expert Lester R. Brown (see his Guest Essay on the website for this chapter) puts it, "No economy, however technologically advanced, can survive the collapse of its environmental support systems."

Using these guidelines to implement the four principles of sustainability could result in an *environmental revolution* during your lifetime. Figure 1-13 (p. 18) lists some of the shifts involved in bringing about this new cultural revolution.

Can we make such a cultural change? Environmental evidence indicates that we have perhaps 50 years and no more than 100 years to do so. We live during in an exciting and challenging cultural hinge of history that within your lifetime can swing us forward toward sustainability or backward toward unsustainability.

The challenge is to make creative use of our economic and political systems to implement the solutions during your lifetime. One key is to recognize that most economic and political change comes about as a result of individual actions and individuals acting together to bring about change. In other words, *individuals matter.* Research by social scientists suggests that it takes only 5–10% of the population of a country, or of the world, to bring about major social change. Such research also shows that significant social change can occur in a much shorter time than most people think.

Anthropologist Margaret Mead summarized our potential for social change: "Never doubt that a small group of thoughtful, committed citizens can change the world. Indeed, it is the only thing that ever has."

Many people believe we ought to accept an ethical responsibility for sustaining the earth's natural capital by leaving the earth in as good a condition as what we found, if not better. *What an exciting time to be alive!*

Solutions

Principles of Sustainability

How Nature Works		Lessons for Us
Runs on perpetual solar energy.		Rely mostly on direct and indirect solar energy.
Recycles nutrients and wastes. There is little waste in nature.		Prevent and reduce waste and pollution and recycle and reuse resources.
Uses biodiversity to maintain itself and adapt to new environmental conditions.		Preserve biodiversity by protecting ecosystems and preventing premature extinction of species.
Includes limiting factors that control a species' population sizes and resource use.		Recognize nature's limits on population size and resource use and learn to live within these limits.

Figure 1-12 Solutions: implications of the four principles of sustainability derived from observing nature (left) for the long-term sustainability of human societies (right).

Current Emphasis	Sustainability Emphasis
Pollution cleanup	Pollution prevention (cleaner production)
Waste disposal (bury or burn)	Waste prevention and reduction
Protecting species	Protecting where species live (habitat protection)
Environmental degradation	Environmental restoration
Increased resource use	Less wasteful (more efficient) resource use
Population growth	Population stabilization by decreasing birth rates
Depleting and degrading natural capital	Protecting natural capital and living off the biological interest it provides

Figure 1-13 Solutions: some shifts involved in bringing about the *environmental* or *sustainability revolution.*

What's the use of a house if you don't have a decent planet to put it on?

HENRY DAVID THOREAU

CRITICAL THINKING

1. List **(a)** three forms of economic growth that you think are environmentally unsustainable and **(b)** three forms that you think are environmentally sustainable.

2. Explain why you agree or disagree with the following propositions:
- **a.** Stabilizing population is not desirable because without more consumers, economic growth would stop.
- **b.** The world will never run out of resources because we can use technology to find substitutes and to help us reduce resource waste.

3. See if the affluenza bug infects you by indicating whether you agree or disagree with the following statements.
- **a.** I am willing to work at a job I despise so I can buy lots of stuff.
- **b.** When I am feeling down, I like to go shopping to make myself feel better.
- **c.** I would rather be shopping right now.
- **d.** I owe more than a $1,000 on my credit cards.
- **e.** I usually make only the minimum payment on my monthly credit card bills.
- **f.** I am running out of room to store my stuff.

If you agree with two of these statements, you are infected with affluenza. If you agree with more than two, you have a serious case of affluenza. Compare your answers with those of your classmates and discuss the effects of the results on the environment and your feelings of happiness.

4. When you read that at least 19,200 human beings die prematurely each day (13 per minute) from preventable malnutrition and infectious disease, do you **(a)** doubt that it is true, **(b)** not want to think about it, **(c)** feel hopeless, **(d)** feel sad, **(e)** feel guilty, or **(f)** want to do something about this problem?

5. How do you feel when you read that the average American consumes about 35 times more resources than the average citizen of India and that human activities are projected to make the earth's climate warmer: **(a)** skeptical about these statements, **(b)** indifferent, **(c)** sad, **(d)** helpless, **(e)** guilty, **(f)** concerned, or **(g)** outraged? Which of these feelings help perpetuate such problems, and which can help alleviate these problems?

6. Which one or more of the four principles of sustainability are involved in each of the following actions: **(a)** recycling soda cans; **(b)** using a rake instead of leaf blower; **(c)** choosing to have no more than one child;

(d) walking to class instead of driving; (e) taking your own reusable bags to the grocery store to carry things home in; (f) volunteering in a prairie restoration project; (g) planting trees; and (h) lobbying elected officials to require that 20% of your country's electricity be produced by renewable wind power by 2020.

7. Explain why you agree or disagree with each of the following statements: (a) humans are superior to other forms of life, (b) humans are in charge of the earth, (c) all economic growth is good, (d) the value of other species depends only on whether they are useful to humans, (e) because all species eventually become extinct we should not worry about whether our activities cause the premature extinction of a species, (f) all species have an inherent right to exist, (g) nature has an almost unlimited storehouse of resources for human use, (h) technology can solve our environmental problems, (i) I do not believe I have any obligation to future generations, and (j) I do not believe I have any obligation to other species.

8. List two questions that you would like to have answered as a result of reading this chapter.

LEARNING ONLINE

The website for this book contains helpful study aids and many ideas for further reading and research. They include a chapter summary, review questions for the entire chapter, flash cards for key terms and concepts, a multiple-choice practice quiz, interesting Internet sites, references, information about green careers, and a guide for accessing thousands of InfoTrac® College Edition articles. Log into

www.thomsonedu.com/biology/miller

Then choose Chapter 1, and select a learning resource. For access to animations, additional quizzes, chapter outlines and summaries, register and log into

at **www.thomsonedu.com** using the access code card in the front of your book.

2 SCIENCE, MATTER, ENERGY, AND ECOSYSTEMS: CONNECTIONS IN NATURE

When we try to pick out anything by itself, we find it hitched to everything else in the universe.

JOHN MUIR

THE NATURE OF SCIENCE

What Do Scientists Do?

Scientists collect data, form hypotheses, and develop theories, models, and laws about how nature works.

Science is a search for understanding of the natural world. It is based on the assumption that events in the natural world follow orderly cause-and-effect patterns that can be understood through careful observation and experimentation. Figure 2-1 summarizes the scientific process.

There is nothing mysterious about this process. You use it all the time in making decisions. Here is an example of applying the scientific process to an everyday situation:

Observation: You switch on your trusty flashlight and nothing happens.

Question: Why did the light not come on?

Hypothesis: Maybe the batteries are bad.

Test the hypothesis: Put in new batteries and switch on the flashlight.

Result: Flashlight still does not work.

New hypothesis: Maybe the bulb is burned out.

Experiment: Replace bulb with a new bulb.

Result: Flashlight works when switched on.

Conclusion: Second hypothesis is verified.

Here is a more formal outline of steps that scientists often take in trying to understand nature, although not always in the order listed.

- **Ask a question or identify a problem to be investigated.**

- **Collect data related to the question or problem by making observations and measurements.** Scientists often conduct **experiments** to study some phenomenon under known conditions.

- **Develop a hypothesis to explain the data.** Scientists working on a particular problem suggest possible explanations, or **scientific hypotheses,** of what they (or other scientists) observe in nature.

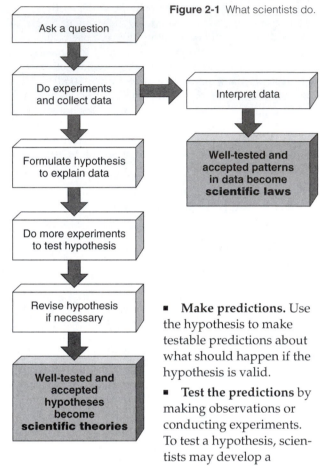

Figure 2-1 What scientists do.

- **Make predictions.** Use the hypothesis to make testable predictions about what should happen if the hypothesis is valid.

- **Test the predictions** by making observations or conducting experiments. To test a hypothesis, scientists may develop a **model**—an approximate or simplified representation or simulation of a system being studied.

- **Accept or reject the hypothesis.** If the new data do not support the hypothesis, come up with another testable explanation. Scientists keep repeating this process until there is general agreement among scientists in their field that a certain hypothesis provides a useful explanation of the data. A well-tested widely accepted scientific hypothesis is called a **scientific theory.**

We often hear about *the* scientific method. In reality, there are many **scientific methods:** ways in which scientists gather data and formulate and test scientific hypotheses, models, theories, and laws. There are no specific rules or guidelines for carrying out a scientific process, which involves reasoning, imagination, and creativity.

Three important features of the scientific process are *skepticism, peer review,* and *reproducibility.* Scientists tend to be highly skeptical of any new data or hypotheses until they can be confirmed or verified. *Peer review* happens when scientists publish details of their methods, results, and reasoning for other scientists working in the same field (their peers) to examine and criticize. Ideally, other scientists will repeat and analyze the work to see if the data can be reproduced and whether the proposed hypotheses are reasonable and useful.

Scientific Theories and Laws: The Most Important Results of Science

A widely tested and accepted scientific hypothesis becomes a scientific theory and a scientific law describes what we find happening in nature over and over again.

If an overwhelming body of observations, measurements, or tests supports a hypothesis, it becomes a *scientific theory. Scientific theories are not to be taken lightly.* They are not guesses, speculations, or suggestions, but are useful explanations of natural phenomena that have a high degree of certainty. They have been tested widely, are supported by extensive evidence, and are largely accepted by the scientific community.

Another important result of science is a **scientific, or natural, law:** a description of what we find happening in nature over and over in the same way. For example, after making thousands of observations and measurements over many decades, scientists discovered the second law of thermodynamics. It says that heat always flows spontaneously from hot matter to cold matter. Scientific laws describe repeated, consistent findings in nature, whereas scientific theories are widely accepted explanations of data, hypotheses, and laws.

Testing Hypotheses

Scientists test hypotheses by using controlled experiments and by running mathematical models on high-speed computers.

Many *variables* or *factors* influence most processes of nature that scientists seek to understand. Scientists conduct *controlled experiments* to try to isolate and study the effect of one variable at a time.

To do such *single-variable analysis,* scientists set up two groups—an *experimental group* in which the chosen variable is changed in a known way, and a *control group* in which the chosen variable is not changed. If the experiment is designed properly, any difference between the two groups should result from the variable that was changed in the experimental group.

Many of the phenomena environmental scientists investigate involve a huge number of interacting variables. Testing each variable with a traditional controlled experiment would be incredibly time consuming, costly, and often impossible. Sometimes this problem is overcome by using *multivariable analysis*—running mathematical models on high-speed computers to analyze the interactions of many variables.

Making Measurements: Uncertainty, Accuracy, and Precision

All scientific measurements involve some uncertainty, which scientists reduce by checking their measuring instruments and repeating their measurements.

When we make a scientific measurement, how do we know whether it is correct? All scientific observations and measurements involve some degree of *uncertainty* because people and measuring devices are not perfect.

Scientists take great pains to reduce the errors in observations and measurements by using standard procedures, testing (calibrating) measuring devices, and repeating their measurements several times, and then finding the average value of these measurements.

In determining the uncertainty involved in a measurement, it is important to distinguish between accuracy and precision. *Accuracy* is the degree to which a measurement conforms to what is accepted as the correct value for the measured quantity, based on careful measurements by many people over a long time. *Precision* is a measure of *reproducibility,* or how closely a series of measurements of the same quantity agree with *one another.*

Frontier Science, Sound Science, and Junk Science

Scientific results fall into three categories: those that have not been confirmed (frontier science), those that have been well tested and widely accepted (sound science), and junk science.

News reports often focus on two things: new so-called scientific breakthroughs, and disputes between scientists over the validity of preliminary and untested data and hypotheses. These preliminary results, called **frontier science,** are often controversial because they have not been widely tested and accepted, and some will be discredited. At the frontier stage, it is normal and healthy for reputable scientists to disagree about the meaning and accuracy of data and the validity of various hypotheses.

By contrast, **sound** science, or **consensus science,** consists of data, theories, and laws that are widely accepted by scientists who are considered experts in the field involved. The results of sound science are based on a self-correcting process of open peer review and reproducibility. To find out what scientists generally agree on, you can seek out reports by scientific bodies such as the U.S. National Academy of Sciences and the British Royal Society, which attempt to summarize consensus among experts in key areas of science.

Junk science consists of scientific results or hypotheses presented as sound science but not having undergone the rigors of the peer-review process, or having been discarded as a result of peer review. Here are some critical thinking questions you can use to uncover junk science.

- Have the data that support the proposed hypotheses been verified? (*Are they reproducible?*)

- Do the conclusions and hypotheses follow logically from the data?

- Does the explanation account for all of the observations? Are there no alternative explanations?

- Are the investigators unbiased in their interpretations of the results? Are they free of a hidden agenda? Is their funding from an unbiased source?

- Have the conclusions been verified by impartial peer review?

- Are the conclusions of the research widely accepted by other experts in this field?

If "yes" is the answer to each of these questions, then the results can be classified as *sound science*. Otherwise, the results may represent *frontier science* that needs further testing and evaluation, or they can be classified as *junk science*.

Limitations of Environmental Science

Inadequate data and scientific understanding limit environmental science and make some of its results controversial.

Before we begin our study of environmental science, we need to recognize some of its limitations, as well as those of science in general. For example, scientists cannot prove anything absolutely because there is always some degree of uncertainty in scientific measurements, observations, and models. Instead, scientists try to establish that a particular model, theory, or law has a very high *probability* (90–99%) of being true.

Secondly, no scientists are capable of complete objectivity about their research results. They are human and cannot be expected to be totally free of bias. However, objectivity is an important goal of science and bias can be minimized and often uncovered by the high standards of evidence required through peer review.

Another limitation especially important to environmental science involves validity of data. There is no way to measure accu-

rately how many metric tons of soil are eroded worldwide, for example. Instead, we use statistical sampling and methods to estimate such numbers. However, such environmental data should not be dismissed as "only estimates" because they can indicate important trends.

Another limitation is that most environmental problems are difficult to understand completely because they involve many variables and highly complex interactions. Much progress has been made, but we still know too little about how the earth works, its current state of environmental health, and the impacts of our activities. Different people can take the same information and reach completely different conclusions and believe that they are right. Critical thinking can uncover and evaluate such differences so that people can move beyond them and work together to solve problems.

Finally, no scientist can predict the results of any action with absolute certainty. Any action in a complex system has multiple, unintended, and often unpredictable effects. Most of the environmental problems we face today are unintended results of activities designed to increase the quality of human life (Figure 2-2).

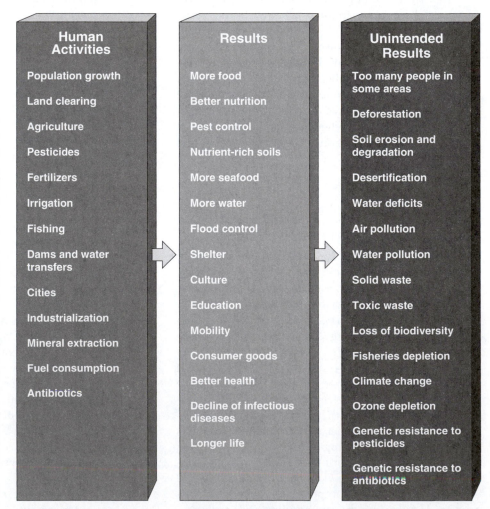

Figure 2-2 Natural capital degradation: human activities designed to improve the quality of life have had a number of unintended harmful environmental effects.

MATTER AND ENERGY

Elements and Compounds

Matter exists in chemical forms as elements and compounds.

Matter is anything that has mass (the amount of material in an object) and takes up space. There are two chemical forms of matter. One is **elements**: the distinctive building blocks of matter that make up every material substance. The other consists of **compounds**: two or more different elements held together in fixed proportions by attractive forces called *chemical bonds*.

To simplify things, chemists represent each element by a one- or two-letter symbol. Examples used in this book are hydrogen (H), carbon (C), oxygen (O), nitrogen (N), phosphorus (P), sulfur (S), chlorine (Cl), fluorine (F), bromine (Br), sodium (Na), calcium (Ca), lead (Pb), mercury (Hg), arsenic (As), and uranium (U).

Nature's Building Blocks: Atoms, Ions, Molecules and Cells

Atoms, ions, and molecules are the building blocks of matter.

If you had a super microscope capable of looking at individual elements and compounds, you could see they are made up of three types of building blocks. The first is an **atom**: the smallest unit of matter that exhibits the characteristics of an element.

The second is an **ion**: an electrically charged atom or combination of atoms. Examples encountered in this book include *positive* hydrogen ions (H^+), sodium ions (Na^+), calcium ions (Ca^{2+}), and ammonium ions (NH_4^+) and *negative* chloride ions (Cl^-), nitrate ions (NO_3^-), sulfate ions (SO_4^{2-}), and phosphate ions (PO_4^{3-}).

A third building block is a **molecule**: a combination of two or more atoms of the same or different elements held together by chemical bonds. Molecules are the building blocks of *compounds*.

Chemists use a shorthand **chemical formula** to show the number of atoms (or ions) of each type in a compound. The formula contains the symbols for each of the elements present and uses subscripts to represent the number of atoms or ions of each element in the compound's basic structural unit. Examples of compounds and their formulas encountered in this book are water (H_2O, read as "H-two-O"), oxygen (O_2), ozone (O_3), nitrogen (N_2), nitrous oxide (N_2O), nitric oxide (NO), hydrogen sulfide (H_2S), carbon monoxide (CO), carbon dioxide (CO_2), nitrogen dioxide (NO_2), sulfur dioxide (SO_2), ammonia (NH_3), sulfuric acid (H_2SO_4), nitric acid (HNO_3), methane (CH_4), and glucose ($C_6H_{12}O_6$).

Table sugar, vitamins, plastics, aspirin, penicillin, and most of the chemicals in your body are **organic compounds** that contain at least two carbon atoms combined with each other and with atoms of one or more other elements such as hydrogen, oxygen, nitrogen, sulfur, phosphorus, chlorine, and fluorine. One exception, methane (CH_4), has only one carbon atom. All other compounds are called **inorganic compounds.**

The millions of known organic (carbon-based) compounds include the following:

- *Hydrocarbons:* compounds of carbon and hydrogen atoms. An example is methane (CH_4), the main component of natural gas, and the simplest organic compound.

- *Chlorinated hydrocarbons:* compounds of carbon, hydrogen, and chlorine atoms. An example is the insecticide DDT ($C_{14}H_9Cl_5$).

- *Simple carbohydrates* (simple sugars): certain types of compounds of carbon, hydrogen, and oxygen atoms. An example is glucose ($C_6H_{12}O_6$), which most plants and animals break down in their cells to obtain energy.

Larger and more complex organic compounds, called *polymers*, consist of a number of basic structural or molecular units (*monomers*) linked by chemical bonds, somewhat like cars linked in a freight train. The four major types of organic polymers—the building materials of life—are

- *complex carbohydrates* consisting of two or more monomers of simple sugars linked together (such as starches that plants use to store energy and also provide energy for animals that feed on plants)

- *proteins* formed by linking together monomers of amino acids, important to living organisms for storing energy, maintaining immune systems, building body tissues and creating hormones

- *nucleic acids* (such as DNA and RNA used in reproduction) made by linked sequences of monomers called nucleotides

- and *lipids*, large organic compounds (such as fats and oils for storing energy and waxes for structure) that do not dissolve in water

 Examine atoms—their parts, how they work, and how they bond together to form molecules—at ThomsonNOW.

Genes consist of specific sequences of nucleotides in a DNA molecule. These coded units of genetic information about specific traits are passed on from parents to offspring during reproduction. **Chromosomes** are combinations of genes that make up a single DNA molecule, together with a number of proteins. Each chromosome typically contains thousands of genes. Genetic information coded in your chromosomal DNA is what makes you different from an oak leaf, an alligator, or a flea, and from your parents.

Finally, these building blocks combine to form the fundamental unit of all living things, the **cell:** a minute compartment containing chemicals necessary for life and within which most of the processes of life take place. Organisms may consist of a single cell (bacteria, for instance) or plants and animals that contain huge numbers of cells. The relationships of genetic material to cells are depicted in Figure 2-3.

What Are Atoms Made Of?

Each atom has a tiny nucleus containing protons and, in most cases, neutrons and one or more electrons whizzing around somewhere outside the nucleus.

If you increased the magnification of your super microscope, you would find that each different type of atom contains a certain number of *subatomic particles.* There are three types of these atomic building blocks: positively charged **protons** (p), uncharged **neutrons** (n), and negatively charged **electrons** (e).

Each atom consists of an extremely small center, or **nucleus** and one or more electrons in rapid motion somewhere outside the nucleus. The nucleus contains one or more protons and, in most cases, neutrons. Atoms are incredibly small. More than 3 million hydrogen atoms could sit side by side on the period at the end of this sentence.

Each atom has an equal number of positively charged protons inside its nucleus and negatively charged electrons whirling around outside its nucleus. Because these electrical charges cancel one another, *the atom as a whole has no net electrical charge.*

Each element has its own specific **atomic number,** equal to the number of protons in the nucleus of each of its atoms. The simplest element, hydrogen (H), has only 1 proton in its nucleus, so its atomic number is 1. Carbon (C), with 6 protons, has an atomic number of 6, whereas uranium (U), a much larger atom, has 92 protons and an atomic number of 92.

Because electrons have so little mass compared with the mass of a proton or a neutron, *most of an atom's mass is concentrated in its nucleus.* The mass of an atom is described in terms of its **mass number:** the total number of neutrons and protons in its nucleus. For example, a hydrogen atom with 1 proton and no neutrons in its nucleus has a mass number of 1, and an atom of uranium with 92 protons and 143 neutrons in its nucleus has a mass number of 235 (92 + 143 = 235).

All atoms of an element have the same number of protons in their nuclei. But they may have different numbers of uncharged neutrons in their nuclei, and thus may have different mass numbers. Various forms

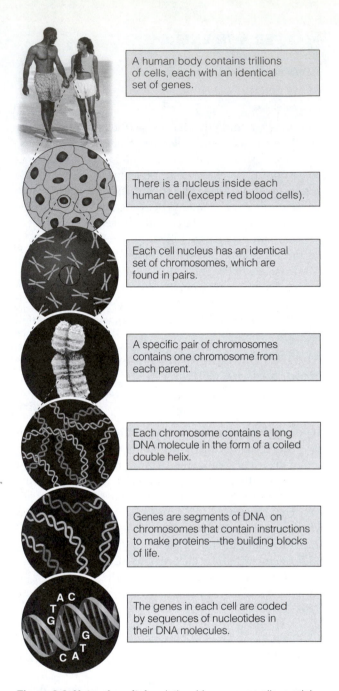

A human body contains trillions of cells, each with an identical set of genes.

There is a nucleus inside each human cell (except red blood cells).

Each cell nucleus has an identical set of chromosomes, which are found in pairs.

A specific pair of chromosomes contains one chromosome from each parent.

Each chromosome contains a long DNA molecule in the form of a coiled double helix.

Genes are segments of DNA on chromosomes that contain instructions to make proteins—the building blocks of life.

The genes in each cell are coded by sequences of nucleotides in their DNA molecules.

Figure 2-3 Natural capital: relationships among cells, nuclei, chromosomes, DNA, and genes.

of an element having the same atomic number but a different mass number are called **isotopes** of that element. Scientists identify isotopes by attaching their mass numbers to the name or symbol of the element. For example, hydrogen has three isotopes: hydrogen-1 (H-1, with 1 proton and no neutrons in its nucleus), hydrogen-2 (H-2, common name *deuterium*, with 1 proton and 1 neutron in its nucleus), and hydrogen-3 (H-3, common name *tritium*, with 1 proton and 2 neutrons in its nucleus). A natural sample of an element contains a mixture of its isotopes in a fixed proportion or percentage abundance by weight.

Matter Quality

Matter can be classified as having high or low quality depending on how useful it is to us as a resource.

The atoms, ions, and molecules that make up matter are found in four *physical states,* the most common of which on earth are solid, liquid, and gas. For example, water exists as solid ice, liquid water, and water vapor. A fourth state of matter, called *plasma,* is a high-energy mixture of positively charged ions and negatively charged electrons. The sun and all stars consist mostly of plasma, which makes it the most abundant form of matter in the universe.

Matter quality is a measure of how useful a form of matter is to us as a resource, based on its availability and concentration, as shown in Figure 2-4. **High-quality matter** is concentrated, usually is found near the earth's surface, and has great potential for use as a matter resource. **Low-quality matter** is dilute, often is deep underground or dispersed in the ocean or the atmosphere, and usually has little potential for use as a material resource.

An aluminum can is a more concentrated, higher-quality form of aluminum than aluminum ore containing the same amount of aluminum. That is why it takes less energy, water, and money to recycle an aluminum can than to make a new can from aluminum ore.

Material efficiency, or **resource productivity,** is the total amount of material needed to produce each unit of goods or services. Business expert Paul Hawken and physicist Amory Lovins contend that resource productivity in developed countries could be improved by 75–90% within two decades using existing technologies.

What Is Energy?

Energy is the work needed to move matter and the heat that flows from hot to cooler samples of matter.

Energy is the capacity to do work and transfer heat. Work is performed when an object such as a grain of sand, this book, or a giant boulder is moved over some distance. Work, or matter movement, is needed to burn natural gas to heat a house or cook food.

There are two major types of energy. One is **kinetic energy,** possessed by matter because of the matter's mass and its speed or velocity. Examples of this energy in motion are wind (a moving mass of air), flowing streams, heat flowing from warmer to colder matter, and electricity (flowing electrons).

The second type is **potential energy,** which is stored and potentially available for use. Examples of this stored energy are this book held in your hand, an

High Quality   **Low Quality**

Solid — Gas

Salt — Solution of salt in water

Coal — Coal-fired power plant emissions

Gasoline — Automobile emissions

Aluminum can — Aluminum ore

Figure 2-4 Examples of differences in matter quality. High-quality matter (left-hand column) is fairly easy to extract and is concentrated; low-quality matter (right-hand column) is more difficult to extract and is more dispersed than high-quality matter.

unlit match, the chemical energy stored in gasoline molecules, and the nuclear energy stored in the nuclei of atoms.

Potential energy can be changed to kinetic energy. Drop this book on your foot and the book's potential energy, from when you held it, changes into kinetic energy. When you burn gasoline in a car engine, the potential energy stored in the chemical bonds of its molecules changes into mechanical (kinetic) energy that propels the car.

 Witness how kinetic and potential energy might be used by a Martian at ThomsonNOW.

Within these two major types, energy exists in a number of different forms and can be changed from one form to another. Examples are *electrical energy* from the flow of electrons, *mechanical energy* used to move or lift matter, *heat* when energy flows from a hot to a colder body, *chemical energy* stored in the chemical bonds holding matter together, and *nuclear* energy stored in the nuclei of atoms.

Another important form of energy is **electromagnetic radiation**—energy traveling in the form of a *wave* as a result of the changes in electric and magnetic fields. Electromagnetic radiation occurs in different *wavelengths* (distance between successive peaks or troughs in the waves) and *energy content,* as shown in Figure 2-5. Such radiation, which includes light, travels through space at the *speed of light*—about 300,000 kilometers per second (186,000 miles per second).

 Find out how color, wavelengths, and energy intensities of visible light are related at ThomsonNOW.

Energy Quality

Energy can be classified as having high or low quality depending on how useful it is to us as a resource.

Energy quality is a measure of an energy source's ability to do useful work. **High-quality energy** is concentrated and can perform much useful work. Examples are electricity, the chemical energy stored in coal and gasoline, concentrated sunlight, and the nuclei of uranium-235, used as fuel in nuclear power plants.

By contrast, **low-quality energy** is dispersed and has little ability to do useful work. An example is heat dispersed in the moving molecules of a large amount of matter (such as the atmosphere or a large body of water) so that its temperature is low.

Physical and Chemical Changes in Matter

Matter can change from one physical form to another or change its chemical composition.

When a sample of matter undergoes a **physical change,** its chemical composition does not change. A piece of aluminum foil cut into small pieces is still aluminum foil. When solid water (ice) is melted or liquid water is boiled, none of the H_2O molecules involved are altered; instead, the molecules are organized in different spatial (physical) patterns.

In a **chemical change,** or **chemical reaction,** there is a change in the chemical composition of the elements or compounds. Chemists use shorthand chemical equations to represent what happens in a chemical reaction. For example, when coal burns completely, the solid carbon (C) it contains combines with oxygen gas (O_2) from the atmosphere to form the gaseous compound carbon dioxide (CO_2). We can represent this chemical reaction in shorthand form as $C + O_2 \longrightarrow CO_2 +$ energy.

Energy is given off in this reaction, making coal a useful fuel. The reaction also shows how the complete burning of coal (or any of the carbon-containing compounds in wood, natural gas, oil, and gasoline) gives off carbon dioxide gas, which helps warm the lower atmosphere.

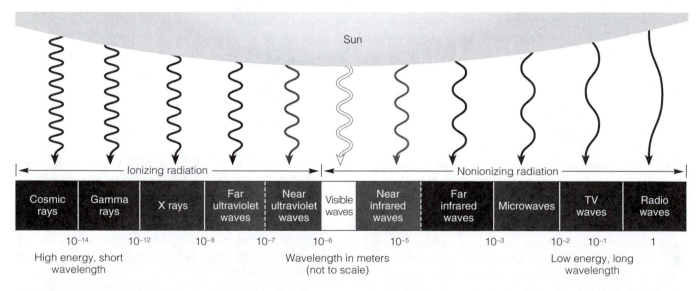

ThomsonNOW **Active Figure 2-5** The *electromagnetic spectrum:* the range of electromagnetic waves, which differ in wavelength (distance between successive peaks or troughs) and energy content. *See an animation based on this figure and take a short quiz on the concept.*

The Law of Conservation of Matter: There Is No "Away"

When a physical or chemical change occurs, no atoms are created or destroyed.

We may change various elements and compounds from one physical or chemical form to another, but we can never create or destroy any of the atoms involved in any physical or chemical change. All we can do is rearrange them into different spatial patterns (physical changes) or different combinations (chemical changes). This statement, based on many thousands of measurements, is known as the **law of conservation of matter.**

This law means there is no "away" as in "to throw away." *Everything we think we have thrown away remains here with us in some form.* We can make the environment cleaner and convert some potentially harmful chemicals into less harmful physical or chemical forms. But *the law of conservation of matter means we will always face the problem of what to do with some quantity of wastes and pollutants.*

The First Law of Thermodynamics: We Cannot Create or Destroy Energy

In a physical or chemical change, we can change energy from one form to another, but we can never create or destroy any of the energy involved.

Scientists have observed energy being changed from one form to another in millions of physical and chemical changes. But they have never been able to detect the creation or destruction of any energy (except in nuclear changes). The results of their experiments have been summarized in the **law of conservation of energy,** also known as the **first law of thermodynamics:** *In all physical and chemical changes, energy is neither created nor destroyed, but it may be converted from one form to another.*

This scientific law tells us that when one form of energy is converted to another form in any physical or chemical change, *energy input always equals energy output.* No matter how hard we try or how clever we are, we cannot get more energy out of a system than we put in; in other words, *we cannot get something for nothing in terms of energy quantity.* This is one of Mother Nature's basic rules that we have to live with.

The Second Law of Thermodynamics: Energy Quality Always Decreases

Whenever energy is changed from one form to another we always end up with less usable energy than we started with.

Because the first law of thermodynamics states that energy can be neither created nor destroyed, we may be tempted to think there will always be enough energy. Yet if we fill a car's tank with gasoline and drive around or use a flashlight battery until it is dead, something has been lost. If it is not energy, what is it? The answer is *energy quality,* the amount of energy available that can perform useful work.

Countless experiments have shown that when energy is changed from one form to another, a decrease in energy quality always occurs. The results of these experiments have been summarized in what is called the **second law of thermodynamics:** *When energy is changed from one form to another, some of the useful energy is always degraded to lower-quality, more dispersed, less useful energy.* This degraded energy usually takes the form of heat given off at a low temperature to the surroundings (environment). There it is dispersed by the random motion of air or water molecules and becomes even less useful as a resource.

In other words, *we cannot break even in terms of energy quality because energy always goes from a more useful to a less useful form when energy is changed from one form to another.* No one has ever found a violation of this fundamental scientific law. It is another one of Mother Nature's rules.

Consider three examples of the second law of thermodynamics in action. *First,* when a car is driven, only about 20–25% of the high-quality chemical energy available in its gasoline fuel is converted into mechanical energy (to propel the vehicle) and electrical energy (to run its electrical systems). The remaining 75–80% is degraded to low-quality heat that is released into the environment and eventually lost into space. Thus, most of the money you spend for gasoline is not used to get you anywhere.

Second, when electrical energy flows through filament wires in an incandescent light bulb, it is changed into about 5% useful light and 95% low-quality heat that flows into the environment. In other words, this so-called *light bulb* is really a *heat bulb.* Good news: Scientists have developed compact fluorescent bulbs that are four times more efficient, and even more efficient bulbs are on the way.

Third, in living systems, solar energy is converted into chemical energy (food molecules) and then into mechanical energy (moving, thinking, and living). During each of these conversions, high-quality energy is degraded and flows into the environment as low-quality heat. Trace the flows and energy conversions in Figure 2-6 (p. 28).

The second law of thermodynamics also means *we can never recycle or reuse high-quality energy to perform useful work.* Once the concentrated energy in a serving of food, a liter of gasoline, a lump of coal, or a chunk of uranium is released, it is degraded to low-quality heat that is dispersed into the environment.

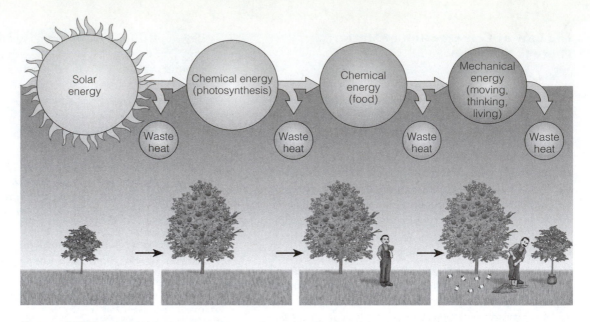

ThomsonNOW™ Active Figure 2-6 The second law of thermodynamics in action in living systems. Each time energy changes from one form to another, some of the initial input of high-quality energy is degraded, usually to low-quality heat that is dispersed into the environment. *See an animation based on this figure and take a short quiz on the concept.*

 See examples of how the first and second laws of thermodynamics apply in our world at ThomsonNOW.

Energy efficiency, or **energy productivity,** is a measure of how much useful work is accomplished by a particular input of energy into a system. *Good news:* There is plenty of room for improving energy efficiency. Scientists estimate that only about 16% of the energy used in the United States ends up performing useful work. The remaining 84% is either unavoidably wasted because of the second law of thermodynamics (41%) or unnecessarily wasted (43%).

EARTH'S LIFE-SUPPORT SYSTEMS: FROM ORGANISMS TO THE BIOSPHERE

What Is Ecology?

Ecology is a study of connections in nature.

Ecology (from the Greek words *oikos*, "house" or "place to live," and *logos*, "study of") is the study of how organisms interact with one another and with their nonliving environment. In effect, it is a study of *connections in nature*—the house for the earth's life. To enhance their understanding of nature, scientists classify matter into levels of organization from atoms to cells to the biosphere. Ecologists focus on trying to understand the interactions among organisms, populations, communities, ecosystems, and the biosphere (Figure 2-7).

An **organism** is any form of life. It is the most fundamental unit of ecology. The **cell** is the basic unit of life in organisms. Organisms may consist of a single cell (bacteria, for instance) or many cells. Look in the mirror. What you see is about 10 trillion cells divided into about 200 different types.

Organisms can be classified into **species,** groups of organisms that resemble one another in appearance, behavior, chemistry, and genetic makeup. Organisms that reproduce sexually by combining cells from both parents are classified as members of the same species if, under natural conditions, they can potentially breed with one another and produce live, fertile offspring.

How many species are on the earth? We do not know. Estimates range from 4 million to 100 million—most of them microorganisms too small to be seen with the naked eye. A best guess is that we share the planet with 10–14 million other species. So far, biologists have identified and named about 1.4 million species, most of them insects (see following Case Study).

Case Study: What Species Rule the World?

Multitudes of tiny microbes such as bacteria, protozoa, fungi, and yeast help keep us alive.

They are everywhere and there are trillions of them. Billions are found inside your body, on your body, in a handful of soil, and in a cup of river water.

These mostly invisible rulers of the earth are *microbes,* or *microorganisms,* catchall terms for many thousands of species of bacteria, protozoa, fungi, and yeasts—most too small to be seen with the naked eye.

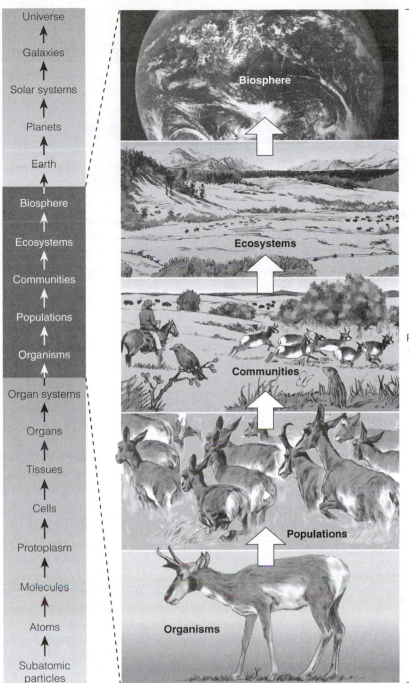

| Universe |
| Galaxies |
| Solar systems |
| Planets |
| Earth |
| Biosphere |
| Ecosystems |
| Communities |
| Populations |
| Organisms |
| Organ systems |
| Organs |
| Tissues |
| Cells |
| Protoplasm |
| Molecules |
| Atoms |
| Subatomic particles |

Biosphere

Ecosystems

Communities

Populations

Organisms

Realm of ecology

ThomsonNOW Active Figure 2-7
Natural capital: levels of organization of matter in nature. Ecology focuses on five of these levels. *See an animation based on this figure and take a short quiz on the concept.*

Microbes do not get the respect they deserve. Most of us think of them as threats to our health in the form of infectious bacteria or "germs," fungi that cause athlete's foot and other skin diseases, and protozoa that cause diseases such as malaria. But these harmful microbes are in the minority.

You are alive because of multitudes of microbes toiling away mostly out of sight. Soil bacteria convert nitrogen gas in the atmosphere into forms that plants can take up from the soil as nutrients. They also help produce foods such as bread, cheese, yogurt, vinegar, tofu, soy sauce, beer, and wine. Bacteria and fungi in the soil decompose organic wastes into nutrients that can be taken up by plants that we and most other animals eat. Without these wee creatures we would be up to our eyeballs in waste matter.

Microbes, especially bacteria, help purify the water you drink by breaking down wastes. Bacteria in your intestinal tract break down the food you eat. Some microbes in your nose prevent harmful bacteria from reaching your lungs. You have ten times more bacteria cells in your body than you have human cells.

Pick up a pinch of soil. It has up to 10 billion bacteria, many thousands of them species that scientists have not identified. A milliliter of ocean water contains about 1 million bacteria. Other bacteria are the source of disease-fighting antibiotics, including penicillin, erythromycin, and streptomycin. Scientists are working on using microbes to develop new medicines and fuels, and genetic engineers are developing microbes that can extract metals from ores, and help clean up polluted water and soils. Some microbes help control plant diseases and populations of insect species that attack our food crops. Relying more on these microbes for pest control can reduce the use of potentially harmful chemical pesticides. We spend much more money on learning about the moon and Mars than in understanding the microbes that sustain us and all forms of life.

Case Study: Have You Thanked the Insects Today?

Insects play vital roles in sustaining life on the earth by pollinating plants that provide food for us and other animals and by eating insects we classify as pests.

Insects have a bad reputation. We classify many insect species as *pests* because they compete with us for food, spread human diseases such as malaria, and invade our lawns, gardens, and houses. Some people have "bugitis," fearing all insects and thinking the only good bug is a dead bug. They fail to recognize the vital roles insects play in helping sustain life on earth.

Many of the earth's plant species depend on insects to pollinate their flowers. Without the natural service of pollination, plants cannot reproduce sexually, and no plant species would be around for long. Without pollinating insects, very few fruits and vegetables would be available for us.

Insects that eat other insects help control the populations of at least half the species of insects we call pests. This free pest-control service is an important part of the natural capital that helps sustain us.

Insects have been around for at least 400 million years and are phenomenally successful forms of life. Some insects can reproduce at an astounding rate. For example, a single female housefly and her offspring can theoretically produce about 5.6 trillion flies in only one year!

Insects can rapidly evolve new genetic traits, such as resistance to pesticides. They also have an exceptional ability to evolve into new species when faced with new environmental conditions, and they are very resistant to extinction. The environmental lesson is that although insects can thrive without newcomers such as us, we and most other land organisms would perish without them.

Populations, Communities, and Ecosystems

Members of a species interact in groups called populations; populations of different species living and interacting in an area form a community; and a community interacting with its physical environment of matter and energy is known as an ecosystem.

A **population** is a group of interacting individuals of the same species occupying a specific area. Examples are all sunfish in a pond, white oak trees in a forest, and people in a country. In most natural populations, individuals vary slightly in their genetic makeup, which is why they do not all look or act alike. This is called a population's **genetic diversity.**

The place where a population (or an individual organism) normally lives is its **habitat.** It may be as large as an ocean or as small the intestine of a termite.

A **community**, or **biological community**, consists of all the populations of the different species living and interacting in an area—a complex network of plants, animals, and microorganisms. An **ecosystem** is a community of different species interacting with one another and with their physical environment of matter and energy, struggling to get enough organic and inorganic compounds or **nutrients** to stay alive and reproduce, and each playing a different role in the flow of energy and cycling of nutrients.

Ecosystems can range in size from a puddle of water to a stream, a patch of woods, an entire forest, or a desert. Ecosystems can be natural or artificial (human created). Examples of artificial ecosystems are crop fields, farm ponds, and reservoirs. All of the earth's ecosystems together make up what we call the **biosphere** within which local ecosystems are connected by energy flowing across their fuzzy boundaries. The key ecological lesson from studying the biosphere is that *everything is linked to everything else.*

 Learn more about how the earth's life is organized on five levels in the study of ecology at ThomsonNOW.

The Earth's Life-Support Systems: Four Spheres

The earth is made up of interconnected spherical layers that contain air, water, soil, minerals, and life.

We can think of the earth as being made up of several spherical layers, as diagrammed in Figure 2-8. Study this figure carefully. The **atmosphere** is a thin envelope or membrane of air around the planet. Its inner layer, the **troposphere,** extends only about 17 kilometers (11 miles) above sea level. It contains most of the planet's air, mostly nitrogen (78%) and oxygen (21%).

The next layer, stretching 17–48 kilometers (11–30 miles) above the earth's surface, is the **stratosphere.** Its lower portion contains enough ozone (O_3) to filter out most of the sun's harmful ultraviolet radiation. This allows life to exist on land and in the surface layers of bodies of water.

The **hydrosphere** consists of the earth's water. It is found as *liquid water* (both surface and underground), *ice* (polar ice, icebergs, and ice in frozen soil layers called *permafrost*), and *water vapor* in the atmosphere. The **lithosphere** is the earth's crust and upper mantle; the crust contains nonrenewable fossil fuels and minerals we use as well as renewable soil chemicals (nutrients) needed for plant life.

All parts of the biosphere are interconnected. Any change in one of the biosphere's components or processes can have a ripple effect on other parts. If the earth were an apple, the biosphere would be no thicker

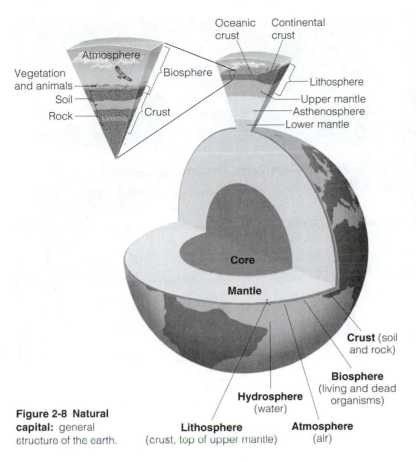

Figure 2-8 Natural capital: general structure of the earth.

Oceanic crust
Continental crust
Atmosphere
Vegetation and animals
Biosphere
Soil
Lithosphere
Rock
Upper mantle
Crust
Asthenosphere
Lower mantle

Core

Mantle

Crust (soil and rock)

Biosphere (living and dead organisms)

Hydrosphere (water)

Lithosphere (crust, top of upper mantle)

Atmosphere (air)

than the apple's skin. *The goal of ecology is to understand the interactions in this thin, life-supporting global membrane of air, water, soil, and organisms.*

What Sustains Life on Earth?

Solar energy, the cycling of matter, and gravity sustain the earth's life.

Life on the earth depends on three interconnected factors shown in Figure 2-9.

■ The *one-way flow of high-quality energy* from the sun through materials and living things in their feeding interactions, into the environment as low-quality energy (mostly heat dispersed into air or water molecules at a low temperature), and eventually back into space as heat. No round-trips are allowed because energy cannot be recycled.

■ The *cycling of matter* (the atoms, ions, or molecules needed for survival by living organisms) through parts of the biosphere. Because the earth is closed to significant inputs of matter from space, the earth's essentially fixed supply of nutrients must be recycled again and again for life to continue. All nutrient trips in ecosystems are round-trips.

■ *Gravity*, which allows the planet to hold on to its atmosphere and causes the downward movement of chemicals in the matter cycles.

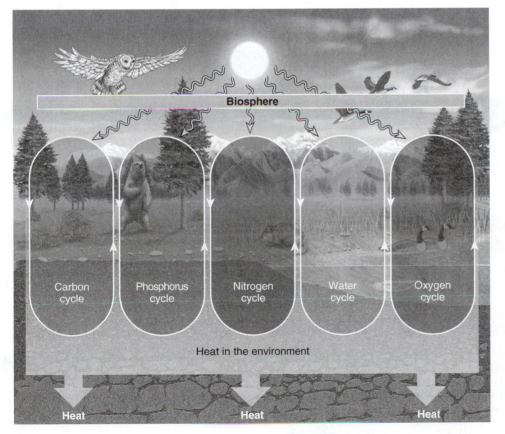

Biosphere

Carbon cycle
Phosphorus cycle
Nitrogen cycle
Water cycle
Oxygen cycle

Heat in the environment

Heat   Heat   Heat

ThomsonNOW™ Active Figure 2-9 Natural capital: life on the earth depends on the *one-way flow of energy* (wavy arrows) from the sun through the biosphere, the *cycling of crucial elements* (solid lines around ovals), and *gravity*, which keeps atmospheric gases from escaping into space and helps recycle nutrients through air, water, soil, and organisms. This simplified model depicts only a few of the many cycling elements. *See an animation based on this figure and take a short quiz on the concept.*

What Happens to Solar Energy Reaching the Earth?

Solar energy flowing through the biosphere warms the atmosphere, evaporates and recycles water, generates winds, and supports plant growth.

About one-billionth of the sun's output of energy reaches the earth—a tiny sphere in the vastness of space—in the form of electromagnetic waves—mostly as visible light (Figure 2-5). Much of this energy is either reflected away or absorbed by chemicals in its atmosphere (Figure 2-10). Trace the flows in this diagram.

Most of the solar radiation making it through the atmosphere hits the earth and is degraded into longer-wavelength infrared radiation. This infrared radiation interacts with so-called *greenhouse gases* (such as water vapor, carbon dioxide, methane, nitrous oxide, and ozone) in the troposphere. As this infrared radiation intersects with these gaseous molecules, it increases their kinetic energy, helping warm the troposphere and the earth's surface. Without this **natural greenhouse effect** (Figure 2-11), the earth would be too cold to support life as we know it and you would not be around to read this book.

Learn more about the flow of energy—from sun to earth and within the earth's systems—at ThomsonNOW.

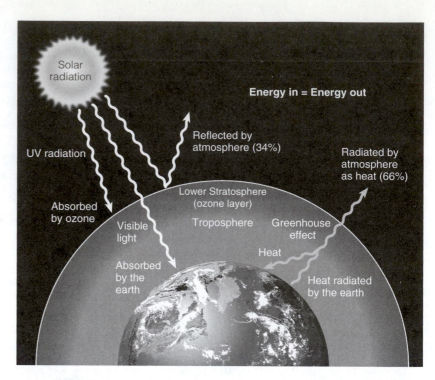

ThomsonNOW Active Figure 2-10 **Solar capital:** flow of energy to and from the earth. *See an animation based on this figure and take a short quiz on the concept.*

Why Is the Earth So Favorable for Life?

The earth's temperature range, distance from the sun, and size result in conditions that are favorable for life as we know it.

Life on the earth as we know it needs a certain temperature range. It depends on the liquid water that dominates the earth's surface. Temperature is crucial be-

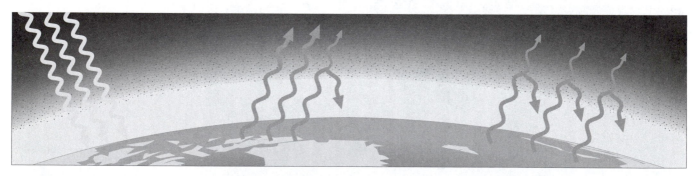

(a) Rays of sunlight penetrate the lower atmosphere and warm the earth's surface.

(b) The earth's surface absorbs much of the incoming solar radiation and degrades it to longer-wavelength infrared (IR) radiation, which rises into the lower atmosphere. Some of this IR radiation escapes into space as heat and some is absorbed by molecules of greenhouse gases and emitted as even longer wavelength IR radiation, which warms the lower atmosphere.

(c) As concentrations of greenhouse gases rise, their molecules absorb and emit more infrared radiation, which adds more heat to the lower atmosphere.

Figure 2-11 Natural capital: the *natural greenhouse effect*. Without the atmospheric warming provided by this natural effect, the earth would be a cold and mostly lifeless planet. According to the widely accepted greenhouse theory, when concentrations of greenhouse gases in the troposphere rise, the average temperature of the troposphere rises. (Modified by permission from Cecie Starr, *Biology: Concepts and Applications*, 4th ed., Pacific Grove, Calif.: Brooks/Cole, © 2000)

cause most life on the earth needs average temperatures between the freezing and boiling points of water.

The earth's orbit is the right distance from the sun to provide these conditions. If the earth were much closer, it would be too hot—like Venus—for water vapor to condense to form rain. If it were much farther away, its surface would be so cold—like Mars—that its water would exist only as ice. The earth also spins; if it did not, the side facing the sun would be too hot and the other side too cold for water-based life to exist.

The earth is also the right size; that is, it has enough gravitational mass to keep its iron and nickel core molten and to keep the light gaseous molecules (such as N_2, O_2, CO_2, and H_2O) in its atmosphere from flying off into space.

On a time scale of millions of years, the earth is enormously resilient and adaptive. During the 3.7 billion years since life arose, the average surface temperature of the earth has remained within the narrow range of 10–20°C (50–68°F), even with a 30–40% increase in the sun's energy output. In short, this remarkable planet that we live on is just right for life as we know it.

Major Components of Ecosystems

Ecosystems consist of nonliving (abiotic) and living (biotic) components.

Two types of components make up the biosphere and its ecosystems. One type, called **abiotic,** consists of nonliving components such as water, air, nutrients, and solar energy. The other type, called **biotic,** consists of biological components such as *producers* (mostly plants and floating algae phytoplankton), *consumers* (animals that get their food by eating plants or other animals), and *decomposers* (mostly bacteria that break down the dead remains of plants and animals and recycle them into the soil or water for reuse by producers).

Figure 2-12 is a greatly simplified diagram of some of the biotic and abiotic components in a terrestrial ecosystem. Look carefully at these components and how they are connected to one another through the consumption habits of organisms.

Each population in an ecosystem has a **range of tolerance** to variations in its physical and chemical environment (Figure 2-13, p. 34). Individuals within a population may also have slightly different tolerance ranges for temperature or other factors because of small differences in genetic makeup, health, and age. For example, a trout population may do best within a narrow band of temperatures (*optimum level* or *range*), but a few individuals can survive above and below that band. However, if the water becomes too hot or too cold, none of the trout can survive.

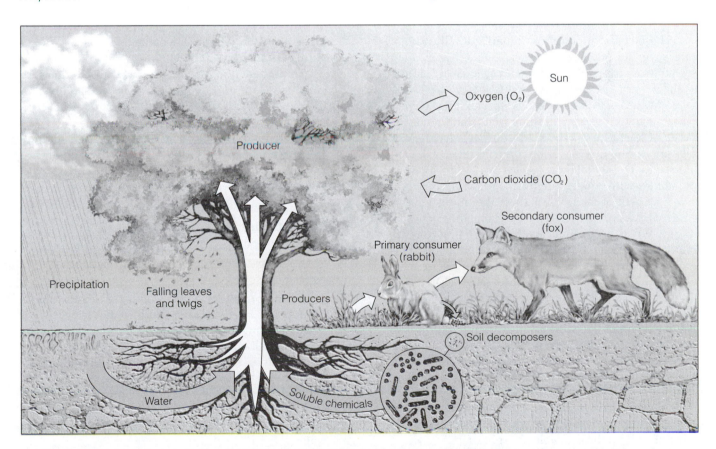

ThomsonNOW Active Figure 2-12 **Natural capital:** major components of an ecosystem in a field.
See an animation based on this figure and take a short quiz on the concept.

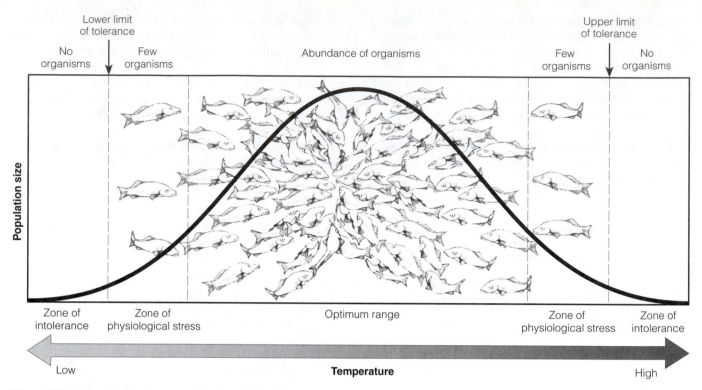

Figure 2-13 Natural capital: range of tolerance for a population of organisms, such as fish, to an abiotic environmental factor—in this case, temperature. These restrictions keep particular species from taking over an ecosystem by keeping their population size in check.

These observations are summarized in the **law of tolerance:** *The existence, abundance, and distribution of a species in an ecosystem are determined by whether the levels of one or more physical or chemical factors fall within the range tolerated by that species.* A species may have a wide range of tolerance to some factors and a narrow range of tolerance to others. Most organisms are least tolerant during juvenile or reproductive stages of their life cycles. Highly tolerant species can live in a variety of habitats with widely different conditions.

A variety of factors can affect the number of organisms in a population. However, sometimes one factor, known as a **limiting factor,** is more important in regulating population growth than other factors. This ecological principle, related to the law of tolerance, is called the **limiting factor principle:** *Too much or too little of any abiotic factor can limit or prevent growth of a population, even if all other factors are at or near the optimum range of tolerance.*

On land, precipitation often is the limiting factor. Lack of water in a desert limits plant growth. Soil nutrients also can act as a limiting factor on land. Suppose a farmer plants corn in phosphorus-poor soil. Even if water, nitrogen, potassium, and other nutrients are at optimum levels, the corn will stop growing when it uses up the available phosphorus.

Too much of an abiotic factor can also be limiting. For example, adding too much water or too much fer-tilizer is a common mistake made by many beginning gardeners and can kill plants.

Important limiting factors for aquatic life zones include temperature, sunlight, nutrient availability, and **dissolved oxygen (DO) content**—the amount of oxygen gas dissolved in a given volume of water at a particular temperature and pressure. Another limiting factor in aquatic life zones is **salinity**—the amounts of various inorganic minerals or salts dissolved in a given volume of water.

Major Biological Components of Ecosystems

Some organisms in ecosystems produce food and others consume food.

The earth's organisms either produce or consume food. **Producers,** sometimes called **autotrophs** (self-feeders), make their own food from compounds and energy obtained from their environment.

On land, most producers are green plants. In freshwater and marine life zones, algae and plants are the major producers near shorelines. In open water, the dominant producers are *phytoplankton*—mostly microscopic organisms that float or drift in the water.

Most producers capture sunlight to make complex compounds (such as glucose, $C_6H_{12}O_6$) by **photosyn-**

thesis. Although hundreds of chemical changes take place during photosynthesis, the overall reaction can be summarized as follows:

carbon dioxide + water + **solar energy** ⟶ glucose + oxygen

All other organisms in an ecosystem are **consumers,** or **heterotrophs** ("other-feeders") that get their energy and nutrients by feeding on other organisms or their remains. **Primary consumers** or **herbivores,** such as rabbits and zooplankton, eat producers. **Secondary consumers** or **carnivores,** such as foxes and fish, feed on herbivores. **Tertiary,** or **third,** and **higher level consumers** are carnivores that feed on other carnivores. **Omnivores** play dual roles by feeding on both plants and animals. Examples are pigs, rats, foxes, bears, cockroaches, and humans.

Decomposers (mostly certain types of bacteria and fungi) are specialized consumers that recycle organic matter in ecosystems. They do this by breaking down (*biodegrading*) dead organic material or **detritus** ("di-TRI-tus," meaning "debris") to get nutrients. This releases the resulting simpler inorganic compounds into the soil and water, where producers can take them up as nutrients.

Detritivores include decomposers and *detritus feeders* that directly consume detritus. Hordes of these waste eaters and degraders can transform a fallen tree trunk into a powder and finally into simple inorganic molecules that plants can absorb as nutrients (Figure 2-14). *In natural ecosystems, there is little or no waste.* One organism's wastes serve as resources for others, as the nutrients that make life possible are recycled again and again.

Producers, consumers, and decomposers use the chemical energy stored in glucose and other organic compounds to fuel their life processes. In most cells this energy is released by **aerobic respiration,** which uses oxygen to convert organic nutrients back into carbon dioxide and water. The net effect of the hundreds of steps in this complex process is represented by the following reaction:

glucose + oxygen ⟶ carbon dioxide + water + **energy**

Although the detailed steps differ, the net chemical change for aerobic respiration is the opposite of that for photosynthesis.

The survival of any individual organism depends on the *flow of matter and energy* through its body.

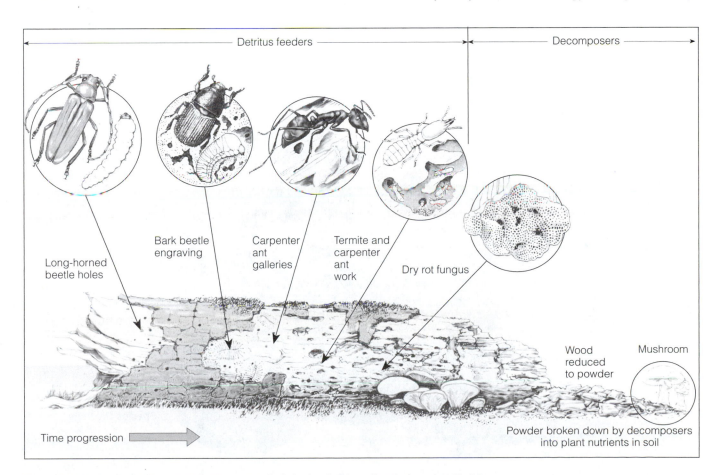

Figure 2-14 **Natural capital:** some detritivores, called *detritus feeders,* directly consume tiny fragments of this log. Other detritivores, called *decomposers* (mostly fungi and bacteria), digest complex organic chemicals in fragments of the log, converting them into simpler inorganic nutrients that can be used again by producers.

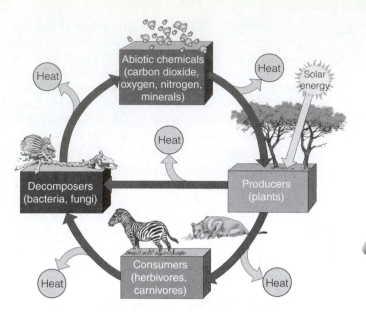

However, an ecosystem as a whole survives primarily through a combination of *matter recycling* and *one-way energy flow*. Figure 2-15 sums up most of what is going on in our planetary home.

Decomposers complete the cycle of matter by breaking down detritus into inorganic nutrients that can be reused by producers. These waste eaters and nutrient recyclers provide this crucial ecological service and never send us a bill. Without decomposers, the entire world would be knee-deep in plant litter, dead animal bodies, animal wastes, and garbage, and most life as we know it would no longer exist.

Thomson NOW! Explore the components of ecosystems, how they interact, the roles of bugs and plants, and what a fox will eat at ThomsonNOW.

ThomsonNOW™ Active Figure 2-15 Natural capital: the main structural components of an ecosystem (energy, chemicals, and organisms). Matter recycling and the flow of energy—first from the sun, then through organisms, and finally into the environment as low-quality heat—links these components. *See an animation based on this figure and take a short quiz on the concept.*

Biodiversity: A Crucial Resource

A vital renewable resource is the biodiversity found in the earth's variety of genes, species, ecosystems, and ecosystem processes.

Biological diversity, or **biodiversity,** is one of the earth's most important renewable resources. It includes four components, as shown in Figure 2-16.

Biodiversity involves several of the levels of organization of life shown in Figure 2-7. They include:

Functional Diversity
The biological and chemical processes such as energy flow and matter recycling needed for the survival of species, communities, and ecosystems.

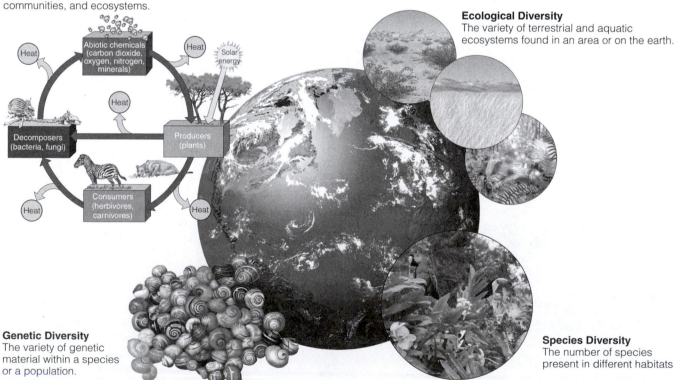

Ecological Diversity
The variety of terrestrial and aquatic ecosystems found in an area or on the earth.

Genetic Diversity
The variety of genetic material within a species or a population.

Species Diversity
The number of species present in different habitats

Figure 2-16 Natural capital: the major components of the earth's *biodiversity*—one of the earth's most important renewable resources. Some people also include *human cultural diversity* as part of the earth's biodiversity. Each human culture has developed various ways to deal with changing environmental conditions.

- **Genetic diversity:** differences in genes as represented by DNA among the individuals in a particular species

- **Species diversity:** the number or variety of species found in the world or in a particular area

- **Ecological diversity:** the number and variety of ecosystems in the world or in a particular area—including *community diversity,* the number and variety of biotic communities in an area, and *habitat diversity,* the number and variety of habitats in an area

- **Functional diversity:** the biological and chemical processes such as energy flow and matter cycling needed for the survival of species, communities, and ecosystems (Figure 2-15)

The earth's biodiversity is the biological wealth or capital that helps keep us alive and supports our economies. It supplies us with food, wood, fibers, energy, raw materials, industrial chemicals, and medicines—all of which pour hundreds of billions of dollars into the world economy each year. It also helps preserve the quality of the air and water, maintain the fertility of soils, dispose of wastes, and control populations of pests that attack crops and forests.

Yet, there is considerable scientific evidence that we are destroying and degrading some of the earth's biodiversity at an increasing rate as the ecological footprints of more and more people spread across the planet's surface. In 2005, the Millennium Ecosystem Assessment estimated that 12% of birds, 25% of mammals, and at least 32% of amphibians will be threatened with extinction in this century. The Assessment also indicated that current extinction rates are 100 to 10,000 times higher than the world's natural rate of extinction, typical of Earth's long-term history, because of human activities. We explore the causes and potential solutions for this problem in later chapters.

Biodiversity is a renewable resource as long we live off the biological income it provides instead of using up the natural capital that supplies this income. Understanding, protecting, and sustaining biodiversity is a major theme of ecology and of this book. Biodiversity is our lifeline!

ENERGY FLOW IN ECOSYSTEMS

Food Chains and Food Webs: Who Eats and Decomposes Whom

Food chains and webs show how eaters, the eaten, and the decomposed are connected to one another in an ecosystem.

All organisms, whether dead or alive, are potential sources of food for other organisms. A caterpillar eats a leaf, a robin eats the caterpillar, and a hawk eats the robin. Decomposers consume the leaf, caterpillar, robin, and hawk after they die. As a result, *there is little waste in natural ecosystems.*

The sequence of organisms, each of which is a source of food for the next, is called a **food chain.** It determines how energy and nutrients move from one organism to another through the ecosystem (Figure 2-17).

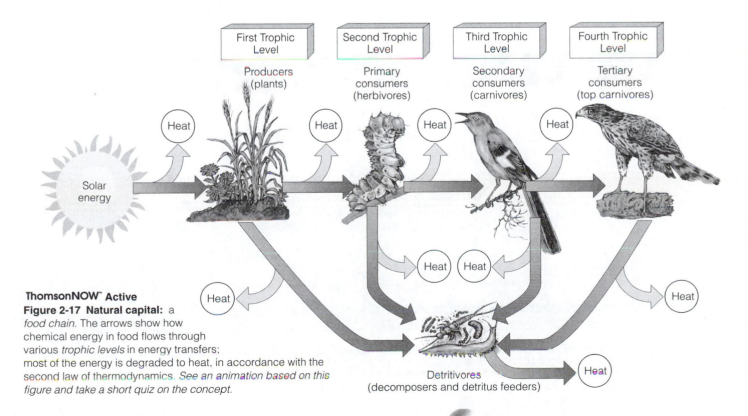

**ThomsonNOW™ Active
Figure 2-17 Natural capital:** a *food chain.* The arrows show how chemical energy in food flows through various *trophic levels* in energy transfers; most of the energy is degraded to heat, in accordance with the second law of thermodynamics. *See an animation based on this figure and take a short quiz on the concept.*

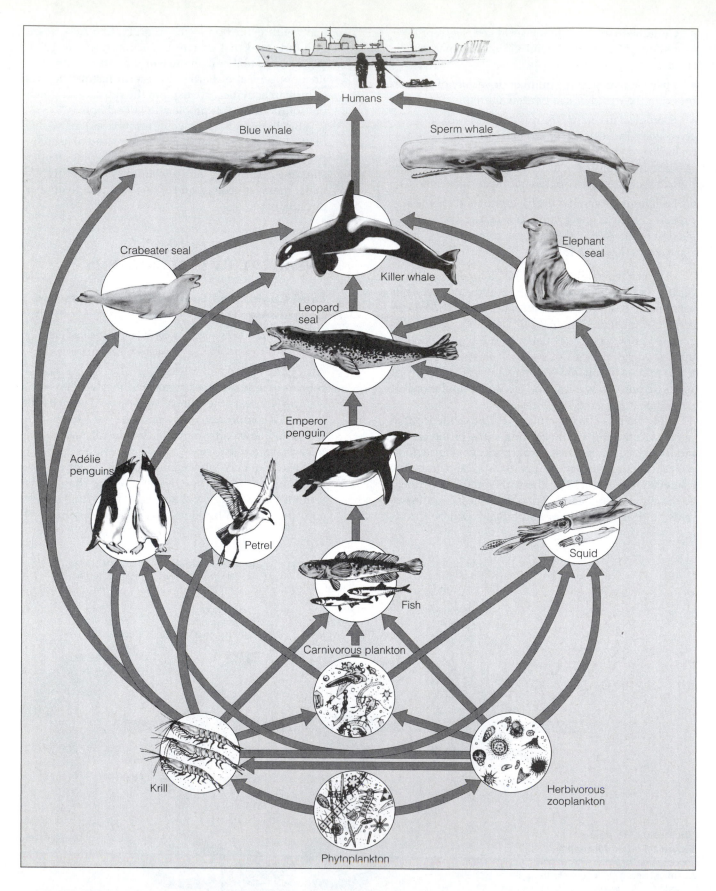

ThomsonNOW™ Active Figure 2-18 Natural capital: a greatly simplified *food web* in the Antarctic. Many more participants in the web, including an array of decomposer organisms, are not depicted here. *See an animation based on this figure and take a short quiz on the concept.*

Ecologists assign each organism in an ecosystem to a feeding level, also called a **trophic level,** depending on whether it is a producer or a consumer and on what it eats or decomposes. Producers belong to the first trophic level, primary consumers to the second trophic level, secondary consumers to the third, and so on. Detritivores and decomposers process detritus from all trophic levels.

Real ecosystems are more complex than this. Most consumers feed on more than one type of organism, and most organisms are eaten by more than one type of consumer. Because most species participate in several different food chains, the organisms in most ecosystems form a complex network of interconnected food chains called a **food web** (Figure 2-18). Trace the flows of matter and energy within this web. Trophic levels can be assigned in food webs just as in food chains. A food web shows how eaters, the eaten, and the decomposed are connected to one another. It is a map of life's interdependence.

Energy Flow in an Ecosystem: Losing Energy in Food Chains and Webs

There is a decrease in the amount of energy available to each succeeding organism in a food chain or web.

Each trophic level in a food chain or web contains a certain amount of **biomass,** the dry weight of all organic matter contained in its organisms. In a food chain or web, chemical energy stored in biomass is transferred from one trophic level to another.

The percentage of usable energy transferred as biomass from one trophic level to the next is called **ecological efficiency.** It ranges from 2–40% (that is, a loss of 60–98%) depending on the types of species and the ecosystem involved, but 10% is typical.

Assuming 10% ecological efficiency (90% loss) at each trophic transfer, if green plants in an area manage to capture 10,000 units of energy from the sun, then only about 1,000 units of energy will be available to support herbivores and only about 100 units will be available to support carnivores.

The more trophic levels, or steps in a food chain or web, the greater the cumulative loss of usable energy as energy flows through the various trophic levels. The **pyramid of energy flow** in Figure 2-19 illustrates this energy loss for a simple food chain, assuming a 90% energy loss with each transfer. How does this diagram help explain why there are not very many tigers in the world?

Energy flow pyramids explain why the earth can support more people if they eat at lower trophic levels by consuming grains, vegetables, and fruits directly rather than passing such crops through another trophic level and eating grain-eaters such as cattle.

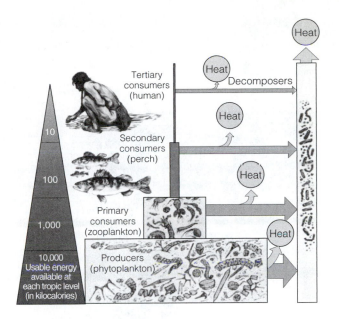

ThomsonNOW **Active Figure 2-19 Natural capital:** generalized *pyramid of energy flow* showing the decrease in usable energy available at each succeeding trophic level in a food chain or web. In nature, ecological efficiency varies from 2% to 40%, with 10% efficiency being common. This model assumes a 10% ecological efficiency (90% loss in usable energy to the environment, in the form of low-quality heat) with each transfer from one trophic level to another. QUESTION: *Why is calling this the "pyramid of energy" scientifically inaccurate? See an animation based on this figure and take a short quiz on the concept.*

The large loss of energy between successive trophic levels also explains why food chains and webs rarely have more than four or five trophic levels. In most cases, too little energy is left after four or five transfers to support organisms feeding at these high trophic levels. This explains why there are so few top carnivores such as eagles, hawks, tigers, and great white sharks. It also explains why such species usually are the first to suffer when the ecosystems that support them are disrupted, and why these species are so vulnerable to extinction. Do you think humans are on this list?

 Examine how energy flows among organisms at different trophic levels and through food webs in rain forests, prairies, and other ecosystems at ThomsonNOW.

Productivity of Producers: The Rate Is Crucial

Different ecosystems use solar energy to produce and use biomass at different rates.

The *rate* at which an ecosystem's producers convert solar energy into chemical energy as biomass is the ecosystem's **gross primary productivity (GPP).** However, to stay alive, grow, and reproduce, an ecosystem's producers must use some of the biomass they produce for their own respiration. **Net primary productivity**

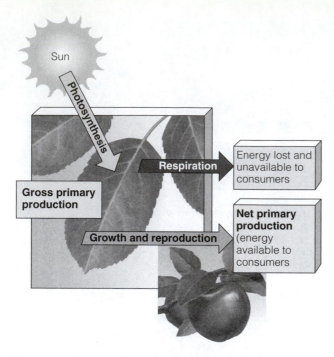

Figure 2-20 Natural capital: distinction between gross primary productivity and net primary productivity. A plant uses some of its gross primary productivity to survive through respiration. The remaining energy is available to consumers.

(NPP), is the *rate* at which producers use photosynthesis to store energy *minus* the *rate* at which they use some of this stored energy through aerobic respiration, as shown in Figure 2-20. In other words, NPP = GPP − R, where R is energy used in respiration. NPP is a measure of how fast producers can provide the food needed by other organisms (consumers) in an ecosystem.

Various ecosystems and life zones differ in their NPP, as graphed in Figure 2-21. Looking at this graph, what are nature's three most productive and three least productive systems? Despite its low net primary productivity, there is so much open ocean that it produces more of the earth's NPP per year than any of the other ecosystems and life zones shown in Figure 2-21.

As we have seen, producers are the source of all food in an ecosystem. Only the biomass represented by NPP is available as food for consumers and they use only a portion of this. Thus, *the planet's NPP ultimately limits the number of consumers (including humans) that can survive on the earth.* This is an important lesson from nature.

Peter Vitousek, Stuart Rojstaczer, and other ecologists estimate that humans now use, waste, or destroy about 27% of the earth's total potential NPP and 10–55% of the NPP of the planet's terrestrial ecosys-

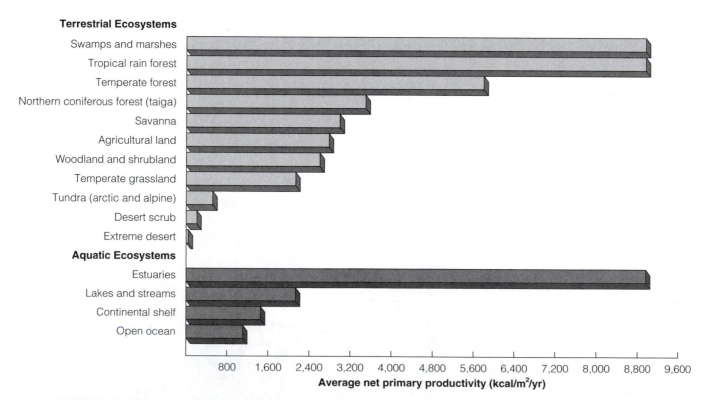

Figure 2-21 Natural capital: estimated annual average *net primary productivity* per unit of area in major life zones and ecosystems, expressed as kilocalories of energy produced per square meter per year (kcal/m²/yr). QUESTION: *What are nature's three most productive and three least productive systems?* (Data from *Communities and Ecosystems,* 2nd ed., by R. H. Whittaker, 1975. New York: Macmillan)

tems. They contend that this is the main reason why we are crowding out or eliminating the habitats and food supplies of a growing number of other species.

It is tempting to conclude from Figure 2-21 that we could feed the world's hungry millions by harvesting plants in highly productive estuaries, swamps, and marshes. But people cannot eat most plants that grow in these areas and these plants provide vital food sources for aquatic species that provide us and other consumers with protein.

We might also conclude that we could clear highly productive tropical rain forests and plant food crops. But most nutrients in rain forests are stored in the vegetation rather than in the soil, where they need to be to grow crops. When the trees are removed, frequent rains and growing and harvesting crops would deplete the nutrient-poor soils making them useful for only a short time without massive and expensive applications of fertilizers.

The earth's vast open oceans provide the largest percentage of the earth's net primary productivity, but harvesting the widely dispersed, tiny floating producers in the open ocean would take much more fossil fuel and other types of energy than the food energy we would get. In addition, this would disrupt the food webs of the open ocean.

SOIL: A RENEWABLE RESOURCE

What Is Soil and Why Is It Important?

Soil is a slowly renewed resource that provides most of the nutrients needed for plant growth and helps purify water.

Soil is a thin covering over most land that is a complex mixture of eroded rock, mineral nutrients, decaying organic matter, water, air, and billions of living organisms, most of them microscopic decomposers (Figure 2-22).

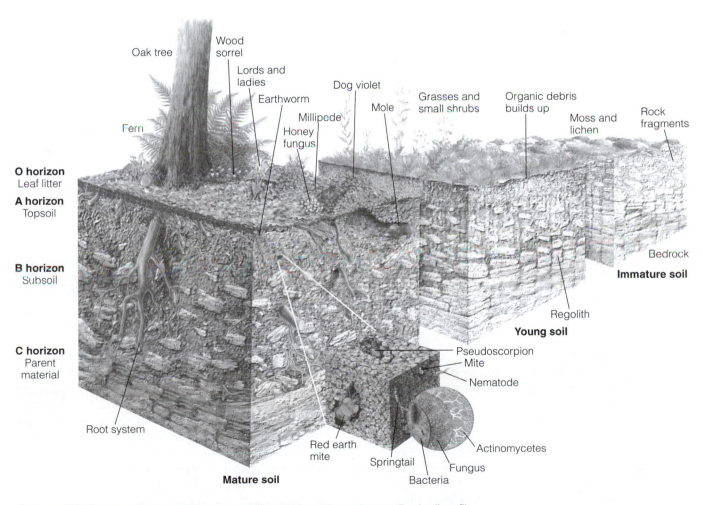

ThomsonNOW™ Active Figure 2-22 Natural capital: soil formation and generalized soil profile. Horizons, or layers, vary in number, composition, and thickness, depending on the type of soil. *See an animation based on this figure and take a short quiz on the concept.* (Used by permission of Macmillan Publishing Company from Derek Elsom, *Earth*, New York. Macmillan, 1992. Copyright © 1992 by Marshall Editions Developments Limited.)

Soil is a renewable resource but it is renewed very slowly. Depending mostly on climate, the formation of just 1 centimeter (0.4 inch) can take from 15 years to hundreds of years.

Soil is the base of life on land because it provides most of the nutrients needed for plant growth. Indeed, you are mostly made of soil nutrients imported into your body by the food you eat. Soil is also the earth's primary filter that cleanses water as it passes through. It also helps decompose and recycle biodegradable wastes and is a major component of the earth's water recycling and water storage processes.

Yet since the beginning of agriculture, human activities have led to rapid soil erosion, which can convert this renewable resource into a nonrenewable resource. Entire civilizations have collapsed because they mismanaged the topsoil that supported their populations. Studies indicate that one-third to one-half of the world's croplands are losing topsoil faster than it is being renewed by natural processes.

Layers in Mature Soils

Most soils developed over a long time consist of several layers containing different materials.

Mature soils (soils that have developed over a long time) are arranged in a series of horizontal layers called **soil horizons,** each with a distinct texture and composition that varies with different types of soils. A cross-sectional view of the horizons in a soil is called a **soil profile.** Most mature soils have at least three of the possible horizons (Figure 2-22). Think of them as floors in the building of life underneath your feet.

The top layer is the *surface litter layer,* or *O horizon.* It consists mostly of freshly fallen undecomposed or partially decomposed leaves, twigs, crop wastes, animal waste, fungi, and other organic materials. Normally, it is brown or black.

The *topsoil layer,* or *A horizon,* is a porous mixture of partially decomposed organic matter, called **humus,** and some inorganic mineral particles. It is usually darker and looser than deeper layers. A fertile soil that produces high crop yields has a thick topsoil layer with lots of humus. This helps topsoil hold water and nutrients taken up by plant roots.

The roots of most plants and most of a soil's organic matter are concentrated in a soil's two upper layers. As long as vegetation anchors these layers, soil stores water and releases it in a nourishing trickle.

The two top layers of most well-developed soils teem with bacteria, fungi, earthworms, and small insects that interact in complex food webs. Bacteria and other decomposer microorganisms found by the billions in every handful of topsoil break down some of its complex organic compounds into simpler inorganic compounds soluble in water. Soil moisture carrying these dissolved nutrients is drawn up by the roots of plants and transported through stems and into leaves as part of the earth's chemical cycling processes.

The color of its topsoil tells us a lot about how useful a soil is for growing crops. Dark-brown or black topsoil is nitrogen-rich and high in organic matter. Gray, bright yellow, or red topsoils are low in organic matter and need nitrogen enrichment to support most crops.

The spaces, or pores, between the solid organic and inorganic particles in the upper and lower soil layers contain varying amounts of air (mostly nitrogen and oxygen gas) and water. Plant roots need the oxygen for cellular respiration.

Some of the precipitation that reaches the soil percolates through the soil layers and occupies many of the soil's pores. This downward movement of water through soil is called **infiltration.** As the water seeps down, it dissolves various minerals and organic matter in upper layers and carries them to lower layers in a process called **leaching.**

Most of the world's crops are grown on soils exposed when grasslands and deciduous forests are cleared. Five important soil types, each with a distinct profile, are shown in Figure 2-23.

Soils vary in their content of *clay* (very fine particles), *silt* (fine particles), *sand* (medium-size particles), and *gravel* (coarse to very coarse particles). The relative amounts of the different sizes and types of these mineral particles determine **soil texture.**

To get an idea of a soil's texture, take a small amount of topsoil, moisten it, and rub it between your fingers and thumb. A gritty feel means it contains a lot of sand. A sticky feel means a high clay content, and you should be able to roll it into a clump. Silt-laden soil feels smooth, like flour. A loam topsoil, best suited for plant growth, has a texture between these extremes—a crumbly, spongy feeling—with many of its particles clumped loosely together.

A numerical scale of **pH** values is used to compare the acidity and alkalinity in water solutions (Figure 2-24, p. 44). The pH of a soil influences the uptake of soil nutrients by plants.

 Compare soil profiles from grassland, desert, and three types of forests at ThomsonNOW.

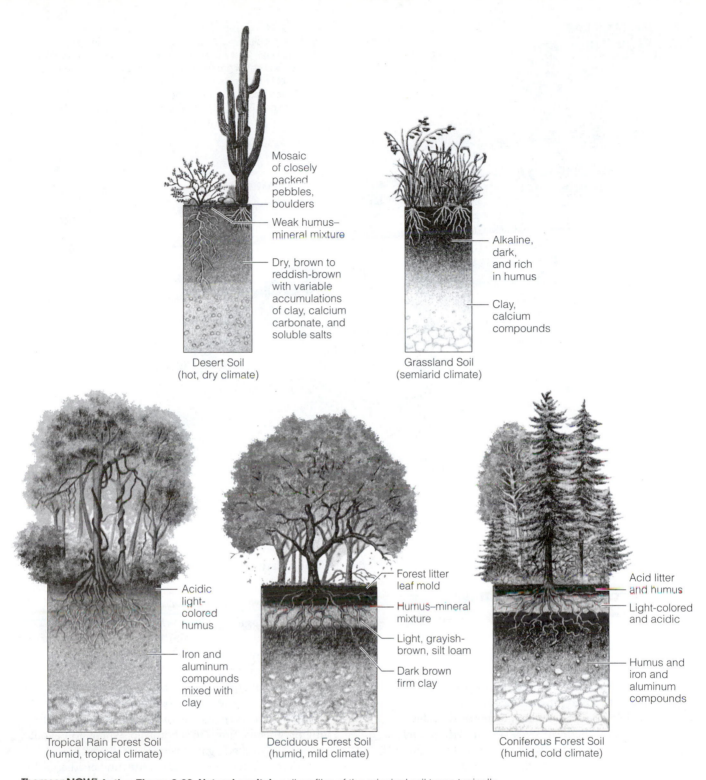

Mosaic of closely packed pebbles, boulders

Weak humus–mineral mixture

Dry, brown to reddish-brown with variable accumulations of clay, calcium carbonate, and soluble salts

Desert Soil
(hot, dry climate)

Alkaline, dark, and rich in humus

Clay, calcium compounds

Grassland Soil
(semiarid climate)

Acidic light-colored humus

Iron and aluminum compounds mixed with clay

Tropical Rain Forest Soil
(humid, tropical climate)

Forest litter leaf mold

Humus–mineral mixture

Light, grayish-brown, silt loam

Dark brown firm clay

Deciduous Forest Soil
(humid, mild climate)

Acid litter and humus

Light-colored and acidic

Humus and iron and aluminum compounds

Coniferous Forest Soil
(humid, cold climate)

ThomsonNOW **Active Figure 2-23 Natural capital:** soil profiles of the principal soil types typically found in five types of terrestrial ecosystems. *See an animation based on this figure and take a short quiz on the concept.*

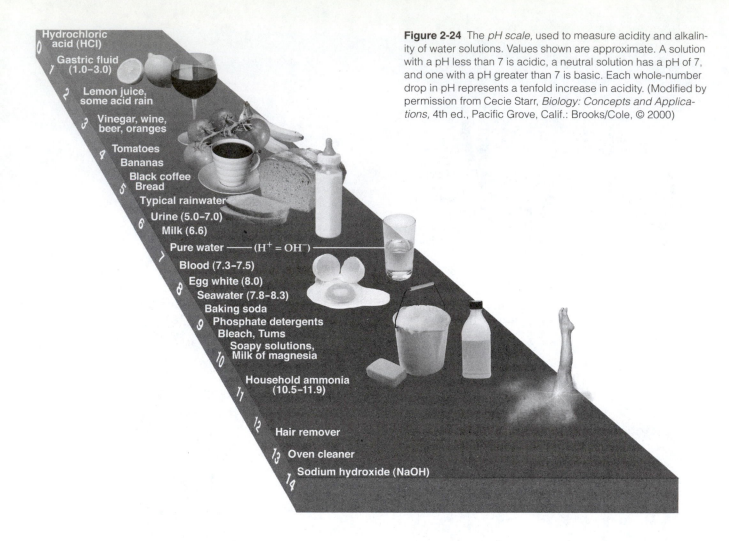

Figure 2-24 The *pH scale*, used to measure acidity and alkalinity of water solutions. Values shown are approximate. A solution with a pH less than 7 is acidic, a neutral solution has a pH of 7, and one with a pH greater than 7 is basic. Each whole-number drop in pH represents a tenfold increase in acidity. (Modified by permission from Cecie Starr, *Biology: Concepts and Applications*, 4th ed., Pacific Grove, Calif.: Brooks/Cole, © 2000)

MATTER CYCLING IN ECOSYSTEMS

Nutrient Cycles: Global Recycling

Global cycles recycle nutrients through the earth's air, land, water, and living organisms and, in the process, connect past, present, and future forms of life.

Nutrients are the elements and compounds that organisms need to live, grow, and reproduce. These substances move through air, water, soil, rock, and living organisms in cycles called **biogeochemical cycles** (literally, life–earth–chemical cycles) or **nutrient cycles.** These cycles, driven directly or indirectly by incoming solar energy and gravity, include the carbon, oxygen, nitrogen, phosphorus, and hydrologic (water) cycles (Figure 2-9).

The earth's chemical cycles also connect past, present, and future forms of life. Some of the carbon atoms in your skin may once have been part of a leaf, a dinosaur's skin, or a layer of limestone rock. Your grandmother, Plato, or a hunter–gatherer who lived 25,000 years ago may have inhaled some of the oxygen molecules you just inhaled.

The Water Cycle

A vast global cycle collects, purifies, distributes, and recycles the earth's fixed supply of water.

The **hydrologic cycle,** or **water cycle,** which collects, purifies, and distributes the earth's fixed supply of water, is shown in Figure 2-25. Trace the flows and paths in this diagram.

Solar energy evaporates water found on the earth's surface into the atmosphere. Some of this water returns to the earth as rain or snow, passes through living organisms, flows into bodies of water, and eventually is evaporated again to continue the cycle.

Some of the fresh water returning to the earth's surface as precipitation in this cycle becomes locked in glaciers. But most precipitation falling on terrestrial ecosystems becomes *surface runoff*. This water flows into streams and lakes, which eventually carry water back to the oceans, where it can be evaporated to cycle again.

Only about 0.02% of the earth's vast water supply is available to us as liquid freshwater in accessible groundwater deposits and in lakes, rivers, and streams.

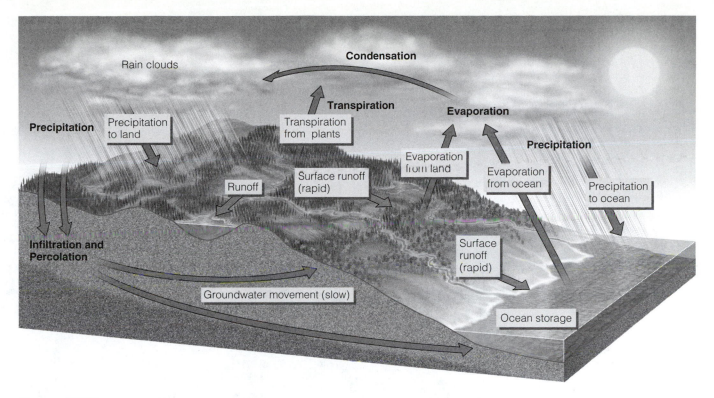

Rain clouds

Condensation

Precipitation

Precipitation to land

Transpiration

Transpiration from plants

Evaporation

Precipitation

Runoff

Surface runoff (rapid)

Evaporation from land

Evaporation from ocean

Precipitation to ocean

Infiltration and Percolation

Surface runoff (rapid)

Groundwater movement (slow)

Ocean storage

ThomsonNOW™ Active Figure 2-25 Natural capital: simplified model of the *hydrologic cycle. See an animation based on this figure and take a short quiz on the concept.*

The rest is too salty for us to use, is tied up as ice, or is too deep underground to extract.

Besides replenishing streams and lakes, surface runoff also causes soil erosion, which moves soil and weathered rock fragments from one place to another. Water is thus the primary sculptor of the earth's landscape. Because water dissolves many nutrient compounds, it is a major medium for transporting nutrients within and between ecosystems.

Throughout the hydrologic cycle, many natural processes act to purify water. Evaporation and subsequent precipitation act as a natural distillation process that removes impurities dissolved in water. Water flowing above ground through streams and lakes and below ground in *aquifers* is naturally filtered and purified by chemical and biological processes. Thus, *the hydrologic cycle also can be viewed as a cycle of natural renewal of water quality.*

Effects of Human Activities on the Water Cycle

We alter the water cycle by withdrawing large amounts of freshwater, clearing vegetation, eroding soils, polluting surface and underground water, and contributing to climate change.

During the past 100 years, we have been intervening in the earth's current water cycle in four major ways.

First, we withdraw large quantities of freshwater from streams, lakes, and underground sources.

Second, we clear vegetation from land for agriculture, mining, road and building construction, and other activities and sometimes cover the land with buildings, concrete, or asphalt. This increases runoff and flooding, reduces infiltration that recharges groundwater supplies, and accelerates soil erosion and landslides. *Third,* we modify water quality by adding nutrients (such as phosphates and nitrates found in fertilizers) and other pollutants.

Fourth, scientists have found that the earth's water cycle is speeding up as a result of a warmer climate caused partially by human inputs of carbon dioxide and other greenhouse gases into the atmosphere. This could change global precipitation patterns that affect the severity and frequency of droughts, floods, and storms. It can also intensify global warming by adding more water vapor—a powerful greenhouse gas—into the troposphere.

The Carbon Cycle: Part of Nature's Thermostat

Carbon cycles through the earth's air, water, soil, and living organisms.

Carbon, the basic building block of carbohydrates, fats, proteins, DNA, and other organic compounds

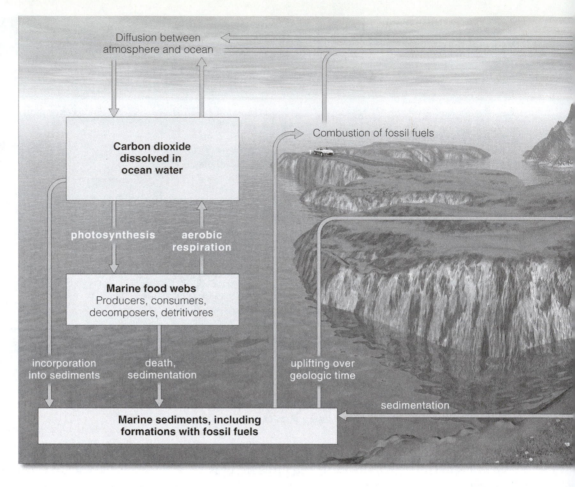

ThomsonNOW™ Active

Figure 2-26 Natural capital:
simplified model of the global
carbon cycle. Carbon moves
through both marine ecosystems
(left side) and terrestrial ecosystems (right side). Carbon reservoirs are shown as boxes;
processes that change one form
of carbon to another are shown in
unboxed print. *See an animation
based on this figure and take a
short quiz on the concept.* (From
Cecie Starr, *Biology: Concepts
and Applications,* 4th ed., Pacific
Grove, Calif.: Brooks/Cole,
© 2000)

necessary for life, is circulated through the biosphere
by the **carbon cycle,** as shown in Figure 2-26. Trace the
flows and paths in this diagram.

The carbon cycle is based on carbon dioxide (CO_2)
gas, which makes up 0.038% of the volume of the troposphere and is also dissolved in water. Carbon dioxide is a key component of nature's thermostat. If the
carbon cycle removes too much CO_2 from the atmosphere, the atmosphere will cool; if the cycle generates
too much, the atmosphere will get warmer. Thus, even
slight changes in the carbon cycle can affect climate
and ultimately the types of life that can exist on various parts of the planet.

Terrestrial producers remove CO_2 from the atmosphere, and aquatic producers remove it from the water. They then use photosynthesis to convert CO_2 into
complex carbohydrates such as glucose ($C_6H_{12}O_6$).

The cells in oxygen-consuming producers, consumers, and decomposers then carry out aerobic respiration. This process breaks down glucose and other
complex organic compounds and converts the carbon
back to CO_2 in the atmosphere or water for reuse by
producers. This linkage between *photosynthesis* in producers and *aerobic respiration* in producers, consumers,
and decomposers circulates carbon in the biosphere
and is a major part of the global carbon cycle. Oxygen

and hydrogen, the other elements in carbohydrates,
cycle almost in step with carbon.

Some carbon atoms take a long time to recycle.
Over millions of years, buried deposits of dead plant
matter and bacteria are compressed between layers of
sediment, where they form carbon-containing *fossil fuels* such as coal and oil (Figure 2-26). This carbon is not
released to the atmosphere as CO_2 for recycling until
these fuels are extracted and burned, or until long-
term geological processes expose these deposits to air.
In only a few hundred years, we have extracted and
burned fossil fuels that took millions of years to form.
This is why fossil fuels are nonrenewable resources on
a human time scale.

Effects of Human Activities on the Carbon Cycle

*Burning fossil fuels and clearing photosynthesizing vegetation
faster than it is replaced can increase the average temperature
of the atmosphere by adding excess carbon dioxide to the
atmosphere.*

Since 1800—and especially since 1950—we have been
intervening in the earth's carbon cycle in two ways
that add carbon dioxide to the atmosphere. *First,* in
some areas we clear trees and other plants that absorb

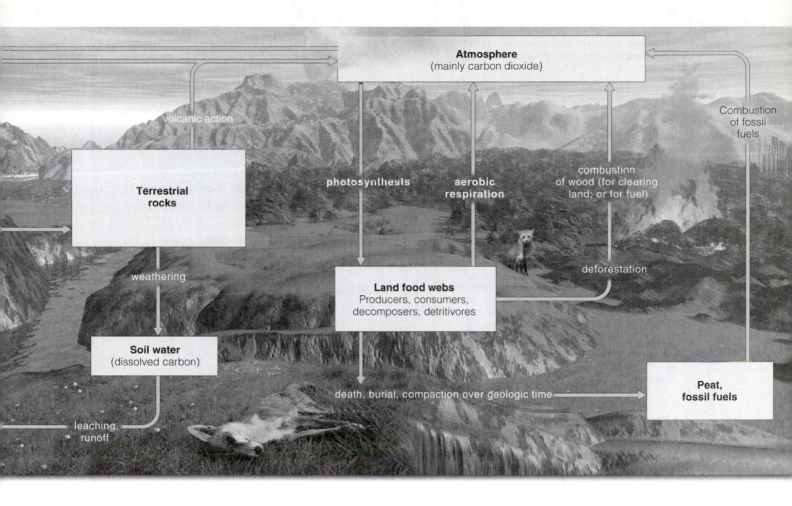

Atmosphere
(mainly carbon dioxide)

volcanic action

Combustion of fossil fuels

Terrestrial rocks

photosynthesis

aerobic respiration

combustion of wood (for clearing land; or for fuel)

weathering

deforestation

Land food webs
Producers, consumers, decomposers, detritivores

Soil water
(dissolved carbon)

death, burial, compaction over geologic time

Peat, fossil fuels

leaching, runoff

CO_2 through photosynthesis faster than they can grow back. *Second*, we add large amounts of CO_2 by burning fossil fuels and wood.

Computer models of the earth's climate systems suggest that increased concentrations of atmospheric CO_2 and other gases we are adding to the troposphere could enhance the planet's *natural greenhouse effect* that helps warm the lower atmosphere (troposphere) and the earth's surface (Figure 2-11). The resulting *global warming* could disrupt global food production and wildlife habitats, alter temperature and precipitation patterns, and raise the average sea level in various parts of the world.

The Nitrogen Cycle: Bacteria in Action

Different types of bacteria help recycle nitrogen through the earth's air, water, soil, and living organisms.

Nitrogen is the atmosphere's most abundant element, with chemically nonreactive nitrogen gas (N_2) making up 78% of the volume of the troposphere. However, N_2 cannot be absorbed and used directly as a nutrient by multicellular plants or animals.

Fortunately, atmospheric electrical discharges in the form of lightning and certain types of bacteria in aquatic systems, in the soil, and in the roots of some

plants can convert N_2 into compounds useful as nutrients for plants and animals as part of the **nitrogen cycle,** depicted in Figure 2-27 (p. 48). Trace the flows and paths in this diagram.

The nitrogen cycle consists of several major steps. In *nitrogen fixation,* specialized bacteria convert gaseous nitrogen (N_2) to ammonia (NH_3) that can be used by plants.

Ammonia not taken up by plants may undergo *nitrification.* In this process, specialized aerobic bacteria convert most of the ammonia in soil to *nitrite ions* (NO_2^-), which are toxic to plants, and *nitrate ions* (NO_3^-), which are easily taken up by plants as a nutrient. Animals, in turn, get their nitrogen by eating plants or plant-eating animals.

Plants and animals return nitrogen-rich organic compounds to the environment as wastes, cast-off particles, and dead bodies. In the *ammonification* step, vast armies of specialized decomposer bacteria convert this detritus into simpler nitrogen-containing inorganic compounds such as ammonia and water-soluble salts containing ammonium ions (NH_4^+).

Nitrogen leaves the soil in the *denitrification* step in which other specialized bacteria in waterlogged soil and in the bottom sediments of lakes, oceans, swamps, and bogs convert NH_3 and NH_4^+ back into nitrite and

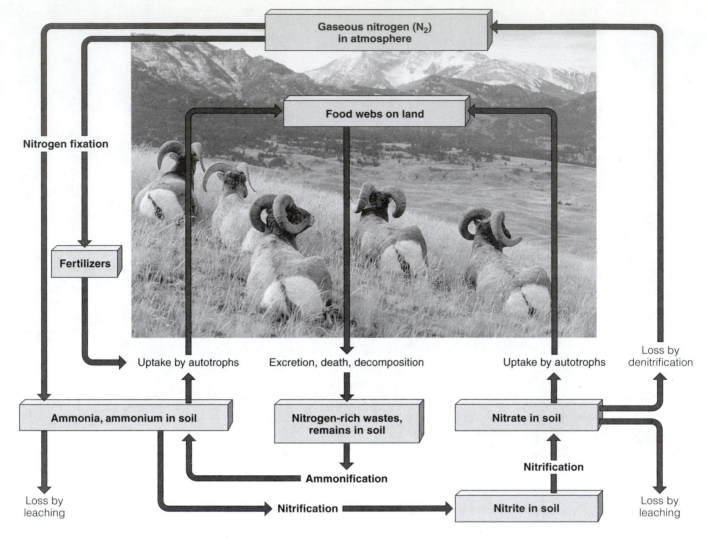

ThomsonNOW™ Active Figure 2-27 Natural capital: simplified model of the *nitrogen cycle* in a terrestrial ecosystem. Nitrogen reservoirs are shown as boxes; processes changing one form of nitrogen to another are shown in unboxed print. *See an animation based on this figure and take a short quiz on the concept.* (Adapted from Cecie Starr, *Biology: Today and Tomorrow*, Pacific Grove, CA: Brooks/Cole © 2005)

nitrate ions and then into nitrogen gas (N_2) and nitrous oxide gas (N_2O). These gases are released to the troposphere to begin the cycle again.

Effects of Human Activities on the Nitrogen Cycle

We add large amounts of various nitrogen-containing compounds to the earth's air and water and remove nitrogen from the soil.

We intervene in the nitrogen cycle in several ways. *First*, we add large amounts of nitric oxide (NO) into the troposphere when we burn any fuel. In the atmosphere, this gas can be converted to nitrogen dioxide gas (NO_2) and nitric acid (HNO_3), which can return to the earth's surface as damaging *acid deposition*, commonly called *acid rain*—more on this in Chapter 11.

Second, we add nitrous oxide (N_2O) to the troposphere through the action of anaerobic bacteria on livestock wastes and commercial inorganic fertilizers applied to the soil. This gas can warm the troposphere and deplete ozone in the stratosphere.

Third, nitrate ions in inorganic fertilizers can leach through the soil and contaminate drinking water. Since 1960, flows of nitrogen in terrestrial ecosystems have doubled, largely in the form of synthetic nitrogen fertilizer.

Fourth, we release large quantities of nitrogen stored in soils and plants as gaseous compounds into the troposphere through destruction of forests, grasslands, and wetlands.

Fifth, we upset aquatic ecosystems by adding excess nitrates in agricultural runoff and discharges from municipal sewage systems—more on this in Chapter 9.

Sixth, we remove nitrogen from topsoil when we harvest nitrogen-rich crops, irrigate crops, and burn or clear grasslands and forests before planting crops.

The Phosphorus Cycle

Phosphorus cycles fairly slowly through the earth's water, soil, and living organisms.

Phosphorus is a key component of DNA, RNA, and energy storage molecules in cells that circulates through water, the earth's crust, and living organisms in the **phosphorus cycle,** depicted in Figure 2-28. Trace the flows and paths in this diagram.

Unlike carbon and nitrogen, very little phosphorus circulates in the atmosphere because soil conditions do not allow bacteria to convert chemical forms of phosphorus to gaseous forms of phosphorus. The phosphorus cycle is slow, and on a short human time scale much phosphorus flows one way from the land to the oceans.

Phosphorous typically is found as phosphate salts containing phosphate ions (PO_4^{3-}) in terrestrial rock formations and ocean bottom sediments. Phosphate can be lost from the cycle for long periods when it washes from the land into streams and rivers and is carried to the ocean. There it can be deposited as sediment on the sea floor and remain for millions of years. Someday geological uplift processes can expose these seafloor deposits from which phosphate can be eroded to start the cyclical process again.

Because most soils contain little phosphate, it is often the *limiting factor* for plant growth on land unless phosphorus (as phosphate salts mined from the earth) is applied to the soil as a fertilizer. Phosphorus also limits the growth of producer populations in many freshwater streams and lakes because phosphate salts are only slightly soluble in water.

Effects of Human Activities on the Phosphorus Cycle

We remove large amounts of phosphate from the earth to make fertilizer, reduce phosphorus in tropical soils by clearing forests, and add excess phosphates to aquatic systems.

We intervene in the earth's phosphorus cycle in three ways. *First,* we mine large quantities of phosphate rock to make commercial inorganic fertilizers and detergents. *Second,* we reduce the available phosphate in tropical soils when we cut down tropical forests. *Third,* we disrupt aquatic systems with phosphates from runoff of animal wastes and fertilizers and discharges from sewage treatment systems.

Thomson™ NOW! Learn more about the water, carbon, nitrogen, and phosphorus cycles using interactive animations at ThomsonNOW.

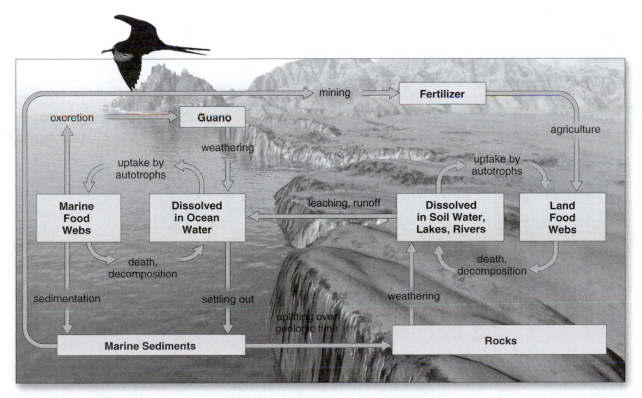

Figure 2-28 Natural capital: simplified model of the *phosphorus cycle.* Phosphorus reservoirs are shown as boxes; processes that change one form of phosphorus to another are shown in unboxed print. (From Cecie Starr and Ralph Taggart, *Biology: The Unity and Diversity of Life,* 9th ed., Belmont, Calif.: Wadsworth, © 2001)

What Is the Rock Cycle? Really Slow Recycling

The earth's three types of rock are transformed and recycled by a rock cycle that is the planet's slowest cyclical process.

Rock is any material that makes up a large, natural, continuous part of the earth's crust. Based on the way it forms, rock is placed in three broad classes. One is **igneous rock,** formed below or on the earth's surface when molten rock material (magma) wells up from the earth's upper mantle or deep crust, cools, and hardens into rock. Examples are granite (formed underground) and lava rock (formed above ground when molten lava cools and hardens).

A second type is **sedimentary rock,** formed from sediment when existing rocks are weathered and eroded into small pieces and carried by wind and water to be deposited in a lake or river. As deposited layers from weathering and erosion become buried and compacted, the resulting pressure causes their particles to bond together to form sedimentary rocks such as sandstone and shale.

The third type is **metamorphic rock,** produced when an existing rock is subjected to high temperatures (which may cause it to melt partially), high pressures, chemically active fluids, or a combination of these agents. Examples are anthracite (a form of coal), slate, and marble.

Rocks are constantly exposed to various physical and chemical conditions that can change them over time. The interaction of processes that change rocks from one type to another is called the **rock cycle** (Figure 2-29).

The rock cycle recycles the earth's three types of rocks over millions of years and is the slowest of the earth's cyclic processes. It concentrates the planet's nonrenewable mineral resources on which we depend. Without the incredibly slow rock cycle, you would not exist.

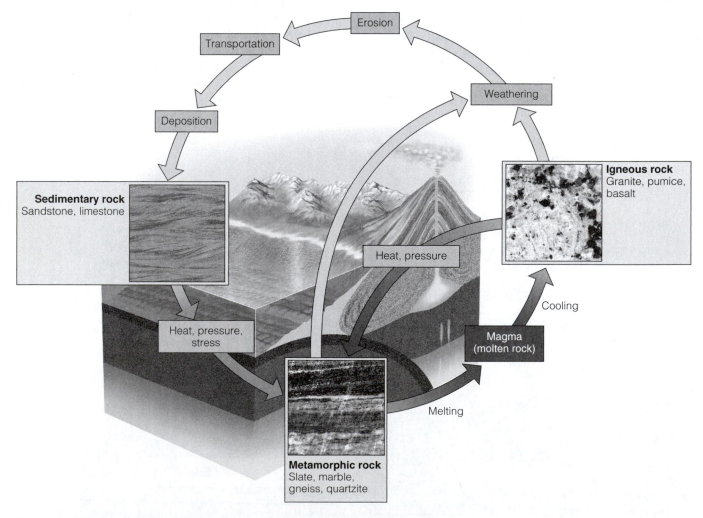

Figure 2-29 Natural capital: the *rock cycle* is the slowest of the earth's cyclic processes. The earth's materials are recycled over millions of years by three processes: *melting*, *erosion*, and *metamorphism*, which produce *igneous, sedimentary*, and *metamorphic* rocks. Rock from any of these classes can be converted to rock of either of the other two classes, or can be recycled within its own class.

All things come from earth, and to earth they all return.

MENANDER (342–290 BC)

CRITICAL THINKING

1. Respond to the following statements:
   a. Scientists have not absolutely proven that anyone has ever died from smoking cigarettes.
   b. The greenhouse theory—that certain gases, such as water vapor and carbon dioxide, warm the troposphere—is not a reliable idea because it is only a scientific theory.

2. Try to find an advertisement or article describing some aspect of science in which **(a)** the concept of scientific proof is misused, **(b)** the term *theory* is used when it should have been *hypothesis,* or **(c)** a sound or consensus scientific finding is dismissed or downplayed because it is "only a theory."

3. A tree grows and increases its mass. Explain why this is not a violation of the law of conservation of matter.

4. If there is no "away," as in "to throw away," why is the world not filled with waste matter?

5. Use the second law of thermodynamics to explain why a barrel of oil can be used only once as a fuel.

6. **(a)** A bumper sticker asks, "Have you thanked a green plant today?" Give two reasons for appreciating a green plant. **(b)** Trace the sources of the materials that make up the bumper sticker, and decide whether the sticker itself is a sound application of the slogan. **(c)** Explain how decomposers help keep you alive.

7. Using the second law of thermodynamics, explain why **(a)** there is such a sharp decrease in usable energy as energy flows through a food chain or web and **(b)** many poor people in developing countries live on a mostly vegetarian diet.

8. What would happen to an ecosystem if **(a)** all its decomposers and detritus feeders were eliminated or **(b)** all its producers were eliminated?

9. List two questions that you would like to have answered as a result of reading this chapter.

LEARNING ONLINE

The website for this book contains helpful study aids and many ideas for further reading and research. They include a chapter summary, review questions for the entire chapter, flash cards for key terms and concepts, a multiple-choice practice quiz, interesting Internet sites, references, information about green careers, and a guide for accessing thousands of InfoTrac® College Edition articles. Log into

www.thomsonedu.com/biology/miller

Then choose Chapter 2, and select a learning resource. For access to animations, additional quizzes, chapter outlines and summaries, register and log into

at **www.thomsonedu.com** using the access code card in the front of your book.

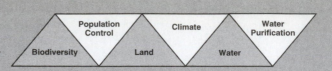

In looking at nature . . . never forget that every single organic being around us may be said to be striving to increase its numbers.

CHARLES DARWIN, 1859

EVOLUTION AND ADAPTATION

What Is Evolution?

Evolution is the change in a population's genetic makeup over time.

According to scientific evidence, populations of organisms adapt to changes in environmental conditions through **biological evolution,** or **evolution**—the change in a population's genetic makeup through successive generations. Note that *populations, not individuals, evolve by becoming genetically different.*

According to the **theory of evolution,** all species descended from earlier, ancestral species. In other words, life comes from life. This widely accepted scientific theory explains how life has changed over the past 3.7 billion years (Figure 3-1) and why life is so diverse today.

Religious and other groups offer other explanations, but the theory of evolution is the accepted *scientific* explanation of how life on the earth has developed. Most of the evidence for this theory comes from **fossils:** mineralized or petrified replicas of skeletons, bones, teeth, shells, leaves, and seeds, or impressions of such items found in rocks. Evidence about the earth's early history also comes from chemical analysis and measurements of elements in primitive rocks and fossils. Analysis of material in cores drilled out of buried ice and comparisons of the DNA of past and current organisms offer still more information.

 Get a detailed look at early evolution—the roots of the tree of life—at ThomsonNOW.

Biological Evolution through Mutation: Changes in the Gene Pool

A population's gene pool changes over time when beneficial changes or mutations in its DNA molecules are passed on to offspring.

Because biological evolution involves the change in a population's genetic makeup through successive generations, *populations—not individuals—evolve by becoming genetically different.* A population's **gene pool** consists of all of the genes (Figure 2-3) in its individuals. In modern scientific terms, *biological evolution* is a change in a population's gene pool over time.

The first step in microevolution is the development of *genetic variability* in a population. Such genetic variety occurs through **mutations:** random changes in the structure or number of DNA molecules in a cell that can be inherited by offspring. Mutations can occur in two major ways. One is by exposure of DNA to external agents such as radioactivity, X rays, and natural and human-made chemicals (called *mutagens*). The other results from random mistakes that sometimes occur in coded genetic instructions when DNA molecules are copied each time a cell divides and whenever an organism reproduces.

Mutations can occur in any cells, but only those in reproductive cells are passed on to offspring. Some mutations are harmless but most are lethal. *Every so often, a mutation is beneficial.* The result is new genetic traits that give that individual and its offspring better chances for survival and reproduction under existing environmental conditions or when such conditions change.

Natural Selection and Adaptation: Leaving More Offspring with Beneficial Genetic Traits

Some members of a population may have genetic traits that enhance their ability to survive and produce offspring with such traits.

The next step in biological evolution is **natural selection.** It occurs when some individuals of a population have genetically based traits that increase their chances of survival and their ability to produce offspring with the same traits.

Three conditions are necessary for evolution of a population by natural selection to occur. *First,* there must be enough *genetic variability* for a trait to exist in a population. *Second,* the trait must be *heritable,* meaning it can be passed from one generation to another. *Third,* the trait must lead to **differential reproduction.** This means it must enable individuals with the trait to leave more offspring than other members of the population. Note that natural selection acts on individuals, but evolution occurs in populations.

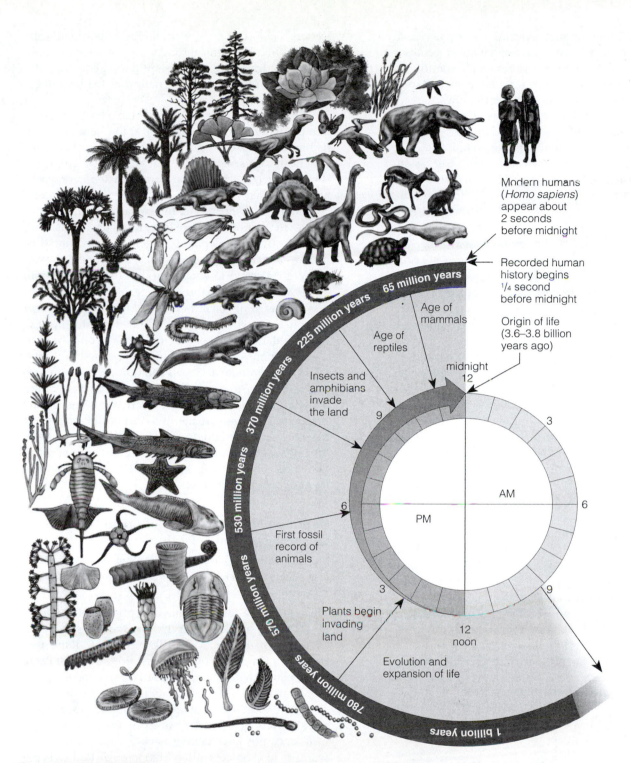

Figure 3-1 Natural capital: greatly simplified overview of the biological evolution of life on the earth, which was preceded by about 1 billion years of chemical evolution. Microorganisms (mostly bacteria) that lived in water dominated the early span of biological evolution on the earth, between about 3.7 billion and 1 billion years ago. Plants and animals evolved first in the seas. Fossil and recent DNA evidence suggests that plants began invading the land some 780 million years ago, and animals began living on land about 370 million years ago. Humans arrived on the scene only a very short time ago—equivalent to less than an eye blink of the earth's roughly 3.7-billion-year history of biological evolution.

An **adaptation,** or **adaptive trait,** is any heritable trait that enables organisms to survive and reproduce better under prevailing environmental conditions. Natural selection tends to preserve beneficial adaptations in populations and discard harmful ones.

Structural adaptations include *coloration* (allowing more individuals to hide from predators or to sneak up on prey), *mimicry* (looking like a poisonous or dangerous species), *protective cover* (shell, thick skin, bark, or thorns), and *gripping mechanisms* (hands with opposable thumbs). *Physiological adaptations* include the ability to hibernate during cold weather and give off chemicals that poison or repel prey. *Behavioral adaptations* include the ability to fly to a warmer climate during winter.

When faced with a change in environmental conditions, a population of a species has three possibilities: *adapt* to the new conditions through natural selection, *migrate* (if possible) to an area with more favorable conditions, or *become extinct.*

The process of biological evolution by natural selection can be summarized as follows: *Genes mutate, individuals are selected, and populations evolve to be better adapted to survive and reproduce under existing environmental conditions.*

New evidence indicates that some new species can arise through *hybridization.* It occurs when individuals of two distinct species crossbreed to produce an individual or *hybrid* that in some cases has a better ability to survive than conventional offspring of the two parent species.

Biologists are also finding examples of species exchanging genes through *horizontal gene transfer.* Such gene swapping can occur when one species feeds upon, infects, or comes into close contact with another species (such as a bacterium or virus) and transfers bits of genetic information. Hybridization and gene transfers and the resulting adaptations can occur rapidly compared to the thousands to millions of years required for conventional evolution through natural selection.

 How many moths can you eat? Find out and learn more about adaptation at ThomsonNOW.

Limits on Adaptation through Conventional Natural Selection

A population's ability to adapt to new environmental conditions is limited by its gene pool and how fast it can reproduce.

Will adaptations in the not-too-distant future allow our skin to become more resistant to the harmful effects of ultraviolet radiation, our lungs to cope with air pollutants, and our livers to better detoxify pollutants?

The answer is *no* because of the two limits on adaptations in nature through conventional natural se-

lection. *First,* a change in environmental conditions can lead to adaptation through natural selection only for genetic traits already present in the gene pool of a population. You must have genetic dice to play the genetic dice game.

Second, even if a beneficial heritable trait is present in a population, the population's ability to adapt may be limited by its reproductive capacity. Populations of genetically diverse species that reproduce quickly—such as weeds, mosquitoes, rats, bacteria, or cockroaches—often adapt to a change in environmental conditions in a short time. In contrast, populations of species that cannot produce large numbers of offspring rapidly—such as elephants, tigers, sharks, and humans—take a long time (typically thousands or even millions of years) to adapt through natural selection. You have to be able to throw the genetic dice fast.

Here is some *bad news* for most members of a population. Even when a favorable genetic trait is present in a population, most of the population would have to die or become sterile so that individuals with the trait could predominate and pass the trait on through natural selection. Thus, most players get kicked out of the genetic dice game before they can win. This means that most members of the human population would have to die for hundreds of thousands of generations for a new genetic trait to emerge through natural selection. This is hardly a desirable solution to the environmental problems we face.

Common Myths about Biological Evolution through Natural Selection

Evolution through natural selection is about leaving the most descendants, and there is no master plan leading to genetic perfection.

There are three common misconceptions about evolution. One is that the "survival of the fittest" means "survival of the strongest." To biologists, *fitness* is a measure of reproductive success, not strength. Thus, the fittest individuals are those that leave the most descendants.

Another misconception is that organisms develop certain traits because they need or want them. A giraffe does not have a long neck because it needs or wants to feed on vegetation high in trees. Rather, some ancestor had a gene for long necks that gave it an advantage over other members of its population in getting food, that giraffe produced more offspring with long necks, and those offspring had a better chance of survival and producing more offspring because of easier access to food with their long necks.

Third is the misconception that evolution involves some grand plan of nature in which species become more perfectly adapted. From a scientific standpoint, no plan or goal of perfection has been identified in the evolutionary process.

Geologic Processes, Climate Change, Catastrophes, and Evolution

Tectonic plate movements, volcanic eruptions, earthquakes, climate change, and catastrophes can wipe out existing species and increase the formation of new ones.

Throughout the earth's history, continents have split and joined as their underlying solid plates, called *tectonic plates*, have drifted very slowly back and forth across the planet's surface. This process has had two important effects on the evolution and location of life on the earth. *First*, the locations of continents and oceanic basins greatly influence the earth's climate and thus help determine where plants and the animals can live. *Second*, the movement of continents has enhanced evolution by allowing species to move, adapt to new environments, and form new species through natural selection. When continents join together, populations can disperse to new areas and evolve to adapt to new environmental conditions. And when continents separate, populations must evolve under isolated conditions.

Volcanic eruptions and earthquakes can also affect biological evolution by destroying habitats and reducing or wiping out populations. On the other hand, deposits of lava can yield a soil that can provide habitats for some species. And earthquakes can separate populations, which can lead to the formation of new species

as each isolated population changes genetically in response to new environmental conditions.

Throughout its long history, the earth's climate has changed drastically, sometimes cooling and covering much of the earth with ice and sometimes heating up, melting ice, and drastically raising sea levels. These changes have major effects on evolution by determining where different types of plants and animals can live and thrive. Life on the earth has endured even more dramatic catastrophes from asteroids and other extraterrestrial bodies striking the earth and from major upheavals of the earth's crust. Such impacts have caused widespread destruction of ecosystems and wiped out large numbers of species, opening up opportunities for the evolution of new species.

ECOLOGICAL NICHES AND ADAPTATION

What Is an Ecological Niche? How Species Coexist

Each species in an ecosystem has a specific role or way of life.

If asked what role a certain species such as an alligator plays in an ecosystem, an ecologist would describe its **ecological niche,** or simply **niche** (pronounced

---

Cockroaches: Nature's Ultimate Survivors

Cockroaches, the bugs many people love to hate, have been around for about 350 million years and are one of the great success stories of evolution. They are so successful because they are *generalists*.

The earth's 3,500 cockroach species can eat almost anything, including algae, dead insects, fingernail clippings, electrical cords, glue, paper, and soap. They can also live and breed almost anywhere except in polar regions.

Some cockroach species can go for a month without food, survive for a month on a drop of water from a dishrag, and withstand massive doses of radiation. One species can survive being frozen for 48 hours.

They usually can evade their predators, and a human foot in hot

pursuit, because most species have antennae that can detect minute movements of air, sensors in their knee joints to sense vibration, and rapid response times (faster than you can blink). Some even have wings. They also have compound eyes that allow them to see in almost all directions at once. Each eye has about 2,000 lenses, compared to one in each of your eyes.

They also have high reproductive rates. In only a year, a single Asian cockroach (especially prevalent in Florida) and its young can add about 10 million new cockroaches to the world. Their high reproductive rate also helps them to quickly develop genetic resistance to almost any poison we throw at them.

Most cockroaches also sample food before it enters their mouths and learn to shun foul-tasting poisons. They also clean up after themselves by eating their own dead and, if food is scarce enough, their living.

About 25 species of cockroaches live in homes. They can carry viruses and bacteria that cause diseases such as hepatitis, polio, typhoid fever, plague, and salmonella. They can also cause people to have allergic reactions ranging from watery eyes to severe wheezing. About 60% of Americans suffering from asthma are allergic to live or dead cockroaches.

However, cockroaches play an important role in nature's food webs. They make tasty meals for birds and lizards.

Critical Thinking

If you could, would you exterminate all cockroach species? What might be some ecological consequences of doing this?

"nitch"). It is a species' way of life or role in an ecosystem and includes everything that affects its survival and reproduction.

A species' ecological niche includes the *adaptations* or *adaptive traits* its members have acquired through evolution. It also includes its range of tolerance for various physical and chemical conditions, such as temperature, the types and amounts of resources it uses, how it interacts with other living and nonliving components of its ecosystems, and the role it plays in the energy flow and matter cycling in an ecosystem.

The ecological niche of a species is different from its **habitat,** the physical location where its populations live. Ecologists often say that a niche is like a species' occupation, whereas habitat is like its address.

Generalist and Specialist Species: Broad and Narrow Niches

Some species have broad ecological roles and others have narrower or more specialized roles.

Scientists use the niches of species to classify them broadly as *generalists* or *specialists*. **Generalist species** have broad niches. They can live in many different places, eat a variety of foods, and tolerate a wide range of environmental conditions. Flies, cockroaches (Spotlight, p. 55), mice, rats, white-tailed deer, raccoons, coyotes, copperheads, starlings, and humans, and many weeds such as crabgrass and dandelions are generalist species.

On the other hand, *tiger salamanders* are **specialists** because their niches are narrow. They can breed only in fishless ponds where their larvae will not be eaten. Other specialists are the *red-cockaded woodpeckers,* which carve nest holes almost exclusively in old (at least 75-year-

old) longleaf pines, and China's highly endangered *giant pandas,* which feed almost exclusively on various types of bamboo. Some shorebirds also occupy specialized niches, feeding on crustaceans, insects, and other organisms on sandy beaches and their adjoining coastal wetlands (Figure 3-2).

Is it better to be a generalist than a specialist? It depends. When environmental conditions are fairly constant, as in a tropical rain forest (Figure 3-2), specialists have an advantage because they have fewer competitors. But under rapidly changing environmental conditions, the generalist usually is better off than the specialist.

SPECIATION, EXTINCTION, AND BIODIVERSITY

How Do New Species Evolve? Moving Out and Moving On

A new species can arise when members of a population are isolated from other members so long that changes in their genetic makeup prevent them from producing fertile offspring if they get together again.

Under certain circumstances natural selection can lead to an entirely new species. In this process, called **speciation,** two species arise from one. For sexually reproducing species, a new species is formed when some members of a population can no longer breed with other members to produce fertile offspring.

The most common mechanism of speciation (especially among sexually reproducing animals) takes

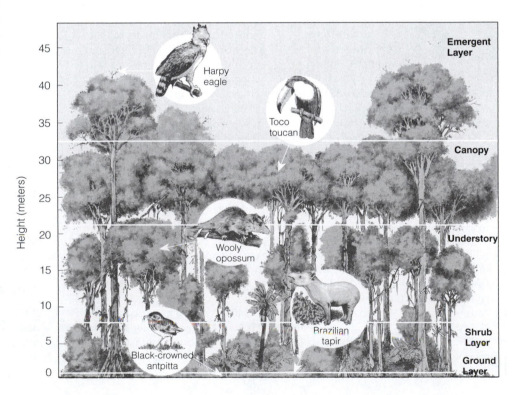

Figure 3-2 Natural capital: stratification of specialized plant and animal niches in various layers of a tropical rain forest. The presence of these specialized niches enables species to avoid or minimize competition for resources and results in the coexistence of a great variety of species.

place in two phases: geographic isolation and reproductive isolation. **Geographic isolation** occurs when different groups of the same population of a species become physically isolated from one another for long periods. For example, part of a population may migrate in search of food and then begin living in another area with different environmental conditions. Populations also may become separated by a physical barrier (such as a mountain range, stream, lake, or road), by a change such as a volcanic eruption or earthquake, or when a few individuals are carried to a new area by wind or water

In **reproductive isolation,** mutation and natural selection operate independently in the gene pools of geographically isolated populations. If this process continues long enough, members of the geographically and reproductively isolated populations of sexually reproducing species may become so different in genetic makeup that if they get together again they cannot produce live, fertile offspring. Then one species has become two, and speciation has occurred (Figure 3-3).

For some rapidly reproducing organisms, this type of speciation may occur within hundreds of years. For most species, it takes from tens of thousands to millions of years. However, some species can speed up the process of evolution by forming new species through hybridization and lateral gene transfer.

 Learn more about different types of speciation and ways in which they occur at ThomsonNOW.

Extinction: Lights Out

A species becomes extinct when its populations cannot adapt to changing environmental conditions.

After speciation, the second process affecting the number and types of species on the earth is **extinction,** in which an entire species ceases to exist. When environmental conditions change drastically enough, a species must evolve (become better adapted), move to a more favorable area if possible, or become extinct.

Extinction is the ultimate fate of all species, just as death is for all individual organisms. Biologists estimate that 99.9% of all the species that have ever existed are now extinct; humans will probably not escape this ultimate fate.

Some species are highly adaptive and can tolerate significant environmental change or migrate to other areas. Other species become extinct when environmental conditions change so rapidly or severely that they cannot genetically adapt to the change.

Species that are found in only one area are called **endemic species** and are especially vulnerable to extinction. They are found on islands and in other unique small areas, especially in tropical rain forests where most species are highly specialized (Figure 3-2). Taking down just one or two large trees in a tropical forest can doom certain specialized species to extinction.

As local environmental conditions change, a certain number of species disappear at a low rate, called **background extinction.** Based on the fossil record and analysis of ice cores, biologists estimate that the average annual background extinction rate is one to five species for each million species on the earth.

In contrast, **mass extinction** is a significant rise in extinction rates above the background level. It is a catastrophic, widespread (often global) event (such as climate change) in which large groups of existing species (perhaps 25–70%) are wiped out in a geological period lasting up to 5 million years. Fossil and geological evidence indicates that the earth's species have experienced five mass extinctions (20–60 million years apart) during the past 500 million years

Scientists have also identified periods of **mass depletion** in which extinction rates were much higher than normal but not high enough to classify as a mass

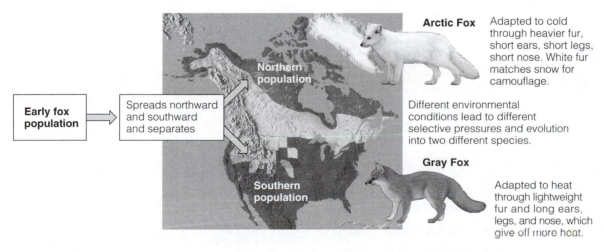

Early fox population → Spreads northward and southward and separates

Northern population

Southern population

Arctic Fox Adapted to cold through heavier fur, short ears, short legs, short nose. White fur matches snow for camouflage.

Different environmental conditions lead to different selective pressures and evolution into two different species.

Gray Fox Adapted to heat through lightweight fur and long ears, legs, and nose, which give off more heat.

Figure 3-3 *Geographic isolation* can lead to reproductive isolation, divergence of gene pools, and speciation.

extinction. A mass extinction or mass depletion crisis for some species is an opportunity for other species that can fill unoccupied niches or newly created ones. The existence of millions of species today means that speciation, on average, has kept ahead of extinction, especially during the last 250 million years (Figure 3-4).

Evidence shows that the earth's mass extinctions and depletions have been followed by periods of recovery called **adaptive radiations** in which numerous new species evolve to fill new or vacated niches in changed environments. Fossil records suggest that it takes 1–10 million years for adaptive radiations to rebuild biological diversity after a mass extinction or depletion.

Effects of Human Activities on the Earth's Biodiversity: Are We a Wise Species?

The scientific consensus is that human activities are decreasing the earth's biodiversity.

Speciation minus extinction equals *biodiversity*, the planet's genetic raw material for future evolution in response to changing environmental conditions. Extinction is a natural process. But much evidence indicates that humans have become a major force in the premature extinction of species.

According to biologists Stuart Primm and Edward O. Wilson, during the 20th century, extinction rates increased by 100–1,000 times the natural background extinction rate. As human population and resource consumption increase over the next 50–100 years, we are expected to take over more and more of the earth's surface and net primary productivity (NPP). According to Wilson and Pimm, this may cause the premature extinction of at least one-fifth of the earth's current species by 2030 and up to half by the end of this century. This could constitute a new *mass depletion* and possibly a new *mass extinction*. Wilson says that if we make an "all-out effort to save the biologically richest parts of the world, the amount of loss can be cut at least by half."

On our short time scale, such major losses cannot be recouped by formation of new species; it took millions of years after each of the earth's past mass extinctions and depletions for life to recover to the previous level of biodiversity. We are also destroying or degrading ecosystems such as tropical forests, coral reefs, and wetlands that are centers for future speciation. See the Guest Essay on this topic by Norman Myers on the website for this chapter.

Each species here today represents a long chain of evolution, and each of these species plays a unique ecological role in the earth's communities and ecosystems. These species, communities, and ecosystems also are essential for future evolution as populations of species continue to adapt to changes in environmental conditions by changing their genetic makeup.

THE FUTURE OF EVOLUTION

Artificial Selection: Picking the Traits We Like

We have learned how to selectively breed members of populations and to use genetic engineering to produce plants and animals with certain genetic traits.

We have used **artificial selection** to change the genetic characteristics of populations with similar genes. In this process, we select one or more desirable genetic traits in the population of a plant or animal, such as a type of wheat, fruit, or dog. Then we use *selective breeding* to end up with populations of the species contain-

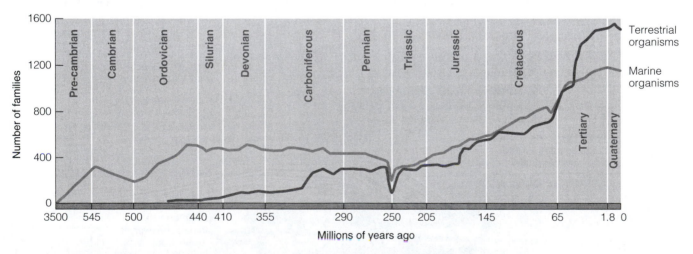

Figure 3-4 Natural capital: changes in the earth's biodiversity over geological time. The biological diversity of life on land and in the oceans has increased dramatically over the last 3.5 billion years, especially during the past 250 million years. During the last 1.8 million years this increase has leveled off.

ing large numbers of individuals with the desired traits.

Artificial selection has yielded food crops with higher yields, cows that give more milk, trees that grow faster, and a variety of different types of dogs and cats. But traditional crossbreeding is a slow process. And it can combine traits only from species that are close to one another genetically.

Now scientists are using genetic engineering to speed up our ability to manipulate genes. **Genetic engineering,** or **gene splicing,** is the alteration of an organism's genetic material, through adding, deleting, or changing segments of its DNA to produce desirable traits or eliminate negative ones. It enables scientists to transfer genes between different species that would not interbreed in nature. For example, genes from a fish species can be put into a tomato or strawberry.

The resulting organisms are called **genetically modified organisms (GMOs)** or **transgenic organisms.** Figure 3-5 (p. 60) outlines the steps involved in developing a genetically modified or transgenic plant.

Compared to traditional crossbreeding, gene splicing takes about half as much time to develop a new crop or animal variety and costs less. Traditional crossbreeding involves mixing the genes of similar types of organisms through breeding. But genetic engineering allows us to transfer traits between different types of organisms without breeding them.

Scientists have used gene splicing to develop modified crop plants, genetically engineered drugs, and pest-resistant plants. They have also created genetically engineered bacteria to clean up spills of oil and other toxic pollutants.

Genetic engineers have also learned how to produce a *clone,* or genetically identical version, of an individual in a population. Scientists have made clones of individuals in populations of domestic animals such as sheep and cows and may someday be able to clone humans—a possibility that excites some people and horrifies others.

Researchers envision using genetically engineered animals to act as biofactories for producing drugs, vaccines, antibodies, hormones, industrial chemicals such as plastics and detergents, and human body organs. This new field is called *biopharming.*

Some Concerns about the Genetic Revolution

Genetic engineering has great promise for improving the human condition, but it is an unpredictable process and raises a number of privacy, ethical, legal, and environmental issues.

The hype about genetic engineering suggests that its results are controllable and predictable. In reality genetic engineering is messy and unpredictable. Genetic engi-

neers can insert a gene into the nucleus of a cell, but with current technology they do not know whether the cell will incorporate the new gene into its DNA. They also do not know where the new gene will be located in the DNA molecule's structure and what effects this will have on the organism.

Thus, *conventional genetic engineering is a trial and error process* with many failures and unexpected results. Indeed, the average success rate of genetic engineering experiments is only about 1%. However, new techniques and advances in synthetic biology could overcome some of these problems.

Applying our increasing genetic knowledge is filled with great promise, but it raises some serious ethical and privacy issues. For example, some people have genes that make them more likely to develop certain genetic diseases or disorders. We now have the power to detect these genetic deficiencies, even before birth.

This raises some important issues. If gene therapy is developed for correcting these deficiencies, who will get it? Will it be mostly for the rich? Will this mean more abortions of genetically defective fetuses? Will health insurers refuse to insure people with certain genetic defects that could lead to health problems? Will employers refuse to hire them?

Soon we may enter the age of *designer babies* where people will choose the traits they want in their offspring from a genetic shopping list. To some this opens up a number of potential problems. What will be the beneficial and harmful environmental effects of such a change in the reproductive process?

Some people dream of a day when our genetic prowess could eliminate death and aging altogether. As one's cells, organs, or other parts wear out or are damaged, they would be replaced with new ones. These replacement parts might be grown in genetic engineering laboratories or biopharms. Or people might choose to have a clone available for spare parts.

This raises a number of questions. Is it moral to do this? Who decides? Who regulates this? Will genetically designed humans and clones have the same legal rights as people?

What might be the environmental impacts of such genetic developments on population, resource use, pollution, and environmental degradation? If everyone could live with good health as long as they wanted for a price, sellers of body makeovers would encourage customers to line up. Each of these wealthy, long-lived people could have an enormous ecological footprint for perhaps centuries.

X *HOW WOULD YOU VOTE?* Should we legalize the production of human clones if a reasonably safe technology for doing so becomes available? Cast your vote online at www.thomsonedu.com/biology/miller.

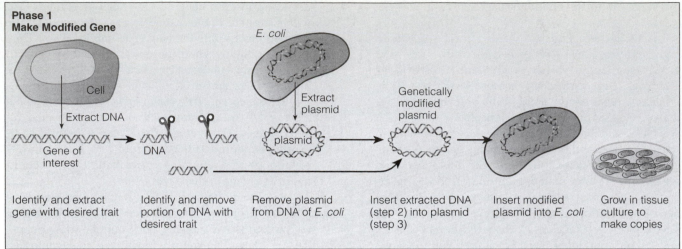

Phase 1
Make Modified Gene

E. coli

Cell

Extract DNA

Gene of
interest

DNA

Extract
plasmid

plasmid

Genetically
modified
plasmid

Identify and extract
gene with desired trait

Identify and remove
portion of DNA with
desired trait

Remove plasmid
from DNA of *E. coli*

Insert extracted DNA
(step 2) into plasmid
(step 3)

Insert modified
plasmid into *E. coli*

Grow in tissue
culture to
make copies

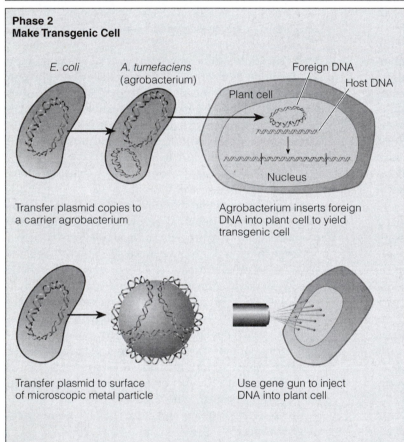

Phase 2
Make Transgenic Cell

E. coli

A. tumefaciens
(agrobacterium)

Foreign DNA

Host DNA

Plant cell

Nucleus

Transfer plasmid copies to
a carrier agrobacterium

Agrobacterium inserts foreign
DNA into plant cell to yield
transgenic cell

Transfer plasmid to surface
of microscopic metal particle

Use gene gun to inject
DNA into plant cell

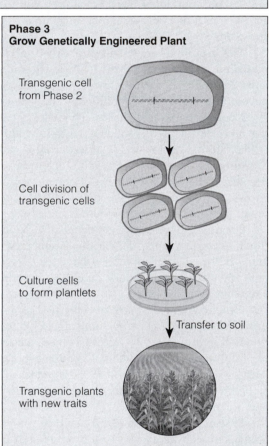

Phase 3
Grow Genetically Engineered Plant

Transgenic cell
from Phase 2

Cell division of
transgenic cells

Culture cells
to form plantlets

Transfer to soil

Transgenic plants
with new traits

Figure 3-5 *Genetic engineering.* Steps in genetically modifying a plant.

BIOMES: CLIMATE AND LIFE ON LAND

Why Do Different Organisms Live in Different Places? Think Climate

Different climates lead to different communities of organisms, especially vegetation.

Biologists have classified the terrestrial (land) portion of the biosphere into **biomes** ("BY-ohms"). They are large regions such as forests, deserts, and grasslands characterized by a distinct climate and specific forms of life, especially vegetation, adapted to it.

Why is one area of the earth's land surface a desert, another a grassland, and another a forest? Why do different types of deserts, grasslands, and forests exist?

The general answer to these questions is, because of differences in **climate**—a region's long-term atmospheric conditions, typically over decades. *Average* *temperature* and *average precipitation* are the two main factors determining a region's climate. Figure 3-6 shows the global distribution of biomes. Study this figure carefully and identify the type of biome you live in.

On maps such as the one in Figure 3-6, biomes are presented as having sharp boundaries and being covered with the same general types of vegetation. In reality, *biomes are not uniform.* They consist of a *mosaic of patches,* with somewhat different biological communities but with similarities unique to the biome. These patches occur mostly because the resources plants and animals need are not uniformly distributed and because human activities remove and alter natural vegetation. Go to a natural area in or near where you live and see if you can find patches with different vegetation.

For plants, precipitation generally is the limiting factor that determines whether a land area is *desert* (with low precipitation and sparse, widely spaced, mostly low vegetation), *grassland* (with enough precipitation to support grass but not large stands of

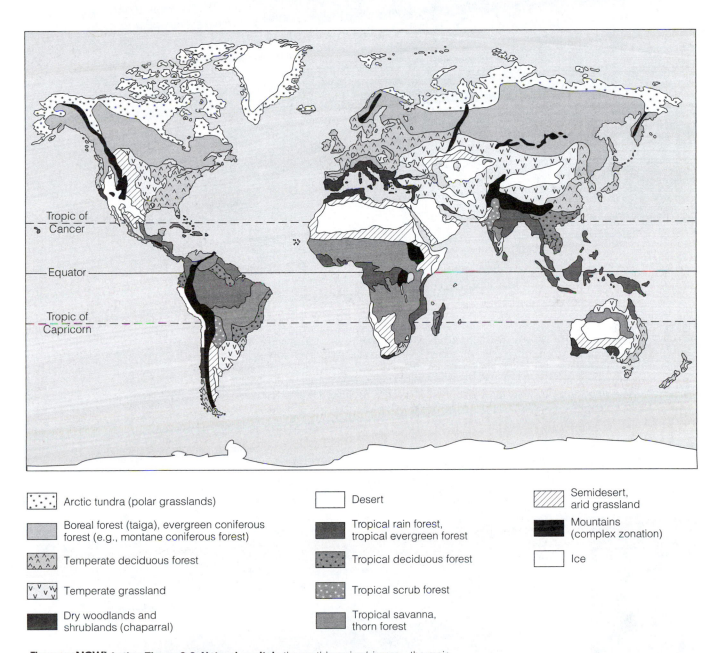

Arctic tundra (polar grasslands)		Desert		Semidesert, arid grassland	
Boreal forest (taiga), evergreen coniferous forest (e.g., montane coniferous forest)		Tropical rain forest, tropical evergreen forest		Mountains (complex zonation)	
Temperate deciduous forest		Tropical deciduous forest		Ice	
Temperate grassland		Tropical scrub forest			
Dry woodlands and shrublands (chaparral)		Tropical savanna, thorn forest			

ThomsonNOW™ Active Figure 3-6 Natural capital: the earth's major *biomes*—the main types of natural vegetation in various undisturbed land areas—result primarily from differences in climate. Each biome contains many ecosystems whose communities have adapted to differences in climate, soil, and other environmental factors. Humans have removed or altered much of this natural vegetation in some areas for farming, livestock grazing, lumber and fuel wood, mining, and construction. *See an animation based on this figure and take a short quiz on the concept.*

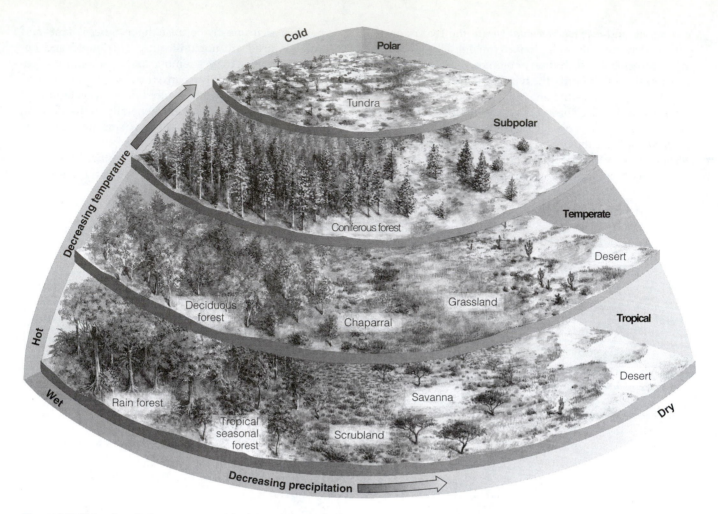

Figure 3-7 Natural capital: average precipitation and average temperature, acting together as limiting factors over a period of 30 or more years, determine the type of desert, grassland, or forest biome in a particular area. Although the actual situation is much more complex, this simplified diagram explains how climate determines the types and amounts of natural vegetation found in an area left undisturbed by human activities. (Used by permission of Macmillan Publishing Company, from Derek Elsom, *The Earth,* New York: Macmillan, 1992. Copyright ©1992 by Marshall Editions Developments Limited)

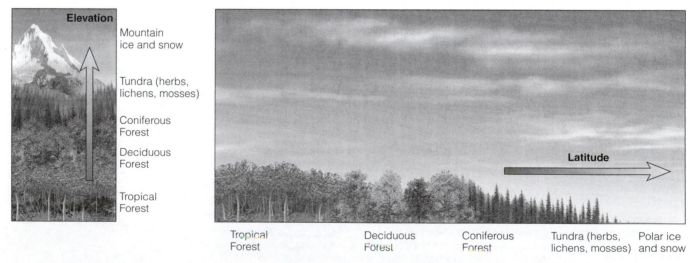

Figure 3-8 Natural capital: generalized effects of altitude (left) and latitude (right) on climate and biomes. Parallel changes in vegetation type occur when we travel from the equator to the poles or from lowlands to mountaintops.

trees), or *forest* (with enough precipitation to support stands of various tree species and smaller forms of vegetation).

Average annual precipitation and temperature are the most important factors in producing tropical, temperate, or polar deserts, grasslands, and forests, as shown in Figure 3-7.

Figure 3-8 shows how climate and vegetation vary with **latitude** (distance from the equator) and **altitude** (elevation above sea level). If you climb a tall mountain from its base to its summit, you can observe changes in plant life similar to those you would encounter in traveling from the equator to the earth's poles.

Find a map showing all the world's biomes and zoom in for details at ThomsonNOW.

What Impacts Do Humans Have on Deserts, Grasslands, and Forests?

Human activities are having major environmental impacts on most of the world's deserts, grasslands, and forests.

Figure 3-9 shows some major components and interactions in a temperate desert biome. Study this figure and note the types of species that live in this biome and how they are connected to one another.

Figure 3-10 (p. 64) shows major human impacts on deserts. Look closely at this figure and consider what effects your lifestyle might have on desert biomes. Deserts take a long time to recover from disturbances because they lack water and have slow plant growth, low species diversity, and slow nutrient cycling (because of little bacterial activity in their soils). Desert

| Producer to primary consumer | Primary to secondary consumer | Secondary to higher-level consumer | ◆ All producers and consumers to decomposers |

Figure 3-9 Natural capital: some components and interactions in a *temperate desert ecosystem.* When these organisms die, decomposers break down their organic matter into minerals that plants use. Arrows indicate transfers of matter and energy between producers, primary consumers (herbivores), secondary or higher-level consumers (carnivores), and decomposers. Organisms are not drawn to scale.

Deserts

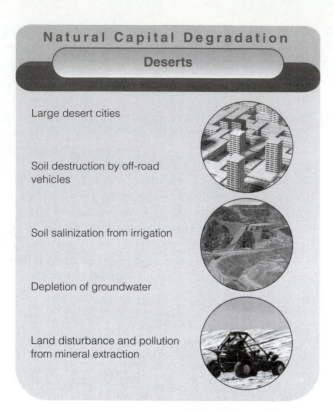

Large desert cities

Soil destruction by off-road vehicles

Soil salinization from irrigation

Depletion of groundwater

Land disturbance and pollution from mineral extraction

Figure 3-10 Natural capital degradation: major human impacts on the world's deserts. QUESTION: *What are the direct and indirect effects of your lifestyle on deserts?*

Grasslands

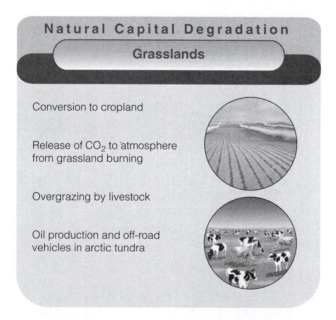

Conversion to cropland

Release of CO_2 to atmosphere from grassland burning

Overgrazing by livestock

Oil production and off-road vehicles in arctic tundra

Figure 3-11 Natural capital degradation: major human impacts on the world's grasslands. Some 70% of Brazil's tropical savanna—once the size of the Amazon—has been cleared and converted to the world's biggest grain growing area. QUESTION: *What are the direct and indirect effects of your lifestyle on grasslands?*

vegetation destroyed by livestock overgrazing and off-road vehicles may take decades to grow back.

Figure 3-11 lists major human impacts on grasslands. What effects might your lifestyle have on grass-

Forests

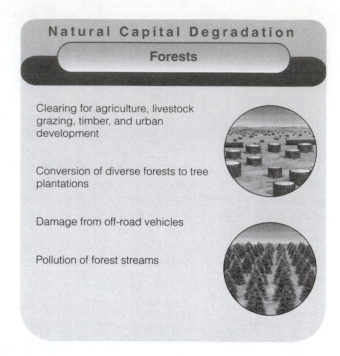

Clearing for agriculture, livestock grazing, timber, and urban development

Conversion of diverse forests to tree plantations

Damage from off-road vehicles

Pollution of forest streams

Figure 3-12 Natural capital degradation: major human impacts on the world's forests. QUESTION: *What are the direct and indirect effects of your lifestyle on forests?*

land biomes? Because of their thick and fertile soils, temperate grasslands are widely used to grow crops. But plowing disrupts their soils and leaves them vulnerable to wind and water erosion.

Figure 3-12 lists major human impacts on the world's forests. Large areas of the world's temperate forests have been cleared to grow crops and build urban areas. Tropical forests are also being cleared rapidly for agriculture, timber, and mining. Consider the possible effects of your lifestyle on forests.

LIFE IN WATER ENVIRONMENTS

Aquatic Life Zones

Life exists in freshwater and ocean aquatic life zones.

Scientists divide the watery parts of the biosphere into **aquatic life zones,** each containing numerous ecosystems. Aquatic life zones are classified into two major types: *freshwater life zones* (such as lakes, streams, and inland wetlands) and *ocean* or *marine life zones* (such as coral reefs, coastal estuaries, and the deep ocean). Waters in the marine zones contain higher concentrations of dissolved salts than those of freshwater zones.

Most aquatic life zones can be divided into three layers: *surface, middle,* and *bottom.* A number of environmental factors determine the types and numbers of organisms found in these layers. Examples are *temperature, access to sunlight for photosynthesis, dissolved oxygen content*, and *availability of nutrients* such as carbon (as dissolved CO_2 gas), nitrogen (as NO_3^-), and phosphorus (mostly as PO_4^{3-}) for producers.

Major Saltwater Life Zones: The Ocean Planet

Oceans occupy almost three-fourths of the earth's surface and consist of a coastal zone and the open sea.

A more accurate name for Earth would be *Ocean* because saltwater oceans cover about 71% of the planet's surface (Figure 3-13). Figure 3-14 lists important ecological and economic services provided by these marine systems. We know more about the surface of the moon than about the oceans that cover most of the earth.

Oceans have two major life zones: the **coastal zone** and the **open sea** (Figure 3-15, p. 66). Although it makes up less than 10% of the world's ocean area, the coastal zone contains 90% of all marine species and is the site of most large commercial marine fisheries. This zone has numerous interactions with the land and thus human activities easily affect it.

Most ecosystems found in the coastal zone have a very high net primary productivity per unit of area. This occurs because the zone has ample supplies of sunlight and plant nutrients (flowing from land and distributed by wind and ocean currents).

One highly productive area in the coastal zone is an *estuary*, a partially enclosed area of coastal water where seawater mixes with freshwater and nutrients and pollutants from rivers, streams, and runoff from land. Another consists of *coastal wetlands:* land areas covered with water all or part of the year. They include *mangrove forest swamps* in tropical waters and *salt marshes* in temperate zones.

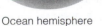

Ocean hemisphere   Land–ocean hemisphere

Figure 3-13 Natural capital: the ocean planet. The salty oceans cover 71% of the earth's surface. About 97% of the earth's water is in the interconnected oceans, which cover 90% of the planet's mostly ocean hemisphere (left) and 50% of its land–ocean hemisphere (right). Freshwater systems cover less than 1% of the earth's surface.

Natural Capital

Marine Ecosystems

Ecological Services	Economic Services
Climate moderation	Food
CO_2 absorption	Animal and pet feed (fish meal)
Nutrient cycling	
Waste treatment and dilution	Pharmaceuticals
Reduced storm impact (mangrove, barrier islands, coastal wetlands)	Harbors and transportation routes
	Coastal habitats for humans
Habitats and nursery areas for marine and terrestrial species	Recreation
	Employment
Genetic resources and biodiversity	Offshore oil and natural gas
	Minerals
Scientific information	Building materials

Figure 3-14 Natural capital: major ecological and economic services provided by marine systems. Scientists estimate that marine systems provide $21 trillion in goods and services per year—70% more than terrestrial ecosystems. QUESTION: *Which two ecological services and which two economic services do you think are the most important?*

Coral reefs form in the shallow coastal zones of warm tropical and subtropical oceans. These beautiful natural wonders are among the world's most diverse and productive ecosystems and are homes for about one-fourth of all marine species.

The open sea is divided into three vertical *zones—euphotic, bathyal,* and *abyssal*—based primarily on the penetration of sunlight (Figure 3-15, p. 66). This vast volume contains only about 10% of all marine species. Average net primary productivity per unit of area is quite low in the open sea (Figure 2-21, bottom, p. 40) except at an occasional equatorial upwelling, where currents bring up nutrients from the ocean bottom. However, because the open sea covers so much of the earth's surface, it makes the largest contribution to the earth's overall net primary productivity.

Figure 3-16 (p. 66) lists major human impacts on marine systems. Does your lifestyle contribute to any of these impacts?

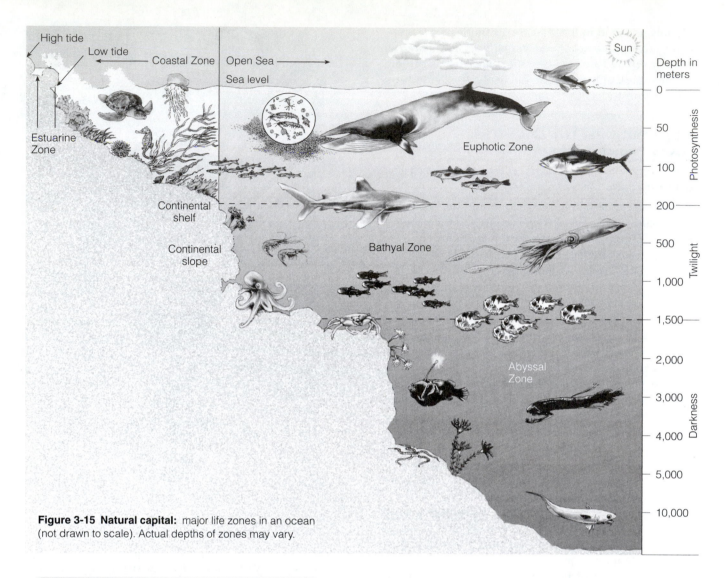

High tide

Low tide

Coastal Zone

Open Sea

Sea level

Sun

Depth in meters

Estuarine Zone

Continental shelf

Continental slope

Euphotic Zone

0

50

100

200

Bathyal Zone

500

1,000

1,500

2,000

Abyssal Zone

3,000

4,000

5,000

10,000

Photosynthesis

Twilight

Darkness

Figure 3-15 Natural capital: major life zones in an ocean (not drawn to scale). Actual depths of zones may vary.

Natural Capital Degradation

Marine Ecosystems

Half of coastal wetlands lost to agriculture and urban development

Over one-third of mangrove forests lost since 1980 to agriculture, development, and aquaculture shrimp farms

About 10% of world's beaches eroding because of coastal development and rising sea level

Ocean bottom habitats degraded by dredging and trawler fishing boats

Over 25% of coral reefs severely damaged and 11% have been destroyed

Major Freshwater Life Zones: Lakes, Wetlands, and Rivers

Freshwater life zones, which cover less than 1% of the earth's surface, provide important ecological and economic services.

Freshwater life zones consist of *standing* bodies of freshwater, such as lakes, ponds, and inland wetlands, and *flowing* systems, such as streams and rivers. These freshwater systems provide a number of important ecological and economic services, summarized in Figure 3-17.

Lakes are large natural bodies of standing freshwater formed when precipitation, runoff, or groundwater seepage fills depressions in the earth's surface. Deep lakes normally consist of distinct zones (Figure 3-18), which provide habitats and niches for different species.

Figure 3-16 Natural capital degradation: major human impacts on the world's marine systems. QUESTION: *Which two of these threats do you think are the most serious?*

Natural Capital

Freshwater Systems

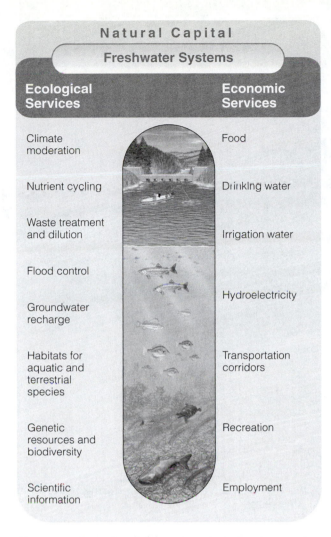

Ecological Services	Economic Services
Climate moderation	Food
Nutrient cycling	Drinking water
Waste treatment and dilution	Irrigation water
Flood control	
Groundwater recharge	Hydroelectricity
Habitats for aquatic and terrestrial species	Transportation corridors
Genetic resources and biodiversity	Recreation
Scientific information	Employment

Figure 3-17 Natural capital: major ecological and economic services provided by freshwater systems. QUESTION: *Which two ecological services and which two economic services do you think are the most important?*

Learn more about the zones of a lake, how its water turns over between seasons, and how lakes differ below their surfaces at ThomsonNOW.

Ecologists classify lakes according to their nutrient content and primary productivity. A newly formed lake generally has a small supply of plant nutrients and is called an **oligotrophic** (poorly nourished) **lake.** Such nutrient-poor lakes are often deep and usually have crystal-clear blue or green water because their low net primary productivity supports few algae and other producers.

Over time, sediment and nutrients wash into an oligotrophic lake, and plants grow and decompose to form bottom sediments. A lake with a large or excessive supply of nutrients (mostly nitrates and phosphates) needed by producers is called a **eutrophic** (well-nourished) **lake.** Such nutrient-rich lakes typically are shallow, have a high net primary productivity, and have murky brown or green water with very poor visibility because of their high content of algae and other producers. Many lakes fall somewhere between the two extremes of nutrient enrichment and are called **mesotrophic lakes.**

Precipitation that does not sink into the ground or evaporate is **surface water.** It becomes **runoff** when it flows into streams. A **watershed,** or **drainage basin,** is the land area that delivers runoff, sediment, and dissolved substances to a stream. Small streams join to form rivers, and rivers flow downhill to the ocean (Figure 3-19, p. 68) as part of the hydrologic cycle.

In many areas, *streams* begin in mountainous or hilly areas that collect and release water falling to the earth's surface as rain or snow that melts during warm

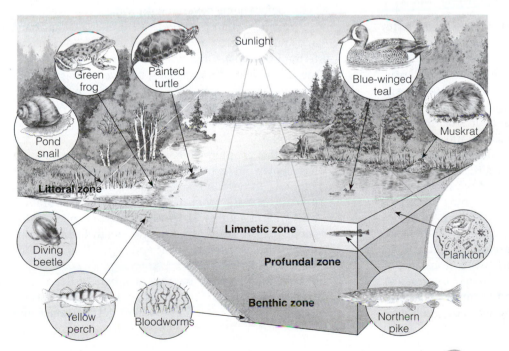

ThomsonNOW™ Active Figure 3-18 Natural capital: distinct zones of life in a fairly deep temperate zone lake. *See an animation based on this figure and take a short quiz on the concept.*

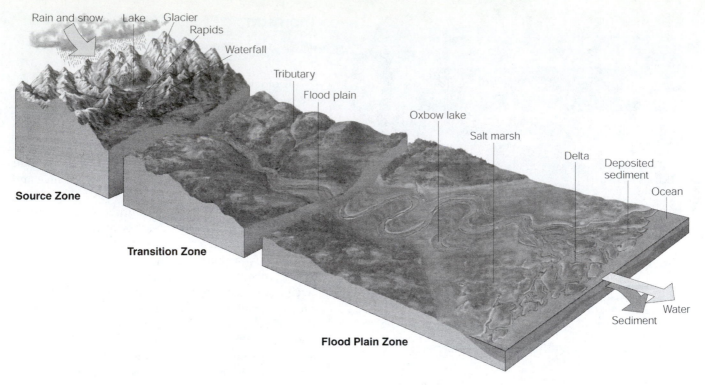

Figure 3-19 Natural capital: three zones in the downhill flow of water: *source zone* containing mountain (headwater) streams; *transition zone* containing wider, lower-elevation streams; and *floodplain zone* containing rivers, which empty into the ocean.

seasons. The downward flow of water from mountain highlands to the sea takes place in three different aquatic life zones with different environmental conditions (Figure 3-19). Because of different environmental conditions in each zone, a river system is a series of different ecosystems with different average depths, flow rates, dissolved oxygen levels, temperatures, and aquatic species.

As streams flow downhill, they become powerful shapers of land. Over millions of years, the friction of moving water levels mountains, cuts deep canyons, and moves rock and soil as sediment to low-lying areas.

We have established dams, power plants, sewage treatment plants, cities, recreation areas, shipping terminals, and farmlands along the shores of rivers and streams, especially in their transition and floodplain zones. This greatly increases the flow of plant nutrients, sediment, and pollutants into these ecosystems. To protect a stream or river system from excessive inputs of nutrients and pollutants, we must protect its *watershed*, the land around it.

Inland wetlands are lands covered with freshwater all or part of the time (excluding lakes, reservoirs, and streams) and located away from coastal areas. They include *marshes* (dominated by grasses), *swamps* (dominated by trees and shrubs), *prairie potholes* (depressions carved out by glaciers), and *flood-plains* (which receive excess water during heavy rains and floods).

Some wetlands are covered with water year-round. Others, called *seasonal wetlands*, usually are under water or soggy for only a short time each year. They include prairie potholes, floodplain wetlands, and bottomland hardwood swamps. Some stay dry for years before being covered with water again.

Inland wetlands provide a number of important and free ecological and economic services such as filtering toxic wastes and pollutants, absorbing and storing excess water from storms, and providing habitats for a variety of species. But we continue to destroy or degrade many of these important systems. What factors in your life contribute to the destruction of inland wetlands?

In this chapter, we have seen that the earth's life has shown an extraordinary ability to diversify by adapting to new biological opportunities and, in the process, creating even more biological opportunities. In the earth's ballet of life that has been playing for about 3.7 billion years, life and death are interconnected. Some species appear and some disappear, but the show goes on.

We only recently became members of this evolutionary ballet. Will we temporarily interrupt the show, take out some of the dancers, and get kicked off the stage? Or will we learn the choreography of the ballet and have a long run? We live in interesting and challenging times.

Nothing in biology makes sense except in the light of evolution.

THEODOSIUS DOBZHANSKY

CRITICAL THINKING

1. **(a)** How would you respond to someone who tells you that he or she does not believe in biological evolution because it is "just a theory"? **(b)** How would you respond to a statement that we should not worry about air pollution because through natural selection the human species will develop lungs that can detoxify pollutants?

2. How would you respond to someone who says that because extinction is a natural process, we should not worry about the loss of biodiversity?

3. An important adaptation of humans is a strong opposable thumb, which allows us to grip and manipulate things with our hands. As a demonstration of the importance of this trait, fold each of your thumbs into the palm of its hand and then tape them securely in that position for an entire day. At the end of the day, make a list of the things you could not do without the use of your thumbs.

4. Explain why you are for or against each of the following: **(a)** requiring labels indicating the use of genetically modified components in any food item, **(b)** using genetic engineering to develop "superior" human beings, and **(c)** using genetic engineering to eliminate aging and death.

5. What type of biome do you live in or near? What have been the effects of human activities over the past 50 years on the characteristic vegetation and animal life normally found in the biome you live in? How is your own lifestyle affecting this biome?

6. You are a defense attorney arguing in court for sparing an undeveloped old-growth tropical rain forest and a coral reef from severe degradation or destruction by development. Write your three most important arguments for the defense of each of these ecosystems. If the judge decides you can save only one of the ecosystems, which one would you choose, and why?

7. List two questions that you would like to have answered as a result of reading this chapter.

LEARNING ONLINE

The website for this book contains helpful study aids and many ideas for further reading and research. They include a chapter summary, review questions for the entire chapter, flash cards for key terms and concepts, a multiple-choice practice quiz, interesting Internet sites, references, information about green careers, and a guide for accessing thousands of InfoTrac® College Edition articles. Log into

www.thomsonedu.com/biology/miller

Then choose Chapter 3, and select a learning resource. For access to animations, additional quizzes, chapter outlines and summaries, register and log into

at **www.thomsonedu.com** using the access code card in the front of your book.

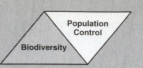

4 COMMUNITY ECOLOGY, POPULATION ECOLOGY, AND SUSTAINABILITY

Animal and vegetable life is too complicated a problem for human intelligence to solve, and we can never know how wide a circle of disturbance we produce in the harmonies of nature when we throw the smallest pebble into the ocean of organic life.

GEORGE PERKINS MARSH

TYPES OF SPECIES

Types of Species in Communities

Communities can contain native, nonnative, indicator, keystone, and foundation species that play different ecological roles.

Ecologists often use labels such as *native, nonnative, indicator, keystone,* or *foundation* to describe the major ecological roles or niches various species play in communities. Any given species may play more than one of these five roles in a particular community.

Native species are those that normally live and thrive in a particular community. Others that migrate into or are deliberately or accidentally introduced into a community are called **nonnative species, invasive species,** or **alien species.**

Many people tend to think of nonnative or invasive species as villains. But most introduced and domesticated species of crops and animals such as chick-

CASE STUDY

Why Are Amphibians Vanishing?

Amphibians (frogs, toads, and salamanders) live part of their lives in water and part on land, and some are classified as indicator species. Frogs, for example, are good indicator species because they are especially vulnerable to environmental disruption at certain points in their life cycle, shown in Figure 4-A.

As tadpoles they live in water and eat plants, and as adults they live mostly on land and eat insects that can expose them to pesticides. Their eggs have no protective shells to block ultraviolet (UV) radiation or pollution. As adults, they take in water and air through their thin, permeable skins, which can readily absorb pollutants from water, air, or soil.

Since 1980, populations of hundreds of the world's estimated 5,743 amphibian species have been vanishing or declining in almost every part of the world, even in protected wildlife reserves and parks. According to the Global Amphibian Assessment, scientists estimate that about 33% of all known amphibian species are

threatened with extinction, and populations of 43% of the species were declining.

No single cause has been identified to explain the amphibian declines. However, scientists have identified a number of factors that can affect frogs and other amphibians at certain points in their life cycles:

- *Habitat loss and fragmentation* (draining and filling of inland wetlands, deforestation, and development)
- *Prolonged drought* (which dries up breeding pools so few tadpoles survive)
- *Pollution* (especially pesticides, which can make frogs vulnerable to bacterial, viral, and fungal diseases and sexual abnormalities)
- *Increases in ultraviolet radiation* (which can harm embryos of amphibians in shallow ponds)
- *Parasites*
- *Overhunting* (especially in Asia and France, where frog legs are a delicacy)
- *Viral and fungal diseases*
- *Natural immigration or deliberate introduction of nonnative*

predators and competitors (such as fish)

A combination of such factors probably is responsible for the decline or disappearance of most amphibian species.

Why should we care if various amphibian species become extinct? Scientists give three reasons. *First,* this trend suggests that environmental health is deteriorating in parts of the world because amphibians are sensitive biological indicators of changes in environmental conditions such as habitat loss and degradation, pollution, UV exposure, and climate change.

Second, adult amphibians play important ecological roles in biological communities. For example, amphibians eat more insects (including mosquitoes) than do birds. In some habitats, extinction of certain amphibian species could lead to extinction of other species, such as reptiles, birds, aquatic insects, fish, mammals, and other amphibians that feed on them or their larvae.

Third, amphibians represent a genetic storehouse of pharmaceutical products waiting to be discovered. Compounds in hundreds

ens, cattle, fish from around the world, and many non-native wild game species are beneficial to us.

However, some nonnative species can thrive and crowd out native species—an example of unintended consequences. Examples are fire ants and rapidly growing plant species such as kudzu.

Indicator Species: Biological Smoke Alarms

Some species can alert us to harmful changes that are taking place in biological communities.

Species that serve as early warnings of damage to a community are called **indicator species.** Birds are excellent biological indicators because they are found almost everywhere and are affected quickly by environmental change such as loss or fragmentation of their habitats and exposure to chemical pesticides. Research indicates that a major factor in the current decline of some species of migratory, insect-eating songbirds in North America is habitat loss or fragmentation. The tropical forests of Latin America and the

Caribbean that are winter habitats for such birds are disappearing rapidly. Their summer habitats in North America also are disappearing or are being fragmented into patches that make the birds more vulnerable to attack by predators and parasites.

Some amphibians (frogs, toads, and salamanders), which live part of their lives in water and part on land, are also classified to be indicator species (Case Study, below). Butterflies are a good indicator species because their association with various plant species makes them vulnerable to habitat loss and fragmentation.

Keystone Species: Major Players

Keystone species help determine the types and numbers of various other species in a community.

A keystone is the wedge-shaped stone placed at the top of a stone archway. Remove this stone and the arch collapses. In some communities, certain species called **keystone species** apparently serve a similar role. They

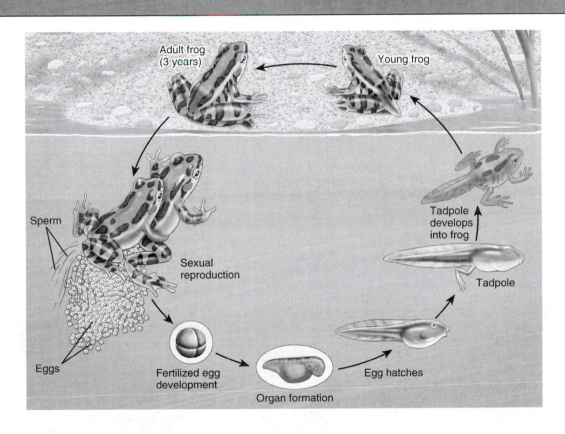

Adult frog (3 years)

Young frog

Sperm

Sexual reproduction

Eggs

Fertilized egg development

Organ formation

Egg hatches

Tadpole

Tadpole develops into frog

Figure 4-A Typical *life cycle of a frog.* Populations of some frog species can decline because of the effects of harmful factors at different points in their life cycle. Such factors include habitat loss, drought, pollution, increased ultraviolet radiation, parasitism, disease, overhunting, and nonnative predators and competitors.

of secretions from amphibian skin have been isolated and some are used as painkillers and antibiotics and in treating burns and heart disease.

The plight of some amphibian indicator species is a warning signal. They do not need us, but we and other species need them.

Critical Thinking

What three things would you do to decrease the loss of amphibian species?

have a much larger effect on the types and abundances of other species in a community than their numbers would suggest. According to this hypothesis, eliminating a keystone species can dramatically alter the structure and function of a community.

Keystone species play critical ecological roles. One is *pollination* of flowering plant species by bees, hummingbirds, bats, and other species. In addition, *top predator* keystone species feed on and help regulate the populations of other species. Examples are the wolf, leopard, lion, alligator, and great white shark (Case Study, below).

The loss of a keystone species can lead to population crashes and extinctions of other species that depend on it for certain ecological services. According to biologist Edward O. Wilson, "The loss of a keystone species is like a drill accidentally striking a power line. It causes lights to go out all over."

X **HOW WOULD YOU VOTE?** Do we have an ethical obligation to protect shark species from premature extinction and treat them humanely? Cast your vote online at www.thomsonedu .com/biology/miller.

Foundation Species: Other Major Players

Foundation species can create and enhance habitats that can benefit other species in a community.

Some ecologists think the keystone species concept should be expanded to include *foundation species,* which play a major role in shaping communities by creating and enhancing habitat that benefits other species.

For example, elephants push over, break, or uproot trees, creating forest openings in the savanna grasslands and woodlands of Africa. This promotes the growth of grasses and other forage plants that ben-

Why Should We Protect Sharks?

CASE STUDY

The world's 370 shark species vary widely in size. The smallest is the dwarf dog shark, about the size of a large goldfish. The largest is the whale shark, the world's largest fish. It can grow to 15 meters (50 feet) long and weigh as much as two full-grown African elephants.

Various shark species, feeding at the top of food webs, cull injured and sick animals from the ocean and thus play an important ecological role. Without such shark species, the oceans would be teeming with dead and dying fish.

Many people—influenced by movies, popular novels, and widespread media coverage of a fairly small number of shark attacks per year—think of sharks as people-eating monsters. However, the three largest species—the whale shark, basking shark, and megamouth shark—are gentle giants. They swim through the water with their mouths open, filtering out and swallowing huge quantities of *plankton* (small, free-floating sea creatures).

Media coverage of shark attacks greatly distorts the danger. Every year, members of a few species, such

as great white, bull, tiger, gray reef, lemon, hammerhead, shortfin mako, and blue sharks, injure 60–100 people worldwide. Since 1990, sharks have killed an average of seven people per year. Most attacks are by great white sharks, which feed on sea lions and other marine mammals and sometimes mistake divers and surfers for their usual prey.

For every shark that injures a person, we kill at least 1 million sharks, a total of about 100 million sharks each year. Sharks are caught mostly for their fins and then thrown back into the water, fins removed, to bleed to death or drown because they can no longer swim.

Shark fins are widely used in Asia as a soup ingredient and as a pharmaceutical cure-all. A fin from a large whale shark can fetch up to $10,000. In high-end Hong Kong restaurants, a single bowl of shark fin soup can cost as much as $100. Ironically, shark fins have been found to contain dangerously high levels of toxic mercury.

Sharks are also killed for their livers, meat, hides, and jaws, and because we fear them. Some sharks die when they are trapped in nets or by lines deployed to catch swordfish, tuna, shrimp, and other commercially important species.

In addition to their important ecological roles, sharks could save human lives. We may learn from them how to fight cancer, which sharks almost never get. Scientists are also studying their highly effective immune systems because they allow wounds to heal without becoming infected.

Sharks are especially vulnerable to overfishing because they grow slowly, mature late, and have only a few young. Today, they are among the most vulnerable and least protected animals on earth.

In 2003, experts at the National Aquarium in Baltimore, Maryland, estimated that populations of some shark species have decreased by 90% since 1992.

In response to a public outcry over depletion of some shark species, the United States and other countries have banned shark hunting in their territorial waters. But such bans are difficult to enforce.

Sharks have been around for more than 400 million years. Sustaining this portion of the earth's biodiversity begins with the knowledge that, although sharks do not need us, we and other species need them.

efit smaller grazing species such as antelope. It also accelerates nutrient cycling rates. Some bats and bird foundation species can regenerate deforested areas and spread fruit plants by depositing plant seeds in their droppings.

SPECIES INTERACTIONS

How Do Species Interact?

Competition, predation, parasitism, mutualism, and commensalism are ways in which species can interact and increase their ability to survive.

When different species in a community have activities or resource needs in common, they may interact with one another. Members of these species may be harmed, helped, or unaffected by the interaction. Ecologists identify five basic types of interactions between species: *interspecific competition, predation, parasitism, mutualism,* and *commensalism.*

The most common interaction between species is *competition* for shared or scarce resources such as space and food. Ecologists call such competition between species **interspecific competition.** When intense competition for limited resources occurs, one of the competing species must migrate (if possible) to another area, shift its feeding habits or behavior through natural selection and evolution, suffer a sharp population decline, or become extinct in that area.

Humans are in competition with many other species for space, food, and other resources. As we convert more and more of the earth's land and aquatic resources and net primary productivity to our uses, we deprive many other species of resources they need to survive.

Reducing or Avoiding Competition: Sharing Resources

Some species evolve adaptations that allow them to reduce or avoid competition for resources with other species.

Over a time scale long enough for evolution to occur, some species competing for the same resources evolve adaptations that reduce or avoid competition. One way this happens is through **resource partitioning.** It occurs when species competing for similar scarce resources evolve more specialized traits that allow them to use shared resources at different times, in different ways, or in different places (Figure 4-1).

Here are some other examples of resource partitioning. When lions and leopards live in the same area, lions take mostly larger animals as prey and leopards take smaller ones. Hawks and owls feed on similar prey, but hawks hunt during the day and owls hunt at night. Some bird species feed on the ground, whereas others seek food in trees and shrubs.

Figure 4-1 Sharing the wealth: *resource partitioning* of five species of insect-eating warblers in the spruce forests of the U.S. state of Maine. Each species minimizes competition with the others for food by spending at least half its feeding time in a distinct portion (shaded areas) of the spruce trees, and by consuming somewhat different insect species. (After R. H. MacArthur, "Population Ecology of Some Warblers in Northeastern Coniferous Forests," *Ecology* 36 (1958): 533–536)

Predator and Prey Species: Eating and Being Eaten

Species called predators feed on all or parts of other species called prey.

In **predation,** members of one species (the *predator*) feed directly on all or part of a living organism of another species (the *prey*). Together, the two kinds of organisms, such as lions (the predator or hunter) and zebras (the prey or hunted), are said to have a **predator–prey relationship.** Such relationships are depicted in Figures 2-17 and 2-18 (pp. 37–38).

At the individual level, members of the prey species are clearly harmed. At the population level, predation plays a role in evolution by natural selection. It can benefit the prey species because predators such as some shark species (Case Study, p. 72) kill the sick, weak, aged, and least fit members of a population. The remaining prey gain better access to food supplies and avoid excessive population growth. Predation also helps successful genetic traits to become more dominant in the prey population through natural selection, which can enhance that species' reproductive success and long-term survival.

Some people tend to view predators with contempt. When a hawk tries to capture and feed on a rabbit, some tend to root for the rabbit. Yet the hawk, like all predators, is merely trying to get enough food to feed itself and its young. In doing so, it plays an important ecological role in controlling rabbit populations.

Parasites: Sponging Off of Others

Although parasites can harm their host organisms, they can promote community biodiversity.

Parasitism occurs when one species (the *parasite*) feeds on part of another organism (the *host*) usually by living on or in the host. In this relationship, the parasite benefits and the host is harmed.

Parasitism can be viewed as a special form of predation. But unlike a conventional predator, a parasite usually is much smaller than its host (prey) and rarely kills its host. Also, most parasites remain closely associated with, draw nourishment from, and may gradually weaken their host over time.

Tapeworms, disease-causing microorganisms, and some other parasites live *inside* their hosts. Other parasites attach themselves to the *outside* of their hosts. Examples are ticks, fleas, mosquitoes, mistletoe plants, and sea lampreys that use their sucker-like mouths to attach themselves to their fish hosts and feed on their blood. Some parasites move from one host to another, as fleas and ticks do; others, such as tapeworms, spend their adult lives with a single host.

Some parasites have little contact with their host. For example, North American cowbirds parasitize or take over the nests of other birds by laying their eggs in them and then letting the host birds raise their young.

From the host's point of view, parasites are harmful, but parasites play important ecological roles. For instance, they promote biodiversity by helping keep some species from becoming so plentiful that they eliminate other species.

Mutualism: Win–Win Relationships

Pollination, bacteria in your gut that digest your food, and fungi that help plant roots take up nutrients are examples of species interactions that benefit both species.

In **mutualism,** two species interact in a way that benefits both. Such benefits include having pollen and seeds dispersed for reproduction, being supplied with food, or receiving protection.

For example, honeybees, caterpillars, butterflies, and other insects may feed on a male flower's nectar, picking up pollen in the process, and then pollinate female flowers when they feed on them. And coral reefs survive by a mutualistic relationship between reef-building coral animals and bacteria that live in their tissues.

Figure 4-2 shows two examples of mutualistic relationships that combine *nutrition* and *protection*. One involves birds that ride on the backs of large animals like African buffalo, elephants, and rhinoceroses (Figure 4-2a). The birds remove and eat parasites from the animal's body and often make noises warning the animal when predators approach.

The second example involves clownfish species, which live within sea anemones, whose tentacles sting and paralyze most fish that touch them (Figure 4-2b). The clownfish, which are not harmed by the tentacles, gain protection from predators and feed on the detritus left from the meals of the anemones. The sea anemones benefit because the clownfish protect them from some of their predators.

Another example involves vast armies of bacteria that live in the digestive systems of animals and break down (digest) their food. The bacteria receive a sheltered habitat and food from their host. In turn, they help break down (digest) their host's food. Hundreds of millions of bacteria in your gut help digest the food you eat and help keep you alive. It is tempting to think of mutualism as an example of cooperation between species, but actually it involves each species benefiting by exploiting the other.

Commensalism: Using without Harming

Some species interact in a way that helps one species but has little, if any, effect on the other.

Commensalism is an interaction that benefits one species but has little, if any, effect on the other species.

(a) Oxpeckers and black rhinoceros

(b) Clown fish and sea anemone

Figure 4-2 Natural capital: examples of *mutualism.* **(a)** Oxpeckers (or tickbirds) feed on parasitic ticks that infest large, thick-skinned animals such as the endangered black rhinoceros. **(b)** A clownfish gains protection and food by living among deadly stinging sea anemones and helps protect the anemones from some of their predators.

One example is some kinds of silverfish insects that move along with the columns of army ants to share the food left over during their raids. The army ants receive no apparent harm or benefit from the silverfish.

Another example is plants called *epiphytes* (such as some types of orchids and bromeliads), which attach themselves to the trunks or branches of large trees in tropical and subtropical forests. These *air plants* benefit by having a solid base on which to grow. They also live in an elevated spot that gives them better access to sunlight, water from the humid air and rain, and nutrients falling from the tree's upper leaves and limbs. Their presence apparently does not harm the tree.

 Review the way species can interact and see the results of an experiment on species interaction at ThomsonNOW.

ECOLOGICAL SUCCESSION: COMMUNITIES IN TRANSITION

Ecological Succession: How Communities Change over Time

New environmental conditions allow one group of species in a community to replace other groups.

All communities change their structure and composition in response to changing environmental conditions. The gradual change in species composition of a given area is called **ecological succession.** During succession, *colonizing* or *pioneer species* are the first to arrive. As environmental conditions change, they are replaced by others, and later these species may be replaced by another set of species.

Ecologists recognize two types of ecological succession depending on the conditions present at the beginning of the process. One is **primary succession,** which involves the gradual establishment of biotic communities on essentially lifeless ground where there is no soil in a terrestrial community (Figure 4-3, p. 76) or no bottom sediment in an aquatic community. Examples include bare rock exposed by a retreating glacier or severe soil erosion, newly cooled lava, an abandoned highway or parking lot, or a newly created shallow pond or reservoir.

Primary succession usually takes a long time. One reason is that before a community can become established on land, there must be soil. Depending mostly on the climate, it takes natural processes several hundred to several thousand years to produce fertile soil.

With the other, more common type, called **secondary succession,** a series of communities with different species can develop in places containing soil or bottom sediment. It begins in an area where the natural community of organisms has been disturbed, removed, or destroyed, but the soil or bottom sediment remains. Candidates for secondary succession include abandoned farmlands (Figure 4-4, p. 77), burned or cut forests, heavily polluted streams, and land that has been flooded. Because some soil or sediment is present, new vegetation usually can begin to germinate within a few weeks. Seeds can be present in soils, or they can be carried from nearby plants by wind or by birds and other animals.

During primary or secondary succession, disturbances such as natural or human-caused fires or deforestation can convert a particular stage of succession to an earlier stage. Such disturbances create new conditions that encourage some species and discourage or eliminate others.

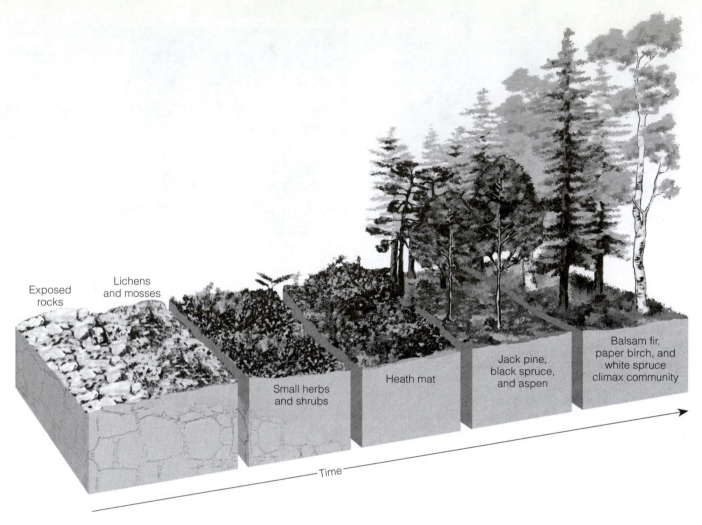

Labels on image:
- Exposed rocks
- Lichens and mosses
- Small herbs and shrubs
- Heath mat
- Jack pine, black spruce, and aspen
- Balsam fir, paper birch, and white spruce climax community
- Time

Figure 4-3 Natural capital: *primary ecological succession* over several hundred years of plant communities on bare rock exposed by a retreating glacier on Isle Royale, Michigan (USA), in northern Lake Superior. The details vary from one site to another.

 Explore the differences between primary and secondary succession at ThomsonNOW.

Can We Predict the Path of Succession toward a Balance of Nature?

Scientists cannot project the course of a given succession or view it as preordained progress toward a stable climax community that is in balance with its environment.

According to the traditional view, succession proceeds along an expected path until a certain stable type of *climax community* occupies an area. Such a community is dominated by a few long-lived plant species and is in balance with its environment. This equilibrium model of succession is what ecologists meant when they talked about the *balance of nature*.

Over the last several decades, many ecologists have changed their views about balance and equilibrium in nature. Under the old balance-of-nature view, a large terrestrial community undergoing succession eventually became covered with an expected type of climax vegetation. But a close look at almost any community reveals that it consists of an ever-changing mosaic of vegetation patches at different stages of succession. And most ecologists now recognize that mature late-successional communities are not in a state of permanent equilibrium. Instead, they are in a state of continual disturbance and change.

The current view is that we cannot predict the course of a given succession or view it as preordained progress toward an ideally adapted climax community. Rather, succession reflects the ongoing struggle by different species for enough light, nutrients, food, and space. This competition allows them to survive and gain reproductive advantages over other species.

Annual weeds

Perennial weeds and grasses

Shrubs and pine seedlings

Young pine forest with developing understory of oak and hickory trees

Mature oak-hickory forest

Time

ThomsonNOW™ Active Figure 4-4 Natural capital: natural ecological restoration of disturbed land. *Secondary ecological succession* of plant communities on an abandoned farm field in the U.S. state of North Carolina. It took 150–200 years after the farmland was abandoned for the area to become covered with a mature oak and hickory forest. A new disturbance such as deforestation or fire would create conditions favoring pioneer species such as annual weeds. In the absence of new disturbances, secondary succession would recur over time, but not necessarily in the same sequence shown here. *See an animation based on this figure and take a short quiz on the concept.*

POPULATION DYNAMICS AND CARRYING CAPACITY

Limits on Population Growth: Biotic Potential versus Environmental Resistance

No population can grow indefinitely because key resources are limited and competitors or predators are usually plentiful.

Four variables—*births, deaths, immigration,* and *emigration*—govern changes in population size. A population increases by birth and immigration (arrival of individuals from outside the population) and decreases by death and emigration (departure of individuals from the population):

$$\text{Population change} = (\text{Births} + \text{Immigration}) - (\text{Deaths} + \text{Emigration})$$

Populations vary in their capacity for growth, also known as their **biotic potential.** The **intrinsic rate of increase (r)** is the rate at which a population would grow if it had unlimited resources. Most populations grow at a rate slower than this maximum.

Individuals in populations with a high rate of growth typically *reproduce early in life, have short generation times* (the time between successive generations), *can reproduce many times* (have a long reproductive life), and *have many offspring each time they reproduce.*

Some species have an astounding biotic potential. Without any controls on its population growth, the descendants of a single female housefly could total about 5.6 trillion flies within about 13 months. If this rapid exponential growth kept up, within a few years these houseflies would cover the earth's entire surface!

Fortunately, this is not realistic because *no population can grow indefinitely*. In the real world, a rapidly

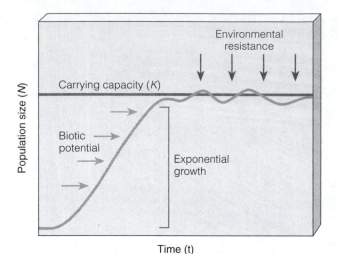

Time (t)

ThomsonNOW Active Figure 4-5 **Natural capital:** no population can grow forever. *Exponential growth* (lower part of the curve) occurs when resources are not limiting a population, which can then grow at its *intrinsic rate of increase (r)* or *biotic potential.* Such exponential growth is converted to *logistic growth,* in which the growth rate decreases as the population becomes larger and faces environmental resistance. Over time, the population size stabilizes at or near the *carrying capacity (K)* of its environment, which results in a sigmoid (S-shaped) population growth curve. Depending on resource availability, the size of a population often fluctuates around its carrying capacity, although a population may temporarily exceed its carrying capacity and suffer a sharp decline or crash in its numbers. *See an animation based on this figure and take a short quiz on the concept.*

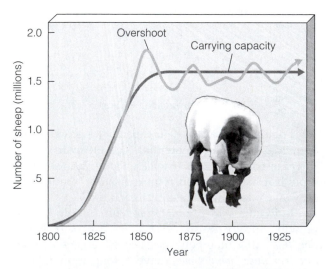

Figure 4-6 Boom and bust: *logistic growth* of a sheep population on the island of Tasmania between 1800 and 1925. After sheep were introduced in 1800, their population grew exponentially thanks to an ample food supply. By 1855, they had overshot the land's carrying capacity. Their numbers then stabilized and fluctuated around a carrying capacity of about 1.6 million sheep.

growing population reaches some size limit imposed by one or more limiting factors, such as light, water, space, or nutrients, or by too many competitors or predators. *There are always limits to population growth in nature.* This important lesson is one of nature's four principles of sustainability.

Environmental resistance consists of all factors that act to limit the growth of a population. Together, biotic potential and environmental resistance determine the **carrying capacity (K)**: the maximum population of a given species that a particular habitat can sustain indefinitely without degrading the habitat. The growth rate of a population decreases as its size nears the carrying capacity of its environment because resources such as food and water begin to dwindle.

Exponential and Logistic Population Growth: J-Curves and S-Curves

With ample resources a population can grow rapidly, but as resources become limited, its growth rate slows and levels off.

A population with few, if any, resource limitations grows exponentially at a fixed rate such as 1% or 2% a year. **Exponential growth** starts slowly and grows faster over time because the base size of the population is increasing. Plotting the number of individuals against time yields a J-shaped growth curve (Figure 4-5, lower part of curve). Whether an exponential growth curve looks steep or "fast" depends on the observation time period.

Logistic growth involves exponential population growth followed by a steady decrease in population growth until the population size levels off (Figure 4-5, top half of curve). This slowdown occurs as the population encounters environmental resistance and approaches the carrying capacity of its environment. After leveling off, a population with this type of growth typically fluctuates slightly above and below the carrying capacity.

A plot of the number of individuals against time yields a sigmoid, or *S*-shaped, logistic growth curve. Figure 4-6 shows such a case involving sheep on the island of Tasmania, south of Australia, in the early 19th century.

 Learn how to estimate a population of butterflies and see a mouse population growing exponentially at ThomsonNOW.

Exceeding Carrying Capacity: Move, Change Habits, or Decline in Size

When a population exceeds its resource supplies, many of its members die unless they can switch to new resources or move to an area with more resources.

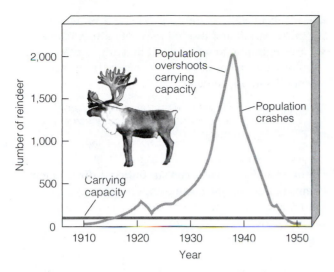

Figure 4-7 Exponential growth, overshoot, and population crash of reindeer introduced to a small island off the southwest coast of Alaska. When 26 reindeer (24 of them female) were introduced in 1910, lichens, mosses, and other food sources were plentiful. By 1935, the herd size had soared to 2,000, overshooting the island's carrying capacity. This led to a population crash, with the herd size plummeting to only 8 reindeer by 1950.

The transition from exponential growth to logistic growth is not always smooth. Some populations use up their resource supplies and temporarily *overshoot*, or exceed, the carrying capacity of their environment. This occurs because of a *reproductive time lag:* the period needed for the birth rate to fall and the death rate to rise in response to resource overconsumption.

In such cases the population suffers a *dieback*, or *crash,* unless the excess individuals can switch to new resources or move to an area with more resources. Such a crash occurred when reindeer were introduced onto a small island off the southwest coast of Alaska (Figure 4-7).

Sometimes when a population exceeds the carrying capacity of an area, it can cause damage that reduces the area's carrying capacity. This means that the area cannot support as many individuals as it once did. For example, overgrazing by cattle on dry western lands in the United States has reduced grass cover in some areas. This has allowed sagebrush—which cattle cannot eat—to move in, thrive, and replace grasses, reducing the land's carrying capacity for cattle.

Humans are not exempt from population overshoot and dieback. Ireland experienced a population crash after a fungus destroyed the potato crop in 1845. About 1 million people died, and 3 million people migrated to other countries.

Technological, social, and other cultural changes have extended the earth's carrying capacity for the human species. We have increased food production and used large amounts of energy and matter resources to make normally uninhabitable areas habitable. How long will we be able to keep doing this on a planet with a finite size, finite resources, and a human population whose size and per capita resource use is growing exponentially? Some say we can keep doing this indefinitely mostly because of our technological ingenuity. Others say that sooner or later we will reach the limits that nature always imposes on populations.

X *How Would You Vote?* Can we continue to expand the earth's carrying capacity for humans? Cast your vote online at www.thomsonedu.com/biology/miller.

Reproductive Patterns: Opportunists and Competitors

While some species have a large number of small offspring and give them little parental care, other species have a few larger offspring and take care of them until they can reproduce.

Species use different reproductive patterns to help ensure their survival. At one extreme are species that reproduce early and put most of their energy into reproduction. These species have many, usually small, offspring and give them little or no parental care or protection. They overcome the massive loss of their offspring by producing so many that a few will survive to reproduce many more offspring to begin this reproductive pattern again. Examples include algae, bacteria, rodents, annual plants (such as dandelions), and most insects.

Such species tend to be *opportunists*. They reproduce and disperse rapidly when conditions are favorable or when a disturbance opens up a new habitat or niche for invasion, as in the early stages of ecological succession. Environmental changes caused by disturbances can allow opportunist species to gain a foothold. However, once established, their populations may crash because of unfavorable environmental conditions or invasion by more competitive species. This helps explain why most opportunist species go through irregular and unstable boom-and-bust cycles in their population size.

At the other extreme are *competitor* species that tend to reproduce late in life and have a small number of offspring with fairly long life spans. Typically, the offspring of such species develop inside their mothers (where they are safe), are born fairly large, mature slowly, and are cared for and protected by one or both parents until they reach reproductive age. This reproductive pattern results in a few big and strong individuals that can compete for resources and reproduce a few young to begin their reproductive pattern again.

Such species tend to do well in competitive conditions when their population size is near the carrying capacity of their environment. Their populations typically follow a logistic growth curve.

Most large mammals (such as elephants, whales, and humans), birds of prey, and large and long-lived plants (such as the saguaro cactus, redwood trees, and most tropical rain forest trees) are competitor species. Many of these species—especially those with long generation times and low reproductive rates like elephants, rhinoceroses, and sharks—are prone to extinction.

Most competitor species thrive best in communities with fairly constant environmental conditions. In contrast, opportunists thrive in habitats that have experienced disturbances such as a tree falling, a forest fire, or the clearing of a forest or grassland for raising crops.

The reproductive pattern of a species may give it a temporary advantage, but *the availability of suitable habitat for individuals of a population in a particular area is what determines its ultimate population size.* Regardless of how fast a species can reproduce, there can be no more dandelions than there is dandelion habitat and no more zebras than there is zebra habitat in a particular area.

HUMAN IMPACTS ON NATURAL SYSTEMS: LEARNING FROM NATURE

Humans Impacts on Natural Ecosystems: Our Big Footprints

We have used technology to alter much of the rest of nature to meet our growing needs and wants in ways that threaten the survival of many other species and that could reduce the quality of life for our own species.

To survive and provide resources for growing numbers of people, we have modified, cultivated, built on, or degraded a large and increasing area of the earth's natural systems. Excluding Antarctica, our activities have, to some degree, directly affected about 83% of the earth's land surface. Figure 4-8 compares some of the characteristics of natural and human-dominated systems.

We have used technology to alter much of the rest of nature to meet our growing needs

and wants in eight major ways (Figure 4-9). To survive, we must exploit and modify parts of nature. However, we are beginning to understand that any human intrusion into nature has multiple effects, most of them unintended and unpredictable.

 Examine how resources have been depleted or degraded around the world at ThomsonNOW.

We face two major challenges. *First*, we need to maintain a balance between simplified, human-altered communities and the more complex natural communities on which we and other species depend. *Second*, we need to slow down the rates at which we are simplifying, homogenizing, and degrading nature for our purposes. Otherwise, what is at risk is not the resilient earth but rather the quality of life for our own species and the very existence of other species, which are being driven to premature extinction partly due to our activities.

The Precautionary Principle: Look Before You Leap

The precautionary principle, while a subject of debate, may be a good approach to take toward some environmental problems and is the basis for several international treaties.

How can we live more sustainably? Ecologists say, study how nature has survived and adapted for 3.7 billion years and copy this strategy, outlined in Chapter 1 (Figures 1-11, p.17 and 1-12, p. 18). Increasingly, envi-

Figure 4-8 Some typical characteristics of natural and human-dominated systems. Many human activities threaten local ecological processes and some bring about harmful regional and global changes.

Property	Natural Systems	Human-Dominated Systems
Complexity	Biologically diverse	Biologically simplified
Energy source	Renewable solar energy	Mostly nonrenewable fossil fuel energy
Waste production	Little, if any	High
Nutrients	Recycled	Often lost or wasted
Net primary productivity	Shared among many species	Used, destroyed, or degraded to support human activities

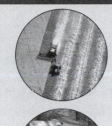

ThomsonNOW™ Active Figure 4-9 Natural capital degradation: major ways humans have altered the rest of nature to meet their growing population and resource needs and wants. QUESTION: *Which three of these items do you think have been the most harmful? See an animation based on this figure and take a short quiz on the concept.*

ronmental scientists and ecologists are also urging that we base our efforts to prevent damage to the earth's life-support system on the **precautionary principle:** when evidence indicates that an activity can seriously harm human health or the environment, we should take precautionary measures to prevent harm even if some of the cause-and-effect relationships have not been fully established scientifically. It is based on the common sense idea behind many adages such as "Better safe than sorry," "Look before you leap," "First, do no harm," and "Slow down for speed bumps."

As an analogy, we know that eating too much of certain types of foods and not getting enough exercise can greatly increase our chances of a heart attack, diabetes, and other disorders. But the exact connections between these health problems, chemicals in various foods, exercise, and genetics are still under study and often debated. We could use this uncertainty and unpredictability as an excuse to continue overeating and

not exercising. However, the wise course is to eat better and exercise more to help *prevent* potentially serious health problems.

Recently, the precautionary principle has formed the basis of several international environmental treaties. One example is the global treaty developed by 122 countries in 2000 to ban or phase out 12 *persistent organic pollutants.*

However, some analysts point out that we should be selective in applying the precautionary principle. We need to project possible unintended effects of our activities as carefully as possible. But we can never know all of the unintended effects of our actions and technologies. Thus, we must be willing to take some risks. Otherwise, we would stifle creativity and innovation and severely limit the development of new technologies and products.

Solutions: How Can We Develop More Sustainable Economies?

We can live more sustainably by shifting from high-throughput to low-throughput economies.

Most of today's advanced industrialized countries have **high-throughput economies** that attempt to boost economic growth by increasing the flow of matter and energy resources through their economic systems (Figure 4-10, p. 82). These resources flow through the economies of such societies to planetary *sinks* (air, water, soil, organisms), where pollutants and wastes can accumulate to harmful levels.

What happens if more and more people continue to use and waste more and more energy and matter resources at an increasing rate? In other words, what happens if most of the world's people get infected with the affluenza virus?

The law of conservation of matter and the two laws of thermodynamics (p. 27) tell us that eventually this consumption will exceed the capacity of the environment to dilute and degrade waste matter and absorb waste heat. This could lead to environmental and economic unsustainability. However, these scientific laws do not tell us how close we are to reaching such limits.

A temporary solution to this problem is to convert a linear high-throughput economy into a circular **matter-recycling-and-reuse economy** that mimics nature (Figure 1-11, p. 17). This means recycling and reusing most of our matter outputs within the economy instead of dumping them into the environment.

Changing to a matter-recycling-and-reuse economy is an important way to buy some time. But this does not allow more and more people to use more and more resources indefinitely, even if all of them were somehow perfectly recycled and reused; the two laws of thermodynamics tell us that recycling and reusing matter resources always requires using high-quality

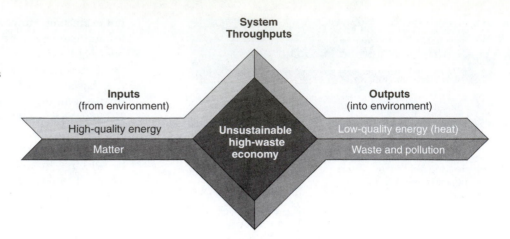

Figure 4-10 The *high-throughput economies* of most developed countries are based on continually increasing the rates of energy and matter flow. This practice produces valuable goods and services but also converts high-quality matter and energy resources into waste, pollution, and low-quality heat.

System Throughputs

Inputs (from environment)

High-quality energy

Matter

Unsustainable high-waste economy

Outputs (into environment)

Low-quality energy (heat)

Waste and pollution

energy (which cannot be recycled) and adds waste heat to the environment.

The three scientific laws governing matter and energy changes and the four principles of sustainability derived from nature (Figure 1-11, p. 17) suggest that the best long-term solution to our environmental and resource problems is to shift from an economy based on maximizing matter and energy flow (high-throughput)

to a more sustainable **low-throughput (low-waste) economy,** as summarized in Figure 4-11.

 Compare how energy is used in high- and low-throughput economies at ThomsonNOW.

In the next chapter, we will apply the principles of population dynamics and sustainability discussed in

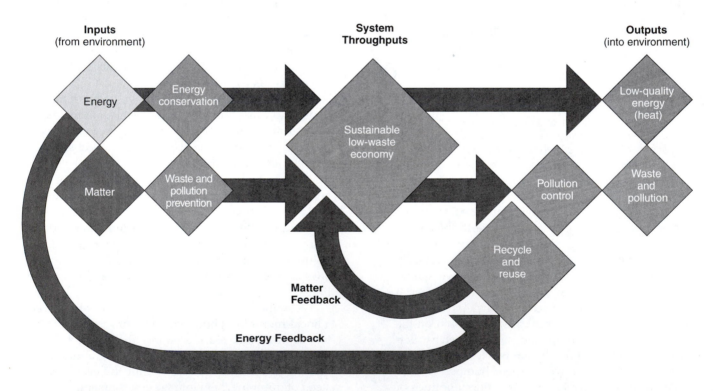

ThomsonNOW **Active Figure 4-11 Solutions:** lessons from nature. A *low-throughput economy,* based on energy flow and matter recycling, works with nature to reduce the throughput of matter and energy resources. This is done by **(1)** reusing and recycling most nonrenewable matter resources, **(2)** using renewable resources no faster than they are replenished, **(3)** using matter and energy resources efficiently, **(4)** reducing unnecessary consumption, **(5)** emphasizing pollution prevention and waste reduction, and **(6)** controlling population growth.

this chapter to the growth of the human population. In the two chapters after that, we apply them to understanding and sustaining the earth's terrestrial and aquatic biodiversity.

We cannot command nature except by obeying her.

SIR FRANCIS BACON

CRITICAL THINKING

1. How would you reply to someone who argues we should not worry about our effects on natural systems because succession will heal the wounds of human activities and restore the balance of nature?

2. How would you determine whether a particular species found in a given area is a keystone species?

3. Explain why a simplified community such as a cornfield usually is much more vulnerable to harm from insects and plant diseases than a more complex, natural community such as a grassland. Does this mean that we should never convert a grassland to a cornfield? Explain. What restrictions, if any, would you put on such conversions?

4. A bumper sticker reads "Nature always bats last and owns the stadium." What does this mean in ecological terms? What is its lesson for the human species?

5. Explain why you agree or disagree with the notion of using the precautionary principle in approaching some environmental problems. List two environmental problems that might warrant such an approach, even if you disagree in general with the principle and reject its use for these problems. Explain your position for or against using the principle to address each problem.

6. List two questions that you would like to have answered as a result of reading this chapter.

LEARNING ONLINE

The website for this book contains helpful study aids and many ideas for further reading and research. They include a chapter summary, review questions for the entire chapter, flash cards for key terms and concepts, a multiple-choice practice quiz, interesting Internet sites, references, information about green careers, and a guide for accessing thousands of InfoTrac® College Edition articles. Log into

www.thomsonedu.com/biology/miller

Then choose Chapter 4, and select a learning resource. For access to animations, additional quizzes, chapter outlines and summaries, register and log into

at **www.thomsonedu.com** using the access code card in the front of your book.

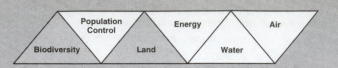

The problems to be faced are vast and complex, but come down to this: 6.4 billion people are breeding exponentially. The process of fulfilling their wants and needs is stripping earth of its biotic capacity to produce life; a climactic burst of consumption by a single species is overwhelming the skies, earth, waters, and fauna.

PAUL HAWKEN

HUMAN POPULATION GROWTH

Population Growth in the Past

We have kept the human population growing by expanding into ecosystems throughout the world, developing agricultural and industrial production innovations to expand the food supply, and using modern hygiene and medical advances to significantly lower death rates.

For most of history, the human population grew slowly, but about 200 years ago, human population growth took off (Figure 1-3, p. 7). Three major reasons explain this population increase.

First, humans developed the ability to expand into diverse new habitats and different climate zones. *Second*, the emergence of early and modern agriculture allowed more people to be fed. *Third*, we developed sanitation systems, antibiotics, and vaccines to help control infectious disease agents. As a result, death rates dropped sharply below birth rates.

Because of these three factors, the human population has experienced rapid exponential growth reflected in the characteristic J-curve (Figure 4-5, left side, p. 78). About 10,000 years ago when agriculture began, there were about 5 million humans on the planet; now there are 6.7 billion of us.

Population Growth Today: Rapid but Slowing

The rate at which the world's population is increasing has slowed, but the population is still growing rapidly.

During 2006, about 81 million people were added to the world's population. At a growth rate of 1.3% a year, we share the earth and its resources with about 222,000 more people each day, or 2.4 more people every time your heart beats.

The rate of the world's annual population growth dropped by almost half between 1963 and 2006, from 2.2% to 1.3%. This is *good news*, but during the same period, the population base doubled from 3.2 billion to 6.7 billion. This drop in the rate of population increase is somewhat like learning that the truck heading straight toward you has slowed from 100 kilometers per hour (kph) to 45 kph while its weight has doubled.

An exponential growth rate of 1.3% may seem small. But adding 81 million people per year to the world's population is roughly equal to adding another New York City every month or another Germany every year. Also, there is a big difference between population growth rates in developed and developing countries. In 2006, the population of developed countries was growing at a rate of 0.1%. That of the developing countries was 1.5%—fifteen times faster.

Where Are We Headed?

We don't know how long we can continue increasing the earth's carrying capacity for humans.

Scientific studies of populations of other species tell us that *no population can continue growing indefinitely*. In other words, there are limits to the earth's long-term carrying capacity for humans and other species.

No one knows how close we are to the environmental limits that sooner or later will control the population size of the human species. But there is growing evidence that we are steadily degrading the natural capital that keeps us and other species alive and supports our economies.

How many of us are likely to be here in 2050? Answer: 7.2–10.6 billion people, depending mostly on projections about the average number of babies that women are likely to have. The medium projection is 8.9 billion people. About 97% of this growth is projected to take place in developing countries, where acute poverty (living on less than $1 per day) is a way of life for about 1.4 billion people.

The human population may stabilize during this century as it moves from a J-shaped curve of exponential growth to an S-shaped curve of logistic growth (Figure 4-5, p. 78). The key question is the following: *Can the world provide an adequate standard of living for the medium projection of 8.9 billion people in 2050 without causing widespread environmental damage?*

FACTORS AFFECTING HUMAN POPULATION SIZE

Birth Rates and Death Rates: Entrances and Exits

Population increases because of births and immigration and decreases through deaths and emigration.

Demography is the study of the size, composition, and distribution of human populations and changes in these characteristics. Specialists in this field, called *demographers*, study how human populations grow or decline through the interplay of three factors: *births*, *deaths*, and *migration*. **Population change** is calculated by subtracting the number of people leaving a population (through death and emigration) from the number entering it (through birth and immigration) during a specified period of time (usually a year):

$$\text{Population change} = (\text{Births} + \text{Immigration}) - (\text{Deaths} + \text{Emigration})$$

When births plus immigration exceed deaths plus emigration, population increases; when the reverse is true, population declines.

Instead of using the total numbers of births and deaths per year, demographers use the **birth rate,** or **crude birth rate** (the number of live births per 1,000 people in a population in a given year), and the **death rate,** or **crude death rate** (the number of deaths per 1,000 people in a population in a given year). Figure 5-1 shows the crude birth and death rates for various groupings of countries in 2006.

What three countries have the largest numbers of people? Number 1 is China with 1.3 billion people in 2006, or one of every five people in the world. Number 2 is India with 1.1 billion people, or one of every six people. Together China and India have 37% of the world's population. The United States, with 300 million people in 2006, has the world's third largest population but only 4.5% of its people.

 Can you guess what regions of the world will likely have the greatest population increases by 2025? Find out at ThomsonNOW.

Declining Fertility Rates: Fewer Babies per Woman

The average number of children that a woman bears has dropped sharply since 1950, but is not low enough to stabilize the world's population in the near future.

Fertility is the number of births that occur to an individual woman. Two types of fertility rates affect a country's population size and growth rate. The first type,

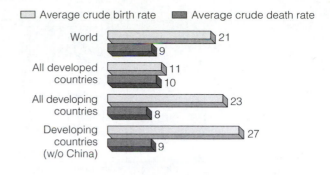

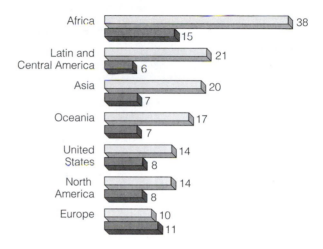

Figure 5-1 Average crude birth and death rates for various groupings of countries in 2006. (Data from Population Reference Bureau)

replacement-level fertility, is the number of children a couple must bear to replace themselves. It is slightly higher than two children per couple (2.1 in developed countries and as high as 2.5 in some developing countries), mostly because some female children die before reaching their reproductive years.

Does reaching replacement-level fertility mean an immediate halt to population growth? No, because so many future parents are alive. If each of today's couples had an average of 2.1 children and their children also had 2.1 children, the world's population would continue to grow for 50 years or more (assuming death rates do not rise).

The second type of fertility rate is the **total fertility rate (TFR):** the average number of children a woman typically has during her reproductive years. TFRs have dropped sharply since 1950. Today, the average woman in the world has half as many children as her counterpart in 1972. In 2006, the average global TFR was 2.7 children per woman: 1.6 in developed countries (down from 2.5 in 1950) and 2.9 in developing countries (down from 6.5 in 1950). This drop in the TFR in developing countries is impressive, but is still far above the replacement level of 2.1.

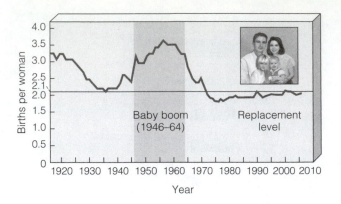

Figure 5-2 Total fertility rates for the United States between 1917 and 2006. (Data from Population Reference Bureau and U.S. Census Bureau)

Case Study: Fertility and Birth Rates in the United States

Population growth in the United States has slowed but is not close to leveling off.

The population of the United States has grown from 76 million in 1900 to 300 million in 2006, despite oscillations in the country's TFR (Figure 5-2). In 1957, the peak of the baby boom after World War II, the TFR reached 3.7 children per woman. Since then it has generally declined, remaining at or below replacement level since 1972.

The drop in the TFR has led to a decline in the rate of population growth in the United States. But the country's population is still growing faster than that of any other developed country and is not close to leveling off. Nearly 3.0 million people were added to the U.S. population in 2006. About 56% of this growth was

the result of more births than deaths and the rest came from legal and illegal immigration.

According to U.S. Census Bureau's medium projections, the U.S. population is likely to increase from 300 million in 2006 to 420 million by 2050 and reach 571 million by 2100. In contrast, population growth has slowed in other major developed countries since 1950 and most are expected to have declining populations after 2010. Because of a high per capita rate of resource use, each addition to the U.S. population has an enormous environmental impact (Figure 1-10, p. 14).

In addition to the almost fourfold increase in population growth, some amazing changes in lifestyles took place in the United States during the 20th century (Figure 5-3), which led to dramatic increases in per capita resource use and a much larger U. S. ecological footprint.

Factors Affecting Birth Rates and Fertility Rates

The number of children women have is affected by many factors, including the cost of raising children, educational and employment opportunities for women, and availability of contraceptives and abortions.

Many factors affect a country's average birth rate and TFR. One is the *importance of children as a part of the labor force*. Proportions of children working tend to be higher in developing countries—especially in rural areas, where children begin working to help raise crops at an early age.

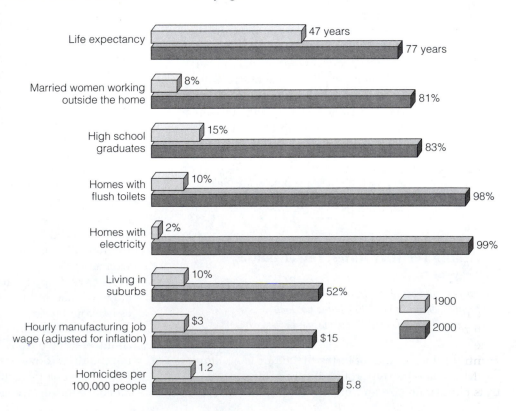

Figure 5-3 Some major changes that took place in the United States between 1900 and 2000. QUESTION: *Which two of these changes do you think were the most important?* (Data from U.S. Census Bureau and Department of Commerce)

Another economic factor is the *cost of raising and educating children*. Birth and fertility rates tend to be lower in developed countries, where raising children is much more costly because they do not enter the labor force until they are in their late teens or 20s. In the United States, it costs about $250,000 to raise a middle-class child from birth to age 18.

The *availability of private and public pension systems* affects a couple's decision on how many children to have. Pensions eliminate parents' need to have a lot of children to help support them in old age.

Urbanization plays a role. People living in urban areas usually have better access to family planning services and tend to have fewer children than those living in rural areas where children are needed to perform essential tasks.

Another important factor is the *educational and employment opportunities available for women*. TFRs tend to be low when women have access to education and paid employment outside the home. In developing countries, women with no education generally have two more children than women with a secondary school education.

Another factor is the *infant mortality rate*. In areas with low infant mortality rates, people tend to have a smaller number of children because fewer children die at an early age.

Average age at marriage (or, more precisely, the average age at which women have their first child) also plays a role. Women normally have fewer children when their average age at marriage is 25 or older.

Birth rates and TFRs are also affected by the *availability of legal abortions*. Each year about 190 million women become pregnant. The United Nations and the World Bank estimate that about 46 million of these women get abortions: 26 million of them legal and 20 million illegal (and often unsafe).

The *availability of reliable birth control methods* allows women to control the number and spacing of the children they have. *Religious beliefs, traditions,* and *cultural norms* also play a role. In some countries, these factors favor large families and strongly oppose abortion and some forms of birth control.

Factors Affecting Death Rates

Death rates have declined because of increased food supplies, better nutrition, advances in medicine, improved sanitation and personal hygiene, and safer water supplies.

The rapid growth of the world's population over the past 100 years is not the result of a rise in the crude birth rate. Instead, it has been caused largely by a decline in crude death rates, especially in developing countries.

More people started living longer and fewer infants died because of increased food supplies and distribution, better nutrition, medical advances such as immunizations and antibiotics, improved sanitation, and safer water supplies (which curtailed the spread of many infectious diseases).

Two useful indicators of the overall health of people in a country or region are **life expectancy** (the average number of years a newborn infant can expect to live) and the **infant mortality rate** (the number of babies out of every 1,000 born who die before their first birthday).

Great news. The global life expectancy at birth increased from 48 years to 67 years (77 years in developed countries and 65 years in developing countries) between 1955 and 2006. It is projected to reach 74 by 2050. Between 1900 and 2006, life expectancy in the United States increased from 47 to 78 years and is projected to reach 82 years by 2050.

However, in the world's poorest countries, life expectancy is 49 years or less. In many African countries, life expectancy is expected to fall further because of more deaths from AIDS.

Infant mortality is viewed as the best single measure of a society's quality of life because it reflects a country's general level of nutrition and health care. A high infant mortality rate usually indicates insufficient food (undernutrition), poor nutrition (malnutrition), and a high incidence of infectious disease (usually from contaminated drinking water and weakened disease resistance from undernutrition and malnutrition).

Good news. Between 1965 and 2006, the world's infant mortality rate dropped from 20 per 1,000 live births to 6.3 in developed countries and from 118 to 59 in developing countries.

Bad news. Annually, at least 7.6 million infants (most in developing countries) die of preventable causes during their first year of life—an average of 21,000 mostly unnecessary infant deaths per day. This is equivalent to 55 jumbo jets, each loaded with 400 infants under age 1, crashing each day with no survivors!

The U.S. infant mortality rate declined from 165 in 1900 to 6.7 in 2006. This sharp decline was a major factor in the marked increase in U.S. average life expectancy during this period. Still, some 46 countries had lower infant mortality rates than did the United States in 2006. If the U.S. infant mortality rate was as low as that of Singapore (ranked no. 1) in 2006, this would have saved the lives of 18,900 American children.

Three factors that keep the U.S. infant mortality rate higher than it could be are inadequate health care for poor women during pregnancy and for their babies after birth, drug addiction among pregnant women, and a high birth rate among teenagers.

Table 5-1 (p. 88) summarizes important terms and concepts related to human population change.

Table 5-1 Summary of Important Terms and Concepts

population change	the number of people entering a population (through birth and immigration) minus the number of people leaving the population (through death and emigration) during a specified period
birth rate	(or crude birth rate) the number of live births per 1,000 people in a population in a given year
death rate	(or crude death rate) the number of deaths per 1,000 people in a population in a given year
replacement-level fertility	the number of children a couple must bear to replace themselves
total fertility rate (TFR)	the average number of children a woman typically has during her reproductive years
life expectancy	the average number of years a newborn infant can expect to live
infant mortality rate	the number of babies out of every 1,000 born who die before their first birthday
age structure	the distribution of males and females in each age group in a population

Case Study: U.S. Immigration (Economics and Politics)

Immigration has played, and continues to play, a major role in the growth and cultural diversity of the U.S. population.

Since 1820, the United States admitted almost twice as many immigrants and refugees as all other countries combined. However, the number of legal immigrants (including refugees) has varied during different periods because of changes in immigration laws and rates of economic growth (Figure 5-4). Currently, legal and illegal immigration account for about 41% of the country's annual population growth.

Between 1820 and 1960, most legal immigrants to the United States came from Europe. Since 1960, most have come from Latin America (53%), Asia (25%), and Europe (14%). In 2006, Latinos (two-thirds of them from Mexico) made up 14% of the U.S. population, making them America's largest minority. By 2050, Latinos are projected to account for one of every four people in the United States and one of every 3 by 2100.

In 1995, the U.S. Commission on Immigration Reform recommended reducing the number of legal immigrants from about 900,000 to 700,000 per year for a transition period and then to 550,000 a year. Some analysts want to limit legal immigration to about 20% of the country's annual population growth. They would accept immigrants only if they can support

themselves, arguing that providing immigrants with public services makes the United States a magnet for the world's poor.

In 2006, an estimated 12 million illegal immigrants lived in the United States (up from 8.4 million in 2000), and as many as 800,000 more are believed to arrive each year. There is strong support for sharply reducing illegal immigration. But some are concerned that a crackdown on the country's illegal immigrants could lead to discrimination against some of the country's legal immigrants.

Proponents of reducing immigration argue that it would allow the United States to stabilize its population sooner and help reduce the country's enormous environmental impact. Polls show that almost 60% of the U.S. public strongly support reducing immigration.

Others oppose reducing current levels of legal immigration. They argue that this would diminish the historic role of the United States as a place of opportunity for the world's poor and oppressed. In addition, legal and illegal immigrants pay taxes, take many menial and low-paying jobs that other Americans shun, open businesses, and create jobs.

In 2004, the government estimated that illegal immigrants paid $7 billion in Social Security taxes, even though they will not receive benefits. Moreover, according to the U.S. Census Bureau, after 2020, higher immigration levels will be needed to supply enough workers as baby boomers retire.

X *HOW WOULD YOU VOTE?* Should legal immigration into the United States (or the country where you live) be reduced? Cast your vote online at **www.thomsonedu.com/biology/miller**.

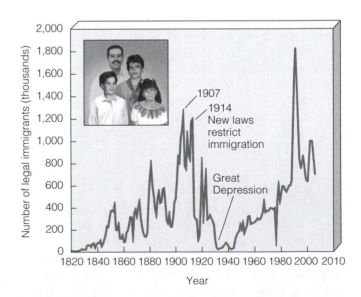

Figure 5-4 Legal immigration to the United States, 1820–2003. The large increase in immigration since 1989 resulted mostly from the Immigration Reform and Control Act of 1986, which granted legal status to illegal immigrants who could show they had been living in the country for several years. (Data from U.S. Immigration and Naturalization Service and the Pew Hispanic Center)

POPULATION AGE STRUCTURE

Age-Structure Diagrams

The number of people in young, middle, and older age groups determines how fast populations grow or decline.

As mentioned earlier, even if the replacement-level fertility rate of 2.1 were magically achieved globally tomorrow, the world's population would keep growing for at least another 50 years (assuming no large increase in the death rate). This results mostly from the **age structure:** the distribution of males and females in each age group of the world's population.

Demographers construct a diagram of population age structure by plotting the percentages or numbers of males and females in the total population in each of three age categories: *prereproductive* (ages 0–14), *reproductive* (ages 15–44), and *postreproductive* (ages 45 and up). Figure 5-5 presents generalized age-structure diagrams for countries with rapid, slow, zero, and negative population growth rates. Which of these figures best represents the country where you live?

Effects of Age Structure on Population Growth: Teenagers Rule

The number of people under age 15 is the major factor determining a country's future population growth.

Any country with many people below age 15 (represented by a wide base in Figure 5-5, left) has a powerful built-in momentum to increase its population size

unless death rates rise sharply. The number of births will rise even if women have only one or two children because a large number of girls will soon be moving into their reproductive years.

What is perhaps the world's most important population statistic? *Twenty-nine percent of the people on the planet were under 15 years old in 2006.* These 1.9 billion young people are poised to move into their prime reproductive years. In developing countries the number is even higher: 32% (42% in Africa), compared with 17% in developed countries. We live in a *demographically divided world,* as shown by population data for the United States, Brazil, and Nigeria (Figure 5-6, p. 90).

Using Age-Structure Diagrams to Make Population and Economic Projections

Changes in the distribution of a country's age groups have long-lasting economic and social impacts.

Between 1946 and 1964, the United States had a *baby boom* that added 79 million people to its population. Over time, this group looks like a bulge moving up through the country's age structure (Figure 5-7, p. 90).

 Examine how the baby boom affects the U.S. age structure over several decades at ThomsonNOW.

Baby boomers now make up nearly half of all adult Americans. As a result, they dominate the population's demand for goods and services. They also play

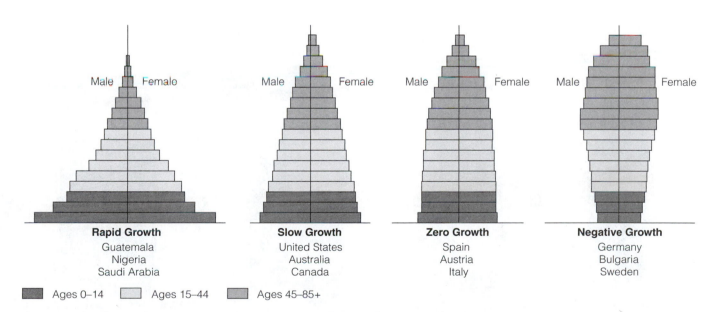

Rapid Growth	Slow Growth	Zero Growth	Negative Growth
Guatemala	United States	Spain	Germany
Nigeria	Australia	Austria	Bulgaria
Saudi Arabia	Canada	Italy	Sweden

■ Ages 0–14   □ Ages 15–44   ▨ Ages 45–85+

ThomsonNOW™ Active Figure 5-5 Generalized population age structure diagrams for countries with rapid (1.5–3%), slow (0.3–1.4%), zero (0–0.2%), and negative population growth rates (a declining population). Populations with a large proportion of its people in the prereproductive ages of 1–14 (at left) have a large potential for rapid population growth. *See an animation based on this figure and take a short quiz on the concept.* (Data from Population Reference Bureau)

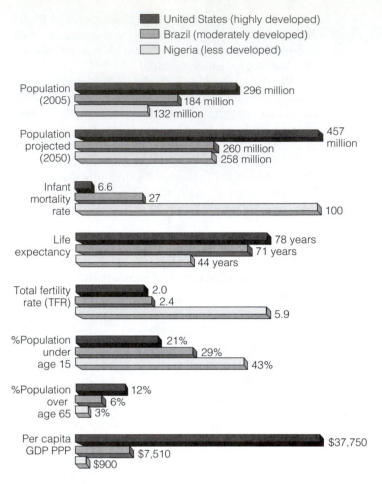

Legend:
- United States (highly developed)
- Brazil (moderately developed)
- Nigeria (less developed)

Population (2005):
- 296 million
- 184 million
- 132 million

Population projected (2050):
- 457 million
- 260 million
- 258 million

Infant mortality rate:
- 6.6
- 27
- 100

Life expectancy:
- 78 years
- 71 years
- 44 years

Total fertility rate (TFR):
- 2.0
- 2.4
- 5.9

%Population under age 15:
- 21%
- 29%
- 43%

%Population over age 65:
- 12%
- 6%
- 3%

Per capita GDP PPP:
- $37,750
- $7,510
- $900

Figure 5-6 Global connections: comparison of key demographic indicators for three countries, one highly developed (United States), one moderately developed (Brazil), and one less developed (Nigeria) in 2006. (Data from Population Reference Bureau)

an increasingly important role in deciding who gets elected and what laws are passed. Baby boomers who created the youth market in their teens and 20s are now creating the 50-something market and will soon move on to create a 60-something market. In 2011, the first baby boomers will turn 65, and the number of Americans over age 65 will consequently grow sharply through 2029.

According to some analysts, the retirement of baby boomers is likely to create a shortage of workers in the United States unless immigrant workers replace some of these retirees. Retired baby boomers are likely to use their political clout to force the smaller number of people in the baby-bust generation that followed them (Figure 5-7) to pay higher income, health-care, and Social Security taxes. One of the biggest public policy questions in the United States is related to the country's age structure. How will the country balance the needs of seniors with those of other age groups, at levels of taxation that don't strangle the economy?

In other respects, the baby-bust generation should have an easier time than the baby-boom generation. Fewer people will be competing for educational opportunities, jobs, and services. Also, labor shortages may drive up their wages, at least for jobs requiring education or technical training beyond high school. However, this may not happen if American-owned companies operating at the global level (multinational companies) continue to export many jobs to other countries.

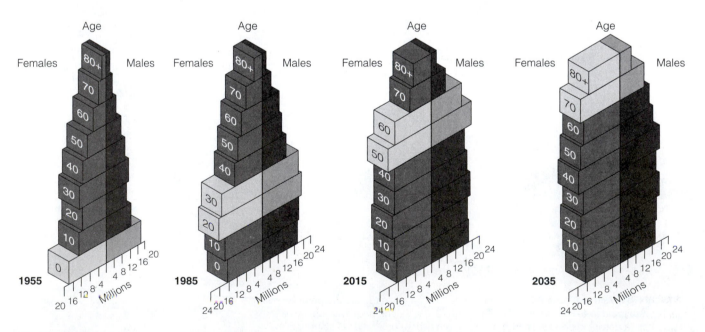

ThomsonNOW™ Active Figure 5-7 Tracking the baby boom. U.S. population by age and sex, 1900, 2000, and 2050 (projected). *See an animation based on this figure and take a short quiz on the concept.*

As these projections indicate, any booms or busts in the age structure of a population create social and economic changes that ripple through a society for decades.

Benefits of Slowing Population Growth

Slowing population growth can have economic and social benefits.

By 2006, 40 countries had populations that were either stable (annual growth rates at or below 0.3%) or declining. All, except Japan, are in Europe. By 2050, according to the UN, the population size of most developed countries (but not the United States) will have stabilized.

Slower population growth can offer economic benefits. As the number of children declines, as happened in many Asian nations in the 1960s, this frees up more human and monetary resources for investment and adult consumption. Over the next decade, most Middle Eastern countries could benefit from a similar *demographic economic dividend.*

Rapid Population Decline from Declining Fertility Rates: Demographic Liabilities

Rapid population decline can lead to long-lasting economic and social problems.

As the age structure of the world's population changes and the percentage of people age 60 or older increases, more countries will begin experiencing population declines. If population decline is gradual, its harmful effects usually can be managed.

But rapid population decline, like rapid population growth, can lead to severe economic and social problems. A country undergoing rapid population decline because of a "baby bust" or a "birth dearth" has a sharp rise in the proportion of older people. This puts severe strains on government budgets because these older people consume an increasingly larger share of medical care, social security funds, and other costly public services funded by a decreasing number of working taxpayers. Such countries can also face labor shortages unless they rely more on greatly increased automation or immigration of foreign workers. Demographers project that 51 countries, most of them developed, will lose population between 2006 and 2050. Figure 5-8 lists some of the problems associated with rapid population decline.

Population Decline from a Rising Death Rate: The AIDS Tragedy

A large number of deaths from AIDS can disrupt a country's social and economic structure by removing large numbers of young adults from its age structure.

Between 2000 and 2050, AIDS is projected to cause the premature deaths of 278 million people in 53 coun-

- Can threaten economic growth
- Less government revenues with fewer workers
- Less entrepreneurship and new business formation
- Less likelihood for new technology development
- Increasing public deficits to fund higher pension and healthcare costs

Figure 5-8 Some problems with rapid population decline. QUESTION: *Which three of these problems do you think are the most important?*

tries—38 of them in Africa. These premature deaths are almost equal to the entire current population of the United States. Read this paragraph again and think about the enormity of this tragedy.

Unlike hunger and malnutrition, which kills mostly infants and children, AIDS kills many young adults. This change in the young adult age structure of a country has a number of drastic effects. One is a sharp drop in average life expectancy. In 8 African countries, where 16–39% of the adult population is infected with HIV, life expectancy could drop to 34–40 years of age.

Another harmful effect is a loss of a country's most productive young adult workers and trained personnel such as scientists, farmers, engineers, teachers, and government, business, and health-care workers. This causes a sharp drop in the number of productive adults available to support the young and the elderly and to grow food and provide essential services.

Analysts call for the international community—especially developed countries—to develop and fund a massive program to help countries ravaged by AIDS in Africa and elsewhere. The program would have two major goals. *First,* it would reduce the spread of HIV through a combination of improved education and health care. *Second,* it would provide financial assistance for education and health care as well as volunteer teachers and health-care and social workers to help compensate for the missing young adult generation.

SOLUTIONS: INFLUENCING POPULATION SIZE

Is the Earth Overpopulated? An Important Controversy

There is disagreement about whether the world should encourage or discourage population growth.

The projected increase of the human population from 6.6 to 8.9 billion or more between 2004 and 2050 reminds us of the important question: *Can the world provide an adequate standard of living for 2.2 billion more people without causing widespread environmental damage?*

Controversy surrounds this and two related questions: First, is the earth currently overpopulated? Second, what measures, if any, should be taken to slow population growth?

Those who do not believe the earth is overpopulated point out that the average life span of the world's 6.7 billion people is longer today than at any time in the past and is projected to get longer. They say that the world can support billions more people—the world's most valuable resource, they argue, for solving the problems we face and stimulating economic growth through more consumption.

Some view any form of population regulation as a violation of their religious or moral beliefs. Others see it as an intrusion into their privacy and personal freedom to have as many children as they want. Some developing countries and members of minorities in developed countries regard population control as a form of genocide to keep their numbers and power from rising.

Proponents of slowing and eventually stopping population growth have a different view. They point out that we fail to provide the basic necessities for one of every five people on the earth today. If we cannot or will not provide basic support for about 1.4 billion people today, they ask, how will we be able to provide for the projected 2.2 billion more people by 2050? See the Guest Essay by Garrett Hardin on this topic on the website for this chapter.

Proponents of slowing population growth warn of two serious consequences if we do not sharply lower birth rates. *First,* the death rate may increase because of declining health and environmental conditions in some areas—something that is already happening in parts of Africa. *Second,* resource use and environmental degradation may intensify as more consumers increase their already large ecological footprint in developed countries and in rapidly developing countries, such as China and India.

✗ *HOW WOULD YOU VOTE?* Should the population of the country where you live be stabilized as soon as possible? Cast your vote online at www.thomsonedu.com/biology/miller.

The Demographic Transition (Economics)

History indicates that as countries become economically developed, their birth and death rates decline.

Demographers have examined the birth and death rates of western European countries that became industrialized during the 19th century. From these data they developed a hypothesis of population change known as the **demographic transition:** as countries become industrialized, first their death rates and then their birth rates decline. According to this hypothesis, the transition takes place in four distinct stages (Figure 5-9).

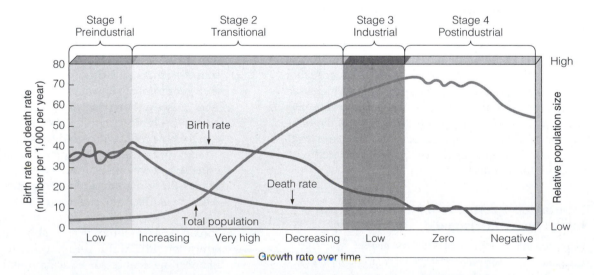

ThomsonNOW **Active Figure 5-9** Generalized model of the *demographic transition.* There is uncertainty about whether this model will apply to some of today's developing countries. *See an animation based on this figure and take a short quiz on the concept.*

First is the *preindustrial stage,* when there is little population growth because harsh living conditions lead to both a high birth rate (to compensate for high infant mortality) and a high death rate.

Next is the *transitional stage,* when industrialization begins, food production rises, and health care improves. Death rates drop and birth rates remain high, so the population grows rapidly (typically 2.5–3% a year).

During the third phase, called the *industrial stage,* the birth rate drops and eventually approaches the death rate as industrialization, medical advances, and modernization become widespread. Population growth continues, but at a slower and perhaps fluctuating rate, depending on economic conditions.

Most developed countries and a few developing countries are now in this third stage. The fertility rates in many developing countries have dropped sharply but are still above replacement-level fertility. Thus, the populations in these countries are still increasing and will continue doing so for several decades.

The last phase is the *postindustrial stage,* when the birth rate declines further, equaling the death rate and reaching zero population growth. If the birth rate falls below the death rate, population size decreases slowly. Forty countries containing about 13% of the world's population have entered this stage and more of the world's developed countries are expected to enter this phase by 2050.

 Explore the effects of economic development on birth and death rates and population growth at ThomsonNOW.

In most developing countries today, death rates have fallen much more than birth rates. In other words, these developing countries are still in the transitional stage, halfway up the economic ladder, with high population growth rates.

Some economists believe developing countries will make the demographic transition over the next few decades with the help of technologies for birth control, communication, and industrial development. Further, they argue, developing countries can learn from mistakes made in developed countries and speed up the transition process.

But some population analysts fear the still-rapid population growth in many developing countries will outstrip economic growth and overwhelm some local life-support systems. This could cause many of these countries to be caught in a *demographic trap* at stage 2, the transition stage. This is now happening as death rates rise in a number of developing countries, especially in Africa. Indeed, countries in Africa being ravaged by the HIV/AIDS epidemic are falling back to stage 1.

Other factors that could hinder the demographic transition in some developing countries are shortages of skilled workers, lack of financial capital, large debts to developed countries, and a drop in economic assistance from developed countries since 1985.

Family Planning: Planning for Babies Works

Family planning has been a major factor in reducing the number of births and abortions throughout most of the world.

Family planning provides educational and clinical services that help couples choose how many children to have and when to have them. Such programs vary from culture to culture, but most provide information on birth spacing, birth control, and health care for pregnant women and infants.

The world's largest source of funds for family planning and reproductive health is the United Nations Population Fund (UNFPA). Since it began in 1969, the UNFPA has provided nearly $6 billion in assistance to governments and nongovernmental organizations in more than 140 countries, some of which have sharply reduced birth, fertility, and abortion rates.

Family planning has helped increase the proportion of married women in developing countries who use modern forms of contraception from 10% of married women of reproductive age in the 1960s to 51% of these women in 2006. Studies also show that family planning is responsible for at least 55% of the drop in TFRs in developing countries, from 6.0 in 1960 to 3.0 in 2006.

One example of a country that sharply reduced its population growth is Thailand. Between 1971 and 2006, it cut its annual population growth rate from 3.2% to 0.8% and its TFR from 6.4 to 1.7 children per family. This was largely due to vigorous government-supported family planning and public relations programs. Also helpful were increased economic roles for women and improved education and health care for women and children.

Despite such successes, several problems remain. *First,* according to John Bongaarts of the Population Council and the UNFPA, 42% of all pregnancies in the developing countries are unplanned and 26% of those end with abortion. *Second,* an estimated 200 million women in developing countries want to limit the number and determine the spacing of their children, but they lack access to family planning services. According to the UNFPA and the Alan Guttmacher Institute, meeting women's current unmet needs for family planning and contraception would prevent *each year:* 23 million unplanned births, 22 million induced abortions, 1.4 million infant deaths, and 142,000 pregnancy-related deaths.

Some analysts call for expanding family planning programs to include teenagers and sexually active unmarried women, who are excluded in many existing programs. For teenagers, many advocate much greater emphasis on abstinence.

Another suggestion is to develop programs that educate men about the importance of having fewer children and taking more responsibility for raising them. Proponents also call for greatly increased research on developing new, more effective, and more acceptable birth control methods for men.

Finally, a number of analysts urge pro-choice and pro-life groups to join forces in greatly reducing unplanned births and abortions, especially among teenagers.

Empowering Women: Ensuring Education, Jobs, and Human Rights

Women tend to have fewer children if they are educated, have a paying job outside the home, and do not have their human rights suppressed.

Three key factors lead women to have fewer and healthier children: education, paying jobs outside the home, and living in societies where their rights are not suppressed.

Women make up roughly half of the world's population. They do almost all of the world's domestic work and child care with little or no pay. Women also provide more unpaid health care than all the world's organized health services combined.

They also do 60–80% of the work associated with growing food, gathering fuelwood, and hauling water in rural areas of Africa, Latin America, and Asia. As one Brazilian woman put it, "For poor women the only holiday is when you are asleep."

Globally, women account for two-thirds of all hours worked but receive only 10% of the world's income, and they own less than 2% of the world's land. In most developing countries, women do not have the legal right to own land or to borrow money. Women also make up 70% of the world's poor and 64% of the world's 800 million illiterate adults.

According to Thorya Obaid, "Many women in the developing world are trapped in poverty by illiteracy, poor health, and unwanted high fertility. All of these contribute to environmental degradation and tighten the grip of poverty. If we are serious about sustainable development, we must break this vicious cycle."

This means giving women everywhere full legal rights and the opportunity to become educated and earn income outside the home. Achieving this would slow population growth, promote human rights and freedom, reduce poverty, and slow environmental degradation—a win–win result.

Empowering women by seeking gender equality will take some major social changes in many countries. This will be difficult to achieve in male-dominated societies but it can be done. An increasing number of women in developing countries are taking charge of their lives and reproductive behavior. They are not waiting around for the slow processes of education and cultural change. As it expands, such bottom-up change by individual women will play an important role in stabilizing population and providing women with basic human rights.

Solutions: Reducing Population Growth

Experience suggests that the best way to slow population growth is a combination of investing in family planning, reducing poverty, and elevating the status of women.

In 1994, the United Nations held its third once-in-a-decade Conference on Population and Development in Cairo, Egypt. One of the conference's goals was to encourage action to stabilize the world's population at 7.8 billion by 2050 instead of the projected 8.9 billion.

The major goals of the resulting population plan, endorsed by 180 governments, are to do the following by 2015:

- Provide universal access to family planning services and reproductive health care
- Improve health care for infants, children, and pregnant women
- Develop and implement national population policies
- Improve the status of women and expand educational and job opportunities for young women
- Provide more education, especially for girls and women
- Increase the involvement of men in child-rearing and family planning
- Sharply reduce poverty
- Sharply reduce unsustainable patterns of production and consumption

The experiences of Japan, Thailand, South Korea, Taiwan, Iran, and China indicate that a country can achieve or come close to replacement-level fertility within a decade or two. Such experiences also suggest that the best way to slow population growth is a combination of *investing in family planning, reducing poverty, and elevating the status of women.*

SLOWING POPULATION GROWTH

Case Study: India

For more than five decades India has tried to control its population growth with only modest success.

The world's first national family planning program began in India in 1952, when its population was nearly 400 million. In 2004, after 54 years of population control efforts, India was the world's second most populous country, with a population of 1.1 billion.

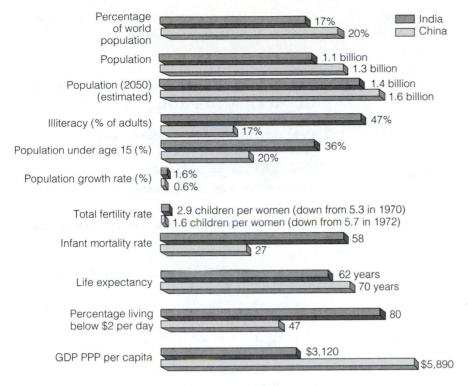

Figure 5-10 Global outlook: Basic demographic data for India and China. (Data from United Nations and Population Reference Bureau)

In 1952, India added 5 million people to its population. In 2006, it added 18.6 million. By 2035, India is projected to be the world's most populous country. Its population is not projected to peak until 2065 at 1.6 billion people. Figure 5-10 compares demographic data for India and China.

India faces a number of already serious poverty, malnutrition, and environmental problems that could worsen as its population continues to grow rapidly. By global standards, about one of every four people in India is poor. And nearly half of India's labor force is unemployed or can find only occasional work.

India currently is self-sufficient in food grain production. Still, about 40% of its population and more than half of its children suffer from malnutrition, mostly because of poverty.

Furthermore, India faces serious resource and environmental problems. With 17% of the world's people, it has just 2.3% of the world's land resources and 2% of the world's forests. About half of the country's cropland is degraded as a result of soil erosion, waterlogging, salinization, overgrazing, and deforestation. In addition, more than two-thirds of India's water is seriously polluted, sanitation services often are inadequate, and many of its major cities suffer from serious air pollution.

India is undergoing rapid economic growth that is expected to accelerate. Its huge and growing middle class is already larger than the entire U.S. population. As these individuals increase their resource use per person, India's ecological footprint will expand and increase the pressure on the country's and the earth's natural capital. On the other hand, economic growth may help slow population growth by accelerating India's ability to make the demographic transition.

Without its long-standing family planning program, India's population and environmental problems would be growing even faster. Still, to its supporters the results of the program have been disappointing for several reasons: poor planning, bureaucratic inefficiency, the low status of women (despite constitutional guarantees of equality), extreme poverty, and lack of administrative and financial support.

The government has provided information about the advantages of small families for years. Yet Indian women still have an average of 2.9 children. Most poor couples believe they need many children to do work and care for them in old age. The strong cultural preference for male children means some couples keep having children until they produce one or more boys. One result is that even though 90% of Indian couples know of at least one modern birth control method, only 46% actually use one. Another consequence is a "bride shortage" because men outnumber women.

Case Study: China

Since 1970 China has used a government-enforced program to cut its birth rate in half and sharply reduce its fertility rate.

Since 1970, China has made impressive efforts to feed its people, bring its population growth under control, and encourage economic growth. Between 1972 and 2006, it cut its crude birth rate in half and trimmed its TFR from 5.7 to 1.6 children per woman (Figure 5-10).

In addition, since 1980 China has moved 300 million people—more than the number of people in the U.S. population—out of poverty and quadrupled the average person's income. China also has a literacy rate of 91% and has used a comprehensive health care system to boost life expectancy to 71 years. It now has more than 300 million new middle-class consumers and is likely to have 600 million such consumers by 2010. By 2020, China could become the world's leading economic power.

To achieve its sharp drop in fertility, China has established the world's most extensive, intrusive, and strict population control program. In the 1960s, government officials concluded that the only alternative to strict population control was mass starvation. China's policy generally limits urban couples to one child and allows rural couples to have two children if the first child is a daughter. Married couples who pledge to have no more than one child receive extra food, larger pensions, better housing, free medical care, salary bonuses, free school tuition for their one child, and preferential treatment in employment when their child enters the job market. Couples who break their pledge lose such benefits.

The government also provides married couples with ready access to free sterilization, contraceptives, and abortion. This helps explain why about 83% of married women in China use modern contraception.

The one-child policy is unpopular and has been enforced mainly in urban areas, with about 30% of the country's population. This uneven enforcement has caused political tension between rural and urban dwellers. Reports of forced abortions and other coercive actions have brought condemnation from the United States and other national governments.

In China (as in India), there is a strong preference for male children because of the lack of a real social security system. A folk saying goes, "Rear a son, protect yourself in old age." Many pregnant women use ultrasound to determine the gender and often get an abortion if the fetus is female. This has lead to a growing *gender imbalance* or "bride shortage" in China's population, with a projected 30–40 million surplus of men by 2020.

China's population control program has sharply reduced birth and fertility rates, and if trends continue, China's population should peak around 2040 and then begin a slow decline. The implications of this are now becoming clear. With fewer children, the average age of China's population is increasing rapidly, which could result in a declining work force, lack of funds for supporting continuing economic development, and fewer children and grandchildren to care for the growing number of elderly people. These concerns and other factors may lead to some relaxation of China's one-child population control policy in the future.

China has 20% of the world's population. But it has only 7% of the world's freshwater and cropland, 4% of its forests, and 2% of its oil. Soil erosion in China is serious and apparently getting worse. In 2005, China's deputy minister of the environment said, "Our raw materials are scarce, we don't have enough land . . . desert areas are expanding . . . half of the water in our seven largest rivers is completely polluted and one-third of the urban population is breathing polluted air."

China's economy is growing at one of the world's highest rates as the country undergoes rapid industrialization. This means that more middle class Chinese will consume more resources per person, increasing China's ecological footprint and the strain on the earth's natural capital.

POPULATION DISTRIBUTION: URBAN GROWTH AND PROBLEMS

Urban Growth: More People and More Poverty

Urban populations are growing rapidly throughout the world and many cities in developing countries have become centers of poverty.

Today, almost half the world's people live in densely populated urban areas. By the end of this century 80–90% of the world's people will probably live in megacities, each containing 10 to 50 million people.

Rural people are *pulled* to urban areas in search of jobs and a better life. Some are also *pushed* from rural areas into urban areas by factors such as poverty, lack of land to grow food, declining agricultural jobs, famine, and war.

Five major trends are important in understanding the problems and challenges of urban growth. First, *the proportion of the global population living in urban areas is increasing.* Between 1850 and 2006, the percentage of people living in urban areas increased from 2% to 49%. According to UN projections, 60% of the world's people will be living in urban areas by 2030. Almost all of this growth will occur in already overcrowded cities in developing countries such as India, Brazil, China, and Mexico (Figure 5-11).

Second, *the number of large cities is mushrooming.* Each week 1 million people are added to the world's urban areas. Today there are 18 *megacities* or *megalopolises* (up from 8 in 1985) each with 10 million or more people. Fifteen of these urban areas are in developing countries (Figure 5-11). Tokyo, Japan, with 35.3 million people, is the world's largest urban area.

Third, *the urban population is increasing rapidly in developing countries.* Between 2006 and 2030, the percentage of people living in urban areas in developing countries is expected to increase from 42% to 56%. In effect, developing countries will need to build the equivalent of a city with more than 1 million people *each week* for the next 25 years.

Fourth, *urban growth is much slower in developed countries than in developing countries.* Still, developed countries, now with 76% urbanization, are projected to reach 84% urbanization by 2030.

Fifth, *poverty is becoming increasingly urbanized as more poor people migrate from rural to urban areas, mostly in developing countries.* The United Nations estimates that at least 1 billion poor people live in crowded and unsanitary slums and shantytowns in or on the outskirts of most cities in developing countries; within 30 years this number may double.

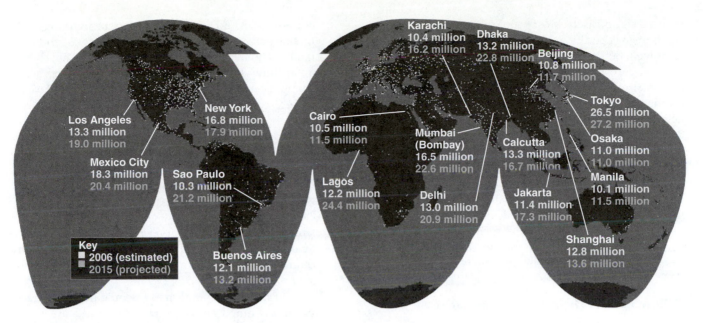

Figure 5-11 Global outlook: major urban areas throughout the world revealed in satellite images of the earth at night showing city lights. Currently, the 49% of the world's people living in urban areas occupy about 2% of the earth's land area. Note that most of the world's urban areas are found along the coasts of continents, and most of Africa and much of the interior of South America, Asia, and Australia are dark at night. This figure also shows the populations of the world's 18 *megacities* (each with 10 or more million people) in 2006 and their projected populations in 2015. All but three are located in developing countries. QUESTION: *Use this figure to list in order the world's five most populous cities in 2006 and the five most populous ones projected for 2015.* (Data from National Geophysics Data Center, National Oceanic and Atmospheric Administration, and United Nations)

Case Study: Urbanization in the United States (Economics)

Eight of every ten Americans live in urban areas, about half of them in sprawling suburbs.

Between 1800 and 2006, the percentage of the U.S. population living in urban areas increased from 5% to 80%. The population has shifted in four phases. First, *people migrated from rural areas to large central cities.* Currently, three-fourths of Americans live in 271 *metropolitan areas* (cities and towns with at least 50,000 people), and nearly half of the country's population lives in consolidated metropolitan areas containing 1 million or more residents (Figure 5-12).

Second, many people *migrated from large central cities to suburbs and smaller cities.* Currently, about 51% of the U.S. population lives in the suburbs and 30% live in central cities.

Third, many people *migrated from the North and East to the South and West.* Since 1980, about 80% of the U.S. population increase has occurred in the South and West, particularly near the coasts. California in the West, with 34.5 million people, is the most populous state, followed by Texas in the Southwest with 21.3 million people. This shift is expected to continue.

Fourth, *some people have migrated from urban areas back to rural areas* since the 1970s, and especially since 1990, resulting in *exurbs*—vast sprawling areas that

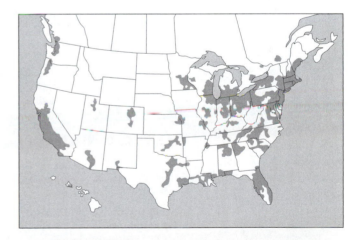

Figure 5-12 Major urban regions in the United States. About 80% of Americans live in urban areas occupying about 1.7% of the land area of the lower 48 states. Nearly half (48%) of Americans live in *consolidated metropolitan areas* with 1 million or more people. These areas are projected to merge into the megalopolises shown as shaded areas in this figure. (Data from U.S. Census Bureau)

have no center and are further from central cities than suburbs are.

During these shifts in the last century, the quality of life for most Americans improved significantly (Figure 5-3). Since 1920, many of the worst urban environmental problems in the United States have been reduced significantly. Most people have better working and housing conditions, and air and water quality have improved. Better sanitation, public water supplies, and medical care have slashed death rates and sickness from malnutrition and transmittable diseases. Concentrating most of the population in urban areas has helped protect the country's biodiversity by reducing the destruction and degradation of wildlife habitat.

However, a number of U.S. cities, especially older ones, have *deteriorating services* and *aging infrastructures* (streets, schools, bridges, housing, and sewers). Many also face *budget crunches* and decreasing public services as businesses and people move to the suburbs and exurbs, reducing revenues from property taxes. And there is *rising poverty* in the centers of many older cities, where unemployment typically is 50% or higher.

Urban Sprawl (Economics and Politics)

When land is available and affordable, urban areas tend to sprawl outward and swallow up surrounding countryside.

A major problem in the United States and some other countries with lots of room for expansion is **urban sprawl.** Growth of low-density development on the edges of cities and towns gobbles up surrounding countryside—frequently prime farmland, forests, and wetlands—and increases dependence on cars. The result is a far-flung hodgepodge of housing developments, shopping malls, parking lots, and office complexes—loosely connected by multilane highways and freeways.

Urban sprawl is the product of ample and affordable land, automobiles, cheap gasoline, and poor urban planning. Figure 5-13 shows some of the undesirable consequences of urban sprawl. Sprawl has increased travel time in automobiles, decreased energy efficiency, increased urban flooding problems, and destroyed prime cropland, forests, open space, and wetlands. It has also led to the economic death of many central cities as people and businesses move to the suburbs.

On the other hand, many people prefer living in suburbs. Compared to central cities, these areas provide lower density, access to larger lot sizes and single-family homes, and often have newer public schools and lower crime rates.

Some cities have redeveloped and rejuvenated downtown areas by building convention and enter-

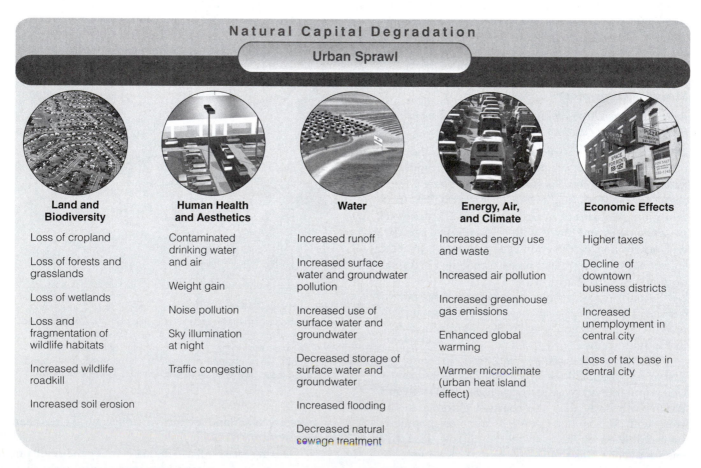

Natural Capital Degradation

Urban Sprawl

Land and Biodiversity	Human Health and Aesthetics	Water	Energy, Air, and Climate	Economic Effects
Loss of cropland	Contaminated drinking water and air	Increased runoff	Increased energy use and waste	Higher taxes
Loss of forests and grasslands	Weight gain	Increased surface water and groundwater pollution	Increased air pollution	Decline of downtown business districts
Loss of wetlands	Noise pollution	Increased use of surface water and groundwater	Increased greenhouse gas emissions	Increased unemployment in central city
Loss and fragmentation of wildlife habitats	Sky illumination at night	Decreased storage of surface water and groundwater	Enhanced global warming	Loss of tax base in central city
Increased wildlife roadkill	Traffic congestion	Increased flooding	Warmer microclimate (urban heat island effect)	
Increased soil erosion		Decreased natural sewage treatment		

Figure 5-13 Natural capital degradation: some undesirable impacts of urban sprawl or car-dependent development. Do you live in an area suffering from urban sprawl? QUESTION: *Which five of these effects do you think are the most harmful?*

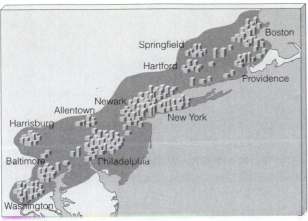

Bowash (Boston to Washington)

Figure 5-14 Natural capital degradation: the U.S. megalopolis of *Bowash*, comprising urban sprawl and coalescence between Boston, Massachusetts, and Washington, D.C.

tainment centers, office buildings, shops, restaurants, hotels, and apartments. This is an encouraging trend but decaying neighborhoods often surround such developments.

As they grow and sprawl outward, separate urban areas merge to form a *megalopolis*. For example, the remaining open space between Boston, Massachusetts, and Washington, D.C., is rapidly urbanizing and coalescing. The result is an almost 800-kilometer-long (500-mile-long) urban area containing about 35 million people, sometimes called *Bowash* (Figure 5-14).

 Examine how the San Francisco Bay area in California (USA) grew in population between 1900 and 1990 at ThomsonNOW.

URBAN ENVIRONMENTAL AND RESOURCE PROBLEMS

Advantages of Urbanization (Economics and Science)

Urban areas can offer more job opportunities and better education and health, and can help protect biodiversity by concentrating people.

Urbanization has many benefits. From an *economic standpoint*, cities are centers of economic development, education, technological developments, and jobs, and have served as centers of industry, commerce, and transportation.

In terms of *health*, urban residents in many parts of the world live longer than do rural residents and have lower infant mortality rates and fertility rates than do rural populations. In addition, urban dwellers generally have better access to medical care, family plan-

ning, education, and social services than do people in rural areas.

Urban areas also have some environmental advantages. For example, recycling is more economically feasible because large concentrations of recyclable materials and per capita expenditures on environmental protection are higher in urban areas. Also, concentrating people in urban areas helps preserve biodiversity by reducing the stress on wildlife habitats.

Disadvantages of Urbanization (Science and Poverty)

Cities are rarely self-sustaining and tend to have mostly harmful impacts on the environment.

Unsustainable systems. Although urban dwellers occupy only about 2% of the earth's land area, they consume about three-fourths of the earth's resources. Because of this and their high waste output (Figure 5-15, p. 100), most of the world's cities are not self-sustaining systems.

Urbanization can help preserve biodiversity in some areas. On the other hand, large areas of land must be disturbed and degraded to provide urban dwellers with food, water, energy, minerals, and other resources. This activity decreases and degrades the earth's biodiversity.

From an environmental standpoint, urban areas are somewhat like gigantic vacuum cleaners, sucking up much of the world's matter, energy, and living resources and spewing out pollution, wastes, and heat. As a consequence, they have large ecological footprints that extend far beyond their boundaries. If you live in a city, you can calculate its ecological footprint by going to the website **www.redefiningprogress.org/**. Also see the Guest Essay on this topic by Michael Cain on this chapter's website.

Lack of Plants. In urban areas, most trees, shrubs, or other plants are destroyed to make way for buildings, roads, and parking lots. Thus, most cities do not benefit from vegetation that would absorb air pollutants, give off oxygen, help cool the air through transpiration, provide shade, reduce soil erosion, muffle noise, provide wildlife habitats, and give aesthetic pleasure. As one observer remarked, "Most cities are places where they cut down most of the trees and then name the streets after them."

Water Problems. As cities grow and their water demands increase, expensive reservoirs and canals must be built and deeper wells drilled. This can deprive rural and wild areas of surface water and deplete groundwater faster than it is replenished.

Flooding also tends to be greater in cities and their suburbs, sometimes because they are built on floodplain areas or along low-lying coastal areas subject to natural flooding. Covering land with buildings,

Inputs

Energy

Food

Water

Raw materials

Manufactured goods

Money

Information

Outputs

Solid wastes

Waste heat

Air pollutants

Water pollutants

Greenhouse gases

Manufactured goods

Noise

Wealth

Ideas

Figure 5-15 Natural capital degradation: urban areas rarely are sustainable systems. The typical city depends on large nonurban areas for huge inputs of matter and energy resources and for disposal of its large outputs of waste matter. According to an analysis by Mathis Wackernagel and William Rees, an area 58 times as large as that of London, England, is needed to supply its residents with resources. They estimate that meeting the needs of all the world's people at the same rate of resource use as that of London would take at least three more earths.

asphalt, and concrete also causes precipitation to run off quickly and overload storm drains. In addition, urban development often destroys or degrades wetlands that act as natural sponges to help absorb excess water. Another threat is that many of the world's largest cities are in coastal areas (Figure 5-11) that could be flooded sometime in this century if sea levels rise due to global warming.

Pollution and Health Problems. Because of their high population densities and high resource consumption, urban dwellers produce most of the world's air pollution, water pollution, and solid and hazardous wastes. Pollutant levels in urban areas are generally higher because pollution is produced in smaller areas and cannot be dispersed and diluted as readily as pollution produced in rural areas. In addition, high population densities in urban areas can increase the spread of *infectious diseases,* especially if adequate drinking water and sewage systems are not available.

Excessive Noise. Most urban dwellers are subjected to *noise pollution:* any unwanted, disturbing, or harmful sound that impairs or interferes with hearing, causes stress, hampers concentration and work efficiency, or causes accidents. Noise levels are measured in decibel-A (dbA) sound pressure units that vary with different human activities (Figure 5-16). Sound pressure becomes damaging at about 85 dbA and painful at around 120 dbA. At 180 dbA it can kill.

Climate and Artificial Light. Cities generally are warmer, rainier, foggier, and cloudier than suburbs and nearby rural areas mostly because of their buildings, pavement, and lack of green space. The enormous amounts of heat generated by cars, factories, furnaces, lights, air conditioners, and heat-absorbing dark roofs and streets in cities create an *urban heat island* surrounded by cooler suburban and rural areas. As cities grow and merge, the heat islands merge and can keep polluted air from being diluted and cleansed.

Also, the artificial light created by cities hinders astronomers from conducting their research and makes it difficult for casual observers to enjoy the night sky. Some species are adversely affected by such *light pollution.* For example, endangered sea turtles lay their eggs on beaches at night and depend on darkness. And migrating birds are lured off course by the lights of high-rise buildings and fatally collide with them.

Poverty and Social Problems. Urban areas can intensify poverty and social problems. Crime rates tend to

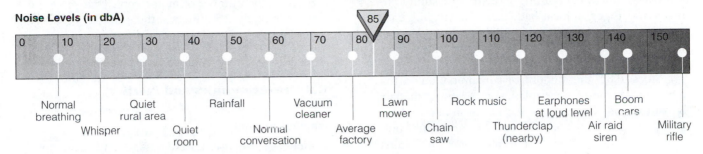

Permanent damage begins after 8-hour exposure

Noise Levels (in dbA)

| 0 | 10 | 20 | 30 | 40 | 50 | 60 | 70 | 80 | 85 | 90 | 100 | 110 | 120 | 130 | 140 | 150 |

Normal breathing — Whisper — Quiet rural area — Quiet room — Rainfall — Normal conversation — Vacuum cleaner — Average factory — Lawn mower — Chain saw — Rock music — Thunderclap (nearby) — Earphones at loud level — Air raid siren — Boom cars — Military rifle

Figure 5-16 Noise levels (in decibel-A sound pressure units) of some common sounds. You are being exposed to a sound level high enough to cause permanent hearing damage if you need to raise your voice to be heard above the racket, if a noise causes your ears to ring, or if nearby speech seems muffled. Prolonged exposure to lower noise levels and occasional loud sounds may not damage your hearing but can greatly increase internal stress. Noise pollution can be reduced by modifying noisy activities and devices, shielding noisy devices or processes, shielding workers or other persons from the noise, moving noisy operations or machinery away from people, and using antinoise (a technology that cancels out one noise with another).

be higher in urban areas than in rural areas. And urban areas are more likely and desirable targets for terrorist acts.

The Urban Poor in Developing Countries (Economics and Poverty)

Most of the urban poor in developing countries live in crowded, unhealthy, and dangerous conditions, but many are better off than the rural poor.

About one of every six people in the world live in crowded and unsanitary conditions in the cities of developing countries. Some live in *slums*—tenements and rooming houses where 3–10 people live in a single room. Others live in *squatter settlements* and *shanty-towns* often on the outskirts of these cities, some perched precariously on steep hillsides subject to landslides. People build shacks from corrugated metal, discarded packing crates, and other scavenged building materials. Others live or sleep on the streets.

Squatters living near the edge of survival in these areas usually lack clean water supplies, sewers, electricity, and roads, and often are subject to severe air and water pollution and hazardous wastes from nearby factories. Their locations may also be especially prone to landslides, flooding, earthquakes, or volcanic eruptions.

Most cities cannot afford to provide squatter settlements and shantytowns with basic services, and their officials fear that improving services will attract even more of the rural poor. Many city governments regularly bulldoze squatter shacks and send police to drive the illegal settlers out. The people then move back in or develop another shantytown somewhere else.

Despite joblessness, squalor, overcrowding, and environmental and health hazards, most squatter and slum residents are better off than the rural poor. With better access to family planning programs, they tend to have fewer children and better access to schools. Many squatter settlements provide a sense of community and a vital safety net of neighbors, friends, and relatives for the poor.

Case Study: Mexico City (Science, Politics, and Poverty)

Mexico City suffers from severe poverty, overcrowding, and air pollution.

Mexico City is an example of an urban area in crisis. About 19 million people—roughly one of every six Mexicans—live there. It is the world's second most populous city, and each year at least 400,000 new residents arrive.

Mexico City suffers from severe air pollution, close to 50% unemployment, deafening noise, overcrowding, traffic congestion, inadequate public transportation, and a soaring crime rate. More than one-third of its residents live in slums called *barrios* or in squatter settlements without running water or electricity.

At least 3 million people have no sewer facilities. This means huge amounts of human waste are deposited in gutters, vacant lots, and open sewers every day, attracting armies of rats and swarms of flies. When the winds pick up dried excrement, a *fecal snow* often falls on parts of the city. Open garbage dumps also contribute dust and bacteria to the atmosphere. This bacteria-laden fallout leads to widespread salmonella and hepatitis infections, especially among children.

Mexico City has one of the world's worst photo-chemical smog problems because of a combination of too many cars and polluting industries, a sunny climate, and topographical bad luck. The city lies in a high-elevation bowl-shaped valley surrounded on three sides by mountains—conditions that trap air pollutants at ground level.

Since 1982, the amount of contamination in the city's air has more than tripled, and breathing that air is said to be roughly equivalent to smoking three packs of cigarettes a day. The city's air and water pollution cause an estimated 100,000 premature deaths per year. Writer Carlos Fuentes has nicknamed it "Makesicko City."

Some environmental progress has been made. The city government has banned cars in its central zone, required catalytic converters on all cars made after 1991, phased out use of leaded gasoline, and replaced old buses and taxis with cleaner ones. The city also bought land for use as green space and planted more than 25 million trees to help absorb pollutants. These measures are working: the percentage of days each year in which air pollution standards are violated has fallen from 50% to 20%.

TRANSPORTATION AND URBAN DEVELOPMENT

Land Availability and Transportation Systems

Land availability determines whether a city must grow vertically or spread out horizontally and whether it relies mostly on mass transportation or the automobile.

The two main types of ground transportation are *individual* (such as cars, motor scooters, bicycles, and walking) and *mass* (mostly buses and rail systems). About 90% of all travel in the world is by foot, bicycle, motor scooter, or bus—mostly because only about 10% of the world's people can afford a car.

Land availability is a key factor determining the types of transportation people use. If a city cannot spread outward, it must grow vertically—upward and downward (below ground)—so it occupies a small land area with a high population density. Most people living in such *compact cities* like Hong Kong, China, and Tokyo, Japan, walk, ride bicycles, or use energy-efficient mass transit.

A combination of cheap gasoline, plentiful land, and a network of highways produces *dispersed cities*. They are found in countries such the United States, Canada, and Australia where ample land often is available for outward expansion. Sprawling cities depend on the automobile for most travel, and have a number of undesirable effects (Figure 5-13).

Today there are about 800 million cars and trucks in the world and each year 44 million more are added. By 2050, the number of motor vehicles is projected to be 3.5 billion—2.5 billion of them in today's developing countries.

Case Study: Motor Vehicles in the United States (Economics and Politics)

Each year Americans drive as far as everyone else in the world combined.

America showcases the advantages and disadvantages of living in a society dominated by motor vehicles. With 4.6% of the world's people, the United States has almost a third of the world's motor vehicles. Over half of American vehicles are gas-guzzling sport utility vehicles (SUVs), pickup trucks, minivans, and other large vehicles.

Mostly because of urban sprawl and convenience, passenger vehicles are used for 98% of all urban transportation and 91% of travel to work in the United States. About 75% of Americans drive to work alone, 5% commute to work on public transit, and 0.4% bicycle to work. Each year Americans drive about the same distance driven by all other drivers in the world, and, in the process, use about 43% of the world's gasoline. According to the American Public Transit Association, if Americans increased their use of mass transit from the current 5% to 10%, this would reduce U.S. dependence on oil by 40%.

Many governments in rapidly industrializing countries such as China plan to develop an automobile-centered transportation system. China plans to reduce its dependence on imported oil by avoiding the trap of relying on fuel-wasting motor vehicles. Instead, by 2020 it plans to become the global leader in producing fuel-efficient cars and the world's largest seller of cars. If this happens, most major U.S. and European car companies could be put out of business. This would lead to a major loss of high-paying jobs, and most U.S. and European citizens would buy motor vehicles from China, helping it become the world's number one economic power.

Trade-offs: Advantages and Disadvantages of Motor Vehicles

Motor vehicles provide personal benefits and help run economies, but they also kill and injure people and animals, pollute the air, promote urban sprawl, and result in traffic jams.

On a personal level, motor vehicles provide mobility and are a convenient and comfortable way to get from one place to another. They also are symbols of power, sex, social status, and success for many people. For some, they also provide escape from an increasingly hectic world.

From an economic standpoint, much of the world's economy is built on producing motor vehicles and supplying roads, services, and repairs for them. In the United States, for example, a quarter of every dollar spent and one of every six nonfarm jobs is connected to the automobile.

Despite their important benefits, motor vehicles have many harmful effects on people and the environment. Almost 18 million people have been killed in automobiles since they were invented. Throughout the world, automobile accidents kill approximately 1.2 million people each year—an average of 3,300 deaths per day—and injure another 15 million people. Each year they also kill about 50 million wild animals and family pets.

In the United States, motor vehicle accidents kill more than 40,000 people a year and injure another 5 million, at least 300,000 of them severely. *Car accidents have killed more Americans than all wars in the country's history.*

Motor vehicles are the world's largest source of air pollution, which causes 30,000–60,000 premature deaths per year in the United States, according to the Environmental Protection Agency. They are also the fastest growing source of climate-changing carbon dioxide emissions—now producing almost one-fourth of them. In addition, they account for two-thirds of the oil used in the United States in the form of gasoline and one-third of the world's oil consumption.

Motor vehicles have helped create urban sprawl. At least a third of urban land worldwide and half in the United States is devoted to roads, parking lots, gasoline stations, and other automobile-related uses. This prompted urban expert Lewis Mumford to suggest that the U.S. national flower should be the concrete cloverleaf.

Another problem is congestion. If current trends continue, U.S. motorists will spend an average of 2 years of their lives in traffic jams. Building more roads may not be the answer. Many analysts agree with the idea, as stated by economist Robert Samuelson, that "cars expand to fill available concrete."

Solutions: Reducing Automobile Use (Economics and Politics)

Although it would not be politically popular, we could reduce reliance on automobiles by having users pay for their harmful effects.

Some environmental scientists and economists suggest that one way to reduce the harmful effects of automobile use is to make drivers pay directly for most of the harmful costs of automobile use—a *user-pays* approach based on honest environmental accounting.

One way to phase in such *full-cost pricing* would be to charge a tax on gasoline that covers the estimated health and environmental costs of driving. Such taxes would amount to about $2 per liter ($7 per gallon) of gasoline in the United States. This would spur the use of more energy-efficient motor vehicles and mass transit, decrease dependence on imported oil and thus increase economic and military security, and reduce pollution and environmental degradation.

Proponents of this approach urge governments to use gasoline tax revenues to help finance mass transit systems, bike paths, and sidewalks and to reduce taxes on income, wages, and wealth to offset the increased taxes on gasoline, thereby making this *tax shift* more politically acceptable.

Most analysts doubt that taxing gasoline would be feasible in the United States, for four reasons. *First,* it faces strong political opposition from two groups: the public, largely unaware of the huge hidden costs they are already paying, and the politically powerful transportation-related industries such as oil and tire companies, road builders, carmakers, and many real estate developers. However, taxpayers might accept sharp increases in gasoline taxes if the extra costs were offset by decreases in taxes on wages and income.

Second, fast, efficient, reliable, and affordable mass transit options and bike paths are not widely available in most of the United States. *Third,* the dispersed nature of most U.S. urban areas makes most people dependent on cars. *Fourth,* most people who can afford cars are virtually addicted to them.

Another way to reduce automobile use and congestion is to raise parking fees and charge tolls on roads, tunnels, and bridges—especially during peak traffic times. And more than 300 cities in Germany, Austria, Italy, Switzerland, and the Netherlands have *car-sharing* networks. Members reserve a car in advance or call the network and are directed to the closest car. They are billed monthly for the time they use a car and the distance they travel. In Berlin, Germany, car sharing has cut car ownership by 75% and car commuting by nearly 90%. Flexcar has been providing car sharing for more than 30,000 members in seven major U.S. metropolitan areas at a rate of $9 an hour.

Solutions: Alternatives to the Car

Alternatives include walking, bicycling, driving scooters, and taking subways, trolleys, trains, and buses.

Several alternatives to car use exist, each with advantages and disadvantages. Examples are *bicycles* (Figure 5-17, p. 104), *mass transit rail systems in urban areas* (Figure 5-18, p. 104), *bus systems in urban areas* (Figure 5-19, p. 104), and *rapid rail systems between urban areas* (Figure 5-20, p. 105).

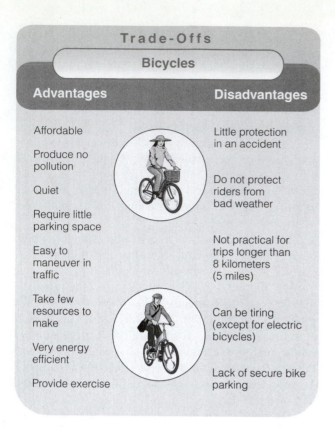

Trade-Offs

Bicycles

Advantages	Disadvantages
Affordable	Little protection in an accident
Produce no pollution	
Quiet	Do not protect riders from bad weather
Require little parking space	
Easy to maneuver in traffic	Not practical for trips longer than 8 kilometers (5 miles)
Take few resources to make	
Very energy efficient	Can be tiring (except for electric bicycles)
Provide exercise	Lack of secure bike parking

Figure 5-17 Trade-offs: advantages and disadvantages of *bicycles*. QUESTION: *What are the single advantage and the single disadvantage that you think are the most important?*

Trade-Offs

Mass Transit Rail

Advantages	Disadvantages
More energy efficient than cars	Expensive to build and maintain
Produces less air pollution than cars	Cost effective only along a densely populated narrow corridor
Requires less land than roads and parking areas for cars	
Causes fewer injuries and deaths than cars	Commits riders to transportation schedules
Reduces car congestion in cities	Can cause noise and vibration for nearby residents

Figure 5-18 Trade-offs: advantages and disadvantages of *mass transit rail systems* in urban areas. QUESTION: *What are the single advantage and the single disadvantage that you think are the most important?*

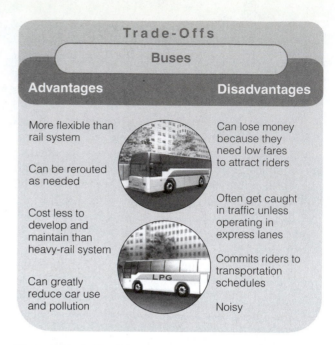

Trade-Offs

Buses

Advantages	Disadvantages
More flexible than rail system	Can lose money because they need low fares to attract riders
Can be rerouted as needed	
Cost less to develop and maintain than heavy-rail system	Often get caught in traffic unless operating in express lanes
	Commits riders to transportation schedules
Can greatly reduce car use and pollution	Noisy

Figure 5-19 Trade-offs: advantages and disadvantages of *bus systems in urban areas*. QUESTION: *What are the single advantage and the single disadvantage that you think are the most important?*

MAKING URBAN AREAS MORE SUSTAINABLE AND DESIRABLE PLACES TO LIVE

Solutions: Smart Growth (Science, Economics, and Politics)

Smart growth can help control growth patterns, discourage urban sprawl, reduce car dependence, and protect ecologically sensitive areas.

Smart growth is emerging as a means to encourage more environmentally sustainable development that requires less dependence on cars, controls and directs sprawl, and reduces wasteful resource use. Figure 5-21 (p. 106) lists popular smart growth tools used to prevent and control urban growth and sprawl.

Some communities are going further and using the principles of *new urbanism* to develop entire villages and recreate mixed-use neighborhoods within existing cities. These principles include:

- *walkability,* with stores and other services within a 10-minute walk of home

- *mixed-use and diversity*—a mix of pedestrian-friendly shops, offices, apartments, and homes and people of different ages, classes, cultures, and races

- *quality urban design* emphasizing beauty, aesthetics, and architectural diversity

Trade-Offs

Rapid Rail

Advantages	Disadvantages
Can reduce travel by car or plane	Expensive to run and maintain
Ideal for trips of 200–1,000 kilometers (120–620 miles)	Must operate along heavily used routes to be profitable
Much more energy efficient per rider over the same distance than a car or plane	Cause noise and vibration for nearby residents

Figure 5-20 Trade-offs: advantages and disadvantages of rapid-rail systems between urban areas. QUESTION: *What are the single advantage and the single disadvantage that you think are the most important?*

- *environmental sustainability* based on development with minimal environmental impact

- *smart transportation* with well-designed train and bus systems connecting neighborhoods, towns, and cities

The Ecocity Concept (Science, Economics, Politics, and Ethics)

An ecocity allows people to walk, bike, or take mass transit for most of their travel, and it recycles and reuses most of its wastes, grows much of its own food, and protects biodiversity by preserving surrounding land.

According to most environmentalists and urban planners, the primary problem is not urbanization but our failure to make cities more sustainable and livable. They call for us to make new and existing urban areas more self-reliant, sustainable, and enjoyable places to live through good ecological design. See the Guest Essay on this topic by David Orr on the website for this chapter.

A more environmentally sustainable city, called an *ecocity* or *green city*, emphasizes the following goals built around the four principles of sustainability:

- Use solar and other locally available renewable energy sources

- Use energy and matter resources efficiently

- Prevent pollution and reduce waste

- Reuse and recycle at least 60% of all municipal solid waste

- Protect and encourage biodiversity by preserving surrounding land and protecting and restoring natural systems and wetlands

- Build just, equitable, and vibrant communities that promote cultural and economic diversity

An ecocity is people-oriented, not car-oriented. Its residents are able to walk, bike, or use low-polluting mass transit for most of their travel. Its buildings, vehicles, and appliances meet high energy-efficiency standards. Trees and plants adapted to the local climate and soils are planted throughout to provide shade and beauty, supply wildlife habitats, and reduce pollution, noise, and soil erosion. Small organic gardens and a variety of plants adapted to local climate conditions often replace monoculture grass lawns.

Abandoned lots, industrial sites, and polluted creeks and rivers are cleaned up and restored. Nearby forests, grasslands, wetlands, and farms are preserved. Much of an ecocity's food comes from nearby organic farms, solar greenhouses, community gardens, and small gardens on rooftops, in yards, and in window boxes. People designing and living in ecocities take seriously the advice U.S. urban planner Lewis Mumford gave more than three decades ago: "Forget the damned motor car and build cities for lovers and friends."

The ecocity is not just a futuristic dream. Examples of cities that have attempted to become more environmentally sustainable and livable include Curitiba, Brazil (Case Study, p. 106); Waitakere City, New Zealand; Leicester, England; Portland, Oregon; Davis, California; Olympia, Washington; and Chattanooga, Tennessee.

According to its proponents, urban areas that fail to become more livable and ecologically sustainable over the next few decades are inviting economic depression and increased unemployment, pollution, and social tension. Sweden is leading the way with 70

Figure 5-21 Solutions: *smart growth tools* used to help prevent and control urban growth and sprawl. QUESTION: *Which five of these tools do you think are the most important ways to prevent or control urban sprawl?*

Solutions

Smart Growth Tools

Limits and Regulations

Limit building permits

Urban growth boundaries

Greenbelts around cities

Public review of new development

Zoning

Encourage mixed use

Concentrate development along mass transportation routes

Promote high-density cluster housing developments

Planning

Ecological land-use planning

Environmental impact analysis

Integrated regional planning

State and national planning

Protection

Preserve existing open space

Buy new open space

Buy development rights that prohibit certain types of development on land parcels

Taxes

Tax land, not buildings

Tax land on value of actual use (such as forest and agriculture) instead of highest value as developed land

Tax Breaks

For owners agreeing legally to not allow certain types of development (conservation easements)

For cleaning up and developing abandoned urban sites (brownfields)

Revitalization and New Growth

Revitalize existing towns and cities

Build well-planned new towns and villages within cities

"eco-municipalities." And China is planning to develop 10 model ecocities.

Case Study: The Ecocity Concept in Curitiba, Brazil

One of the world's most livable and sustainable major cities is Curitiba, Brazil.

Curitiba is known as the ecological capital of Brazil. This beautiful city with its clean air, tree-lined streets, and lack of litter decided in 1969 to focus on mass transit rather than on the car and now has the world's best bus system. Each day it carries about 70% of its more than 2 million people throughout the city along express lanes dedicated to buses. Only high-rise apartment buildings are allowed near major bus routes, and each building must devote the bottom two floors to stores, which reduces the need for residents to travel.

Bike paths run throughout most of the city. Cars are banned from 49 blocks of the city's downtown area, which has a network of pedestrian walkways connected to bus stations, parks, and bike paths. Because Curitiba relies less on automobiles, it uses less energy per person and has less air pollution, greenhouse gas emissions, and traffic congestion than most comparable cities.

Volunteers have planted more than 1.5 million trees, none of which can be cut down without a permit, and two trees must be planted for each one cut down.

The city recycles roughly 70% of its paper and 60% of its metal, glass, and plastic, which is sorted by households for collection three times a week. Recovered materials are sold mostly to the more than 500 major industries in Curitiba that must meet strict pollution standards. Most of these industries are in an industrial park outside the city limits. A major bus line runs to the park, but many of the workers live nearby and can walk or bike to work.

The city uses old buses as roving classrooms to give the poor basic skills needed for jobs. Other retired buses have become health clinics, soup kitchens, and day care centers, which are open 11 hours a day and are free for low-income parents.

The poor receive free medical, dental, and child care, and 40 feeding centers are available for street children. People can exchange recyclable garbage for food, bus tokens, and school supplies. The city has a *build-it-yourself* system that gives each poor family a plot of land, building materials, two trees, and an hour's consultation with an architect.

In Curitiba, virtually all households have electricity, drinking water, and trash collection. About 95% of its citizens can read and write, and 83% of the adults have at least a high school education. All schoolchildren study ecology. Polls show that 99% of the city's inhabitants would not want to live anywhere else.

This internationally acclaimed model of urban planning and sustainability is the brainchild of architect and former college teacher Jaime Lerner, who has served as the city's mayor three times since 1969.

An exciting challenge during this century will be to reshape existing cities and design new ones like Curitiba that are more livable and environmentally sustainable.

The city is not an ecological monstrosity. It is rather the place where both the problems and the opportunities of modern technological civilization are most potent and visible.

PETER SELF

CRITICAL THINKING

1. Why is it rational for a poor couple in a developing country such as India to have four or five children? What changes might induce such a couple to consider their behavior irrational?

2. Choose what you consider to be a major local, national, and global environmental problem, and describe the role of population growth in this problem. Compare your answer with those of your classmates.

3. Suppose that all women in the world today began bearing children at replacement-level fertility rates of 2.1 children per woman. Explain why this would not im-

mediately stop global population growth. About how long would it take for population growth to stabilize (assuming death rates do not rise)?

4. Do you think the population of **(a)** your own country and **(b)** the area where you live is too high? Explain.

5. Should everyone have the right to have as many children as they want? Explain.

6. Some people believe the most important goal is to sharply reduce the rate of population growth in developing countries, where 97% of the world's population growth is expected to take place. Some people in developing countries contend that the most serious environmental problems result from high levels of resource consumption per person in developed countries, which use about 80% of the world's resources. What is your view on this issue? Explain.

7. If you own a car or hope to own one, what conditions, if any, would encourage you to rely less on the automobile and to travel to school or work by bicycle, on foot, by mass transit, or by a carpool or vanpool?

8. Congratulations! You are in charge of the world. List the three most important features of your **(a)** population policy and **(b)** urban policy.

9. List two questions that you would like to have answered as a result of reading this chapter.

LEARNING ONLINE

The website for this book contains helpful study aids and many ideas for further reading and research. They include a chapter summary, review questions for the entire chapter, flash cards for key terms and concepts, a multiple-choice practice quiz, interesting Internet sites, references, information about green careers, and a guide for accessing thousands of InfoTrac® College Edition articles. Log into

www.thomsonedu.com/biology/miller

Then choose Chapter 5, and select a learning resource. For access to animations, additional quizzes, chapter outlines and summaries, register and log into

at **www.thomsonedu.com** using the access code card in the front of your book.

6

SUSTAINING BIODIVERSITY: THE ECOSYSTEM APPROACH

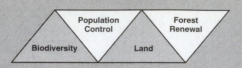

Forests precede civilizations, deserts follow them.

FRANÇOIS-AUGUSTE-RENÉ DE CHATEAUBRIAND

HUMAN IMPACTS ON BIODIVERSITY

Global Outlook: Uses of the World's Land

Over half the world's land consists of forests, woodlands, rangeland, and pasture and 11% is used to grow food.

About 30% of the earth's land is covered by forests and woodlands. Rangeland and pasture occupy about 26% and crops about 11% of the land. The remaining third consists of deserts, tundra, rock, and ice that are unsuitable for most human activities.

We have protected about 12% of the earth's land in parks, wildlife refuges, wilderness, and various types of nature reserves. But more than half of such lands exist as protected areas mostly on paper, without needed funds, tools, or mechanisms to guarantee protection. In other words, *we have reserved about 95% of the world's land for our use.* This is the major reason that the world's biodiversity is under assault.

Effects of Human Activities on Global Biodiversity (Science and Economics)

We have depleted and degraded some of the earth's biodiversity and these threats are expected to increase.

Figure 6-1 summarizes the ways in which many human activities decrease biodiversity. According to biodiversity expert Edward O. Wilson, "The natural world is everywhere disappearing before our eyes—cut to pieces, mowed down, plowed under, gobbled up, replaced by human artifacts."

Consider a few examples of how human activities have decreased and degraded the earth's terrestrial biodiversity. According to a 2002 study on the impact of the human ecological footprint on the earth's land, we have disturbed to some extent at least half, and probably 83%, of the earth's land surface (excluding Antarctica and Greenland). Most of this has been done by filling in wetlands and converting grasslands and forests to crop fields and urban areas. By some estimates, humans use, waste, or destroy about 10–55% of the net primary productivity of the planet's terrestrial ecosystems.

In the United States, at least 95% of the virgin forests in the lower 48 states have been logged for

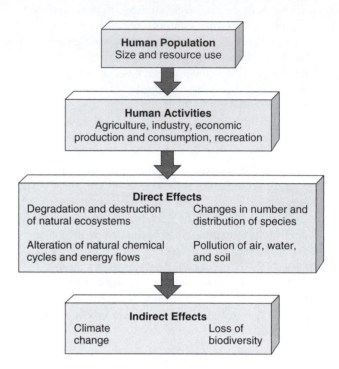

Figure 6-1 Natural capital degradation: major connections between human activities and the earth's biodiversity.

lumber and to make room for agriculture, housing, and industry. In addition, 98% of tallgrass prairie in the Midwest and Great Plains has disappeared, and 99% of California's native grassland and 85% of its original redwood forests are gone.

Human activities are also degrading the world's *aquatic biodiversity.* About half the world's wetlands and half the wetlands in the United States were lost during the last century. An estimated 27% of the world's diverse coral reefs have been severely damaged. By 2050, another 70% may be severely damaged or eliminated.

About three-fourths of the world's 200 commercially valuable marine fish species are either overfished or fished to their estimated sustainable yield. According to a 2004 study by the United Nations Environment Programme (UNEP), stocks of large, open-ocean fish such as tuna, swordfish, and marlin have declined by 90% since 1900.

Human activities also contribute to the *premature extinction of species.* Biologists estimate that the current global extinction rate of species is at least 100 times and probably 1,000 to 10,000 times what it was before

The Ecosystem Approach

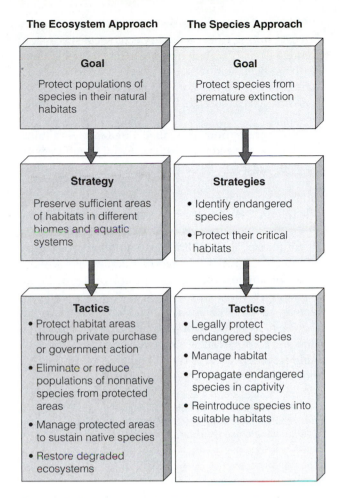

Goal
Protect populations of species in their natural habitats

↓

Strategy
Preserve sufficient areas of habitats in different biomes and aquatic systems

↓

Tactics
- Protect habitat areas through private purchase or government action
- Eliminate or reduce populations of nonnative species from protected areas
- Manage protected areas to sustain native species
- Restore degraded ecosystems

The Species Approach

Goal
Protect species from premature extinction

↓

Strategies
- Identify endangered species
- Protect their critical habitats

↓

Tactics
- Legally protect endangered species
- Manage habitat
- Propagate endangered species in captivity
- Reintroduce species into suitable habitats

Figure 6-2 Solutions: goals, strategies, and tactics for protecting biodiversity.

humans existed. These threats to the world's biodiversity are projected to increase sharply by 2018.

Figure 6-2 outlines the goals, strategies, and tactics for preserving and restoring the terrestrial ecosystems and aquatic systems that provide habitats and resources for the world's species (as discussed in this chapter), and preventing the premature extinction of species (as discussed in Chapter 7).

Why Should We Care about Biodiversity? (Science, Economics, and Ethics)

Biodiversity should be protected from degradation by human activities because it exists and because of its usefulness to us and other species.

Biodiversity researchers contend that we should act to preserve the earth's overall diversity because its genes, species, ecosystems, and ecological processes have two types of value. One is **intrinsic value**—that these components of biodiversity exist, regardless of their use to us. The other is **instrumental value**—our measure of their usefulness to us.

MANAGING AND SUSTAINING FORESTS

Economic and Ecological Services Provided by Forests

Forests provide a number of ecological and economic services and researchers have attempted to estimate their total monetary value.

About 30% of the earth's land surface (excluding Greenland and Antarctica) is covered by forests with at least 10% tree cover. The world's forests provide many important ecological and economic services (Figure 6-3).

For example, forests play an important role in regulating the earth's climate. Trees temporarily remove carbon dioxide from the troposphere, through photosynthesis, and store it as carbon compounds in their tissues as part of the global carbon cycle (Figure 2-26, p. 46). Trees also affect global climate by helping regulate water flows among the oceans, land, and troposphere as part of the global water cycle (Figure 2-25, p. 45).

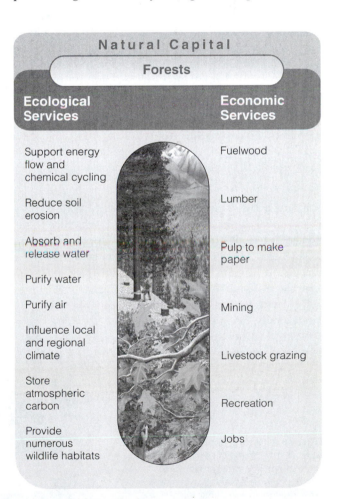

Natural Capital

Forests

Ecological Services
- Support energy flow and chemical cycling
- Reduce soil erosion
- Absorb and release water
- Purify water
- Purify air
- Influence local and regional climate
- Store atmospheric carbon
- Provide numerous wildlife habitats

Economic Services
- Fuelwood
- Lumber
- Pulp to make paper
- Mining
- Livestock grazing
- Recreation
- Jobs

Figure 6-3 Natural capital: major ecological and economic services provided by forests. Forests provide habitats for about two-thirds of the world's people. They also are home for 300 million people and more than 1.6 billion people depend on forests for their livelihood. The global trade in forest products amounts to about $270 billion a year.

Scientists and economists have worked on estimating the economic value of ecological services provided by forests and other ecosystems. For example, in 1997 a team of ecologists, economists, and geographers—led by ecological economist Robert Costanza of the University of Vermont—estimated the monetary worth of the earth's natural capital and biological income to be close to the economic value of all of the goods and services produced throughout the world each year. Based on these estimates, *biodiversity is the world's biggest financial asset.* According to this study, the world's forests provide us with ecological services worth at least $4.7 trillion per year—hundreds of times more then the economic value of forests.

The authors of this study warn that unless the financial value of biodiversity is factored into decisions on how we use forests and other ecological resources, they will be destroyed or degraded for short-term economic gain. And the researchers noted that their estimates could easily be too low by a factor of 10 to 1 million because of a lack of data.

According to ecological economist Robert Costanza, "We have been cooking the books for a long time by leaving out the worth of nature." And biologist David Suzuki warns, "Our economic system has been constructed under the premise that natural services are free. We can't afford that luxury any more."

Types of Forests (Science)

Some forests have not been disturbed by human activities, others have grown back after being cut, and some consist of planted stands of a particular tree species.

Forest managers and ecologists classify forests into three major types based on age and structure. One type consists of **old-growth forests:** uncut forests or regenerated forests that have not been seriously disturbed by human activities or natural disasters for at least several hundred years. Old-growth forests are storehouses of biodiversity because they provide ecological niches for a multitude of wildlife species.

A second type is a **second-growth forest:** a stand of trees resulting from secondary ecological succession (Figure 4-4, p. 77). They develop after the trees in an area have been removed by *human activities* (such as clear-cutting for timber or conversion to cropland) or by *natural forces* (such as fire, hurricanes, or volcanic eruption).

A **tree plantation,** also called a **tree farm,** is a third type. It is a managed tract with uniformly aged trees of one or two species that are harvested by clear-cutting as soon as they become commercially valuable. The land is then replanted and clear-cut again in a regular cycle (Figure 6-4). Currently, about 63% of the world's forests are secondary-growth forests, 22% are old-growth forests, and 5% are tree plantations (that produce about one-fifth of the world's commercial wood).

According to a 2001 study by the Worldwide Fund for Nature (WWF), intensive but sustainable management on just 5% of already deforested lands could more than meet the world's current and future demand for commercial wood and fiber. This trade-off would leave the world's remaining old-growth forests untouched.

X *HOW WOULD YOU VOTE?* Should there be a global effort to sharply reduce the cutting of old-growth forests? Cast your vote online at www.thomsonedu.com/biology/miller.

Global Outlook: Extent of Deforestation

Human activities have reduced the earth's forest cover by 20–50%, and deforestation is continuing at a fairly rapid rate, except in North America, Europe, and China.

Surveys by the World Resources Institute (WRI) indicate that over the past 8,000 years human activities

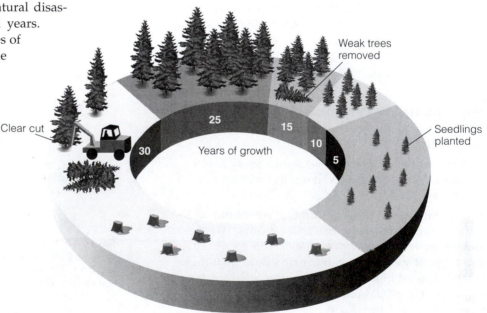

Clear cut

Weak trees removed

Seedlings planted

30 Years of growth 25 15 10 5

Figure 6-4 Natural capital degradation: Short (25- to 30-year) rotation cycle of cutting and regrowth of a monoculture tree plantation in modern industrial forestry. In tropical countries, where trees can grow more rapidly year-round, the rotation cycle can be 6–10 years. Old-growth or secondary forests are clear-cut to provide land for growing most tree plantations.

have reduced the earth's original forest cover by at least 20% and probably by 50%. In addition, surveys by the UN Food and Agricultural Organization (FAO) in 2005 and the WRI indicate that the global rate of forest cover loss between 1990 and 2005 was between 0.2% and 0.5% a year and that at least another 0.1–0.3% of the world's forests were degraded.

If estimates are correct, the world's forests are being cleared or degraded at a rate of 0.3–0.8% a year, with much higher rates in some areas. Over four-fifths of these losses take place in the tropics, mostly in South America and Africa. The WRI estimates that if current deforestation rates continue, about 40% of the world's remaining intact forests will have been logged or converted to other uses within 10–20 years, if not sooner.

Cutting down large areas of forests has important short-term economic benefits (Figure 6-3, right). But it reduces biodiversity and the ecological services such forests provide and has a number of other harmful environmental effects (Figure 6-5).

Learn more about how deforestation can affect the drainage of a watershed and disturb its ecosystem at ThomsonNOW.

There are two encouraging trends. *First,* the total area of forests in North America, Europe, and China increased slightly between 1990 and 2005 because of tree planting and natural reforestation from secondary ecological succession on cleared forest areas and abandoned croplands. *Second,* some of the cut areas in tropical forests have increased tree cover from regrowth and planting of tree plantations.

Harvesting Trees (Science and Economics)

Trees can be harvested individually from diverse forests, or an entire forest stand can be cut down in one or several phases.

Environmental Degradation
Deforestation

- Decreased soil fertility from erosion
- Runoff of eroded soil into aquatic systems
- Premature extinction of species with specialized niches
- Loss of habitat for migratory species such as birds and butterflies
- Regional climate change from extensive clearing
- Releases CO_2 into atmosphere from burning and tree decay
- Accelerates flooding

Figure 6-5 Natural capital degradation: harmful environmental effects of deforestation that can reduce the ecological services provided by forests. QUESTION: *In what ways might your lifestyle contribute directly or indirectly to deforestation?*

The first step in forest management is to build roads for access and timber removal. Even carefully designed logging roads have a number of harmful effects (Figure 6-6)—increased erosion and sediment runoff into waterways, habitat fragmentation, and biodiversity loss. Logging roads also expose forests to invasion by nonnative pests, diseases, and wildlife species. And they open once-inaccessible forests to farmers, miners, ranchers, hunters, and off-road vehicle users. In addition, logging roads on public lands in the United States disqualify the land for protection as wilderness.

Once loggers reach a forest area, they use various methods to harvest the trees (Figure 6-7, p. 112). In **selective cutting,** intermediate-aged or mature trees in

Figure 6-6 Natural capital degradation: Building roads into previously inaccessible forests paves the way to fragmentation, destruction, and degradation.

(a) Selective cutting

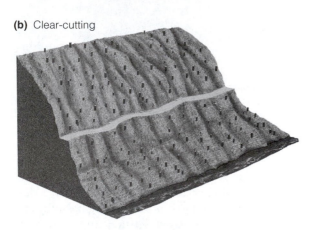

(b) Clear-cutting

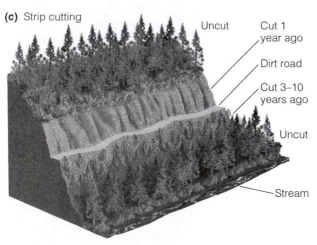

(c) Strip cutting

Uncut

Cut 1 year ago

Dirt road

Cut 3–10 years ago

Uncut

Stream

Figure 6-7 Tree harvesting methods.

an uneven-aged forest are cut singly or in small groups (Figure 6-7a). Selective cutting reduces crowding, removes diseased trees, encourages growth of younger trees, maintains a stand of trees of different species and ages, helps protect the site from soil erosion and wind damage, and allows a forest to be used for multiple purposes.

Some tree species that grow best in full or moderate sunlight are all removed in a single clear-cut (Fig-

ure 6-7b). Figure 6-8 lists the advantages and disadvantages of clear-cutting.

A clear-cutting variation that can allow a more sustainable timber yield without widespread destruction is *strip-cutting* (Figure 6-7c). It involves clear-cutting a strip of trees along the contour of the land, with the corridor narrow enough to allow natural regeneration within a few years. After regeneration, loggers cut another strip above the first, and so on. This allows clear-cutting of a forest in narrow strips over several decades with minimal damage.

Solutions: Managing Forests More Sustainably

We can use forests more sustainably by respecting the economic value of their ecological services, harvesting trees no faster than they are replenished, and protecting old-growth and vulnerable areas.

Biodiversity researchers and a growing number of foresters have suggested several ways to achieve more sustainable forest management (Figure 6-9).

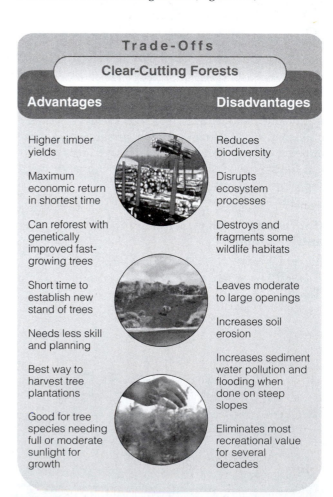

Trade-Offs

Clear-Cutting Forests

Advantages	Disadvantages
Higher timber yields	Reduces biodiversity
Maximum economic return in shortest time	Disrupts ecosystem processes
Can reforest with genetically improved fast-growing trees	Destroys and fragments some wildlife habitats
Short time to establish new stand of trees	Leaves moderate to large openings
Needs less skill and planning	Increases soil erosion
Best way to harvest tree plantations	Increases sediment water pollution and flooding when done on steep slopes
Good for tree species needing full or moderate sunlight for growth	Eliminates most recreational value for several decades

Figure 6-8 Trade-offs: advantages and disadvantages of clear-cutting forests. QUESTION: *What are the single advantage and the single disadvantage that you think are the most important?*

Solutions

Sustainable Forestry

- Conserve biodiversity along with water and soil resources

- Grow more timber on long rotations

- Rely more on selective cutting and strip cutting

- No clear-cutting, seed tree, or shelterwood cutting on steeply sloped land

- No fragmentation of remaining large blocks of forest

- Sharply reduce road building into uncut forest areas

- Leave most standing dead trees and fallen timber for wildlife habitat and nutrient recycling

- Certify timber grown by sustainable methods

- Include ecological services of trees and forests in estimating economic value

Figure 6-9 Solutions: ways to manage forests more sustainably. QUESTION: *Which two of these solutions do you think are the most important?*

Conservation biologists suggest four ways to encourage greater protection of the world's remaining forests. *First*, include estimated economic values of ecological services in all forest use decisions. If we did this, most of the world's old-growth and second-growth forests would not be cut because the monetary value of their ecological services far outweighs the monetary value of their short-term economic services. The ecological services of old-growth and most second-growth forests could be sustained indefinitely by selectively harvesting trees no faster than they are replenished and by using these forests primarily for recreation and as centers of biodiversity.

Second, protect enough forest so that the rate of forest loss and degradation by human and natural factors in a particular area is balanced by the rate of forest renewal.

Third, identify and protect forest areas that are centers of biodiversity and that are threatened by economic development. *Fourth*, establish and use methods to certify timber that has been grown sustainably, as discussed below.

Solutions: Certifying Sustainably Grown Timber

Organizations have developed standards for certifying that timber has been harvested sustainably and that wood products have been produced from sustainably harvested timber.

Collins Pine owns and manages a large area of productive timberland in northeastern California. Since 1940, the company has used selective cutting to help maintain ecological, economic, and social sustainability of its timberland.

Since 1993, Scientific Certification Systems (SCS) has evaluated the company's timber production. SCS is part of the nonprofit Forest Stewardship Council (FSC). It was formed in 1993 to develop a list of environmentally sound practices for use in certifying timber and products made from such timber.

Each year, SCS evaluates Collins's landholdings to ensure that: cutting has not exceeded long-term forest regeneration; roads and harvesting systems have not caused unreasonable ecological damage; soils are not damaged; downed wood (boles) and standing dead trees (snags) are left to provide wildlife habitat; and the company is a good employer and a good steward of its land and water resources.

According to the FSC, between 1995 and 2005, the area of the world's forests that meets its international certification standards grew tenfold. The countries with the largest areas of FSC-certified forests are, in order, Sweden, Poland, the United States, and Canada. Despite this progress, by 2005, only about 6% of the world's forested area was certified.

FOREST RESOURCES AND MANAGEMENT IN THE UNITED STATES

U.S. Forests: Encouraging News

U.S. forests cover more area than they did in 1920, more wood is grown than is cut, and the country has set aside large areas of protected forests.

Forests cover about 30% of the U.S. land area, providing habitats for more than 80% of the country's wildlife species and supplying about two-thirds of the nation's surface water.

Forests (including tree plantations) in the United States cover more area than they did in 1920. Many of the old-growth forests that were cleared or partially cleared between 1620 and 1960 have grown back naturally through secondary ecological succession as fairly diverse second-growth (and in some cases third-growth) forest in every region of the United States, except much of the West. In 1995, environmental writer Bill McKibben cited forest regrowth in the United States—especially in the East—as "the great environmental story of the United States, and in some ways the whole world."

Also, every year more wood is grown in the United States than is cut and each year the total area planted with trees increases. And in 2006, protected forests made up about 40% of the country's total forest area, mostly in the *National Forest System*, which comprises 155 national forests and 22 national grasslands, all managed by the U.S. Forest Service (USFS).

On the other hand, since the mid-1960s, an increasing area of the nation's remaining old-growth and fairly diverse second-growth forests has been clear-cut and often replaced with biologically simplified tree plantations. According to biodiversity researchers, this reduces overall forest biodiversity and disrupts ecosystem processes such as energy flow and chemical cycling. Some environmentally concerned citizens have protested the cutting of ancient trees and forests (Individuals Matter, below).

Logging in U.S. National Forests (Science and Politics)

There has been an ongoing debate over whether U.S. national forests should be managed primarily for timber, their ecological services, recreation, or a mix of these uses.

For decades, controversy has swirled around the use of resources in the national forests. Timber companies push to cut as much of this timber in these forests as possible at low prices. Biodiversity experts and many environmental scientists believe that national forests should be managed primarily to provide recreation

and to sustain biodiversity, water resources, and other ecological services.

The Forest Service's timber-cutting program loses money because revenue from timber sales does not cover the costs of road building, timber sale preparation, administration, and other overhead costs. Because of such government subsidies, timber sales from U.S. federal lands have lost money for taxpayers in 97 of the last 100 years.

Yet, in 2003, the U.S. Congress passed the *Healthy Forests Restoration Act*. Under this law, timber companies are allowed to cut down economically valuable medium-size and large trees in 71% of the total area of the national forests without environmental review in return for clearing away smaller, more fire-prone trees and underbrush.

Critics of this act say that its benefits could be achieved at a much lower cost to taxpayers by giving grants to communities that are vulnerable to wildfires for creating buffer zones around their communities and protecting homes and buildings in their areas. This would be a less environmentally destructive solution than widespread removal of medium- and large-size trees in many of the country's national forests. Critics also say that once these trees have been cut, it will be too late to halt the resulting loss of biodiversity

INDIVIDUALS MATTER

Butterfly in a Redwood Tree

"Butterfly" is the nickname given to Julia Hill. This young woman spent two years of her life on a small platform near the top of a giant redwood tree in California to protest the clear-cutting of a forest of these ancient trees, some of them more than 1,000 years old.

She and other protesters were illegally occupying these trees as a form of *nonviolent civil disobedience* used decades ago by Mahatma Gandhi in his efforts to end the British occupation of India. Butterfly had never participated in any environmental protest or act of civil disobedience.

She went to the site to express her belief that it was wrong to cut down these ancient giants for short-term economic gain, even if you

own them. She planned to stay only for a few days.

But after seeing the destruction and climbing one of these magnificent trees, she ended up staying in the tree for two years to bring publicity to what was happening and help save the surrounding trees. She became a media symbol of the protest and during her stay used a cell phone to communicate with members of the mass media throughout the world to help develop public support for saving the trees.

Can you imagine spending two years of your life in a tree on a platform not much bigger than a king-sized bed 55 meters (180 feet) above the ground and enduring high winds, intense rainstorms, snow, and ice? And Butterfly was not living in a quiet pristine forest. All round her was the noise of trucks, chainsaws, and helicopters trying to scare her into returning to the ground.

Although Butterfly lost her courageous battle to save the surrounding forest, she persuaded Pacific Lumber MAXXAM to save her tree (called Luna) and a 60-meter (200-foot) buffer zone around it. Not too long after she descended from the tree, someone used a chainsaw to seriously damage it, and cables and steel plates have been used to preserve it.

But maybe Butterfly and the earth did not lose. A book she wrote about her stand, and her subsequent travels to campuses all over the world, have inspired a number of young people to stand up for protecting biodiversity and other environmental causes.

Butterfly was leading by following in the tradition of Gandhi, who said, "My life is my message." Would you spend a day or a week of your life protesting something that you believed to be wrong?

and increased threat of fire. Some call this law the *Unhealthy Forest Restoration Act* and view it mostly as a costly giveaway to politically powerful timber companies under the guise of fire prevention. They call for repealing or drastically modifying this law.

✗ How Would You Vote? Do you support repealing or modifying the Healthy Forests Restoration Act? Cast your vote online at www.thomsonedu.com/biology/miller.

Logging in national forests has its advantages and disadvantages. According to a 2000 study by the accounting firm Econorthwest, recreation, hunting, and fishing in national forests add ten times more money to the national economy and provide seven times more jobs than does extraction of timber and other resources.

✗ How Would You Vote? Should logging in U.S. national forests be banned? Cast your vote online at www.thomsonedu.com/biology/miller.

TROPICAL DEFORESTATION

Global Outlook: Clearing and Degradation of Tropical Forests

Large areas of ecologically and economically important forests are being cleared and degraded at a fast rate.

Tropical forests cover about 6% of the earth's land area—roughly the area of the lower 48 states. Climatic and biological data suggest that mature tropical forests once covered at least twice as much area as they do today, with most of the destruction occurring since 1950. Satellite scans and ground-level surveys used to estimate forest destruction indicate that large areas of tropical forests are being cut rapidly in parts of South America (especially Brazil), Africa, and Asia.

Studies indicate that more than half of the world's species of terrestrial plants and animals live in tropical rain forests. Each time a tract of tropical rain forest is cleared, several species—some with possible medical or other benefits to humans—may be lost forever. Brazil has about 40% of the world's remaining tropical rain forest and an estimated 30% of the world's terrestrial plant and animal species in the vast Amazon basin, which is about two-thirds the size of the continental United States.

In 1970, only 1% of the Amazon basin had been deforested. By 2005, Brazil's government estimated that 16% had been deforested or degraded and converted mostly to tropical grassland (savanna). Between 2002 and 2005, the rate of deforestation increased sharply—mostly to make way for cattle ranching and large plantations for growing crops such as soybeans, used for cattle feed. In 2004, researchers at the Smithsonian Tropical Research Institute estimated loggers and farmers cleared and burned an area equivalent to a loss of 11 football fields a minute!

In 2004, Imazon, a leading Brazilian environmental group, said that satellite photos show that land occupation and deforestation covers some 47% of Brazil's Amazon—much larger than official government estimates. In 2005, the Amazon was suffering from its worst drought in forty decades. Meteorologists attribute the drought largely to warmer ocean water, caused mostly by global warming, which is increased by tropical deforestation in a vicious circle of ecological connections.

Estimates of global tropical forest loss vary because of the difficulty of interpreting satellite images and different definitions of forest and deforestation. Also, some countries hide or exaggerate deforestation rates for political and economic reasons. However, the projected rate is high enough to lose or degrade half of the world's remaining tropical forests in 35–117 years. Most biologists believe that cutting and degrading most remaining old-growth tropical forests is a serious global environmental problem because of the important ecological and economic services provided by forests (Figure 6-3).

Causes of Tropical Deforestation and Degradation (Economics and Politics)

The primary causes of tropical deforestation and degradation are population growth, poverty, environmentally harmful government subsidies, debts owed to developed countries, and failure to value their ecological services.

Tropical deforestation results from a number of interconnected primary and secondary causes (Figure 6-10, p. 116). Population growth and poverty combine to drive subsistence farmers and the landless poor to tropical forests, where they try to grow enough food to survive. Government subsidies can accelerate deforestation by making timber harvesting and cattle grazing cheap compared to the economic value of the ecological services the forests provide.

Governments in Indonesia, Mexico, and Brazil also encourage the poor to colonize tropical forests by giving them title to land that they clear. This can help reduce poverty but can lead to environmental degradation unless the new settlers are taught how to use such forests more sustainably, which is rarely done.

In addition, international lending agencies encourage developing countries to borrow huge sums of money from developed countries to finance projects such as roads, mines, logging operations, oil drilling, and dams in tropical forests. Another cause is failure to value the ecological services of forests (Figure 6-3, left).

The depletion and degradation of a tropical forest begins when a road is cut deep into the forest interior for logging and settlement. Loggers then use selective cutting to remove the best timber. This topples many

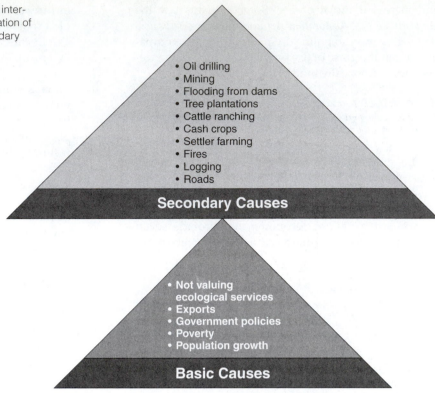

Figure 6-10 Natural capital degradation: major interconnected causes of the destruction and degradation of tropical forests. The importance of specific secondary causes varies in different parts of the world.

Secondary Causes
- Oil drilling
- Mining
- Flooding from dams
- Tree plantations
- Cattle ranching
- Cash crops
- Settler farming
- Fires
- Logging
- Roads

Basic Causes
- Not valuing ecological services
- Exports
- Government policies
- Poverty
- Population growth

other trees because of their shallow roots and the network of vines connecting trees in the forest's canopy. Although timber exports to developed countries contribute significantly to tropical forest depletion and degradation, domestic use accounts for more than 80% of the trees cut in developing countries.

After the best timber has been removed, timber companies often sell the land to ranchers. Within a few years, they typically overgraze it and sell it to settlers who have migrated to the forest hoping to grow enough food to survive. Then they move their land-degrading ranching operations to another forest area. After a few years of crop growing and rain erosion, the nutrient-poor tropical soil is depleted of nutrients. Then the settlers move on to newly cleared land.

According to a 2005 report by the UN Food and Agricultural Organization, cattle ranching is the main cause of forest destruction in Brazil and other parts of Central and South America. Raising soybeans to feed Brazil's rapidly growing beef, poultry, and pork industries is also leading to the destruction of the country's forests.

In some areas—especially Africa and Latin America—large sections of tropical forest are cleared for raising cash crops such as sugarcane, bananas, pineapples, strawberries, and coffee mostly for export to developed countries. Tropical forests are also cleared for mining and oil drilling and to build dams on rivers that flood large areas of the forest.

Healthy rain forests do not burn. But increased logging, settlements, grazing, and farming along roads built in these forests results in fragments of forest that dry out. This makes such areas easier to ignite by lightning or to be burned by farmers and ranchers. In addition to destroying and degrading biodiversity, this releases large amounts of carbon dioxide into the atmosphere.

A 2005 study found that widespread burning of tropical forest areas in the Amazon is changing weather patterns by raising temperatures and reducing rainfall. This is converting deforested areas into grassland or savanna. Models project that at current burning and deforestation rates, 20–30% of the Amazon will turn into savanna in the next fifty years.

Solutions: Reducing Tropical Deforestation and Degradation

There are a number of ways to slow and reduce the deforestation and degradation of tropical forests.

Analysts have suggested various ways to protect tropical forests and use them more sustainably (Figure 6-11). One method is to help new settlers in tropical forests learn how to practice small-scale sustainable agriculture and forestry. Another method is to sustainably harvest some of the renewable resources such as fruits and nuts in tropical rain forests.

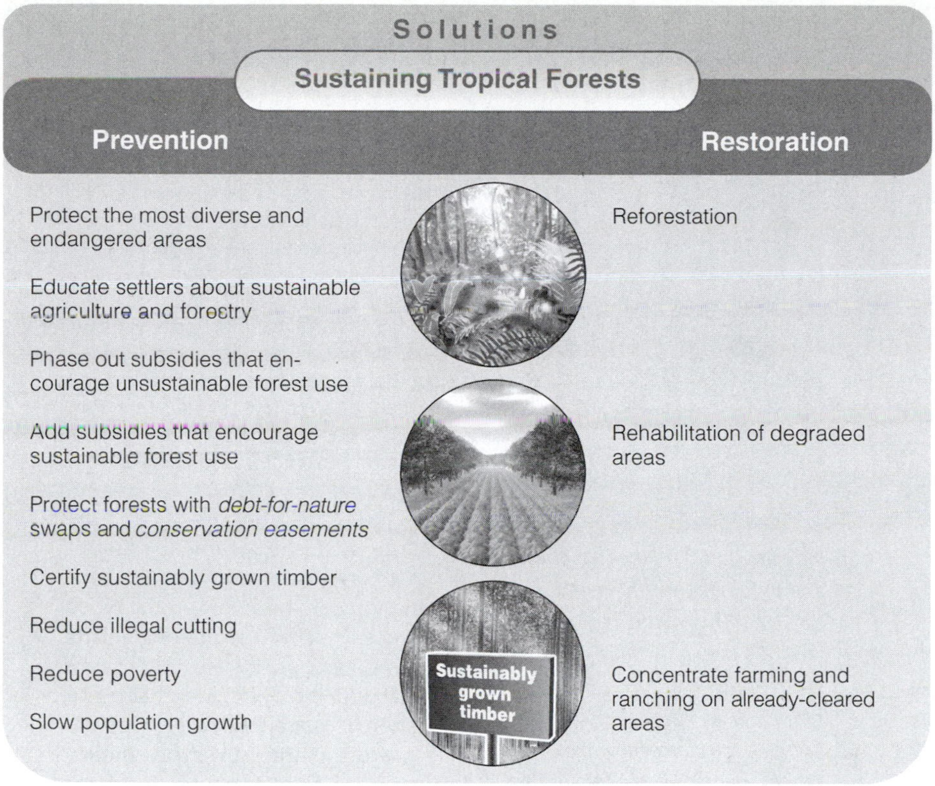

Figure 6-11 Solutions: ways to protect tropical forests and use them more sustainably. QUESTION: *Which three of these solutions do you think are the most important?*

Debt-for-nature swaps make it financially profitable for countries to protect tropical forests. In such a swap, participating countries act as custodians of protected forest reserves in return for foreign aid or debt relief. In a similar newer strategy called *conservation concession,* nations are paid for concessions that preserve their resources. In this win–win approach, the nation gets the money, yet does not have to pay its debts by selling off its resources.

Loggers can also use gentler methods for harvesting trees. For example, cutting canopy vines (lianas) before felling a tree can reduce damage to neighboring trees by 20–40%, and using the least obstructed paths to remove the logs can halve the damage to other trees. In addition, governments and individuals can mount efforts to reforest and rehabilitate degraded tropical forests and watersheds (Individuals Matter, p. 118). Other suggestions include clamping down on illegal logging and promoting the use of ecologically and economically valuable trees (see Solutions, below).

Solutions: The Incredible Neem Tree (Science)

The neem tree could eventually benefit almost everyone on the earth.

Imagine a single plant that could quickly reforest degraded land, supply fuelwood and lumber in dry areas, provide natural alternatives to toxic pesticides, be used to treat numerous diseases, and help control human population growth. The *neem tree,* a broadleaf evergreen member of the mahogany family, is that plant.

This remarkable tropical species, native to India and Burma, is ideal for reforestation because it can grow to maturity in only 5–7 years. It grows well in poor soil in semiarid lands such as those in Africa, providing abundant fuelwood, lumber, and lamp oil.

It also contains various natural pesticides. Chemicals from its leaves and seeds can repel or kill more than 200 insect pest species. Extracts from neem seeds and leaves can fight bacterial, viral, and fungal infections. Villagers call the tree a "village pharmacy" because its chemicals can relieve so many different health problems. People also use the tree's twigs as an antiseptic toothbrush and the oil from its seeds to make toothpaste and soap.

That is not all. Neem-seed oil evidently acts as a strong spermicide and may help in the development of a much-needed male birth control pill. According to a study by the U.S. National Academy of Sciences, the neem tree "may eventually benefit every person on the planet."

Despite its numerous advantages, ecologists caution against widespread planting of the neem tree outside its native range. As a nonnative species, it could take over and displace native species because of its rapid growth and resistance to pests.

Starting with a small tree nursery in her back yard, Wangari Maathai (Figure 6-A) founded the Green Belt Movement in 1977. The goals of this highly regarded women's self-help group are to organize poor women in rural Kenya to plant and protect millions of trees to combat deforestation and provide fuelwood. By 2004, the 50,000 members of this grassroots group had established 6,000 village nurseries and planted and protected more than 30 million trees.

The success of this project has sparked the creation of similar programs in more than 30 other African countries. This inspiring leader has said,

I don't really know why I care so much. I just have something in-side me that tells me that there is a problem and I have got to do something about it. And I'm sure it's the same voice that is speaking to everyone on this planet, at least everybody who seems to be concerned about the fate of the world, the fate of this planet.

Figure 6-A Wangari Maathai was the first Kenyan woman to earn a Ph.D. (in anatomy) and to head an academic department (veterinary medicine) at the University of Nairobi. In 1977, she organized the internationally acclaimed Green Belt Movement. For her work in protecting the environment, she has received many honors, including the Goldman Prize, the Right Livelihood Award, the UN Africa Prize for Leadership, and the 2004 Nobel Peace Prize. After years of being harassed, beaten, and jailed for opposing government policies, she was elected to Kenya's parliament as a member of the Green Party in 2002. In 2003, she was appointed Assistant Minister for Environment, Natural Resources, and Wildlife.

In 2004, she became the first African woman and first environmentalist to be awarded the Nobel Peace Prize for her lifelong efforts. Within an hour of learning that she had won the prize, Maathai planted a tree. In her speech accepting the award she urged everyone in the world to plant a tree as a symbol of commitment and hope.

NATIONAL PARKS

Global Outlook: Threats to National Parks

Countries have established more than 1,100 national parks, but most are threatened by human activities.

Today, more than 1,100 national parks larger than 10 square kilometers (4 square miles) each are located in more than 120 countries. But according to a 1999 study by the World Bank and the World Wildlife Fund, only 1% of the parks in developing countries receive protection.

Local people invade most parks in search of wood, cropland, game animals, and other natural products for their daily survival. Loggers, miners, and wildlife poachers (who kill animals to obtain and sell items such as rhino horns, elephant tusks, and furs) also operate in many of these parks. Park services in developing countries typically have too little money and too few personnel to fight these invasions, either by force or through education.

Another problem is that most national parks are too small to sustain many large-animal species. Also, many parks suffer from invasions by nonnative species that can reduce the populations of native species and cause ecological disruption.

Case Study: Stresses on U.S. National Parks

National parks in the United States face many threats.

The U.S. national park system, established in 1912, has 58 national parks, sometimes called the country's *crown jewels*. State, county, and city parks supplement these national parks. Most state parks are located near urban areas and have about twice as many visitors per year as the national parks.

Popularity is one of the biggest problems for national and state parks in the United States. During the summer, users entering the most popular U.S. national and state parks often face hour-long backups and experience noise, congestion, eroded trails, and stress instead of peaceful solitude. In some parks, noisy dirt bikes, dune buggies, snowmobiles, and other off-road vehicles degrade the aesthetic experience for many visitors, destroy or damage fragile vegetation, and disturb wildlife.

Many visitors expect parks to have grocery stores, laundries, bars, golf courses, video arcades, and other facilities found in urban areas. U.S. Park Service rangers spend an increasing amount of their time on law enforcement and crowd control instead of conservation management and education. Many overworked and underpaid rangers are leaving for better-paying jobs.

Predators such as wolves all but vanished from various parks mostly because of excessive hunting and poisoning by ranchers and federal officials. In some parks, populations of the species that wolves once controlled have exploded, destroying vegetation and crowding out other species. To help correct this situation, the U.S. Fish and Wildlife Service established a controversial program to reintroduce the gray wolf into the Yellowstone National Park area (Case Study, below).

Many parks suffer damage from the migration or deliberate introduction of nonnative species. European wild boars (imported to North Carolina in 1912 for hunting) threaten vegetation in part of the Great Smoky Mountains National Park. Nonnative mountain goats in Washington's Olympic National Park trample native vegetation and accelerate soil erosion. While some nonnative species have moved into parks, some economically valuable native species of animals and plants (including many threatened or endangered species) are killed or removed illegally by poachers in almost half of U.S. national parks.

Nearby human activities that threaten wildlife and recreational values in many national parks include mining, logging, livestock grazing, coal-burning power plants, water diversion, and urban development. Polluted air, drifting hundreds of kilometers, kills ancient trees in California's Sequoia National Park and often blots out the awesome views at Arizona's Grand Canyon. According to the National Park Service, air pollution affects scenic views in national parks more than 90% of the time. In addition, the U.S. General Accounting Office reports that the national parks need at least $6 billion for long overdue repair of trails, buildings, and other infrastructure.

Reintroducing Wolves to Yellowstone

CASE STUDY

At one time, the gray wolf ranged over most of North America. But between 1850 and 1900, an estimated 2 million wolves were shot, trapped, and poisoned by ranchers, hunters, and government employees. The idea was to make the West and the Great Plains safe for livestock and for big-game animals prized by hunters.

It worked. When Congress passed the U.S. Endangered Species Act in 1973, only a few hundred gray wolves remained outside of Alaska, primarily in Minnesota and Michigan.

Ecologists recognize the important role this keystone predator species once played in parts of the West and the Great Plains. These wolves culled herds of bison, elk, caribou, and mule deer, and kept down coyote populations. They also provided uneaten meat for scavengers such as ravens, bald eagles, ermines, and foxes.

In recent years, herds of elk, moose, and antelope have expanded. Their larger numbers have devastated some vegetation, increased erosion, and threatened the niches of other wildlife species. Reintroducing a keystone species such as the gray wolf into a terrestrial ecosystem is one way to help sustain the biodiversity of the ecosystem and prevent further environmental degradation.

In 1987, the USFWS proposed reintroducing gray wolves into the Yellowstone ecosystem. This brought angry protests. Some objections came from ranchers who feared the wolves would attack their cattle and sheep; one enraged rancher said that the idea was "like reintroducing smallpox." Other protests came from hunters who feared the wolves would kill too many big-game animals, and from mining and logging companies who worried the government would halt their operations on wolf-populated federal lands.

Since 1995, federal wildlife officials have caught gray wolves in Canada and relocated them in Yellowstone National Park and northern Idaho. By 2005, there were about 850 gray wolves in these two areas.

With wolves around, elk are gathering less near streams and rivers. Their diminished presence has spurred the growth of aspen, cottonwoods, and willow trees. This stabilizes the banks of streams, which lowers the water temperature and makes the habitat better for trout. Beavers seeking willow and aspen have returned. And leftovers of elk killed by wolves are an important food source for grizzly bears and other scavengers.

The wolves have cut coyote populations in half. This has increased populations of smaller animals such as ground squirrels and gophers hunted by coyotes. This provides more food for eagles and hawks. Between 1995 and 2002, the wolves also killed 792 sheep, 278 cattle, and 62 dogs in the Northern Rockies.

Critical Thinking

Do you support the program to reestablish populations of the gray wolf in the Yellowstone ecosystem? Explain. What are the major drawbacks of this program?

National Parks

- Integrate plans for managing parks and nearby federal lands

- Add new parkland near threatened parks

- Buy private land inside parks

- Locate visitor parking outside parks and use shuttle buses for entering and touring heavily used parks

- Increase funds for park maintenance and repairs

- Survey wildlife in parks

- Raise entry fees for visitors and use funds for park management and maintenance

- Limit number of visitors to crowded park areas

- Increase number and pay of park rangers

- Encourage volunteers to give visitor lectures and tours

- Seek private donations for park maintenance and repairs

Figure 6-12 Solutions: suggestions for sustaining and expanding the national park system in the United States. QUESTION: *Which two of these solutions do you think are the most important?* (Data from Wilderness Society and National Parks and Conservation Association)

Figure 6-12 lists suggestions that various analysts have made for sustaining and expanding the national park system in the United States.

NATURE RESERVES

Protecting Land from Human Exploitation (Science, Economics, and Politics)

Ecologists believe that we should protect more land to help sustain the earth's biodiversity but powerful economic and political interests oppose doing this.

Most ecologists and conservation biologists believe the best way to preserve biodiversity is through a worldwide network of protected areas. Currently, about 12% of the earth's land area has been protected strictly or partially in nature reserves, parks, wildlife refuges, wilderness, and other areas. In other words, we have reserved 88% of the earth's land for ourselves, and most of the remaining 12% we have pro-

tected is ice, tundra, or desert where most people do not want to live.

And this 12% figure is misleading because no more than 5% of the earth's land is strictly protected from potentially harmful human activities. In other words, we have reserved 95% of the earth's land for ourselves, and most of the remaining area consists of ice, tundra, or desert.

Conservation biologists call for protection of at least 20% of the earth's land area in a global system of biodiversity reserves that includes multiple examples of all the earth's biomes. Doing this will require action and funding by national governments (Case Study, p. 121), private groups, and cooperative ventures involving governments, businesses, and private conservation groups.

Private groups play an important role in establishing wildlife refuges and other reserves to protect biological diversity. For example, since its founding by a group of professional ecologists in 1951, the *Nature Conservancy*—with more than 1 million members worldwide—has created the world's largest system of private natural areas and wildlife sanctuaries in 30 countries.

In the United States, private, nonprofit *land trust groups* have protected large areas of land. Members pool their financial resources and accept tax-deductible donations to buy and protect farmlands, woodlands, and urban green spaces. However, the area of land in development is four times the land area being protected by land trusts.

Most developers and resource extractors oppose protecting even the current 12% of the earth's remaining undistributed ecosystems. They contend that most of these areas contain valuable resources that would add to economic growth.

Ecologists and conservation biologists disagree. They view protected areas as islands of biodiversity that help sustain all life and economies, and serve as centers of future evolution. See Norman Myer's Guest Essay on this topic on the website for this chapter.

X *HOW WOULD YOU VOTE?* Should at least 20% of the earth's land area be strictly protected from economic development? Cast your vote online at www.thomsonedu.com/biology/miller.

Whenever possible, conservation biologists call for using the *buffer zone concept* to design and manage nature reserves. This means protecting an inner core of a reserve by two buffer zones in which local people can extract resources in ways that are sustainable and that do not harm the inner core. Doing this can enlist local people as partners in protecting a reserve from unsustainable uses. The United Nations uses this principle in establishing its global network of 425 biosphere reserves in 95 countries (Figure 6-13).

Case Study: Costa Rica—A Global Conservation Leader (Science, Politics, and Economics)

Costa Rica has devoted a larger proportion of land than has any other country to conserving its significant biodiversity.

Tropical forests once completely covered Central America's Costa Rica, which is smaller in area than the U.S. state of West Virginia and about one-tenth the size of France. Between 1963 and 1983, politically powerful ranching families cleared much of the country's forests to graze cattle. They exported most of the beef produced to the United States and western Europe.

Despite such widespread forest loss, tiny Costa Rica is a superpower of biodiversity, with an estimated 500,000 plant and animal species. A single park in Costa Rica is home to more bird species than all of North America.

In the mid-1970s, Costa Rica established a system of reserves and national parks that by 2004 included about a quarter of its land—6% of it in reserves for indigenous peoples. Costa Rica now devotes a larger proportion of land to biodiversity conservation than does any other country.

The country's parks and reserves are consolidated into eight *megareserves* designed to sustain about 80% of Costa Rica's biodiversity (Figure 6-14). Each reserve

Biosphere Reserve

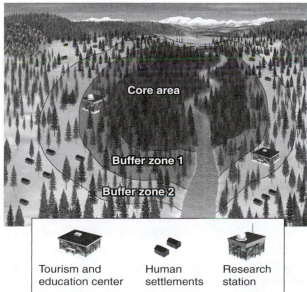

Figure 6-13 Solutions: a model *biosphere reserve*. Each reserve contains a protected inner core surrounded by two buffer zones that local and indigenous people can use for sustainable logging, food growing, cattle grazing, hunting, fishing, and ecotourism.

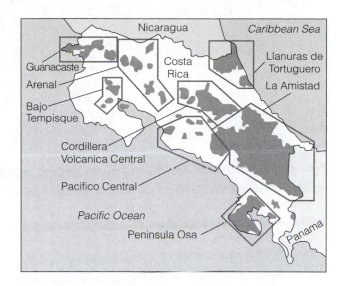

Figure 6-14 Solutions: Costa Rica has consolidated its parks and reserves into eight *megareserves* designed to sustain about 80% of the country's rich biodiversity.

contains a protected inner core surrounded by two buffer zones that local and indigenous people can use for sustainable logging, food growing, cattle grazing, hunting, fishing, and ecotourism.

Costa Rica's biodiversity conservation strategy has paid off. Today, the country's largest source of income is its $1-billion-a-year tourism business—almost two-thirds of it from ecotourists.

To reduce deforestation the government has eliminated subsidies for converting forest to cattle grazing land. And it pays landowners to maintain or restore tree coverage. The goal is to make sustaining forests profitable. It has worked. Costa Rica has gone from having one of the world's highest deforestation rates to having one of the lowest.

Global Biodiversity Hot Spots (Science and Stewardship)

We can prevent or slow down losses of biodiversity by concentrating efforts on protecting hot spots where significant biodiversity is under immediate threat.

In reality, few countries are physically, politically, or financially able to set aside and protect large biodiversity reserves. To protect as much of the earth's remaining biodiversity as possible, conservation biologists use an *emergency action* strategy that identifies and quickly protects *biodiversity hot spots*. These "ecological arks" are areas especially rich in plant and animal species that are found nowhere else and are in great danger of extinction or serious ecological disruption.

Figure 6-15 (p. 122) shows 34 hot spots. They cover a little more than 2% of the earth's land surface and

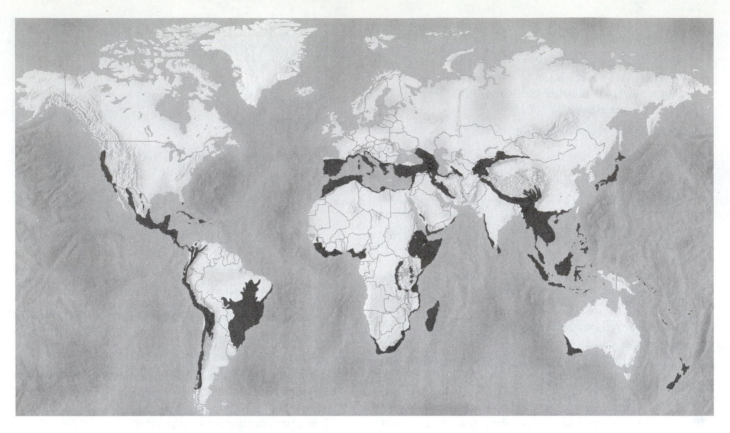

ThomsonNOW™ Active Figure 6-15 Endangered natural capital: 34 *hot spots* identified by ecologists as important but endangered centers of biodiversity that contain a large number of endemic plant and animal species found nowhere else. Identifying and saving these critical habitats is a vital emergency response, although this makes it no less important to protect and sustain biodiversity elsewhere. *See an animation based on this figure and take a short quiz on the concept.* (Data from Center for Applied Biodiversity Science at Conservation International)

contain 52% of the world's plant species and 36% of all terrestrial vertebrates. And they are the only homes for more than one-third of the planet's known terrestrial plant and animal species. According to Norman Myers: "I can think of no other biodiversity initiative that could achieve so much at a comparatively small cost, as the hot spots strategy." Edward O. Wilson, one of the leading authorities on biodiversity, has described the hot spot approach as "the most important contribution to conservation biology of the last century."

 Learn more about hot spots around the world, what is at stake there, and how they are threatened at ThomsonNOW.

Natural Capital: Wilderness

Wilderness is land legally set aside in a large enough area to prevent or minimize harm from human activities.

One way to protect undeveloped lands from human exploitation is by legally setting them aside as un-

developed land called **wilderness.** Hikers and campers can visit such areas but they cannot stay. U.S. President Theodore Roosevelt summarized what we should do with wilderness: "Leave it as it is. You cannot improve it."

The U.S. Wilderness Society estimates that a wilderness area should contain at least 4,000 square kilometers (1,500 square miles). Otherwise, it can be affected by air, water, and noise pollution from nearby human activities.

Wild places are areas where people can experience the beauty of nature and observe natural biological diversity. They can also enhance the mental and physical health of visitors by allowing them to get away from noise, stress, development, and large numbers of people. Wilderness preservationist John Muir advised us,

Climb the mountains and get their good tidings. Nature's peace will flow into you as the sunshine into the trees. The winds will blow their freshness into you, and the storms their energy, while cares will drop off like autumn leaves.

Even those who never use the wilderness areas may want to know they are there, a feeling expressed by novelist Wallace Stegner:

Save a piece of country . . . and it does not matter in the slightest that only a few people every year will go into it. This is precisely its value. . . . We simply need that wild country available to us, even if we never do more than drive to its edge and look in. For it can be a means of reassuring ourselves of our sanity as creatures, a part of the geography of hope.

Some critics oppose protecting wilderness for its scenic and recreational value for a small number of people. They believe this is an outmoded policy that keeps some areas of the planet from being economically useful to humans.

To most biologists, the most important reasons for protecting wilderness and other areas from exploitation and degradation are to *preserve their biodiversity* as a vital part of the earth's natural capital and to *protect them as centers for evolution* in response to mostly unpredictable changes in environmental conditions. In other words, wilderness is a biodiversity and wildness bank and an eco-insurance policy. Some analysts also believe wilderness should be preserved because the wild species it contains have a right to exist and play their roles in the earth's ongoing saga of biological evolution and ecological processes, without human interference.

Case Study: Wilderness Protection in the United States (Science and Politics)

Only a small percentage of the land area of the United States has been protected as wilderness.

In the United States, conservationists have been trying to save wild areas from development since 1900. Overall, they have fought a losing battle. Not until 1964 did Congress pass the Wilderness Act. It allowed the government to protect undeveloped tracts of public land from development as part of the National Wilderness Preservation System.

The area of protected wilderness in the United States increased tenfold between 1970 and 2000. Still, only about 4.6% of U.S. land is protected as wilderness—almost three-fourths of it in Alaska. Only 1.8% of the land area of the lower 48 states is protected, most of it in the West. In other words, Americans have reserved 98% of the continental United States to be used as they see fit and have protected only 2% as wilderness. According to a 1999 study by the World Conservation Union, the United States ranks 42nd among nations in terms of terrestrial area protected as wilderness, and Canada is in 36th place.

In addition, only 4 of the 413 wilderness areas in the lower 48 states are larger than 4,000 square kilometers (1,500 square miles). Also, the system includes

What Can You Do?
Sustaining Terrestrial Biodiversity

- Adopt a forest.

- Plant trees and take care of them.

- Recycle paper and buy recycled paper products.

- Buy sustainable wood and wood products.

- Choose wood substitutes such as bamboo furniture and recycled plastic outdoor furniture, decking, and fencing.

- Restore a nearby degraded forest or grassland.

- Landscape your yard with a diversity of plants natural to the area.

- Live in town because suburban sprawl reduces biodiversity.

Figure 6-16 Individuals matter: ways to help sustain terrestrial biodiversity.

only 81 of the country's 233 distinct ecosystems. Most wilderness areas in the lower 48 states are threatened habitat islands in a sea of development.

Almost 400,000 square kilometers (150,000 square miles) in scattered blocks of public lands could qualify for designation as wilderness—about 60% of it in the national forests. For more than 20 years, these areas have been temporarily protected while they were evaluated for wilderness protection.

For decades, politically powerful oil, gas, mining, and timber industries have sought entry to these areas to develop resources for increased profits and short-term economic growth. Their efforts paid off in 2005 when the Bush administration ceased protecting roadless areas under consideration for classification as wilderness within the national forest system. The Secretary of the Interior is also allowing states to claim old cow paths and off-road vehicle trails as roads that would disqualify an area from being protected as wilderness.

Figure 6-16 lists some ways you can help sustain the earth's terrestrial biodiversity.

ECOLOGICAL RESTORATION

Rehabilitating and Restoring Damaged Ecosystems (Science)

Scientists have developed a number of ways to rehabilitate and restore degraded ecosystems and create artificial ecosystems.

Almost every natural place on the earth has been affected or degraded to some degree by human activities. However, much of the environmental damage we have inflicted on nature is at least partially reversible through **ecological restoration:** the process of repairing damage caused by humans to the biodiversity and dynamics of natural ecosystems. Examples include replanting forests, restoring grasslands, restoring wetlands, reclaiming urban industrial areas (brownfields), reintroducing native species, removing invasive species, and freeing river flows by removing dams.

Farmer and philosopher Wendell Berry says we should try to answer three questions in deciding whether and how to modify or rehabilitate natural ecosystems. *First,* what is here? *Second,* what will nature permit us to do here? *Third,* what will nature help us do here? An important strategy is to mimic nature and natural processes and, ideally, to let nature do most of the work, usually through secondary ecological succession.

By studying how natural ecosystems recover, scientists are learning how to speed up repair operations using a variety of approaches. They include the following:

- *Restoration:* trying to return a particular degraded habitat or ecosystem to a condition as similar as possible to its natural state.

- *Rehabilitation:* attempting to turn a degraded ecosystem back into a functional or useful ecosystem without trying to restore it to its original condition. Examples include removing pollutants and replanting areas such as mining sites, landfills, and clear-cut forests to reduce soil erosion.

- *Replacement:* replacing a degraded ecosystem with another type of ecosystem. For example, a productive pasture or tree farm may replace a degraded forest.

- *Creating artificial ecosystems:* an example is the creation of artificial wetlands.

Researchers have suggested five basic science-based principles for carrying out most forms of ecological restoration and rehabilitation.

- Identify what caused the degradation (such as pollution, farming, overgrazing, mining, or invading species).

Ecological Restoration of a Tropical Dry Forest in Costa Rica (Science and Stewardship)

CASE STUDY

Costa Rica is the site of one of the world's largest *ecological restoration* projects. In the lowlands of the country's Guanacaste National Park (Figure 6-14), a small tropical dry deciduous forest has been burned, degraded, and fragmented by large-scale conversion to cattle ranches and farms.

Now it is being restored and relinked to the rain forest on adjacent mountain slopes. The goal is to eliminate damaging nonnative grass and cattle and reestablish a tropical dry forest ecosystem over the next 100–300 years.

Daniel Janzen, professor of biology at the University of Pennsylvania and a leader in the field of restoration ecology, has helped galvanize international support and has raised more than $10 million for

this restoration project. He recognizes that ecological restoration and protection of the park will fail unless the people in the surrounding area believe they will benefit from such efforts. Janzen's vision is to make the nearly 40,000 people who live near the park an essential part of the restoration of the degraded forest, a concept he calls *biocultural restoration.*

By actively participating in the project, local residents reap educational, economic, and environmental benefits. Local farmers make money by sowing large areas with tree seeds and planting seedlings started in Janzen's lab. Local grade school, high school, and university students and citizens' groups study the ecology of the park and go on field trips to the park. Its location near the Pan American Highway makes it an ideal area for ecotourism, which stimulates the local economy.

The project also serves as a training ground in tropical forest restoration for scientists from all over the world. Research scientists working on the project give guest classroom lectures and lead some of the field trips.

In a few decades, today's children will be running the park and the local political system. If they understand the ecological importance of their local environment, they are more likely to protect and sustain its biological resources. Janzen believes that education, awareness, and involvement—not guards and fences—are the best ways to restore degraded ecosystems and protect largely intact ecosystems from unsustainable use.

Critical Thinking

Would such an ecological restoration project be possible in the area where you live? Explain.

- Stop the abuse by eliminating or sharply reducing these factors. Examples include removing toxic soil pollutants, adding nutrients to depleted soil, adding new topsoil, and controlling or eliminating disruptive nonnative species.

- If necessary, reintroduce species, especially pioneer, keystone, and foundation species to help restore natural ecological processes, as was done with wolves in the Yellowstone area (Case Study, p. 119).

- Protect the area from further degradation.

- Monitor restoration efforts, assess success, and modify strategies as needed.

One approach to restoring degraded land involves *speeding up natural secondary ecological succession.* Examples include planting young trees in clear-cut forest areas and preventing fires that can set back succession in areas such as degraded tropical seasonal forests (Case Study, at left).

Some analysts worry that environmental restoration could encourage continuing environmental destruction and degradation by suggesting any ecological harm we do can be undone. Restorationists agree that restoration should not be used as an excuse for environmental destruction. But they point out that so far we have been able to protect or preserve no more than about 5% of the earth's land from the effects of human activities. So ecological restoration is badly needed for much of the world's ecosystems that we have damaged. They also point out that if a restored ecosystem differs from the original system, this is better than nothing, and that increased experience will improve the effectiveness of ecological restoration.

> X **HOW WOULD YOU VOTE?** Should we mount a massive effort to restore ecosystems we have degraded even though this will be quite costly? Cast your vote online at www.thomsonedu.com/biology/miller.

SUSTAINING AQUATIC BIODIVERSITY

What Do We Know about Aquatic Biodiversity and Why Should We Care about It? (Science)

We know fairly little about the biodiversity of the world's marine and freshwater systems that provides us with important economic and ecological services.

Although we live on a water planet, we have explored only about 5% of the earth's global ocean and know fairly little about its biodiversity and how it works. We also know fairly little about freshwater biodiversity.

Scientific investigation of poorly understood marine and freshwater aquatic systems is a *research frontier* whose study could result in immense ecological and economic benefits.

However, scientists have established three general patterns of marine biodiversity. *First,* the greatest marine biodiversity occurs in coral reefs, estuaries, and the deep-sea floor. *Second,* biodiversity is higher near coasts than in the open sea because of the greater variety of producers, habitats, and nursery areas in coastal areas. *Third,* biodiversity is higher in the bottom region of the ocean than in the surface region because of the greater variety of habitats and food sources on the ocean bottom.

Why should we care about aquatic biodiversity? The answer is that it helps keep us alive and supports our economies. Marine systems provide a variety of important ecological and economic services (Figure 3-14, p. 65), as do freshwater systems (Figure 3-17, p. 67).

Human Impacts on Aquatic Biodiversity: Our Large Aquatic Footprints (Science)

Human activities have destroyed or degraded a large proportion of the world's coastal wetlands, coral reefs, mangroves, and ocean bottom, and overfished many marine and freshwater species.

The greatest threat to the biodiversity of the world's oceans is loss and degradation of habitats, as summarized in Figure 3-16, p. 66. Revisit this figure. For example, more than one-fourth of the world's diverse coral reefs have been severely damaged, mostly by human activities, and up to half of these reefs may be severely damaged or destroyed by 2050.

The 2005 Millennium Ecosystem Assessment estimated that in just two decades, we have removed more than a third of the world's mangroves. More will be flooded and lost if global warming causes further rises in sea levels.

Many bottom habitats are being degraded and destroyed by dredging operations and trawler boats, which, like giant submerged bulldozers, drag huge nets weighted down with heavy chains and steel plates over ocean bottoms to harvest bottom fish and shellfish. Each year, thousands of trawlers scrape and disturb an area of ocean bottom about 150 times larger than the area of forests clear-cut each year. In 2004, some 1,134 scientists signed a statement urging the United Nations to declare a moratorium on bottom trawling on the high seas.

Studies indicate that about half of the world's 200 commercially valuable marine fish species are fished to their estimated sustainable yield and about a fifth are overfished. According to a 2004 study by the United

Nations Environment Programme (UNEP), stocks of large, open-ocean fish, such as tuna, swordfish, and marlin, have declined by 90% since 1900.

One of the results of the increasingly efficient global hunt for fish is that big fish in many populations of commercially valuable species are becoming scarce. Smaller fish are next as the fishing industry has begun working its way down marine food webs. This can reduce the breeding stock needed for recovery of depleted species, unravel food webs, and disrupt marine ecosystems. And we throw away almost one-third of the fish we catch because they are caught unintentionally in our harvest of commercially valuable species.

According to the World Conservation Union, 34% of the world's 15,500 known marine fish species and 71% of the world's freshwater fish species face extinction within your lifetime. Indeed, *marine and freshwater fish are threatened with extinction by human activities more than any other group of species.*

Why Is It Difficult to Protect Aquatic Biodiversity?

Coastal development, the invisibility of problems, the vastness of the world's oceans, and lack of legal jurisdiction hinder protection of marine biodiversity.

Protecting aquatic biodiversity, especially marine biodiversity, is difficult for several reasons. *First,* the human ecological footprint is expanding so rapidly into aquatic areas that it is difficult to monitor the impacts.

Second, much of the damage to the oceans and other bodies of water is not visible to most people. *Third,* many people incorrectly view the seas as an inexhaustible resource that can absorb an almost infinite amount of waste and pollution.

Finally, most of the world's ocean area lies outside the legal jurisdiction of any country. Thus, it is an open-access resource, subject to overexploitation because of the tragedy of the commons.

Protecting and Sustaining Marine Biodiversity

Marine biodiversity can be sustained by protecting endangered species, establishing protected sanctuaries, managing coastal development, reducing water pollution, and preventing overfishing.

Here are some ways to protect and sustain aquatic biodiversity. First, we can *protect endangered and threatened species,* as discussed in Chapter 7. Second, we can *establish protected marine sanctuaries.* Since 1986, the World Conservation Union has helped establish a global system of *marine protected areas* (MPAs)—areas of ocean partially protected from human activities. There are 1,300 MPAs, nearly 200 of them in U.S. waters. Nearly all MPAs allow dredging, trawler fishing, and other resource extraction activities

Marine reserves are areas where no extraction and alteration of any living or nonliving resource is allowed. Scientific studies show that within fully protected marine reserves, fish populations double, fish size grows by almost a third, fish reproduction triples, and species diversity increases by almost one-fourth. Furthermore, this improvement happens within 2–4 years after strict protection begins and lasts for decades.

According to the 2005 Millennium Ecosystem Assessment less than 0.3% of the world's ocean area consists of fully protected marine reserve waters. In the United States, the total area of fully protected marine habitat is only about 130 square kilometers (50 square miles). In other words, *we have failed to strictly protect 99.7% of the world's ocean area from human exploitation.*

Also, many current marine sanctuaries are too small to protect most of the species within them and do not provide adequate protection from pollution that flows from the land into coastal waters. In 2004, a joint study by the World Wildlife Fund International and Great Britain's Royal Society for Protection of Birds called for fully protecting 30% of the world's oceans as marine reserves.

Another approach involves using *zoning rules* to protect and manage marine resources for entire bodies of water. A 2004 study by the Pew Oceans Commission recommended establishing many more protected marine reserves in U.S. coastal waters and connecting them with protected corridors so fish can move back and forth between such areas. As of 2006, the federal government had not implemented this recommendation.

We can also establish *integrated coastal management programs* in which groups of fishers, scientists, conservationists, citizens, business interests, developers, and politicians work together to identify shared problems and goals. Then they attempt to develop workable and cost-effective solutions that preserve biodiversity and environmental quality while meeting economic and social needs. Australia's huge Great Barrier Reef Marine Park is managed this way. Currently, more than 100 integrated coastal management programs are being developed throughout the world, including one for the Chesapeake Bay in the United States.

Another important strategy is to protect existing coastal and inland wetlands from being destroyed or degraded. This would involve steering development away from existing wetlands, restoring degraded wetlands, and trying to prevent and control invasions by nonnative species. Freshwater lakes and rivers would benefit from the same precautions.

We can also manage marine fisheries to prevent overfishing. Figure 6-17 lists measures that analysts have suggested should be taken for managing global fisheries more sustainably and protecting marine biodiversity. Most of these approaches rely on some sort

Figure 6-17 Solutions: ways to manage fisheries more sustainably and protect marine biodiversity. QUESTION: *Which four of these solutions do you think are the most important?*

of government regulation. Finally, we can regulate and prevent aquatic pollution, as discussed in Chapter 9.

WHAT CAN WE DO?

Solutions: Establishing Priorities

Biodiversity expert Edward O. Wilson has proposed eight priorities for protecting most of the world's remaining ecosystems and species.

In 2002, Edward O. Wilson, considered to be one of the world's foremost experts on biodiversity, proposed the following priorities for protecting most of the world's remaining ecosystems and species:

- Take immediate action to preserve the world's biological hot spots (Figure 6-15, p.122).

- Keep intact the world's remaining old-growth forests and cease all logging of such forests.

- Complete the mapping of the world's terrestrial and aquatic biodiversity so we know what we have and can make conservation efforts more precise and cost-effective.

- Determine the world's *marine hot spots* and assign them the same priority for immediate action as for those on land.

- Concentrate on protecting and restoring the world's lakes and river systems, which are the most threatened ecosystems of all.

- Ensure that the full range of the earth's terrestrial and aquatic ecosystems is included in a global conservation strategy.

- Make conservation profitable. This involves finding ways to raise the income of people who live in or near nature reserves so they can become partners in their protection and sustainable use.

- Initiate ecological restoration products worldwide to heal some of the damage we have done and increase the share of the earth's land and water allotted to the rest of nature.

According to Wilson, such a conservation strategy would cost about $30 billion per year—an amount that could be provided by a tax of one cent per cup of coffee.

This strategy for protecting the earth's precious biodiversity, upon which we are completely dependent, will not be implemented without bottom-up political pressure on elected officials from individual citizens and groups. It will also require cooperation among key people in government, the private sector, science, and engineering.

We abuse land because we regard it as a commodity belonging to us. When we see land as a community to which we belong, we may begin to use it with love and respect.

ALDO LEOPOLD

CRITICAL THINKING

1. Explain why you agree or disagree with each of the proposals for providing more sustainable use of forests throughout the world, listed in Figure 6-9. p. 113.

2. Should developed countries provide most of the money to preserve remaining tropical forests in developing countries? Explain.

3. Should there be a ban on the use of off-road motorized vehicles and snowmobiles on all public lands in the United States or in the country where you live? Explain.

4. Do you agree or disagree with the notion that native peoples should be allowed to live according to their traditions within wilderness areas while commercial development of those areas is prohibited? Explain.

5. Are you in favor of establishing more wilderness areas in the United States, especially in the lower 48 states (or in the country where you live)? Explain. What might be some drawbacks of doing this?

6. Why do you think the loss of aquatic biodiversity in marine and freshwater systems is not a more prominent issue in the news media? Think of three ways in which your lifestyle may be connected to this loss.

7. Congratulations! You are in charge of the world. List the three most important features of your policies for using and managing **(a)** forests, **(b)** parks, and **(c)** aquatic biodiversity.

8. List two questions that you would like to have answered as a result of reading this chapter.

LEARNING ONLINE

The website for this book contains helpful study aids and many ideas for further reading and research. They include a chapter summary, review questions for the entire chapter, flash cards for key terms and concepts, a multiple-choice practice quiz, interesting Internet sites, references, information about green careers, and a guide for accessing thousands of InfoTrac® College Edition articles. Log into

www.thomsonedu.com/biology/miller

Then choose Chapter 6, and select a learning resource. For access to animations, additional quizzes, chapter outlines and summaries, register and log into

at **www.thomsonedu.com** using the access code card in the front of your book.

7

SUSTAINING BIODIVERSITY: THE SPECIES APPROACH

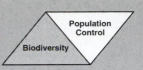

The last word in ignorance is the person who says of an animal or plant: "What good is it?" . . . If the land mechanism as a whole is good, then every part of it is good, whether we understand it or not. . . . Harmony with land is like harmony with a friend; you cannot cherish his right hand and chop off his left.

Aldo Leopold

SPECIES EXTINCTION

Three Types of Species Extinction (Science)

Species can become extinct locally, ecologically, or globally.

Biologists distinguish among three levels of species extinction. *Local extinction* occurs when a species is no longer found in an area it once inhabited but is still found elsewhere in the world. Most local extinctions involve losses of one or more populations of species.

Ecological extinction occurs when so few members of a species are left that it can no longer play its ecological roles in the biological communities where it is found. Finally, *biological extinction* occurs when a species is no longer found anywhere on the earth (Figure 7-1 and Case Study, below). Biological extinction is forever.

Case Study: The Passenger Pigeon—Gone Forever

The passenger pigeon, once the most common bird in North America, became extinct because of human activities, mostly overhunting and loss of habitat.

In 1813, bird expert John James Audubon saw a single flock of passenger pigeons that he estimated was 16 kilometers (10 miles) wide and hundreds of kilometers long and contained perhaps a billion birds. The flock took three days to fly past him and was so dense that it darkened the skies.

By 1914, the passenger pigeon (Figure 7-1, left) had disappeared forever. How could a species that was once the most common bird in North America and probably the world become extinct in only a few decades? The answer is, humans wiped them out. The main reasons for the extinction of this species were uncontrolled commercial hunting and loss of the bird's habitat and food supply as forests were cleared to make room for farms and cities.

Passenger pigeons were good to eat, their feathers made good filling for pillows, and their bones were widely used for fertilizer. They were easy to kill because they flew in gigantic flocks and nested in long, narrow colonies.

Beginning in 1858, passenger pigeon hunting became a big business. Shotguns, traps, artillery, and even dynamite were used. People burned grass or sulfur below their roosts to suffocate the birds. Shooting galleries used live birds as targets. In 1878, one professional pigeon trapper made $60,000 by killing 3 million birds at their nesting grounds near Petoskey, Michigan.

By the early 1880s, only a few thousand birds remained. At that point, recovery of the species was doomed because the females laid only one egg per

Passenger pigeon   Great auk   Dodo   Dusky seaside sparrow   Aepyornis (Madagascar)

Figure 7-1 Lost natural capital: some animal species that have become prematurely extinct largely because of human activities, mostly habitat destruction and overhunting. The Great Auk became extinct in 1844 from overhunting because of its willingness to march up the boardwalks to ships.

nest. On March 24, 1900, a young boy in Ohio shot the last known wild passenger pigeon.

The last passenger pigeon on earth, a hen named Martha after Martha Washington, died in the Cincinnati Zoo in 1914. Her stuffed body is now on view at the National Museum of Natural History in Washington, D.C.

Endangered and Threatened Species—Ecological Smoke Alarms (Science)

An endangered species could soon become extinct and a threatened species is likely to become endangered.

Biologists classify species heading toward biological extinction as either *endangered* or *threatened* (Figure 7-2). An **endangered species** has so few individual survivors that the species could soon become extinct over all or most of its natural range. A **threatened,** or **vulnerable, species** is still abundant in its natural range but because of declining numbers is likely to become endangered in the near future.

Some species have characteristics that make them more vulnerable than others to ecological and biological extinction (Figure 7-3, p. 132). As biodiversity expert Edward O. Wilson puts it, "the first animal species to go are the big, the slow, the tasty, and those with valuable parts such as tusks and skins."

The International Union for the Conservation of Nature and Natural Resources (IUCN) is a coalition of the world's leading conservation groups. Since the 1960s, it has kept *Red Lists* that have become the world standard for listing all threatened species throughout the world. In 2004, the Red List contained 15,589 species at risk of extinction, including one of every four mammal species, one of every seven plant species, and one of every eight bird species. Those compiling the list say it underestimates the true number of threatened species because only a tiny fraction of 1.4 million known species have been assessed.

Estimating Extinction Rates (Science)

Scientists use measurements and models to estimate extinction rates.

Evolutionary biologists estimate that more than 99.9% of all species that have ever existed are now extinct because of a combination of background extinctions, mass extinctions, and mass depletions taking place over thousands to millions of years. Biologists also talk of an *extinction spasm*, wherein large numbers of species are lost over a period of a few centuries or at most 1,000 years.

Biologists trying to catalog extinctions have three problems. *First,* because it takes such a long time for a species to become extinct, it is not easy to document. *Second,* scientists have identified only about 1.4 million

of the world's estimated 4–100 million species. *Third,* scientists know little about most of the species we have identified.

Scientists do the best they can with the tools they have to estimate past and projected future extinction rates. One approach is to study past records that have documented the rate at which mammals and birds have become extinct since humans arrived on earth and compare that with the fossil records of such extinctions prior to the arrival of humans.

Another way that biologists project future extinction rates is to observe how the number of species present increases with the size of an area. This *species–area relationship* suggests that, on average, a 90% loss of habitat causes the extinction of about 50% of the species living in that habitat.

Scientists also use models to estimate the risk that a particular species will become endangered or extinct within a certain time, based on factors such as trends in population size, changes in habitat availability, and interactions with other species. Estimates of future extinction rates vary because of differing assumptions about the earth's total number of species, the proportion of these species that are found in tropical forests, the rate at which tropical forests are being cleared, and the reliability of the methods used to make these estimates.

Effects of Human Activities on Extinction Rates (Science)

Biologists estimate that the current rate of extinction is 100 to 10,000 times the rate before humans arrived.

In due time, all species become extinct but there is considerable evidence that we are hastening the final exit for a growing number of species. Before we came on the scene, the estimated extinction rate was roughly one per million species annually. This amounted to an annual extinction rate of about 0.0001% per year.

Using the methods just described, biologists conservatively estimate that the current rate of extinction is at least 100 to 1,000—and by some estimates 10,000—times the rate before humans arrived on earth. This amounts to an annual extinction rate of 0.01% to 1% per year.

So how many species are being lost prematurely each year? This depends on how many species are on the earth. Assuming that the extinction rate is 0.1%, each year we are losing 5,000 species per year if there are 5 million species, 14,000 if there are 14 million species (biologists' current best guess), and 100,000 if there are 100 million species.

Most biologists would consider the premature loss of 1 million species over 100–200 years an extinction crisis or spasm that, if kept up, would lead to a mass depletion or even a mass extinction. At an extinction

Grizzly bear Kirkland's warbler Knowlton cactus Florida manatee African elephant

Utah prairie dog Swallowtail butterfly Humpback chub Golden lion tamarin Siberian tiger

Giant panda Black-footed ferret Whooping crane Northern spotted owl Blue whale

Mountain gorilla Florida panther California condor Hawksbill sea turtle Black rhinoceros

Figure 7-2 Endangered natural capital: species that are endangered or threatened with premature extinction largely because of human activities. Almost 30,000 of the world's species and 1,260 of those in the United States are officially listed as being in danger of becoming extinct. Most biologists believe the actual number of species at risk is much larger.

Characteristic	Examples
Low reproductive rate	Blue whale, giant panda, rhinoceros
Specialized niche	Blue whale, giant panda, Everglades kite
Narrow distribution	Many island species, elephant seal, desert pupfish
Feeds at high trophic level	Bengal tiger, bald eagle, grizzly bear
Fixed migratory patterns	Blue whale, whooping crane, sea turtles
Rare	Many island species, African violet, some orchids
Commercially valuable	Snow leopard, tiger, elephant, rhinoceros, rare plants and birds
Large territories	California condor, grizzly bear, Florida panther

Figure 7-3 Natural capital loss and degradation: characteristics of species that are prone to ecological and biological extinction. QUESTION: *Which of these characteristics helped lead to the premature extinction of the passenger pigeon?*

rate of 0.1% a year, the time it would take to lose 1 million species would be 200 years if there were a total of 5 million species, 71 years if the total were 14 million species, and 10 years if there were 100 million species. How many years would it take to lose 1 million species for each of these three total species estimates if the extinction rate were 1% a year?

According to researchers Edward O. Wilson and Stuart Primm, at a 1% extinction rate, at least one-fifth of the world's current animal and plant species could be gone by 2030 and half could vanish by the end of this century. In the words of biodiversity expert Norman Myers, "Within just a few human generations, we shall—in the absence of greatly expanded conservation efforts—impoverish the biosphere to an extent that will persist for at least 200,000 human generations or twenty times longer than the period since humans emerged as a species."

Most biologists consider extinction rates of 0.1%–1% to be conservative estimates for several reasons. *First,* both the rate of species loss and the extent of biodiversity loss are likely to increase during the next 50–100 years because of the projected growth of the world's human population and resource use per person. In other words, the size of our already large ecological footprint (Figure 1-6, p. 10) is likely to increase.

Second, current and projected extinction rates are much higher than the global average in parts of the

world that are endangered centers of the world's biodiversity. Conservation biologists urge us to focus our efforts on slowing the much higher rates of extinction in such hot spots (Figure 6-15, p. 122) as the best and quickest way to protect much of the earth's biodiversity from being lost prematurely.

Third, we are eliminating, degrading, and simplifying many biologically diverse environments—such as tropical forests, tropical coral reefs, wetlands, and estuaries—that serve as potential colonization sites for the emergence of new species. Thus, in addition to increasing the rate of extinction, we may also be limiting long-term recovery of biodiversity by reducing the rate of speciation for some types of species. In other words, we are also creating a *speciation crisis.* See the Guest Essay by Norman Myers on this topic on the website for this chapter.

Philip Levin, Donald Levin, and other biologists also argue that the increasing fragmentation and disturbance of habitats throughout the world may increase the speciation rate for rapidly reproducing opportunist species such as weeds, rodents, and cockroaches and other insects. Thus, the real threat to biodiversity from current human activities may not be a permanent decline in the number of species but a long-term erosion in the earth's variety of species and habitats. Such a loss of biodiversity (one of the four principles of sustainability) would reduce the ability of life to adapt to changing conditions by creating new species.

IMPORTANCE OF WILD SPECIES

Why Should We Preserve Wild Species? (Science and Economics)

We should avoid causing the premature extinction of species, if only because of the economic and ecological services they provide.

So what is all the fuss about? If all species eventually become extinct, why should we worry about losing a few more because of our activities? Does it matter that the passenger pigeon, the remaining orangutans, or some unknown plant or insect in a tropical forest becomes prematurely extinct because of human activities?

We know that new species eventually evolve to take the place of ones lost through extinction spasms, mass depletions, or mass extinctions (Figure 3-4, p. 58). So why should we care if we speed up the extinction rate over the next 50–100 years? Because it will take at least 5 million years for natural speciation to rebuild the biodiversity we are likely to destroy during this century!

Conservation biologists and ecologists say we should act now to prevent premature extinction of

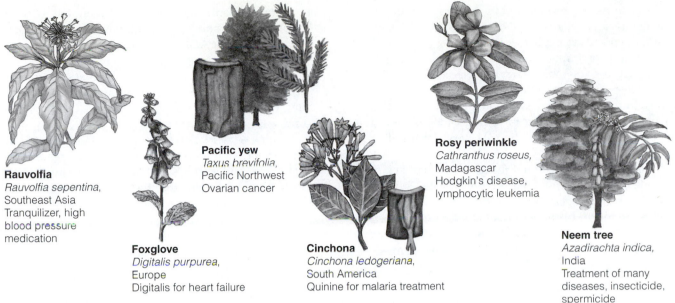

Rauvolfia
Rauvolfia sepentina,
Southeast Asia
Tranquilizer, high
blood pressure
medication

Pacific yew
Taxus brevifolia,
Pacific Northwest
Ovarian cancer

Rosy periwinkle
Cathranthus roseus,
Madagascar
Hodgkin's disease,
lymphocytic leukemia

Foxglove
Digitalis purpurea,
Europe
Digitalis for heart failure

Cinchona
Cinchona ledogeriana,
South America
Quinine for malaria treatment

Neem tree
Azadirachta indica,
India
Treatment of many
diseases, insecticide,
spermicide

Figure 7-4 Natural capital: *nature's pharmacy.* Parts of these and a number of other plants and animals (many of them found in tropical forests) are used to treat a variety of human ailments and diseases. Nine of the ten leading prescription drugs originally came from wild organisms. About 2,100 of the 3,000 plants identified by the National Cancer Institute as sources of cancer-fighting chemicals come from tropical forests. Despite their economic and health potential, fewer than 1% of the estimated 125,000 flowering plant species in tropical forests (and a mere 1,100 of the world's 260,000 known plant species) have been examined for their medicinal properties. Once the active ingredients in the plants have been identified, they can usually be produced synthetically. Many of these tropical plant species are likely to become extinct before we can study them.

species because of their *instrumental value*—their usefulness to us in the form of economic and ecological services (Case Study, at right). For example, species provide economic value in the form of food crops, fuelwood and lumber, paper, and medicine (Figure 7-4).

Another instrumental value is the *genetic information* in species that allows them to adapt to changing environmental conditions and to form new species. Genetic engineers use this information to produce new types of crops (Figure 3-5, p. 60) and foods and edible vaccines for viral diseases such as hepatitis B. Carelessly eliminating many of the species making up the world's vast genetic library is like burning books before we read them. Wild species also provide a way for us to learn how nature works and sustains itself.

The earth's wild plants and animals also provide us with *recreational pleasure.* Each year, Americans spend over three times as many hours watching wildlife—doing nature photography and bird watching, for example—as they spend on watching movies or professional sporting events.

Wildlife tourism, or *ecotourism,* generates at least $500 billion per year worldwide, and perhaps twice as much. Conservation biologist Michael Soulé estimates that one male lion living to age 7 generates $515,000 in tourist dollars in Kenya but only $1,000 if killed for its skin. Similarly, over a lifetime of 60 years, a Kenyan elephant is worth about $1 million in ecotourist rev-

enue—many times more than its tusks are worth when sold illegally for their ivory.

Ecotourism should not cause ecological damage but some of it does. The website for this chapter lists some guidelines for evaluating ecotours.

Case Study: Why Should We Care about Bats? (Science and Economics)

Bats play vital ecological roles, helping to maintain biodiversity, and some are keystone species.

Bats—*the only mammals that can fly*—have two traits that make them vulnerable to extinction. *First,* they reproduce slowly. *Second,* many bat species live in huge colonies in caves and abandoned mines, which people sometimes block. This prevents them from leaving to get food and it can disturb their hibernation.

Bats play important ecological roles. About 70% of the world's 950 known bat species feed on crop-damaging nocturnal insects and other insect pest species such as mosquitoes. This makes them the major nighttime SWAT team for such insects.

In some tropical forests and on many tropical islands, *pollen-eating bats* pollinate flowers, and *fruit-eating bats* distribute plants throughout tropical forests by excreting undigested seeds.

As keystone species, such bats are vital for maintaining plant biodiversity and for regenerating large areas of tropical forest cleared by human activities. If you enjoy bananas, cashews, dates, figs, avocados, or mangos, you can thank bats.

Many people mistakenly view bats as fearsome, filthy, aggressive, rabies-carrying bloodsuckers. But most bat species are harmless to people, livestock, and crops. In the United States, only ten people have died of bat-transmitted disease in four decades of record keeping; more Americans die each year from falling coconuts.

Because of unwarranted fears of bats and lack of knowledge about their vital ecological roles, several bat species have been driven to extinction. Currently, about one-fourth of the world's bat species are listed as endangered or threatened. Conservation biologists urge us to view bats as valuable allies, not as enemies.

The Intrinsic Value of Species (Ethics)

Some believe that each wild species has an inherent right to exist.

Some people believe each wild species has *intrinsic* or *existence* value based on its inherent right to exist and play its ecological roles, regardless of its usefulness to us. According to this view, we have an ethical responsibility to protect species from becoming prematurely extinct as a result of human activities, and to prevent the degradation of the world's ecosystems and its overall biodiversity.

Some people distinguish between the survival rights of plants and those of animals, mostly for practical reasons. Poet Alan Watts once said he was a vegetarian "because cows scream louder than carrots."

Other people distinguish among various types of species. For example, they might think little about getting rid of the world's mosquitoes, cockroaches, rats, or disease-causing bacteria.

Some conservation biologists caution us not to focus primarily on protecting relatively big organisms—the plants and animals we can see and are familiar with. They remind us that the true foundation of the earth's ecosystems and ecological processes are the invisible bacteria, and the algae, fungi, and other *microorganisms* that decompose the bodies of larger organisms and recycle the nutrients needed by all life.

CAUSES OF PREMATURE EXTINCTION OF WILD SPECIES

Habitat Destruction, Degradation, and Fragmentation: Remember HIPPO (Science)

The greatest threat to a species is the loss, degradation, and fragmentation of the place where it lives.

Figure 7-5 shows the basic and secondary causes of the endangerment and premature extinction of wild species. Conservation biologists sometimes summarize

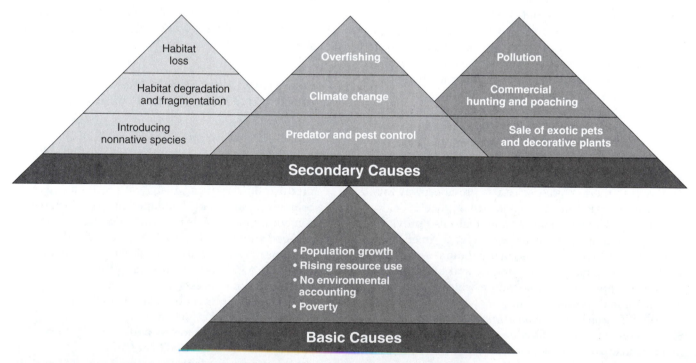

Figure 7-5 Natural capital degradation: underlying and direct causes of depletion and premature extinction of wild species. The major direct cause of wildlife depletion and premature extinction is habitat loss, degradation, and fragmentation. This is followed by the deliberate or accidental introduction of harmful invasive (nonnative) species into ecosystems.

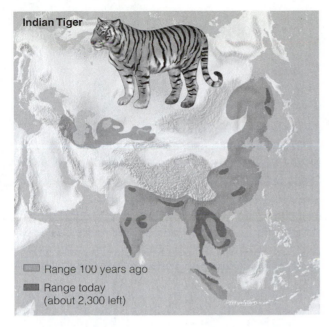

Indian Tiger

☐ Range 100 years ago

■ Range today
(about 2,300 left)

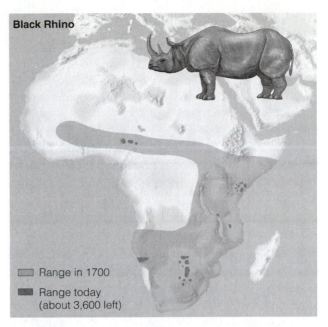

Black Rhino

☐ Range in 1700

■ Range today
(about 3,600 left)

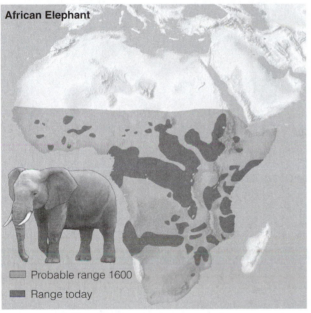

African Elephant

☐ Probable range 1600

■ Range today

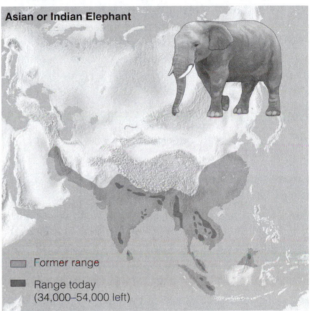

Asian or Indian Elephant

☐ Former range

■ Range today
(34,000–54,000 left)

ThomsonNOW™ Active Figure 7-6 Natural capital degradation: reductions in the ranges of four wildlife species, mostly as the result of habitat loss and hunting. What will happen to these and millions of other species when the world's human population doubles and per capita resource consumption rises sharply in the next few decades? *See an animation based on this figure and take a short quiz on the concept.* (Data from International Union for the Conservation of Nature and World Wildlife Fund)

the most important causes of premature extinction using the acronym **HIPPO** for **H**abitat destruction and fragmentation, **I**nvasive (alien) species, **P**opulation growth (too many people consuming too many resources), **P**ollution, and **O**verharvesting.

According to biodiversity researchers, the greatest threat to wild species is habitat loss (Figure 7-6), degradation, and fragmentation. In other words, many species have a hard time surviving when we take over

their ecological "house" and their food supplies and make them homeless.

Deforestation of tropical forests is the greatest eliminator of species, followed by the destruction and degradation of coral reefs and wetlands, plowing of grasslands, and pollution of streams, lakes, and oceans. Globally, temperate biomes have been affected more by habitat loss and degradation than have tropical biomes because of widespread development in temperate countries over the past 200 years. But development is now shifting to many tropical biomes.

Island species, many of them *endemic species* found nowhere else on earth, are especially vulnerable to extinction when their habitats are destroyed, degraded, or fragmented. Any habitat surrounded by a different one can be viewed as a *habitat island* for most of the species that live there. Most national parks and other protected areas are habitat islands, many of them surrounded by potentially damaging logging, mining, energy extraction, and industrial activities. Freshwater lakes are also habitat islands that are especially vulnerable when nonnative species are introduced.

Habitat fragmentation—from road building, logging, agriculture, and urban development—occurs when a large, continuous area of habitat is reduced in area and divided into smaller, more scattered, and isolated patches or "habitat islands." This can block migration routes and divide populations of a species into smaller and more isolated groups, which are more vulnerable to predators, competitive species, disease, and catastrophic events such as a storm or fire. Also, it creates barriers that limit the abilities of some species to disperse and colonize new areas, get enough to eat, and find mates.

 See how serious the habitat fragmentation problem is for elephants, tigers, and rhinos at ThomsonNOW.

Case Study: A Disturbing Message from the Birds (Science)

Our activities are causing serious declines in the populations of many bird species—mostly from habitat loss, degradation, and fragmentation.

Approximately 70% of the world's 9,800 known bird species are declining in numbers, and about one of every six bird species are threatened with extinction, mostly because of habitat loss, degradation, and fragmentation. Many of the world's bird species are found in South America (Figure 7-7).

A 2004 National Audubon Society study found that 30% of all North American bird species and 70% of those living in grasslands are declining in numbers or are at risk of disappearing. Conservation biologists view this decline with alarm because birds are excellent *environmental indicators*. They live in every climate and biome, respond quickly to environmental changes in their habitats, and are easy to track and count.

Birds also help to control populations of rodents and insects (which decimate many tree species); pollinate a variety of flowering plants, spreading plants throughout their habitats by consuming and excreting plant seeds; and scavenge dead animals. Conservation biologists urge us to listen more carefully to what birds

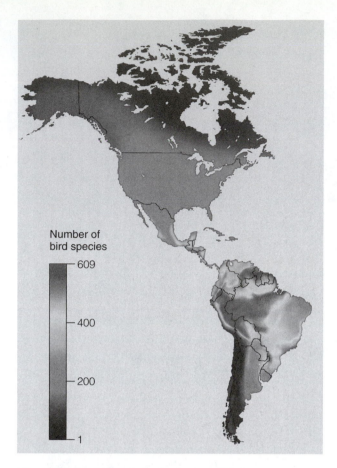

Figure 7-7 Natural capital: distribution of bird species in North America and Latin America. (Data from the Nature Conservancy, Conservation International, World Wildlife Fund, and Environment Canada)

are telling us about the state of the environment for them and for us.

Deliberately Introduced Species (Science and Economics)

Many nonnative species provide us with food, medicine, and other benefits, but a few can wipe out some native species, disrupt ecosystems, and cause large economic losses.

After habitat loss and degradation, the deliberate or accidental introduction of harmful invasive species into ecosystems is the biggest cause of animal and plant extinctions.

Actually, we depend on nonnative organisms for ecosystem services, food, shelter, medicine, and aesthetic enjoyment. According to a 2000 study by ecologist David Pimentel, introduced species such as corn, wheat, rice, other food crops, cattle, poultry, and other livestock provide more than 98% of the U.S. food supply. Similarly, nonnative tree species are grown in about 85% of the world's tree plantations. Some deliberately introduced species have also helped control pests.

The problem is that some introduced species have no natural predators, competitors, parasites, or pathogens to help control their numbers in their new habitats. Such species can reduce or wipe out populations of many native species and trigger ecological disruptions. About 50,000 nonnative species now live in the United States and about one in seven of them have caused ecological and economic harm (Figure 7-8, p. 138). According to biologist Thomas Lovejoy, harmful invader species cost the U.S. public more than $137 billion each year—an average of $16 million per hour!

Nonnative species threaten almost half of the more than 1,260 endangered and threatened species in the United States, and 95% of those in Hawaii, according to the U.S. Fish and Wildlife Service. They are also blamed for about two-thirds of fish extinctions in the United States between 1900 and 2000.

One example of a deliberately introduced nonnative species is the *nutria*, a giant-sized rodent (Figure 7-8). People hoping to sell its fur deliberately transplanted the nutria into the United States and many other countries from its home in South America. Many of these animals escaped, spread rapidly, and have caused serious ecological damage by munching their way through the nation's fragile coastal swamps.

Accidentally Introduced Invasive Species (Science)

A growing number of accidentally introduced species are causing serious economic and ecological damage.

Many unwanted nonnative invaders arrive from other continents as stowaways on aircraft, in the ballast water of tankers and cargo ships, and as hitchhikers on imported products such as wooden packing crates. Cars and trucks can spread seeds of nonnative species imbedded in tire treads. Tourists return home with living plants that may multiply and become invasive or harbor insects that can escape, multiply rapidly, and threaten crops.

In the late 1930s, the extremely aggressive Argentina fire ant (Figure 7-8) was introduced accidentally into the United States in Mobile, Alabama. The ants may have arrived on shiploads of lumber or coffee imported from South America or by hitching a ride in the soil-containing ballast water of cargo ships—a major source of troublesome invasive species.

Without natural predators, fire ants have spread rapidly by land and water (they can float) throughout the South, from Texas to Florida and as far north as Tennessee and Virginia (Figure 7-9, p. 139). They are also found in Puerto Rico and recently have invaded California and New Mexico.

Wherever fire ants have gone, they have sharply reduced or wiped out up to 90% of native ant populations. Bother them, and up to 100,000 ants can swarm out of their nest to attack you with their painful and burning stings. They have killed deer fawns, birds, livestock, pets, and at least 80 people allergic to their venom. These ants have invaded cars (causing accidents by attacking drivers), damaged crops, disrupted phone service and electrical power, caused fires by chewing through underground cables, and cost the United States an estimated $600 million per year. Their large mounds, which look like large boils on the land, can ruin cropland and their painful stings can make backyards uninhabitable.

Widespread pesticide spraying in the 1950s and 1960s temporarily reduced fire ant populations. But this chemical warfare actually hastened the advance of the rapidly multiplying fire ant by reducing populations of many native ant species. Worse, it promoted development of genetic resistance to pesticides in the fire ants through natural selection. In other words, we helped wipe out their competitors and made them genetically stronger.

Researchers at the U.S. Department of Agriculture are experimenting with the use of biological control agents to reduce fire ant populations. Before widespread use of these biological control agents begins, researchers must be sure they will not cause problems for native ant species or become pests themselves.

Solutions: Reducing Threats from Nonnative Species (Science)

Prevention is the best way to reduce the threats from harmful nonnative species because once they have arrived it is difficult and expensive to slow their spread.

Once a nonnative species gets established in an ecosystem, its wholesale removal is almost impossible—somewhat like trying to get smoke back into a chimney. Thus, the best way to limit the harmful impacts of nonnative species is to prevent them from being introduced and becoming established.

There are several ways to do this. One is to identify both the major characteristics that allow species to become successful invaders and the types of ecosystems that are vulnerable to invaders (Figure 7-10, p. 139). Such information can be used to screen out potentially harmful invaders using more ground surveys and satellite observations.

We can also step up inspections of imported goods that are likely to contain invader species. A related strategy is to identify major harmful invader species and pass international laws banning their transfer from one country to another, as is now done for endangered species.

Deliberately Introduced Species

Purple looselife

European starling

African honeybee ("Killer bee")

Nutria

Salt cedar (Tamarisk)

Marine toad (Giant toad)

Water hyacinth

Japanese beetle

Hydrilla

European wild boar (Feral pig)

Accidentally Introduced Species

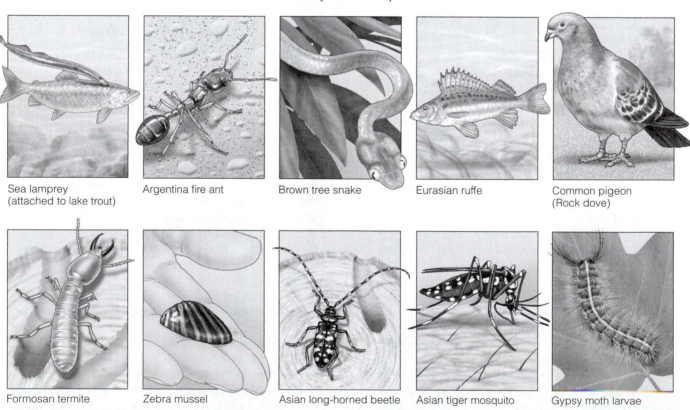

Sea lamprey (attached to lake trout)

Argentina fire ant

Brown tree snake

Eurasian ruffe

Common pigeon (Rock dove)

Formosan termite

Zebra mussel

Asian long-horned beetle

Asian tiger mosquito

Gypsy moth larvae

Figure 7-8 Threats to natural capital: some of the more than 7,100 harmful invasive (nonnative) species that have been deliberately or accidentally introduced into the United States.

We can also require cargo ships to discharge their ballast water and replace it with saltwater at sea before entering ports or to sterilize such water or pump nitrogen into the water to displace dissolved oxygen and kill most invader organisms. We could increase re-

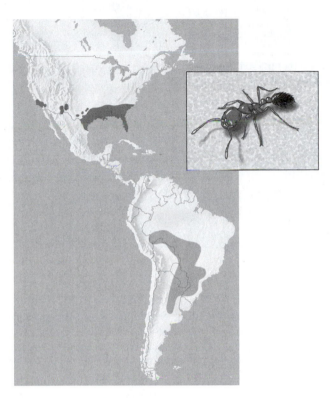

Figure 7-9 Natural capital degradation: the *Argentina fire ant,* introduced accidentally into Mobile, Alabama, in 1932 from South America (lighter shaded area), has spread over much of the southern United States (darker shaded area). This invader is also found in Puerto Rico, New Mexico, and California. (Data from S.D. Porter, Agricultural Research Service, U.S. Department of Agriculture)

Characteristics of Successful Invader Species	Characteristics of Ecosystems Vulnerable to Invader Species
• High reproductive rate, short generation time (r-selected species) • Pioneer species • Long lived • High dispersal rate • Release growth-inhibiting chemicals into soil • Generalists • High genetic variability	• Similar climate to habitat of invader • Absence of predators on invading species • Early successional systems • Low diversity of native species • Absence of fire • Disturbed by human activities

Figure 7-10 Threats to natural capital: some general characteristics of successful invader species and ecosystems vulnerable to invading species.

What Can You Do?
Invasive Species

- Do not allow wild animals to escape.

- Do not spread wild plants to other areas.

- Do not dump the contents of an aquarium into waterways, wetlands, or storm drains.

- When camping use wood near your campsite instead of bringing firewood from somewhere else.

- Do not dump unused bait into the water.

- After dogs visit woods or the water brush them before taking them home.

- After each use clean your vehicle, mountain bike, surfboard, kayaks, canoes, boats, tent, hiking boots, and other gear before heading for home.

- Empty all water from canoes, kayaks, dive gear, and other outdoor equipment before heading home.

- Plant a variety of trees, shrubs, and other plants in your yard to reduce losses from invasive species.

- Do not buy plants from overseas or swap them with others using the Internet.

Figure 7-11 Individuals matter: ways to prevent or slow the spread of harmful invasive species. QUESTION: *Which two of these actions do you think are the most important? Which of these actions do you plan to take?*

search to find and introduce natural predators, parasites, and disease-causing bacteria and viruses to control populations of established invaders.

Finally, each of us can take measures in our own lives to make sure we do not contribute to the growing problem of harmful invasive species. Figure 7-11 shows some of the things you can do to help.

Population Growth, Overconsumption, Pollution, and Climate

Population growth, affluenza, pollution, and climate change have caused the premature extinction of species, and the impact of these factors is growing.

Past and projected human population growth and excessive and wasteful consumption of resources have caused premature extinction of some species. Acting together, these two factors have greatly expanded the human ecological footprint. As humans spread out in this way, we have taken over most of the habitat space and other resources available to other species.

Pollution threatens populations and species in a number of ways. A major extinction threat comes from the unintended effects of pesticides. According to the U.S. Fish and Wildlife Service, each year in the United States, pesticides kill about one-fifth of the country's

Figure 7-12 Natural capital degradation: *bioaccumulation* and *biomagnification.* DDT is a fat-soluble chemical that can accumulate in the fatty tissues of animals. In a food chain or web, the accumulated DDT can be biologically magnified in the bodies of animals at each higher trophic level. The concentration of DDT in the fatty tissues of organisms was biomagnified about 10 million times in this food chain in an estuary near Long Island Sound in New York. If each phytoplankton organism takes up and retains one unit of DDT from the water, a small fish eating thousands of zooplankton (which feed on the phytoplankton) will store thousands of units of DDT in its fatty tissue. Each large fish that eats 10 of the smaller fish will ingest and store tens of thousands of units, and each bird (or human) that eats several large fish will ingest hundreds of thousands of units. Dots represent DDT, and arrows show small losses of DDT through respiration and excretion.

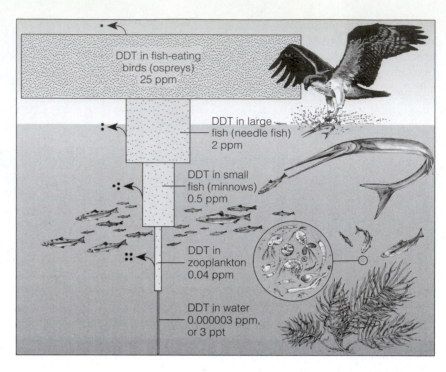

DDT in fish-eating birds (ospreys)
25 ppm

DDT in large fish (needle fish)
2 ppm

DDT in small fish (minnows)
0.5 ppm

DDT in zooplankton
0.04 ppm

DDT in water
0.000003 ppm, or 3 ppt

beneficial honeybee colonies, more than 67 million birds, and 6–14 million fish. They also threaten about one-fifth of the country's endangered and threatened species.

During the 1950s and 1960s, populations of fish-eating birds such as the osprey, cormorant, brown pelican, and bald eagle plummeted. A chemical derived from DDT, when biologically magnified in food webs (Figure 7-12), made the birds' eggshells so fragile they could not reproduce successfully. Also hard hit were such predatory birds as the prairie falcon, sparrow hawk, and peregrine falcon, which help control rabbits, ground squirrels, and other crop eaters. Some *good news* is that since the U.S. ban on DDT in 1972, most of these species have made a comeback.

Most natural climate changes in the past have taken place over long periods of time, giving species time to adapt or evolve into new species in order to cope with changes. But considerable evidence indicates that human activities such as greenhouse gas emissions and deforestation may bring about rapid climate change during this century. This could change the habitats of many species and accelerate the extinction of some. A 2004 study by Conservation International predicted that climate change could drive more than a quarter of all land animals and plants to extinction by the end of this century.

Illegal Killing or Sale of Wild Species (Economics and Ethics)

Some protected species are illegally killed for their valuable parts or are sold live to collectors.

Organized crime has moved into illegal wildlife smuggling because of the huge profits involved—surpassed only by the illegal international trade in drugs and weapons. At least two-thirds of all live animals smuggled around the world die in transit.

Poor people in areas with rich stores of wildlife kill or trap such species in order to make enough money to survive and feed their families. Professional poachers also prey on these species. To poachers, a live *mountain gorilla* is worth $150,000, a *panda* pelt $100,000, a *chimpanzee* $50,000, and an *Imperial Amazon macaw* $30,000. A *rhinoceros horn* is worth as much as $28,600 per kilogram ($13,000 per pound). It is used in dagger handles in the Middle East and as a fever reducer and alleged aphrodisiac in China—the world's largest consumer of wildlife—and other parts of Asia.

In 1950, an estimated 100,000 tigers existed in the world. Despite international protection, only 5,000 to 7,000 tigers remain in the wild, on an ever-shrinking range. Bengal tigers are at risk because a tiger fur sells for $100,000 in Tokyo. With the body parts of a single tiger worth as much as $25,000, it is not surprising that illegal hunting has skyrocketed, especially in India. Without emergency action, few if any tigers may be left in the wild within 20 years.

As commercially valuable species become endangered, their black market demand soars. This increases their chances of premature extinction from poaching. Most poachers are not caught. And the money they can make far outweighs the small risk of being caught, fined, and imprisoned.

Scientists are fighting back. Since 1989, an international agreement banned the sale of ivory between nations, but poachers still kill elephants for their tusks. Scientists are using DNA analysis to trace illegal ivory back to where the elephant was killed, enabling detection and increased protection in poaching hot spots.

Predator and Pest Control (Science, Economics, and Ethics)

Killing predators that bother us or cause economic losses threatens some species with premature extinction.

People sometimes try to exterminate species that compete with them for food and game animals. African farmers kill large numbers of elephants to keep them from trampling and eating food crops. Each year, U.S. government animal control agents shoot, poison, or trap thousands of coyotes, prairie dogs, bobcats, wolves, and other species that prey on livestock, on species prized by game hunters, on crops, or on fish raised in aquaculture ponds.

Since 1929, U.S. ranchers and government agencies have poisoned 99% of North America's prairie dogs because horses and cattle sometimes step into the burrows and break their legs. This has also nearly wiped out the endangered black-footed ferret (Figure 7-2), which preyed on the prairie dog. Only about 600 are left in the wild. Most ranchers and hunters support government pest control programs on their behalf. Some conservationists consider it an unethical taking of life.

Exotic Pets and Decorative Plants (Economics and Ethics)

Legal and illegal trade in wildlife species used as pets or for decorative purposes threatens some species with extinction.

The global legal and illegal trade in wild species for use as pets is a huge and very profitable business. However, for every live animal captured and sold in the pet market, an estimated 50 others are killed.

About 25 million U.S. households have exotic birds as pets, 85% of them imported. More than 60 bird species, mostly parrots, are endangered or threatened because of this wild bird trade. According to the U.S. Fish and Wildlife Service, collectors of exotic birds may pay $10,000 for a threatened hyacinth macaw smuggled out of Brazil; however, during its lifetime, a single macaw left in the wild might yield as much as $165,000 in tourist income. A 1992 study suggested that keeping a pet bird indoors for more than ten years doubles a person's chances of getting lung cancer from inhaling tiny particles of bird dander.

Other wild species whose populations are depleted because of the pet trade include amphibians, reptiles, mammals, and tropical fish (taken mostly from the coral reefs of Indonesia and the Philippines). Divers catch tropical fish by using plastic squeeze bottles of cyanide to stun them. For each fish caught alive, many more die. In addition, the cyanide solution kills the coral animals that create the reef, which is a center for marine biodiversity.

Some exotic plants, especially orchids and cacti, are endangered because they are gathered (often illegally) and sold to collectors to decorate houses, offices, and landscapes. A collector may pay $5,000 for a single rare orchid, and a single, rare, mature crested saguaro cactus can earn cactus rustlers as much as $15,000.

PROTECTING WILD SPECIES: THE LEGAL AND ECONOMIC APPROACHES

Global Outlook and Politics: International Treaties

International treaties have helped reduce the international trade of endangered and threatened species, but enforcement is difficult.

Several international treaties and conventions help protect endangered or threatened wild species. One of the most far reaching is the 1975 *Convention on International Trade in Endangered Species* (CITES). This treaty, now signed by 166 countries, lists some 900 species that cannot be commercially traded as live specimens or wildlife products because they are in danger of extinction. The treaty also restricts international trade of 30,000 other species because they are at risk of becoming threatened.

CITES has helped reduce international trade in many threatened animals, including elephants, crocodiles, and chimpanzees. However, the effects of this treaty are limited because enforcement is difficult and varies from country to country, and convicted violators often pay only small fines. Also, member countries can exempt themselves from protecting any listed species, and much of the highly profitable illegal trade in wildlife and wildlife products goes on in countries that have not signed the treaty.

The Convention on Biological Diversity (CBD), ratified by 188 countries, legally binds signatory governments to reversing the global decline of biological diversity. However, its implementation has been slow because some key countries, including the United States, have not ratified it. Also, it contains no severe penalties or other enforcement mechanisms.

The U.S. Endangered Species Act (Science and Politics)

One of the world's most far-reaching and controversial environmental laws is the U.S. Endangered Species Act passed in 1973.

The *Endangered Species Act of 1973* (ESA, amended in 1982, 1985, 1988, and 2006) was designed to identify and legally protect endangered species in the United

States and abroad. This act is probably the most far-reaching environmental law ever adopted by any nation, which has made it controversial. Canada and a number of other countries have similar laws.

The National Marine Fisheries Service (NMFS) is responsible for identifying and listing endangered and threatened ocean species, and the U.S. Fish and Wildlife Services (USFWS) identifies and lists all other endangered and threatened species. Any decision by either agency to add or remove a species from the list must be based on biological factors alone, without consideration of economic or political factors. However, economic factors can be used in deciding whether and how to protect endangered habitat and in developing recovery plans for listed species.

The ESA also forbids federal agencies (except the Defense Department) to carry out, fund, or authorize projects that would jeopardize an endangered or threatened species or destroy or modify the critical habitat it needs to survive. On private lands, fines up to $100,000 and one year in prison can be imposed to ensure protection of the habitats of endangered species.

The act also makes it illegal for Americans to sell or buy any product made from an endangered or threatened species or to hunt, kill, collect, or injure such species in the United States.

Between 1973 and 2006, the number of U.S. species on the official endangered and threatened lists increased from 92 to about 1,260 species—60% of them plants and 40% animals. According to a 2000 study by the Nature Conservancy, one-third of the country's species are at risk of extinction, and 15% of all species are at high risk. This is far more than the 1,260 species on the ESA list. The study also found that many of the country's rarest and most imperiled species are concentrated in a few hot spots (Figure 7-13).

Getting listed is only half the battle. Until 2006, the ESA required the secretary of the interior to designate and protect the *critical habitat* needed for the survival and recovery of each listed species. So far, critical habitats have been established for only one-third of the species on the ESA list, mostly because of political pressure against designation, a lack of funds, and too little information to designate such areas.

In 2006, the ESA was severely weakened by an amendment eliminating the establishment of a critical habitat—something conservation biologists believe is essential for protecting most species and enhancing their recovery. Since 2001, the government has stopped listing new species and designating critical habitats for listed species unless required to do so by court order.

Encouraging Private Landowners to Protect Endangered Species (Economics and Politics)

Congress has amended the Endangered Species Act to help landowners protect endangered species on their land.

In 1982, Congress amended the ESA to allow the secretary of the interior to use *habitat conservation plans*

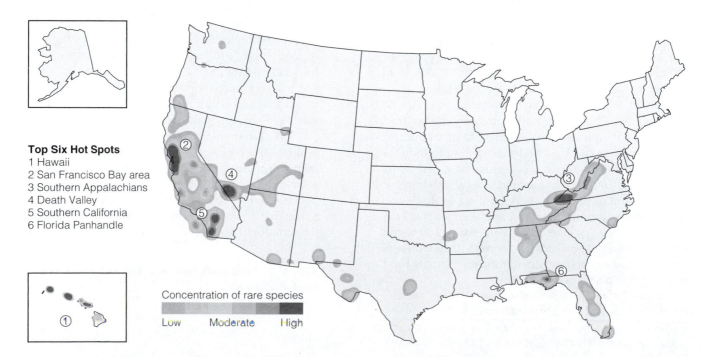

Top Six Hot Spots
1 Hawaii
2 San Francisco Bay area
3 Southern Appalachians
4 Death Valley
5 Southern California
6 Florida Panhandle

Concentration of rare species

Low   Moderate   High

Figure 7-13 Threatened natural capital: biodiversity hot spots in the United States. The shaded areas contain the largest concentrations of rare and potentially endangered species. (Data from State Natural Heritage Programs, the Nature Conservancy, and Association for Biodiversity Information)

(HCP). They are designed to strike a compromise between the interests of private landowners and those of endangered and threatened species, with the goal of not reducing the recovery chances of a protected species.

With an HCP, landowners, developers, or loggers are allowed to destroy some critical habitat in exchange for taking steps to protect members of the species. Such measures might include setting aside a part of the species' habitat as a protected area, paying to relocate the species to another suitable habitat, or paying money to have the government buy suitable habitat elsewhere. Once the plan is approved it cannot be changed, even if new data show that the plan is inadequate to protect a species and help it recover.

There is concern that many of the plans have been approved without enough scientific evaluation of their effects on a species' recovery. Also, many of the plans are political compromises that do not protect the species or that make inadequate provisions for its recovery.

The Future of the Endangered Species Act (Economics and Politics)

Some believe that the Endangered Species Act should be weakened or repealed and others believe it should be strengthened and modified to focus on protecting ecosystems.

Since 1992, Congress has been debating its reauthorization of the ESA with proposals ranging from eliminating the act to weakening it or to strengthening it.

Opponents of the ESA contend that it puts the rights and welfare of endangered plants and animals above those of people; that it has not been effective in protecting endangered species; and that it has caused severe economic losses by hindering development on private land that contains endangered or threatened species. Since 1995, efforts to weaken the ESA have included the following suggested changes:

■ Making protection of endangered species on private land voluntary

■ Having the government compensate landowners if it forces them to stop using part of their land to protect endangered species

■ Making it harder and more expensive to list newly endangered species by requiring government wildlife officials to navigate through a series of hearings and peer-review panels

■ Eliminating the need to designate critical habitats partly because dealing with lawsuits for failure to develop critical habitats takes up most of the limited funds allocated for carrying out the ESA

■ Allowing the secretary of the interior to permit a listed species to become extinct without trying to save it and to determine whether a species should be listed

■ Allowing the secretary of the interior to give any state, county, or landowner permanent exemption from the law, with no requirement for public notification or comment

The 1996 amendments to the act accomplished most of these objectives.

Other critics would go further and do away with this act. But because this is politically unpopular with the American public, most efforts are designed to weaken the act and reduce its meager funding.

Most conservation biologists and wildlife scientists agree that the ESA has some deficiencies and needs to be simplified and streamlined. But they contend that the ESA has not been a failure (Case Study, p. 144).

They also contest the charge that the ESA has caused severe economic losses. Government records show that since 1979, only about 0.05% of the almost 200,000 projects evaluated by the USFWS have been blocked or canceled as a result of the ESA.

A study by the U.S. National Academy of Sciences recommended three major changes to make the ESA more scientifically sound and effective.

■ Greatly increase the meager funding for implementing the act

■ Develop recovery plans more quickly

■ When a species is first listed, establish a core of its survival habitat as critical, as a temporary emergency measure that could support the species for 25–50 years

Most biologists and wildlife conservationists believe the United States needs a new law that emphasizes protecting and sustaining biological diversity and ecological functioning rather than attempting to save individual species. This new ecosystems approach would follow three principles:

■ Find out what species and ecosystems the country has

■ Locate and protect the most endangered ecosystems and species

■ Provide private landowners who agree to help protect specific endangered ecosystems with significant financial incentives (tax breaks and write-offs) and technical help

---

X *HOW WOULD YOU VOTE?* Should the Endangered Species Act be modified to protect the nation's overall biodiversity? Cast your vote online at www.thomsonedu.com/biology/miller.

Accomplishments of the Endangered Species Act (Science, Economics, and Politics)

Critics of the ESA call it an expensive failure because only 37 species have been removed from the endangered list. Most biologists agree that the act needs strengthening and modification. But they disagree that the act has been a failure, for four reasons.

First, species are listed only when they are in serious danger of extinction. This is like setting up a poorly funded hospital emergency room that takes only the most desperate cases, often with little hope for recovery, and saying it should be shut down because it has not saved enough patients.

Second, it takes decades for most species to become endangered or threatened. Thus, it usually takes

decades to bring a species in critical condition back to the point where it can be removed from the list. Expecting the ESA—which has been in existence only since 1973—to quickly repair the biological depletion of many decades is unrealistic.

Third, according to federal data the conditions of more than half the listed species are stable or improving and 99% of the protected species are still surviving. A hospital emergency room taking only the most desperate cases and then stabilizing or improving the condition of more than half of its patients while keeping 99% of them alive would be considered an astounding success.

Fourth, the federal endangered species budget was only $58 million in 2005—about what the Department of Defense spends in a little

more than an hour or 20¢ a year per U.S. citizen. To supporters of the ESA, it is amazing that the ESA has managed to stabilize or improve the conditions of more than half of the listed species on a shoestring budget.

Yes, the act can be improved and federal regulators have sometimes been too heavy handed in enforcing it. But instead of gutting or doing away with this important law, biologists call for it to be strengthened and modified to help protect ecosystems and the nation's overall biodiversity.

Critical Thinking

Should the budget for the Endangered Species Act be drastically increased? Explain.

PROTECTING WILD SPECIES: THE SANCTUARY APPROACH

Wildlife Refuges and Other Protected Areas (Science and Stewardship)

The United States has set aside 542 federal refuges for wildlife, but many refuges are suffering from environmental degradation.

In 1903, President Theodore Roosevelt established the first U.S. federal wildlife refuge at Pelican Island, Florida. Since then, the National Wildlife Refuge System has grown to 542 refuges. More than 35 million Americans visit these places each year to hunt, fish, hike, or watch birds and other wildlife.

More than three-fourths of the refuges serve as vital wetland sanctuaries for protecting migratory waterfowl. About one-fifth of U.S. endangered and threatened species have habitats in the refuge system, and some refuges have been set aside for specific endangered species. These have helped Florida's key deer, the brown pelican, and the trumpeter swan to recover.

Conservation biologists call for setting aside more refuges for endangered plants. They also urge Congress and state legislatures to allow abandoned military lands that contain significant wildlife habitat to become national or state wildlife refuges. However, according to a General Accounting Office study, activi-

ties considered harmful to wildlife occur in nearly 60% of the nation's wildlife refuges.

Gene Banks, Botanical Gardens, and Wildlife Farms (Science, Economics, and Stewardship)

Establishing gene banks and botanical gardens, and using farms to raise threatened species can help protect species from extinction, but these options lack funding and storage space.

Gene or *seed banks* preserve genetic information and endangered plant species by storing their seeds in refrigerated, low-humidity environments. More than 100 seed banks around the world collectively hold about 3 million samples.

Scientists urge the establishment of many more such banks, especially in developing countries. But some species cannot be preserved in gene banks. And the banks are expensive to operate and can be destroyed by accidents.

The world's 1,600 *botanical gardens* and *arboreta* contain living plants, representing almost one-third of the world's known plant species. However, they contain only about 3% of the world's rare and threatened plant species.

We can take pressure off some endangered or threatened species by raising individuals on *farms* for commercial sale. One example is the use of farms in

Florida to raise alligators for their meat and hides. Another example is *butterfly farms* in Papua New Guinea, where many butterfly species are threatened by development activities.

Zoos and Aquariums (Science, Economics, and Stewardship)

Zoos and aquariums can help protect endangered animal species, but efforts lack funding and storage space.

Zoos, aquariums, game parks, and animal research centers are being used to preserve some individuals of critically endangered animal species, with the long-term goal of reintroducing the species into protected wild habitats.

Two techniques for preserving endangered terrestrial species are egg pulling and captive breeding. *Egg pulling* involves collecting wild eggs laid by critically endangered bird species and then hatching them in zoos or research centers. In *captive breeding,* some or all of the wild individuals of a critically endangered species are captured for breeding in captivity, with the aim of reintroducing the offspring into the wild.

Lack of space and money limits efforts to maintain populations of endangered species in zoos and research centers. The captive population of each species must number 100–500 individuals to avoid extinction through accident, disease, or loss of genetic diversity through inbreeding. Recent genetic research indicates that 10,000 or more individuals are needed for an endangered species to maintain its capacity for biological evolution.

However, zoos and research centers contain only about 3% of the world's rare and threatened plant species. Thus, the major conservation role of these facilities will be to help educate the public about the ecological importance of the species they display and the need to protect habitat.

Public aquariums that exhibit unusual and attractive fish and some marine animals such as seals and dolphins also help educate the public about the need to protect such species. In the United States, more than 35 million people visit aquariums each year. However, unlike some zoos, public aquariums have not served as effective gene banks for endangered marine species, especially marine mammals that need large volumes of water.

Instead of seeing zoos and aquariums as sanctuaries, some critics see most of them as prisons for once-wild animals. They also contend that zoos and aquariums can foster the notion that we do not need to preserve large numbers of wild species in their natural habitats. Proponents counter that these facilities play an important role in educating the public about wildlife and the need to protect biodiversity.

Some people criticize zoos and aquariums for putting on shows with animals wearing clothes, riding bicycles, or performing tricks. They see this as fostering the idea that the animals are there primarily to entertain us by doing things people do and, in the process, raising money for their keepers.

Regardless of their benefits and drawbacks, zoos, aquariums, and botanical gardens are not biologically or economically feasible solutions for most of the world's current endangered species and the much larger number of species expected to be threatened over the next few decades.

Reconciliation Ecology (Science and Stewardship)

Reconciliation ecology involves finding ways to share the places that we dominate with other species.

In 2003, ecologist Michael L. Rosenzweig wrote a book entitled *Win–Win Ecology: How Earth's Species Can Survive in the Midst of Human Enterprise* (Oxford University Press). He strongly supports Edward O. Wilson's eight-point program to help sustain the earth's biodiversity (p. 127). He also supports the species protection strategies discussed in this chapter.

But he contends that, in the long run, these approaches will fail for two reasons. *First,* current reserves are devoted to saving only about 7% of the world's terrestrial area. To Rosenzweig, the real challenge is to help sustain wild species in the human-dominated portion of nature that makes up 93% of the planet's terrestrial ecological "cake."

Second, setting aside funds and refuges and passing laws to protect endangered and threatened species are essentially desperate attempts to save species that are in deep trouble. This can help a few species, but the real challenge is learning how to keep more species away from the brink of extinction.

Rosenzweig suggests that we develop a new form of conservation biology, called **reconciliation ecology.** It is the science of inventing, establishing, and maintaining new habitats to conserve species diversity in places where people live, work, or play. In other words, we need to learn how to share the spaces we dominate with other species.

Implementing Reconciliation Ecology (Science and Stewardship)

Some people are finding creative ways to practice reconciliation ecology in their neighborhoods and cities.

Practicing reconciliation ecology begins by looking at the habitats we prefer. Given a choice, most people prefer a grassy and fairly open habitat with a few scattered trees, and many people prefer to live near a stream, lake, river, or ocean. We also love flowers. The problem is that most species do not like what we like or cannot survive in the habitats we prefer. No wonder so few of them live with us.

With restoration ecology, some of our monoculture yards can be replaced with diverse yards using plant species adapted to local climates that are selected to attract certain species. This would help keep down insect pests, save water, and require less use of noisy and polluting lawnmowers.

Communities could have contests and awards for people designing the most biodiverse and species-friendly yards and gardens. Signs could describe the types of ecosystems being mimicked and the species being protected as a way to educate and encourage experiments by other people. Some creative person might be able to design more biologically diverse golf courses and cemeteries. People have worked together to help preserve bluebirds within human-dominated habitats (Case Study, below).

San Francisco's Golden Gate Park is a large oasis of gardens and trees in the midst of a major city. It is a

CASE STUDY

Using Reconciliation Ecology to Protect Bluebirds (Science and Stewardship)

Populations of bluebirds in much of the eastern United States are declining, for two reasons.

First, these birds nest in tree holes of a certain size. Dead and dying trees once provided plenty of these holes. But today timber companies often cut down all of the trees, and many homeowners manicure their property by removing dead and dying trees.

Second, two aggressive, abundant, nonnative bird species—starlings and house sparrows—also like to nest in tree holes and take them away from bluebirds. To make matters worse, starlings eat blueberries that bluebirds need to survive during the winter.

People have come up with a creative way to help save the bluebird. They have designed nest boxes with holes large enough to accommodate bluebirds but too small for starlings. They also found that house sparrows like shallow boxes so they made the bluebird boxes deep enough to make them unattractive nesting sites for the sparrows.

In 1979, the North American Bluebird Society was founded to spread the word and encourage people to use the bluebird boxes on their properties and to keep house cats away from nesting bluebirds. Now bluebird numbers are building back up.

Critical Thinking

See if you can come up with a reconciliation project to help protect threatened bird or other species in your neighborhood or school.

Figure 7-14 Individuals matter: ways to help prevent premature extinction of species. QUESTION: *Which two of these actions do you think are the most important? Which of these actions do you plan to take?*

good example of reconciliation ecology because it was designed and planted by humans who transformed it from a system of sand dunes.

The Department of Defense controls about 10 million hectares (25 million acres) of land in the United States. Biologists propose using some of this land for developing and testing reconciliation ecology projects. Some college campuses and schools might also serve as reconciliation ecology laboratories. How about yours?

In this chapter, we have seen that protecting the species that make up part of the earth's biodiversity from premature extinction is a difficult, controversial, and challenging responsibility. Acting to prevent the premature extinction of species is a key to sustainability. It helps preserve the earth's biodiversity by avoiding disruption of species interactions that help control population sizes and the energy flow and matter cycling in ecosystems. Figure 7-14 lists some things you can do to help prevent the premature extinction of species.

We know what to do. Perhaps we will act in time.

EDWARD O. WILSON

CRITICAL THINKING

1. Discuss your gut-level reaction to the following statement: "Eventually, all species become extinct. Thus, it does not really matter that the passenger pigeon is extinct, and that the whooping crane, and the world's remaining tiger species are endangered mostly because of human activities." Be honest about your reaction, and give arguments for your position.

2. Make a log of your own consumption of all products for a single day. Relate your level and types of consumption to the decline of wildlife species and the increased destruction and degradation of wildlife habitats in the United States (or the country where you live), in tropical forests, and in aquatic ecosystems.

3. Do you accept the ethical position that each species has the inherent right to survive without human interference, regardless of whether it serves any useful purpose for humans? Explain. Would you extend this right to the *Anopheles* mosquito, which transmits malaria, and to infectious bacteria? Explain.

4. Which of the following statements best describes your feelings toward wildlife: (a) As long as it stays in its space, wildlife is OK. (b) As long as I do not need its space, wildlife can stay where it is. (c) I have the right to use wildlife habitat to meet my own needs. (d) When you have seen one redwood tree, elephant, or some other form of wildlife, you have seen them all, so lock up a few of each species in a zoo or wildlife park and do not worry about protecting the rest. (e) Wildlife should be protected.

5. List your three favorite species. Why are they your favorites? Reflect on what your choice of favorite species tells you about your attitudes toward most wildlife.

6. Environmental groups in a heavily forested state want to restrict logging in some areas to save the habitat of an endangered squirrel. Timber company officials argue that the well-being of one type of squirrel is not as important

as the well-being of the many families affected if the restriction causes the company to lay off hundreds of workers. If you had the power to decide this issue, what would you do and why? Can you come up with a compromise?

7. Congratulations! You are in charge of preventing the premature extinction of the world's existing species. What would be three major components of your program to accomplish this goal?

8. List two questions that you would like to have answered as a result of reading this chapter.

LEARNING ONLINE

The website for this book contains helpful study aids and many ideas for further reading and research. They include a chapter summary, review questions for the entire chapter, flash cards for key terms and concepts, a multiple-choice practice quiz, interesting Internet sites, references, information about green careers, and a guide for accessing thousands of InfoTrac® College Edition articles. Log into

www.thomsonedu.com/biology/miller

Then choose Chapter 7, and select a learning resource. For access to animations, additional quizzes, chapter outlines and summaries, register and log into

at **www.thomsonedu.com** using the access code card in the front of your book.

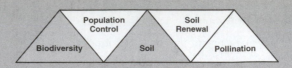

There are two spiritual dangers in not owning a farm. One is the danger of supposing that breakfast comes from the grocery, and the other that heat comes from the furnace.

ALDO LEOPOLD

FOOD SECURITY AND NUTRITION

Food Security (Science, Economics, and Stewardship)

Global food production has stayed ahead of population growth, but one of every six people in developing countries do not have enough land or money to grow or buy the food they need.

Today we produce more than enough food to meet the basic nutritional needs of every person on the earth—if everyone had access to it. Even with this surplus, one of every six people in developing countries is not getting enough to eat.

Most agricultural experts agree that *the root cause of such hunger is poverty*, which prevents poor people from growing or buying enough food. According to the World Bank, one-fifth of the world's people struggle to live on an income of less than $1 (U.S.) per day, and almost half live on less than $2 (U.S.) per day. War and corruption also can deny poor people access to food.

Food security means that every person in a given area has daily access to enough nutritious food to have an active and healthy life. At the national level, poverty can be greatly reduced by government programs that help the poor help themselves, through family planning, education and jobs (especially for women), and small loans to help the poor start a business or buy enough land to grow their own food.

But many developing countries do not produce enough food to feed their people and are too poor to import enough food to provide national food security. This means that developed nations and international lending institutions such as the World Bank must provide technical advice and funding to help such countries become more self-sufficient in meeting their food security needs.

Food security also depends on greatly reducing the harmful environmental effects of agriculture—such as soil erosion and aquifer depletion—at the local, national, and global levels. Unless these and other threats are dealt with, nations cannot become agriculturally self-sufficient and hunger will get worse.

Chronic Hunger and Malnutrition (Science and Poverty)

Some people cannot grow or buy enough food to meet their basic energy needs, and others do not get enough protein and other key nutrients; severe food shortages have also caused mass starvation and many deaths.

To maintain good health and resist disease, we need fairly large amounts of *macronutrients* (such as protein, carbohydrates, and fats), and smaller amounts of *micronutrients* consisting of various vitamins (like A, C, and E) and minerals (such as iron, iodine, and calcium).

People who cannot grow or buy enough food to meet their basic energy needs suffer from **chronic undernutrition,** or **hunger.** Most chronically undernourished children live in developing countries. They are likely to suffer from mental retardation and stunted growth and are susceptible to infectious diseases such as measles and diarrhea, which rarely kill children in developed countries.

Many of the world's poor can afford only to live on a low-protein, high-carbohydrate vegetarian diet consisting only of grains such as wheat, rice, or corn. Many suffer from **malnutrition** resulting from deficiencies of protein, calories, and other key nutrients.

Good news. According to the UN Food and Agriculture Organization (FAO), the average daily food intake in calories per person in the world and in developing countries rose sharply between 1961 and 2000, and is projected to continue rising through 2030. Also, the estimated number of chronically undernourished or malnourished people fell from 918 million in 1970 to 852 million in 2005—about 95% of them in developing countries. Still, this is far from the Millennium Development Goal of reducing the number of hungry and malnourished people to 400 million by 2015.

Despite such progress, one of every six people in developing countries (including about one of every three children younger than age 5) is chronically undernourished or malnourished. In 2005, the FAO estimated that each year nearly 6 million people die prematurely from undernutrition, malnutrition, and increased susceptibility to normally nonfatal infectious diseases (such as measles and diarrhea) because of their weakened condition. This means that each day an average of 16,400 children die prematurely from these causes related to poverty.

Hunger is also a problem in rich developed countries such as the United States. According to the U.S. Department of Agriculture, in 2003 some 35 million Americans (up from 31 million in 1999)—mostly children—went hungry at times, mostly because of poverty.

Not Getting Enough Vitamins and Minerals (Science)

One of every three persons has a deficiency of one or more vitamins and minerals, especially vitamin A, iron, and iodine.

According to the World Health Organization (WHO), about one of every three people suffers from a deficiency of one or more vitamins and minerals, most often in developing countries and involving *vitamin A, iron,* and *iodine.*

According to the WHO, 120–140 million children in developing countries are deficient in vitamin A. Globally, at least 350,000 children under age 6 go blind each year from a lack of vitamin A and up to 80% of them die within a year.

Other nutritional deficiency diseases are caused by the lack of minerals. Too little *iron*—a component of hemoglobin that transports oxygen in the blood—causes *anemia.* According to a 1999 survey by the WHO, one of every three people in the world—mostly women and children in tropical developing countries—suffers from iron deficiency. It causes fatigue, makes infection more likely, and increases a woman's chances of dying in childbirth and an infant's chances of dying from hemorrhage in childbirth.

Elemental *iodine* is essential for proper functioning of the thyroid gland, which produces a hormone that controls the body's rate of metabolism. Iodine is found in seafood and crops grown in iodine-rich soils. Chronic lack of iodine can cause stunted growth, mental retardation, and goiter—a swollen thyroid gland that can lead to deafness. According to the United Nations, about 26 million children suffer brain damage each year from lack of iodine. And 600 million people—mostly in South and Southeast Asia—suffer from goiter.

Solutions: Reducing Childhood Deaths from Hunger and Malnutrition

There are several ways to reduce childhood deaths from nutrition-related causes.

Studies by the United Nations Children's Fund (UNICEF) indicate that one-half to two-thirds of nutrition-related childhood deaths could be prevented at an average annual cost of $5–10 per child by the following measures:

- Immunizing children against childhood diseases such as measles

- Encouraging breast-feeding (except for mothers with AIDS)

- Preventing dehydration from diarrhea by giving infants a mixture of sugar and salt in a glass of water

- Preventing blindness by giving children a vitamin A capsule twice a year at a cost of about 75¢ per child, or fortifying common foods with vitamin A and other micronutrients at a cost of about 10¢ per child annually

- Providing family planning services to help mothers space births at least 2 years apart

- Increasing education for women, with emphasis on nutrition, drinking water sterilization, and child care

Overnutrition: Eating Too Much

Overeating and lack of exercise can lead to reduced life quality, poor health, and premature death.

Overnutrition occurs when food energy intake exceeds energy use and causes excess body fat. Too many calories, too little exercise, or both can cause overnutrition. People who are underfed and underweight and those who are overfed and overweight face similar health problems: *lower life expectancy, greater susceptibility to disease and illness,* and *lower productivity and life quality.*

We live in a world where 1 billion people have health problems because they do not get enough to eat and another 1.2 billion face health problems from eating too much. According to a 2004 study by the International Obesity Task Force, about one of every four people in the world is overweight and one of every twenty is obese.

A 2005 study at Boston University found that about 60% of Americans adults are overweight and 33% are obese (for a total of 93%)—the highest overnutrition rate in any developed country. The $42 billion that Americans spend each year trying to lose weight is 1.8 times more than the $24 billion per year needed to eliminate undernutrition and malnutrition in the world.

FOOD PRODUCTION

The Success of Modern Food Production (Science)

Since 1950, food production from croplands, rangelands, and ocean fisheries has increased dramatically.

Three systems supply most of our food. *Croplands* mostly produce grains and provide about 77% of the world's food, using 11% of the world's land area. *Rangelands* and *pastures* produce meat, mostly from grazing livestock, and supply about 16% of the world's food, using about 29% of the world's land area. *Oceanic fisheries,* and more recently *aquaculture,* supply about 7% of the world's food.

Since 1960, there has been a staggering increase in global food production from all three systems. This occurred because of technological advances such as increased use of tractors and farm machinery and high-tech fishing equipment. Other advances include inorganic chemical fertilizers, irrigation, pesticides, high-yield varieties of wheat, rice, and corn, and methods and devices for raising large numbers of cattle, pigs, chickens, and fish in factory-like conditions.

We face important challenges to increase food production without causing serious environmental harm. Each day, there are about 216,000 more mouths to feed. To feed the world's 8.9 billion people projected to exist in 2050, we must produce and distribute more food than has been produced since agriculture began about 10,000 years ago, and do so in an environmentally sustainable manner.

Can we achieve this goal? Some analysts say we can, mostly by using genetic engineering (Figure 3-5, p. 60). Others have doubts. They are concerned that environmental degradation, pollution, lack of water for irrigation, overgrazing by livestock, overfishing, and loss of vital ecological services may limit future food production.

Plants and Animals That Feed the World (Science)

Wheat, rice, and corn provide more than half of the calories in the food consumed by the world's people.

Of the estimated 30,000 plant species with parts that people can eat, only 14, along with 9 terrestrial animal species, supply an estimated 90% of our global intake of calories. Just three types of grain crops—*wheat, rice,* and *corn*—provide more than half the calories people consume.

Two-thirds of the world's people survive primarily on rice, wheat, and corn, mostly because they cannot afford meat. As incomes rise, most people consume more meat, milk, cheese, and other products of domesticated livestock, which in turn means more grain consumption by those animals.

Fish and shellfish are an important source of food for about 1 billion people, mostly in Asia and in coastal areas of developing countries. But on a global scale, fish and shellfish supply only 7% of the world's food.

Industrial Food Production—High Input Monocultures (Science)

About 80% of the world's food supply is produced by industrialized agriculture.

Industrialized agriculture, or **high-input agriculture,** uses large amounts of fossil fuel energy, water, commercial fertilizers, and pesticides to produce single crops (monocultures) or livestock animals for sale. Practiced on about a fourth of all cropland, mostly in developed countries (Figure 8-1), this form of agriculture has spread since the mid-1960s to some developing countries and now produces about four-fifths of the world's food.

Plantation agriculture is a form of industrialized agriculture used primarily in tropical developing countries. It involves growing *cash crops* (such as bananas, soybeans, sugarcane, cocoa, peanuts, vegetables, and coffee) on large monoculture plantations, mostly for sale in developed countries.

An increasing amount of livestock production in developed countries is industrialized. Large numbers of cattle are brought to densely populated *feedlots,* or *animal factories* where they are fattened up for about 4 months before slaughter. Most pigs and chickens in developed countries spend their lives in densely populated pens and cages and eat mostly grain grown on cropland. Such systems use large amounts of energy and water and produce huge amounts of animal waste that can pollute surface and groundwater and saturate the air with horrible odors.

Croplands, like natural ecosystems, provide the ecological and economic services listed in Figure 8-2. Indeed, agriculture is the world's largest industry, providing a living for one of every five people. It is also the world's most environmentally harmful industry.

Case Study: Industrial Food Production in the United States (Science and Economics)

The United States uses industrialized agriculture to produce about 17% of the world's grain in a very efficient manner.

In the United States, industrialized farming has evolved into *agribusiness* as giant multinational corporations increasingly control the growing, processing, distribution, and sale of food in the United States and in the global marketplace.

In total annual sales, agriculture is bigger than the country's automotive, steel, and housing industries combined. It generates about 18% of the country's gross domestic product and almost one-fifth of all jobs in the private sector, employing more people than any other industry. With only 0.3% of the world's farm labor force, U.S. farms produce about 17% of the world's grain.

Since 1950, U.S. industrialized agriculture has more than doubled the yield of key crops such as wheat, corn, and soybeans without cultivating more land. Such increases in the yield per hectare of key crops have kept large areas of forests, grasslands, and wetlands from being converted to farmland.

Increased yields resulted from a sharp rise in agricultural efficiency. While the U.S. output of crops,

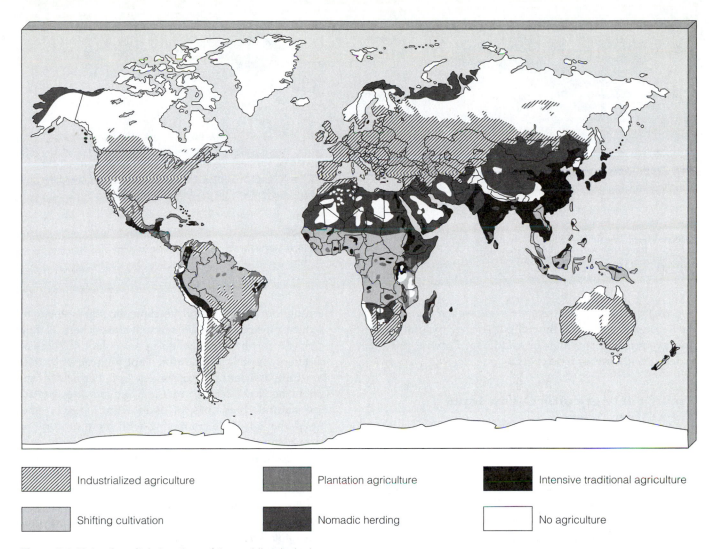

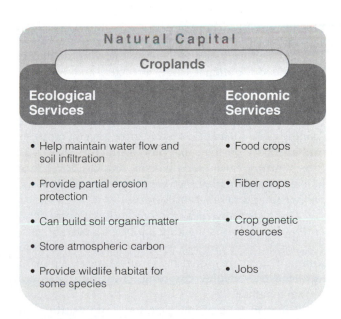	Industrialized agriculture		Plantation agriculture		Intensive traditional agriculture
	Shifting cultivation		Nomadic herding		No agriculture

Figure 8-1 Natural capital: locations of the world's principal types of food production. Excluding Antarctica and Greenland, agricultural systems cover almost one-third of the earth's land surface.

meat, and dairy products has been increasing steadily since 1975, the major inputs of labor and resources—with the exception of pesticides—to produce each unit of that output have fallen steadily since 1950.

This industrialization of agriculture has been made possible by the availability of cheap energy, mostly from oil, used to run farm machinery, process food, and produce commercial fertilizers and pesticides. Putting food on the table consumes about 17% of all commercial energy used in the United States each year (Figure 8-3, p. 152). The input of energy needed to produce a unit of food has fallen considerably so that today, most plant crops in the United States provide more food energy than the energy used to grow them.

However, when we consider the energy used to grow, store, process, package, transport, refrigerate, and cook all plant and animal food, *about 10 units of*

Natural Capital

Croplands

Ecological Services	Economic Services
• Help maintain water flow and soil infiltration	• Food crops
• Provide partial erosion protection	• Fiber crops
• Can build soil organic matter	• Crop genetic resources
• Store atmospheric carbon	
• Provide wildlife habitat for some species	• Jobs

Figure 8-2 Natural capital: ecological and economic services provided by croplands. QUESTION: *Which two ecological and economic services do you think are the most important?*

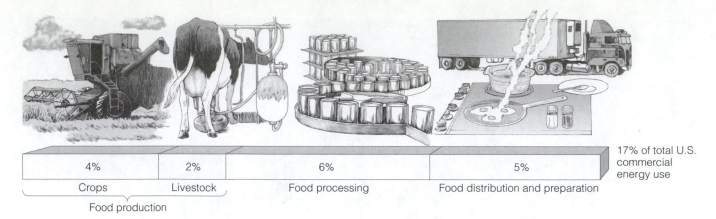

4%	2%	6%	5%	17% of total U.S. commercial energy use
Crops	Livestock	Food processing	Food distribution and preparation	

Food production

Figure 8-3 In the United States, industrialized agriculture uses about 17% of all commercial energy and food travels an average 2,400 kilometers (1,300 miles) from farm to plate. (Data from David Pimentel and Worldwatch Institute)

nonrenewable fossil fuel energy are needed to put 1 unit of food energy on the table. By comparison, every unit of energy from human labor in traditional farming provides 1 to 10 units of food energy.

Traditional Agriculture—Low Input Polyculture (Science)

Many traditional farmers in developing countries use low-input agriculture to produce a variety of crops on each plot of land.

Traditional agriculture consists of two main types, which together are practiced by 2.7 billion people (42% of the world's people) in developing countries, and provide about one-fifth of the world's food supply on about three-fourths of its cultivated land. **Traditional subsistence agriculture** uses mostly human labor and draft animals to produce only enough crops or livestock for a farm family's survival. In **traditional intensive agriculture,** farmers increase their inputs of human and draft-animal labor, fertilizer, and water to produce enough food to feed their families plus some to sell for income.

While some traditional farmers focus on growing a single crop, such as rice, many grow several crops simultaneously on the same plot, a practice known as **interplanting.** Such crop diversity reduces the chance of losing most or all of the year's food supply to pests, bad weather, and other misfortunes.

There are four types of interplanting. **Polyvarietal cultivation** involves planting a plot with several varieties of the same crop. In **intercropping,** two or more different crops are grown at the same time on a plot. In **agroforestry,** or **alley cropping,** crops and trees are grown together.

A fourth type of interplanting is **polyculture,** in which many different plants are planted together. Low-input polyculture has a number of advantages. The crops mature at different times, providing food throughout the year and keeping the soil covered to reduce erosion from wind and water. There is less need for fertilizer and water because root systems at different depths in the soil capture nutrients and moisture efficiently. Insecticides and herbicides are rarely needed because multiple habitats are created for natural predators of crop-eating insects and weeds have trouble competing with the multitude of crop plants.

Recent ecological research found that on average, low-input polyculture produces higher yields per hectare of land than high-input monoculture. For example, a 2001 study by ecologists Peter Reich and David Tilman found that carefully controlled polyculture plots with 16 different species of plants consistently outproduced plots with 9, 4, or only 1 type of plant species.

SOIL EROSION AND DEGRADATION

Erosion and Topsoil (Science)

Soil erosion caused by water, wind, and people lowers soil fertility and can overload nearby bodies of water with eroded sediment.

Some 15–20 centimeters (6–8 inches) of *topsoil* (Figure 2-22, p. 41) is all that stands between much of the world and starvation. Topsoil is a renewable resource because natural processes regenerate it. But this is a slow process—typically taking from many decades to several hundred years to replenish 2.5 centimeters (1 inch) of topsoil, depending mostly on climate and other conditions.

Soil erosion is the movement of soil components, especially surface litter and topsoil, from one place to another by the action of wind or water. When topsoil erodes faster than it forms on a piece of land, it becomes a nonrenewable resource.

Some soil erosion is natural and some is caused by human activities. In undisturbed vegetated ecosystems, the roots of plants help anchor the soil, and usually soil is not lost faster than it forms. Flowing water causes most soil erosion. But wind can remove topsoil rapidly in areas with a dry climate and relatively flat land. This loss of natural capital increases when soil-holding grasses are destroyed through activities such as farming, logging, construction, overgrazing, and off-road vehicle use.

Soil erosion has two major harmful effects. One is *loss of soil fertility* through depletion of plant nutrients in topsoil. The other is *water pollution* in nearby surface waters where eroded soil ends up as sediment, which can kill fish and shellfish and clog irrigation ditches, boat channels, reservoirs, and lakes.

Global Outlook: Soil Erosion

Soil is eroding faster than it is forming on more than a third of the world's cropland.

A joint survey by the United Nations Environment Programme and the World Resources Institute estimated that topsoil is eroding faster than it forms on about 38% of the world's cropland (Figure 8-4). According to a 2000 study by the Consultative Group on International Agricultural Research, soil erosion and degradation has reduced food production on 13% of the world's cropland since 1950 because of human activities. See the Guest Essay on soil erosion by David Pimentel on the website for this chapter.

Some analysts contend that erosion estimates are overstated because they underestimate the abilities of some local farmers to restore degraded land. The FAO also points out that much of the eroded topsoil does not go far and is deposited further down a slope, valley, or plain. In some places, the loss in crop yields in one area could be offset by increased yields elsewhere.

Case Study: Soil Erosion in the United States (Science and Economics)

Soil in the United States is eroding faster than it forms on most cropland, but since 1985, erosion has been cut by about two-thirds.

According to the Natural Resources Conservation Service, soil on cultivated land in the United States is eroding about 16 times faster than it can form. The Great Plains, for example, have lost one-third or more of their topsoil in the 150 years since they were first plowed.

On the other hand, of the world's major food-producing nations, only the United States is sharply reducing some of its soil losses through a combination of planting crops without disturbing the soil and government-sponsored soil conservation programs.

Since 1985, these efforts have cut soil losses on U.S. cropland by 40%. However, effective soil conservation is practiced today on only half of all U.S. agricultural land and on only half of the country's most erodible cropland.

Salinization and Waterlogging of Soils (Science)

Repeated irrigation can reduce crop productivity by salt buildup in the soil and waterlogging of crop plants.

The 20% of the world's cropland that is irrigated produces almost 40% of the world's food. But irrigation

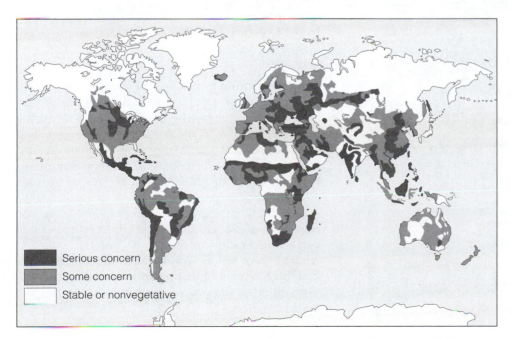

Serious concern
Some concern
Stable or nonvegetative

Figure 8-4 Natural capital degradation: global soil erosion. (Data from UN Environment Programme and the World Resources Institute)

has a downside. Most irrigation water is a dilute solution of various salts, picked up as the water flows over or through soil and rocks. Irrigation water not absorbed into the soil evaporates, leaving behind a thin crust of dissolved salts in the topsoil.

Repeated annual applications of irrigation water in dry climates lead to the gradual accumulation of salts in the upper soil layers—a process called **salinization** (Figure 8-5). It stunts crop growth, lowers crop yields, and eventually kills plants and ruins the land.

The United Nations estimates that severe salinization has reduced yields on about one-fifth of the world's irrigated cropland, and another third has been moderately salinized. The most severe salinization occurs in Asia, especially in China, India, and Pakistan. Salinization affects almost one-fourth of irrigated cropland in the United States, especially in some heavily irrigated western states.

We know how to prevent and deal with soil salinization, as summarized in Figure 8-6. But some of these remedies are expensive and flushing soil with water to wash it results in more saline water for downstream farmers and towns.

Another problem with irrigation is **waterlogging** (Figure 8-5). Farmers often apply large amounts of irri-

Figure 8-6 Solutions: methods for preventing and cleaning up soil salinization. QUESTION: *Which two of these solutions do you believe are the most important?*

gation water to leach salts deeper into the soil. Without adequate drainage, water may accumulate underground and gradually raise the water table. Saline water then envelops the deep roots of plants, lowering their productivity and killing them after prolonged exposure. At least one-tenth of the world's irrigated land suffers from waterlogging, and the problem is getting worse.

SUSTAINABLE AGRICULTURE THROUGH SOIL CONSERVATION

Conservation Tillage

Modern farm machinery can plant crops without disturbing the soil.

Soil conservation involves using a variety of ways to reduce soil erosion and restore soil fertility, mostly by keeping the soil covered with vegetation. Minimizing plowing and tilling is the key to reducing erosion and restoring healthy soil. Many farmers in the United States and South America use **conservation-tillage farming,** which disturbs the soil as little as possible while planting crops.

In 2004, farmers used conservation tillage on about 45% of U.S. cropland. The USDA estimates that using conservation tillage on 80% of U.S. cropland would reduce soil erosion by at least half. Conservation tillage also has great potential to reduce soil erosion and raise crop yields in dry regions in Africa and the Middle East.

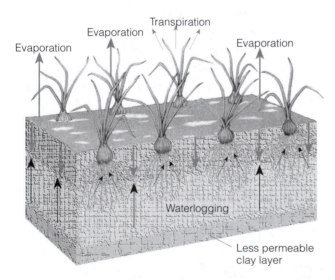

Salinization

1. Irrigation water contains small amounts of dissolved salts.

2. Evaporation and transpiration leave salts behind.

3. Salt builds up in soil.

Waterlogging

1. Precipitation and irrigation water percolate downward.

2. Water table rises.

Figure 8-5 Natural capital degradation: *salinization* and *waterlogging* of soil on irrigated land without adequate drainage can decrease crop yields.

Conservation tillage, however, is not a cure-all. It requires costly machinery, works better in some areas than others, and is more useful with some crops than others.

Other Methods for Reducing Soil Erosion (Science)

Terracing, contour planting, strip cropping, alley cropping, and windbreaks can reduce soil erosion.

Terracing is a way to grow crops on steep slopes without depleting topsoil. It is done by converting the land into a series of broad, nearly level terraces that run across the land's contours (Figure 8-7a). This retains water for crops at each level and reduces soil erosion by controlling runoff.

Contour farming involves plowing and planting crops in rows across the slope of the land rather than up and down (Figure 8-7b). Each row acts as a small dam to help hold soil and to slow water runoff.

Strip cropping (Figure 8-7b) involves planting alternating strips of a row crop (such as corn or cotton) and another crop that completely covers the soil (such as a grass or a grass and legume mixture). The cover crop traps soil that erodes from the row crop, and catches and reduces water runoff.

Another way to reduce erosion is to leave crop residues on the land after the crops are harvested. Farmers can also plant *cover crops* such as alfalfa, clover, or rye immediately after harvest to help protect and hold the soil.

Another method for slowing erosion is *alley cropping* or *agroforestry*, in which one or more crops are planted together in strips or alleys between trees and shrubs that provide shade (Figure 8-7c). This reduces water loss by evaporation and helps retain and slowly release soil moisture—an insurance policy during prolonged drought. The trees also can provide fruit, fuelwood, and trimmings that can be used as mulch (green manure) for the crops and as fodder for livestock.

(a) Terracing

(b) Contour planting and strip cropping

(c) Alley cropping

(d) Windbreaks

Figure 8-7 Solutions: in addition to conservation tillage, soil conservation methods include (a) terracing, (b) contour planting and strip cropping, (c) alley cropping or agroforestry, and (d) windbreaks.

Some farmers establish **windbreaks,** or **shelter-belts,** of trees around crop fields (Figure 8-7d) to reduce wind erosion, help retain soil moisture, supply wood for fuel, increase crop productivity by 5–10%, and provide habitats for birds, pest-eating and pollinating insects, and other animals.

Organic and Inorganic Fertilizers (Science)

Organic and inorganic fertilizers used together can help restore soil nutrients and fertility, but harmful environmental effects of inorganic fertilizers must be controlled.

The best way to maintain soil fertility is through soil conservation. The next best option is to restore some of the plant nutrients that have been washed, blown, or leached out of soil, or removed by repeated crop harvesting.

Fertilizers are used to partially restore lost plant nutrients. Farmers can use **organic fertilizer** from plant and animal materials or **commercial inorganic fertilizer** produced from various minerals.

There are several types of *organic fertilizers.* One is **animal manure:** the dung and urine of cattle, horses, poultry, and other farm animals. It improves soil structure, adds organic nitrogen, and stimulates beneficial soil bacteria and fungi.

A second type of organic fertilizer called **green manure** consists of freshly cut or growing green vegetation plowed into the soil to increase the organic matter and humus available to the next crop. A third type is **compost,** produced when microorganisms in soil break down organic matter, such as leaves, food wastes, paper, and wood, in the presence of oxygen.

Crops such as corn, tobacco, and cotton can deplete nutrients (especially nitrogen) in the topsoil if planted on the same land several years in a row. **Crop rotation** provides one way to reduce these losses. Farmers plant areas or strips with nutrient-depleting crops one year. The next year, they plant the same areas with legumes whose root nodules add nitrogen to the soil. In addition to helping restore soil nutrients, this method reduces erosion by keeping the soil covered with vegetation.

Many farmers (especially in developed countries) rely on *commercial inorganic fertilizers.* The active ingredients typically are inorganic compounds that contain *nitrogen, phosphorus,* and *potassium.* Other plant nutrients also may be present in low or trace amounts. The use of inorganic fertilizers has grown by about 700% since 1960 and now accounts for about one-fourth of the world's crop yield.

Commercial inorganic fertilizers can replace depleted inorganic nutrients, but they do not replace organic matter. Thus, for healthy soil, both inorganic and organic fertilizers should be used.

THE GREEN REVOLUTION AND ITS ENVIRONMENTAL IMPACT

Using Green Revolutions to Increase Food Production (Science)

Since 1950, most of the increase in global food production has come from using high-input agriculture to produce more crops on each unit of land.

Farmers can produce more food by farming more land or getting higher yields per unit of area from existing cropland. Since 1950, about 88% of the increase in global food production has come from increased yields per unit of area of cropland in a process called the **green revolution.**

The green revolution involves three steps. *First,* develop and plant monocultures of selectively bred or genetically engineered high-yield varieties of key crops such as rice, wheat, and corn. *Second,* produce high yields by using large inputs of fertilizer, pesticides, and water. *Third,* increase the number of crops grown per year on a plot of land through *multiple cropping.*

This high-input approach dramatically increased crop yields in most developed countries between 1950 and 1970 in what is called the *first green revolution* (Figure 8-8, dark shading).

A *second green revolution* has been taking place since 1967. Fast-growing dwarf varieties of rice and wheat, specially bred for tropical and subtropical climates, have been introduced into several developing countries (Figure 8-8, lighter shading). Producing more food on less land has the benefit of protecting biodiversity by saving large areas of forests, grasslands, wetlands, and easily eroded mountain terrain from being used to grow food.

Mostly because of the two green revolutions, between 1950 and 1985, world grain production tripled (Figure 8-9, left) and per capita grain production rose by 37% (Figure 8-9, right).

On the other hand, Indian economist Vandana Shiva argues that overall gains in crop yields from green revolution varieties may be much lower than claimed. The yields are based on comparisons between the output per hectare of old and new *monoculture* varieties rather than between the even higher yields per hectare for *polyculture* cropping systems and the new monoculture varieties that often replace polyculture crops.

Problems with Expanding the Green Revolution (Science and Economics)

Lack of water, high costs for small farmers, and physical limits to increasing crop yields hinder expansion of the green revolution.

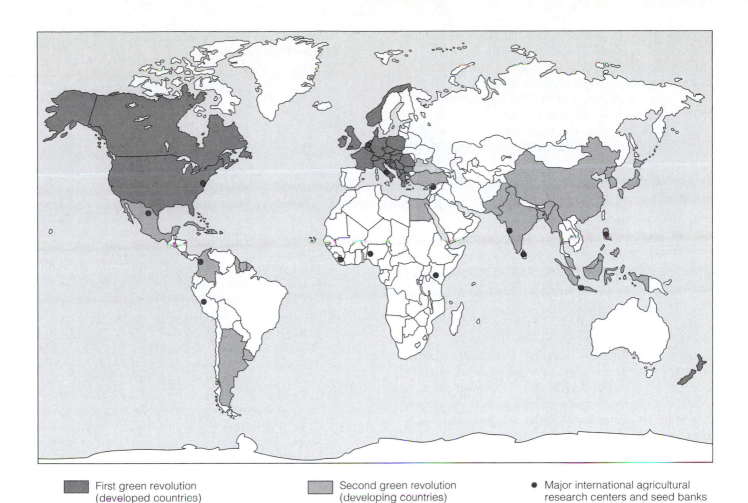

■ First green revolution (developed countries)	■ Second green revolution (developing countries)	● Major international agricultural research centers and seed banks

Figure 8-8 Countries whose crop yields per unit of land area increased during the two green revolutions. The first (dark shading) took place in developed countries between 1950 and 1970; the second (lighter shading) has occurred since 1967 in developing countries with enough rainfall or irrigation capacity. Several agricultural research centers and gene or seed banks (black dots) play a key role in developing high-yield crop varieties.

Total World Grain Production

World Grain Production per Capita

Figure 8-9 Global outlook: total worldwide grain production of wheat, corn, and rice (left), and per capita grain production (right), 1961–2005. In order, the world's three largest grain-producing countries are China, the United States, and India. (Data from U.S. Department of Agriculture, Worldwatch Institute, UN Food and Agriculture Organization, and Earth Policy Institute)

Analysts point to several factors that have limited the success of the green revolutions to date and may continue to do so. Without huge amounts of fertilizer and water, most green revolution crop varieties produce yields that are no higher (and are sometimes lower) than those from traditional strains. And with their high inputs of water, fertilizer, and pesticides, they cost too much for most subsistence farmers in developing countries.

Scientists also point out that continuing to increase fertilizer, water, and pesticide inputs eventually produces no additional increase in crop yields. For example, grain yields rose about 2.1% a year between 1950 and 1990, but then growth dropped to 1.1% per year by 2000 and to 0.5% by 2004—well below the 1.2% growth rate for the world's population. No one knows whether this downward trend will continue.

Can we get around this limitation by irrigating more cropland? About 40% of the world's food production comes from the 20% of its cropland that is irrigated. Between 1950 and 2004, the world's area of irrigated cropland tripled, with most of the growth occurring from 1950 to 1978.

However, the amount of irrigated land per person has been declining since 1978 and is projected to fall much more between 2004 and 2050. One reason is that since 1978, the world's population has grown faster than has irrigated agriculture. Other factors are depletion of underground water supplies (aquifers), inefficient use of irrigation water, and salt buildup in soil on irrigated cropland. In addition, the majority of the world's farmers do not have enough money to irrigate their crops.

Is cultivating more land the answer? Theoretically, clearing tropical forests and irrigating arid land could more than double the world's cropland. But much of this is *marginal land* with poor soil fertility, steep slopes, or both. Cultivation of such land is unlikely to be sustainable.

Much of the world's potentially cultivable land lies in dry areas, especially in Australia and Africa. Large-scale irrigation in these areas would require expensive dam projects, use large inputs of fossil fuel to pump water long distances, and deplete groundwater supplies by removing water faster than it is replenished. It would also require expensive efforts to prevent erosion, groundwater contamination, salinization, and waterlogging, all of which reduce crop productivity.

Furthermore, these potential increases in cropland would not offset the projected loss of almost one-third of today's cultivated cropland caused by erosion, overgrazing, waterlogging, salinization, and urbanization. Such cropland expansion would also reduce wildlife habitats and thus the world's biodiversity. Bottom line: *Many analysts believe that significant expansion of cropland is unlikely over the next few decades for economic and ecological reasons.*

Harmful Environmental Effects of Food Production (Science)

Modern agriculture has a greater harmful environmental impact than any human activity, and these effects may limit future food production.

Modern agriculture has significant harmful effects on air, soil, water, and biodiversity, as Figure 8-10 shows. According to many analysts, agriculture has a greater harmful environmental impact than any human activity.

There is concern that crop yields in some areas may decline because of environmental factors as soil erodes and loses fertility, irrigated soil becomes salty and waterlogged, underground and surface water supplies become depleted and polluted with pesticides and nitrates from fertilizers, and populations of rapidly breeding pests develop genetic immunity to widely used pesticides. We do not know how close we are to such environmental limits.

Another concern is the increasing loss of *agrobiodiversity*—the world's genetic variety of animals, plants, and microorganisms. Scientists estimate that since 1900, we have lost three-fourths of the genetic diversity of agricultural crops. India once planted 30,000 varieties of rice. Now more than 75% of the rice production comes from only 10 varieties. During the last century, 1,000 breeds, or 15%, of the world's cattle and poultry breeds have disappeared. In other words, we are rapidly shrinking the world's genetic "library" needed to increase food yields just when we need it more than ever.

Thus, modern agriculture violates the four principles of sustainability (Figure 1-11, p. 17). It depends heavily on nonrenewable fossil fuels, includes too little recycling of crop and animal wastes, accelerates soil erosion, does not preserve agrobiodiversity, and can disrupt natural species interactions that help control pest population sizes.

THE GENE REVOLUTION

Traditional Crossbreeding and Genetic Engineering (Science)

We can increase crop yields by using crossbreeding to mix the genes of similar types of organisms and by genetic engineering to mix the genes of different organisms.

For centuries, farmers and scientists have used *crossbreeding* through *artificial selection* to develop genetically improved varieties of crop strains. Such selective breeding has had amazing results. Ancient ears of corn were about the size of your little finger and wild tomatoes were once the size of a grape.

But traditional crossbreeding is a slow process, typically taking 15 years or more to produce a commercially valuable new variety, and it can combine

Food Production

Biodiversity Loss	Soil	Water	Air Pollution	Human Health
Loss and degradation of habitat from clearing grasslands and forests and draining wetlands	Erosion	Water waste	Greenhouse gas emissions from fossil fuel use	Nitrates in drinking water
Fish kills from pesticide runoff	Loss of fertility	Aquifer depletion	Other air pollutants from fossil fuel use	Pesticide residues in drinking water, food, and air
Killing of wild predators to protect livestock	Salinization	Increased runoff and flooding from land cleared to grow crops	Greenhouse gas emissions of nitrous oxide from use of inorganic fertilizers	Contamination of drinking and swimming water with disease organisms from livestock wastes
Loss of genetic diversity from replacing thousands of wild crop strains with a few monoculture strains	Waterlogging	Sediment pollution from erosion	Belching of the greenhouse gas methane into the atmosphere by cattle	Bacterial contamination of meat
	Desertification	Fish kills from pesticide runoff	Pollution from pesticide sprays	
		Surface and groundwater pollution from pesticides and fertilizers		
		Overfertilization of lakes and slow-moving rivers from runoff of nitrates and phosphates from fertilizers, livestock wastes, and food processing wastes		

Figure 8-10 Natural capital degradation: major harmful environmental effects of food production. According to a 2002 study by the United Nations, nearly 30% of the world's cropland has been degraded to some degree by soil erosion, salt buildup, and chemical pollution, and 17% has been seriously degraded. QUESTION: *Which item in each of these categories do you think is the most harmful?*

traits only from species that are genetically similar. Resulting varieties remain useful for only 5–10 years before pests and diseases reduce their effectiveness.

Today, scientists are creating a *third green revolution*—actually a *gene revolution*—by using *genetic engineering* to develop genetically improved strains of crops and livestock animals. It involves splicing a gene from one species and transplanting it into the DNA of another species (Figure 3-5, p. 60). Compared to traditional crossbreeding, gene splicing takes about half as long to develop a new crop, cuts costs, and allows the insertion of genes from almost any other organism into crop cells. For example, genetic engineers have developed potatoes that resist disease because they contain chicken genes.

Ready or not, the world is entering the *age of genetic engineering*. More than two-thirds of the food products on U.S. supermarket shelves now contain some form of genetically engineered plants, and that proportion is increasing rapidly. By 2020, more cropland may be devoted to genetically engineered crops than to conventional crossbred crops.

Despite the promise, considerable controversy has arisen over the use of *genetically modified food* (GMF) and other forms of genetic engineering. Its producers and investors see this kind of food as a potentially sustainable way to solve world hunger problems and improve human health. Some critics consider it potentially dangerous "Frankenfood." Figure 8-11 (p. 160) summarizes the projected advantages and disadvantages of this new technology.

Critics recognize the potential benefits of genetically modified crops. But they warn that we know too little about the potential harm to human health and ecosystems from the widespread use of such crops. Also, genetically modified organisms cannot be recalled if they cause some unintended harmful genetic and ecological effects—as some scientists expect.

Most scientists and economists who have evaluated the genetic engineering of crops believe that its enormous potential benefits outweigh the much smaller risks. But critics call for more controlled field experiments, more research and long-term safety testing to better understand the risks, and stricter

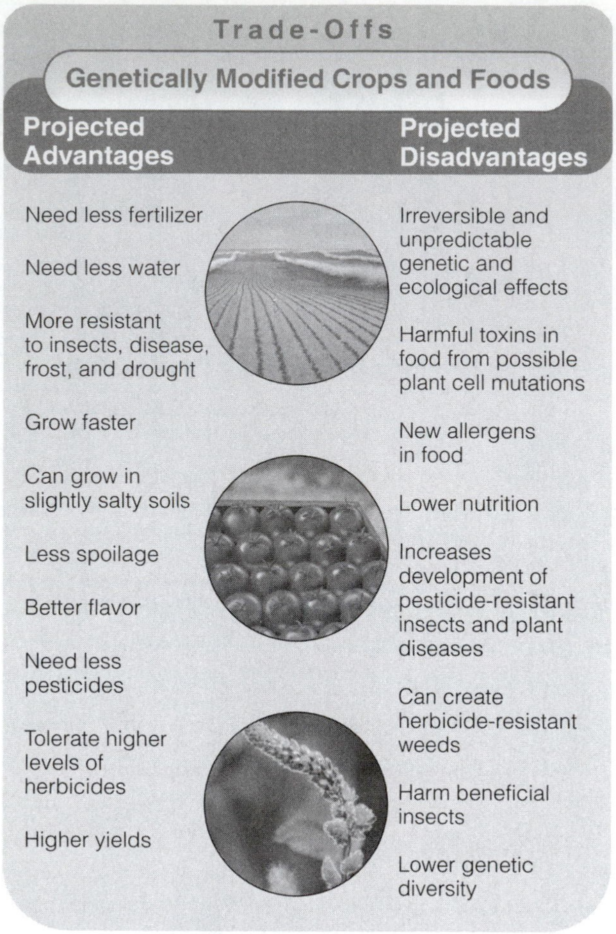

Trade-Offs

Genetically Modified Crops and Foods

Projected Advantages	Projected Disadvantages
Need less fertilizer	Irreversible and unpredictable genetic and ecological effects
Need less water	
More resistant to insects, disease, frost, and drought	Harmful toxins in food from possible plant cell mutations
Grow faster	New allergens in food
Can grow in slightly salty soils	Lower nutrition
Less spoilage	Increases development of pesticide-resistant insects and plant diseases
Better flavor	
Need less pesticides	
Tolerate higher levels of herbicides	Can create herbicide-resistant weeds
Higher yields	Harm beneficial insects
	Lower genetic diversity

Figure 8-11 Trade-offs: projected advantages and disadvantages of genetically modified crops and foods. QUESTION: *What are the single advantage and the single disadvantage that you think are the most important?*

regulation of this rapidly growing technology. A 2004 study by the Ecological Society of America recommended more caution in releasing genetically engineered organisms into the environment.

✗ HOW WOULD YOU VOTE? Do the advantages of genetically engineered foods outweigh their disadvantages? Cast your vote online at www.thomsonedu.com/biology/miller.

Another issue related to GMF arises from court decisions granting seed companies patents (and thus exclusive ownership) of genetically modified crop varieties. The companies have said that saving seeds of their patented varieties is illegal and have successfully sued some farmers for using their seeds. Critics argue that patenting of genetically engineered crop and animal varieties represents a private takeover of a common heritage—the work of all the farmers over the last 10,000 years. And many farmers in developing countries are too poor to buy patented seeds each year and have refused to respect the patent claims on such seeds. Seed companies say they have spent large amounts of

money developing these new varieties and that patents allow them to recoup their expenses and make profits.

Many analysts and consumer advocates believe governments should require mandatory labeling of GMFs. Consumers would then have more information to make informed choices about the foods they buy. Such labeling is required in Japan, Europe, South Korea, Canada, Australia, and New Zealand. Polls show that at least 75% of Americans support mandatory labeling of GMFs.

In the United States, industry representatives and the USDA oppose labeling, claiming that GMFs are not substantially different from foods developed by conventional crossbreeding methods and that labeling would be expensive. Also, they fear—probably correctly—that labeling such foods would hurt sales by arousing suspicion.

✗ HOW WOULD YOU VOTE? Should all genetically engineered foods be so labeled? Cast your vote online at www.thomsonedu.com/biology/miller.

PRODUCING MORE MEAT

Increasing Livestock Production (Science)

Meat is produced by livestock grazing on grass in rangelands and pastures, but factory production in densely packed feedlots produces almost half the world's meat.

Meat and meat products such as milk and cheese are good sources of high-quality protein. Between 1950 and 2004, world meat production increased more than fivefold, and per capita meat production more than doubled. It is likely to more than double again by 2050 as affluence rises and people begin consuming more meat in middle-income developing countries such as China and India.

Two systems are used to raise livestock. In the first approach, livestock graze on grass in unfenced rangelands and enclosed pastures. The second system is industrialized, in which pigs, chickens, and cattle are raised in densely packed *feedlots* and fed grain or meal produced from fish.

Feedlots account for about 43% of the world's beef production, half the pork production, and almost three-fourths of the world's poultry production. In the United States, less than 1% of the cattle slaughtered each year are raised by eating grass from pastures or rangelands. Experts expect such industrialized meat production to expand rapidly. This will increase pressure on the world's grain supply as well as on the world's fish supply. About one-third of the world's fish catch is used to feed livestock.

Cattle and dairy cows also belch methane, accounting for 16% of the world's annual emissions of methane, the second most powerful greenhouse gas after carbon dioxide. Thus, more cattle and dairy cows means more methane and more global warming.

Industrialized meat production subjects workers in meat packing plants where cattle, hogs, chickens, and turkeys are killed, cut up, and packaged to filthy, cold, slippery, and dangerous environments. Meatpacking is one of the most dangerous jobs in the United States; one in three meatpacking workers suffer an injury on the job. Meat workers also suffer from mental health problems because of the stress and danger of their jobs.

Livestock in the United States produce 20 times more waste (manure) than is produced by the country's human population. Manure washing off the land or leaking from lagoons used to store animal wastes can kill fish by depleting dissolved oxygen.

To save grain, some growers feed livestock crop residues mixed with bits and pieces of other animals (meat, bones, feathers) left over after slaughter. This can lead to the formation of certain proteins that destroy the normal proteins in the brains of cattle—an affliction known as *mad cow disease*. It causes animals to stumble, become aggressive, and eventually die. The disease can spread to people who eat infected meat. Since 1986, more than 150 people have died from a human form of this disease. Producing meat in animal feedlots has advantages and disadvantages (Figure 8-12).

Raising cattle on rangelands and pastures is less environmentally destructive that raising them in feedlots as long as the grasslands are not overgrazed.

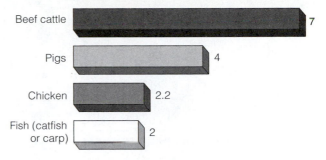

Kilograms of grain needed per kilogram of body weight

Figure 8-13 Efficiency of converting grain into animal protein. Data in kilograms of grain per kilogram of body weight added. (Data from U.S. Department of Agriculture)

Grass-fed cattle require little or no grain, thus eliminating the harmful environmental effects of using fertilizers and pesticides to grow soybeans and corn and the energy costs of shipping these grains to feedlots.

A more sustainable form of meat production and consumption would involve shifting from less grain-efficient forms of animal protein, such as beef and pork, to more grain-efficient ones, such as poultry and herbivorous farmed fish (Figure 8-13).

CATCHING AND RAISING MORE FISH AND SHELLFISH

Harvesting Fish and Shellfish (Science)

After spectacular increases since 1994, the world's total and per capita marine and freshwater fish and shellfish catches have leveled off.

The world's third major food-producing system consists of **fisheries:** concentrations of particular aquatic species suitable for commercial harvesting in a given ocean area or inland body of water. About two-thirds of the world's fish and shellfish harvest comes from the oceans and inland freshwater lakes, rivers, reservoirs, and ponds. The other third comes from using *aquaculture* to raise marine and freshwater fish in ponds and underwater cages.

Figure 8-14 (p. 162) shows the effects of the global efforts to boost the seafood harvest. After increasing threefold between 1950 and 1970, the annual commercial fish catch (marine plus freshwater harvest, but excluding aquaculture) essentially leveled off between 1970 and 1985, and has risen only slightly since then (Figure 8-14, left). After doubling between 1950 and 1970, the fish catch per person leveled off until 1983 and has fallen since then (Figure 8-14, right). Some fisheries scientists project a decline in the global marine fish catch in the future because of overfishing, coastal water pollution, and loss of coastal wetlands, which serve as nurseries for many species of commercially valuable fish and shellfish.

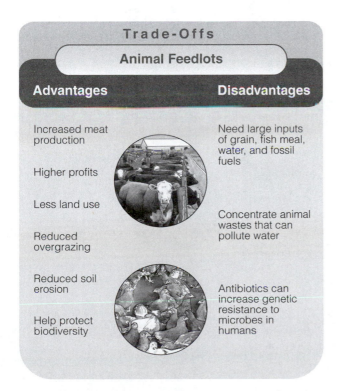

Figure 8-12 Trade-offs: advantages and disadvantages of animal feedlots. QUESTION: *What are the single advantage and the single disadvantage that you think are the most important?*

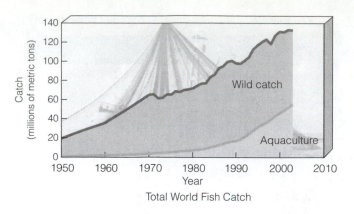

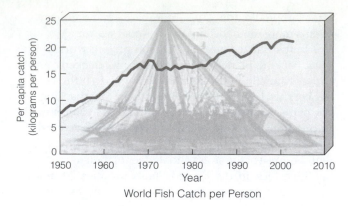

Figure 8-14 Natural capital degradation: world fish catch (left) and world fish catch per person (right), 1950–2003. Estimates for both since 1990 may be about 10% lower than shown here, because it was discovered that China had been inflating its fish catches since 1990. (Data from UN Food and Agriculture Organization, U.S. Census Bureau, and Worldwatch Institute)

Studies indicate that about 70% of the world's 200 commercially valuable marine fish species (33% in U.S. waters) are either overfished or fished to their estimated sustainable yield. According to the Ocean Conservancy, "we are spending the principal of our marine fish resources rather than living off the interest they provide."

Good news. In 1995, fisheries biologists studied population data for 128 depleted fish stocks and concluded that 125 of them could recover with careful management. This involves establishing fishing quotas, restricting use of certain types of fishing gear and methods, limiting the number of fishing boats, closing fisheries during spawning periods, and setting aside networks of no-take reserves.

Aquaculture—Aquatic Feedlots (Science)

Raising large numbers of fish and shellfish in ponds and cages is the world's fastest growing type of food production.

Aquaculture involves raising fish and shellfish for food instead of going out and hunting and gathering them. It is the world's fastest growing type of food production and accounts for about one-third of the global production of fish and shellfish.

There are two basic types of aquaculture: fish farming and fish ranching. **Fish farming,** involves cultivating fish in a controlled environment (often a coastal or inland pond, lake, reservoir, or rice paddy) and harvesting them when they reach the desired size.

Fish ranching involves raising *anadromous species*, such as salmon that live part of their lives in freshwater and part in saltwater. They are held in captivity for the first few years of their lives, usually in fenced-in areas or floating cages in coastal lagoons, estuaries, or the deep ocean. Then the fish are released, and adults are harvested when they return to spawn.

Figure 8-15 lists the major advantages and disadvantages of aquaculture. Some analysts project that

freshwater and saltwater aquaculture production could provide at least half the world's seafood by 2025. But other analysts warn that the harmful environmental effects of aquaculture (Figure 8-15, right) could limit future production.

Figure 8-16 lists some ways to make aquaculture more sustainable and to reduce its harmful environ-

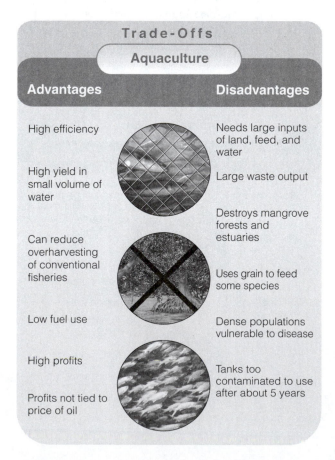

Figure 8-15 Trade-offs: advantages and disadvantages of *aquaculture.* QUESTION: *What are the single advantage and the single disadvantage that you think are the most important?*

Solutions

More Sustainable Aquaculture

- Use less fishmeal feed to reduce depletion of other fish
- Improve management of aquaculture wastes
- Reduce escape of aquaculture species into the wild
- Restrict location of fish farms to reduce loss of mangrove forests and estuaries
- Farm some aquaculture species in deeply submerged cages to protect them from wave action and predators and allow dilution of wastes into the ocean
- Certify sustainable forms of aquaculture

Figure 8-16 Solutions: ways to make aquaculture more sustainable and reduce its harmful environmental effects. QUESTION: *Which two of these solutions do you think are the most important?*

mental effects. However, even under the most optimistic projections, increasing both the wild catch and aquaculture will not increase world food supplies significantly. The reason is that fish and shellfish supply only about 1% of the calories and 6% of the protein in the human diet. A 100% increase would bring those numbers to only 2% and 12%.

SOLUTIONS: TOWARD GLOBAL FOOD SECURITY

Growing More Food in Urban Areas and Cutting Food Waste

People in urban areas could save money by growing more of their food. As much as 70% of the world's food is wasted.

According to the United Nations Development Program, urban gardens provide about 15% of the world's food supply. Farmers in or near 18 of China's largest cities provide urban dwellers with 85% of their vegetables and more than half of their meat and poultry. Food experts believe that people in urban areas could live more sustainably and save money by growing more of their food in empty lots, in backyards, on rooftops and balconies, and by raising fish in tanks.

According to the FAO, as much as 70% of the food produced worldwide is lost through spoilage, inefficient processing and preparation, and plate waste. Affluent countries such as the United States, Canada, and Italy waste 50–60% of their food. Nationwide, U.S. households throw away food worth as much $43 billion a year—almost twice the $24 billion a year needed to eliminate global hunger and malnutrition.

Government Policies and Food Production (Economics and Politics)

Governments can use price controls to keep food prices artificially low, give farmers subsidies to encourage food production, or eliminate food price controls and subsidies and let farmers and fishers respond to market demand.

Agriculture is a financially risky business. Whether farmers have a good year or a bad year depends on factors over which they have little control: weather, crop prices, crop pests and diseases, interest rates, and the global market.

Governments use three main approaches to influence food production:

- *Control prices.* Use price controls to keep food prices artificially low. Consumers are happy, but farmers may not be able to make a living.
- *Provide subsidies.* Give farmers subsidies and tax breaks to keep them in business and encourage them to increase food production. Globally, government price supports and other subsidies for agriculture total more than $350 billion per year (about $100 billion per year in the United States)—more than $666,000 per minute! If government subsidies are too generous and the weather is good, farmers may produce more food than can be sold. The resulting surplus depresses food prices, which reduces the financial incentive for farmers in developing countries to increase domestic food production.
- *Let the marketplace decide.* Another approach is to eliminate most or all price controls and subsidies and let farmers and fishers respond to market demand without government interference. Some analysts urge that any phaseout of farm and fishery subsidies should be coupled with increased aid for the poor and the lower middle class, who would suffer the most from any increase in food prices. Some environmental scientists say that instead of eliminating all subsidies, we should use them to reward farmers and ranchers who protect the soil, conserve water, reforest degraded land, protect and restore wetlands, conserve wildlife, and practice more sustainable agriculture and fishing.

Trying New Foods and Changing Eating Habits

A variety of plants and insects could be used as sources of food, but most consumers are reluctant to try new foods.

Some analysts recommend greatly increased cultivation of less widely known but more nutritious plants to supplement or replace such staples as wheat, rice, and corn. One possibility is the *winged bean*, a fast-growing, plant grown in 50 countries because of its high protein and vitamin content.

Some edible insects—called *microlivestock*—are also important potential sources of protein, vitamins, and minerals in many parts of the world. There are about 1,500 edible insect species. One example is *Mopani*—emperor moth caterpillars—one of several insects eaten in South Africa. Other examples include black ant larvae, served in tacos in Mexico, giant waterbugs crushed into vegetable dip in Thailand, and lightly toasted butterflies, a favorite food in Bali.

Most of these insects are 58–78% protein by weight—three to four times as protein-rich as beef, fish, or eggs. One problem is getting farmers to take the financial risk of cultivating new types of food crops. Another is convincing consumers to try new foods.

PROTECTING FOOD RESOURCES: PEST MANAGEMENT

Natural Pest Control (Science)

Predators, parasites, and disease organisms found in nature control populations of most pest species as part of the earth's free ecological services.

A **pest** is any species that competes with us for food, invades lawns and gardens, destroys wood in houses, spreads disease, invades ecosystems, or is simply a nuisance. Worldwide, only about 100 species of plants (which we call weeds), animals (mostly insects), fungi, and microbes (which can infect crop plants and livestock animals) cause about 90% of the damage to the crops we grow.

In natural ecosystems and many polyculture agroecosystems, *natural enemies* (predators, parasites, and disease organisms) control the populations of about 98% of the potential pest species in keeping with one of the earth's four sustainability principles (Figure 1-11, p. 17).

When we clear forests and grasslands, plant monoculture crops, and douse fields with pesticides, we upset many of these natural population checks and balances. Then we must devise ways to protect our monoculture crops, tree plantations, lawns, and golf courses from insects and other pests that nature once controlled at no charge.

Synthetic Pesticides: Copying Nature

We use chemicals to repel or kill pest organisms as plants have done for millions of years to defend themselves against hungry herbivores.

To help control pest organisms, we have developed a variety of **pesticides**—chemicals to kill or control populations of organisms we consider undesirable. Com-

mon types of pesticides include *insecticides* (insect killers), *herbicides* (weed killers), *fungicides* (fungus killers), and *rodenticides* (rat and mouse killers). *Biocide* is a more accurate name for these chemicals because most pesticides kill other organisms as well as their pest targets.

We did not invent the use of chemicals to repel or kill other species. Plants have been producing chemicals to ward off, deceive, or poison herbivores that feed on them for about 225 million years. This is a never-ending, ever-changing coevolutionary process: herbivores overcome various plant defenses through natural selection; then new plant defenses are favored by natural selection in this ongoing cycle of evolutionary punch and counterpunch.

Since 1950, pesticide use has increased more than 50-fold, and most of today's pesticides are more than ten times as toxic as those used in the 1950s. About three-fourths of these chemicals are used in developed countries, but use in developing countries is soaring.

About one-fourth of pesticide use in the United States is for ridding houses, gardens, lawns, parks, playing fields, swimming pools, and golf courses of pests. According to the U.S. Environmental Protection Agency (EPA), the average lawn in the United States is doused with ten times more synthetic pesticides per hectare than is U.S. cropland. Each year, more than 250,000 people in the United States become ill because of household pesticide use, and such pesticides are a major source of accidental poisonings and deaths of young children.

Many pesticides, called *broad-spectrum agents,* are toxic to many species; others, called *selective,* or *narrow-spectrum agents,* are effective against a narrowly defined group of organisms. Pesticides vary in their *persistence,* the length of time they remain deadly in the environment. In 1962, biologist Rachel Carson warned against relying on synthetic organic chemicals to kill insects and other species we deem pests (Individuals Matter, at right).

Advantages of Modern Synthetic Pesticides (Science and Economics)

Modern pesticides save lives, increase food supplies, increase profits for farmers, work fast, and are safe if used properly.

Proponents of conventional chemical pesticides contend that their benefits outweigh their harmful effects. Conventional pesticides have a number of important benefits.

They save human lives. Since 1945, DDT and other insecticides probably have prevented the premature deaths of at least 7 million people (some say as many as 500 million) from insect-transmitted diseases such as malaria (carried by the *Anopheles* mosquito),

Rachel Carson

Rachel Carson began her professional career as a biologist for the Bureau of U.S. Fisheries (later the U.S. Fish and Wildlife Service). In that capacity, she carried out research on oceanography and marine biology and wrote articles about the oceans and topics related to the environment.

In 1951, she wrote *The Sea Around Us*, which described in easily understandable terms the natural history of oceans and how human activities were harming them. This book sold more than 2 million copies, was translated into 32 languages, and won a National Book Award.

During the late 1940s and throughout the 1950s, DDT and related compounds were increasingly used to kill insects that ate food crops, attacked trees, bothered people, and transmitted diseases such as malaria. In 1958, DDT was sprayed to control mosquitoes near the home and private bird sanctuary of one of Carson's friends. After the spraying, her friend witnessed the agonizing deaths of several birds. She begged Carson to find someone to investigate the effects of pesticides on birds and other wildlife.

Carson decided to look into the issue herself and found that independent research on the environmental effects of pesticides was almost nonexistent. As a well-trained scientist, she surveyed the scientific literature, became convinced that pesticides could harm wildlife and humans, and methodically developed information about the harmful effects of widespread use of pesticides.

In 1962, she published her findings in popular form in *Silent Spring*, the title alluding to the silencing of "robins, catbirds, doves, jays, wrens, and scores of other bird voices" because of their exposure to pesticides. Many scientists, politicians, and policy makers read *Silent Spring*, and the public embraced it.

Chemical manufacturers viewed the book as a serious threat to booming pesticide sales and mounted a campaign to discredit Carson. A parade of critical reviewers and industry scientists claimed her book was full of inaccuracies, made selective use of research findings, and failed to give a balanced account of the benefits of pesticides.

Some critics even claimed that, as a woman, Carson was incapable of understanding such a highly scientific and technical subject. Others charged that she was a hysterical woman and a radical nature lover trying to scare the public in order to sell books.

During these intense attacks, Carson was suffering from terminal cancer. Yet she strongly defended her research and countered her critics. She died in 1964—about 18 months after the publication of *Silent Spring*—without knowing that many historians consider her work an important contribution to the modern environmental movement then emerging in the United States.

bubonic plague (carried by rat fleas), and typhus (carried by body lice and fleas).

They increase food supplies. According to the FAO, 55% of the world's potential human food supply is lost to pests. Without pesticides, these losses would be worse, and food prices would rise.

They increase profits for farmers. Pesticide company officials estimate that every $1 spent on pesticides leads to an increase in U.S. crop yields worth approximately $4. (But studies have shown this benefit drops to about $2 if the harmful effects of pesticides are included).

They work faster and better than alternatives. Pesticides control most pests quickly and at a reasonable cost, have a long shelf life, are easily shipped and applied, and are safe when handled properly by farm workers. When genetic resistance occurs, farmers can use stronger doses or switch to other pesticides.

When used properly, their health risks are very low compared with their benefits. According to Elizabeth Whelan, director of the American Council on Science and Health (ACSH), which presents the position of the pesticide industry, "The reality is that pesticides, when used in the approved regulatory manner, pose no risk to either farm workers or consumers."

Newer pesticides are safer and more effective than many older pesticides. Greater use is being made of botanicals and microbotanicals. Derived originally from plants, they are safer to users and less damaging to the environment than many older pesticides. Genetic engineering is also being used to develop pest-resistant crop strains and genetically altered crops that produce their own pesticides.

Many new pesticides are used at very low rates per unit area compared to older products. Application amounts per hectare for many new herbicides are 1/100 the rates for older ones, and genetically engineered crops could reduce the use of toxic insecticides.

Disadvantages of Modern Synthetic Pesticides (Science)

Pesticides can promote genetic resistance to their effects, wipe out natural enemies of pest species, create new pest species, end up in the environment, and harm wildlife and people.

Opponents of widespread pesticide use believe the harmful effects of pesticides outweigh their benefits.

They cite several serious problems with the use of conventional pesticides:

They accelerate the development of genetic resistance to pesticides by pest organisms. Insects breed rapidly, and within 5–10 years (much sooner in tropical areas) they can develop immunity to widely used pesticides through natural selection and come back stronger than before. Weeds and plant disease organisms also develop genetic resistance, but more slowly than insects do. Since 1945, about 1,000 species of insects and rodents (mostly rats) and 550 types of weeds and plant diseases have developed genetic resistance to one or more pesticides. Even more serious is the ability of species with pesticide-resistant genes to transfer this resistance to pests that have never been exposed to the pesticides.

They can put farmers on a financial treadmill. Because of genetic resistance, many widely used insecticides (such as DDT) no longer do a good job of protecting people from insect-transmitted diseases in some parts of the world. This can put farmers on a *pesticide treadmill,* whereby they pay more and more for a pest control program that often becomes less and less effective.

Some insecticides kill natural predators and parasites that help control the pest populations. Wiping out natural predators can unleash new pests, whose populations their predators had previously held in check, and cause other unexpected effects. Of the 300 most destructive insect pests in the United States, 100 were once minor pests that became major pests after widespread use of insecticides. Mostly because of genetic resistance and reduction of natural predators, pesticide use has not reduced crop losses to pests in the United States. According to a study by David Pimentel, an expert in insect ecology, although the use of synthetic pesticides has increased 33-fold since 1942, 37% of the U.S. food supply is lost to pests today compared to 31% in the 1940s.

Pesticides do not stay put. According to the USDA, only 0.1–2% of the insecticide applied to crops by aerial spraying or ground spraying reaches the target pests. Also, less than 5% of herbicides applied to crops reach the target weeds. In other words, 98–99.9% of the pesticides and more than 95% of the herbicides we apply end up in the air, surface water, groundwater, bottom sediments, food, and nontarget organisms, including humans and wildlife (Figure 7-12, p. 140).

Some pesticides harm wildlife. According to the USDA and the U.S. Fish and Wildlife Service, each year pesticides applied to cropland in the United States wipe out about 20% of U.S. honeybee colonies and damage another 15%. This costs farmers at least $200 million per year because fewer bees are pollinating vital crops. Pesticides also kill more than 67 million birds and 6–14 million fish per year, and according to a 2004 study by the Center for Biological Diversity, menace about one of every three endangered and threatened species in the United States.

Some pesticides can threaten human health. The World Health Organization (WHO) and the UN Environment Programme (UNEP) estimate that each year pesticides seriously poison 3.5 to 5 million agricultural workers in developing countries and at least 300,000 people in the United States. They also cause 20,000–40,000 deaths (about 25 in the United States) per year. Health officials believe the actual number of pesticide-related illnesses and deaths among the world's farm workers probably is greatly underestimated because of poor record-keeping, lack of doctors, inadequate reporting of illnesses, and faulty diagnoses.

According to studies by the National Academy of Sciences, exposure to legally allowed pesticide residues in food causes 4,000–20,000 cases of cancer per year in the United States. Roughly half of these individuals die prematurely. Some scientists are becoming increasingly concerned about possible genetic mutations, birth defects, nervous system disorders (especially behavioral disorders), and effects on the immune and endocrine systems from long-term exposure to low levels of various pesticides. The pesticide industry disputes these claims.

X *HOW WOULD YOU VOTE?* Do the advantages of using synthetic chemical pesticides outweigh their disadvantages? Cast your vote online at www.thomsonedu/biology/miller.

Pesticide Protection Laws in the United States (Politics)

Government regulation has banned a number of harmful pesticides but some scientists call for strengthening pesticide laws.

How well the public in the United States is protected from the harmful effects of pesticides remains a controversial topic. Between 1972 and 2005, the EPA banned or severely restricted the use of 57 active pesticide ingredients, including DDT and most other chlorinated hydrocarbon insecticides.

According to studies by the National Academy of Sciences, federal laws regulating pesticide use in the United States are inadequate and poorly enforced by the EPA, the Food and Drug Administration (FDA), and the USDA. One study by the National Academy of Sciences found that up to 98% of the potential risk of developing cancer from pesticide residues on food grown in the United States would be eliminated if EPA standards were as strict for pre-1972 pesticides as they are for later ones. Another problem is that banned or unregistered pesticides may be manufactured in the

U.S. pesticide companies make and export to other countries pesticides that have been banned or severely restricted—or never even approved—in the United States. Other industrial countries also export banned and unapproved pesticides.

But what goes around can come around. In what environmental scientists call a *circle of poison* or *the boomerang effect*, residues of some of these banned or unapproved chemicals exported to other countries can return to the exporting countries on imported food. The wind can also carry persistent pesticides such as DDT from one country to another.

Environmental and health scientists and have urged Congress—without success—to ban such exports. Supporters of pesticide exports argue that such sales increase economic growth and provide jobs, and that banned pesticides are exported only with the consent of the importing countries. They also contend that if the United States did not export pesticides, other countries would.

In 1998, more than 50 countries developed an international treaty that requires exporting countries to have informed consent from importing countries for exports of 22 pesticides and 5 industrial chemicals. In 2000, more than 100 countries developed an international agreement to ban or phase out the use of 12 especially hazardous persistent organic pollutants (POPs)—9 of them persistent hydrocarbon pesticides such as DDT and other chemically similar pesticides.

Critical Thinking

Should U.S. companies be allowed to export pesticides that have been banned, severely restricted, or not approved for use in the United States? Explain.

United States and exported to other countries (Spotlight, above).

In 2004, the Environmental Working Group evaluated scientific studies of pesticide residues on food grown in other countries and imported to the United States. They found that the six foods most likely to have higher pesticide residues than those allowed by law are, in order, apples, peaches, strawberries, green beans, red raspberries, and pears. Figure 8-17 lists some ways you can reduce your exposure to pesticides.

Other Ways to Control Pests (Science)

A mix of cultivation practices and biological and ecological alternatives to conventional chemical pesticides can help control pests.

Many scientists believe we should greatly increase the use of biological, ecological, and other alternative methods for controlling pests and diseases that affect crops and human health. A number of methods are available.

Fool the pest. A variety of *cultivation practices* can be employed to fake out pest species. Examples are rotating the types of crops planted in a field each year, adjusting planting times so major insect pests either starve or get eaten by their natural predators, and growing crops in areas where their major pests do not exist.

Provide homes for pest enemies. Farmers can increase the use of polyculture, which uses plant diversity to reduce losses to pests. Homeowners can reduce weed invasions by cutting grass no lower than 8 centimeters (3 inches) high. This provides a dense enough cover to keep out crabgrass and many other undesirable weeds.

Implant genetic resistance. Use genetic engineering to speed up the development of pest- and disease-resistant crop strains. But there is controversy over whether the projected advantages of the increasing

What Can You Do?

Reducing Exposure to Pesticides

- Grow some of your food using organic methods.

- Buy organic food.

- Wash and scrub all fresh fruits, vegetables, and wild foods you pick.

- Eat less or no meat.

- Trim the fat from meat.

Figure 8-17 Individuals matter: ways to reduce your exposure to pesticides. QUESTION: *Which three of these actions do you think are the most important? Which things in this list do you do or plan to do?*

use of genetically modified plants and foods outweigh their projected disadvantages (Figure 8-11).

Bring in natural enemies. Use *biological control,* or *biocontrol,* by importing natural predators, parasites, and disease-causing bacteria and viruses to help regulate pest populations. This approach is nontoxic to other species, minimizes genetic resistance, and can save large amounts of money—about $25 for every $1 invested in controlling 70 pests in the United States. However, biological control agents cannot always be mass produced, are often slower acting and more difficult to apply than conventional pesticides, can sometimes multiply and become pests themselves, and must be protected from pesticides sprayed in nearby fields.

Use insect perfumes. Sex attractants (called *pheromones*) can lure pests into traps or attract their natural predators into crop fields (usually the more effective approach). These chemicals attract only one species, work in trace amounts, have little chance of causing genetic resistance, and are not harmful to nontarget species. However, it is costly and time consuming to identify, isolate, and produce the specific sex attractant for each pest or predator.

Bring in the hormones. Hormones are chemicals produced by animals that control developmental processes at different stages of life. Scientists have learned how to identify and use hormones that disrupt an insect's normal life cycle, and prevent it from reaching maturity and reproducing. Insect hormones have the same advantages as sex attractants. But they take weeks to kill an insect, often are ineffective with large infestations of insects, and sometimes break down before they can act. In addition, they must be applied at exactly the right time in the target insect's life cycle. They can sometimes affect the target's predators and other nonpest species and are difficult and costly to produce.

Scald them. Some farmers have controlled some insect pests by spraying them with hot water. This has worked well on cotton, alfalfa, and potato fields and in citrus groves in Florida, and its cost is roughly equal to that of using chemical pesticides.

Case Study: Integrated Pest Management— A Component of Sustainable Agriculture (Science)

An ecological approach to pest control uses an integrated mix of cultivation and biological methods, and small amounts of selected chemical pesticides as a last resort.

Many pest control experts and farmers believe the best way to control crop pests is a carefully designed **integrated pest management (IPM)** program. In this approach, each crop and its pests are evaluated as parts of an ecological system. Then farmers develop a program that includes cultivation and biological and chemical approaches to pest management.

The overall aim of IPM is not to eradicate pest populations but rather to reduce crop damage to an economically tolerable level. Fields are monitored carefully, and when an economically damaging level of pests has been reached, farmers first use biological methods (natural predators, parasites, and disease organisms) and cultivation controls, including vacuuming up harmful bugs. Small amounts of insecticides—mostly based on natural insecticides produced by plants—are applied only as a last resort and in the smallest amount possible. Broad-spectrum, long-lived pesticides are not used and different chemicals are used to slow the development of genetic resistance and to avoid killing predators of pest species. Small plots of *trap crops* are used to lure pests away from the primary crops. They are planted early so they will mature a week or two before the primary crops. When pests arrive, they are heavily sprayed with pesticides and destroyed.

In 1986, the Indonesian government banned 57 of the 66 pesticides used on rice, and phased out pesticide subsidies over a 2-year period. It also launched a nationwide education program to help farmers switch to IPM. The results were dramatic. Between 1987 and 1992, pesticide use dropped by 65%, rice production rose by 15%, and more than 250,000 farmers were trained in IPM techniques. Sweden and Denmark have used IPM to cut their pesticide use in half.

According to a 2003 study by the U.S. National Academy of Sciences, these and other experiences show that a well-designed IPM program can reduce pesticide use and pest control costs by 50–65% without reducing crop yields and food quality. It can also reduce inputs of fertilizer and irrigation water, and slow the development of genetic resistance because pests are assaulted less often and with lower doses of pesticides. Thus IPM is an important form of *pollution prevention* that reduces risks to wildlife and human health.

Despite its promise, IPM, like any other form of pest control, has some disadvantages. It requires expert knowledge about each pest situation and is slower acting than conventional pesticides. Methods developed for a crop in one area might not apply to areas with even slightly different growing conditions. Initial costs may be higher, although long-term costs typically are lower than those of using conventional pesticides.

Widespread use of IPM is hindered by government subsidies for conventional chemical pesticides, opposition by pesticide manufacturers, and a lack of IPM experts.

A 1996 study by the National Academy of Sciences recommended that the United States shift from chemical-based approaches to more ecologically sensitive pest management approaches. Within 5–10 years, such a shift could cut U.S. pesticide use in half, as it has in several other countries.

A growing number of scientists urge the USDA to use three strategies to promote IPM in the United States:

- Add a 2% sales tax on pesticides and use the revenue to fund IPM research and education

- Set up a federally supported IPM demonstration project on at least one farm in every county

- Train USDA field personnel and county farm agents in IPM so they can help farmers use this alternative

The pesticide industry has successfully opposed such measures.

X *HOW WOULD YOU VOTE?* Should governments heavily subsidize a switch to integrated pest management? Cast your vote online at www.thomsonedu.com/biology/miller.

Several UN agencies and the World Bank have joined together to establish an IPM facility. Its goal is to promote the use of IPM by disseminating information and establishing networks among researchers, farmers, and agricultural extension agents involved in IPM.

SOLUTIONS: SUSTAINABLE AGRICULTURE

Low-Input Sustainable Agriculture (Science)

We can produce food more sustainably by reducing resource throughput and working with nature.

There are three main ways to reduce hunger and malnutrition and the harmful environmental effects of agriculture.

- *Slow population growth.*

- *Sharply reduce poverty* so that people can grow or buy enough food for their survival and good health.

- *Develop and phase in systems of sustainable or low-input agriculture over the next few decades.* One component of this is increased use of *organic agriculture.*

Figure 8-18 lists the major components of more sustainable agriculture. Studies have shown that yields of organic crops can be as much as 20% lower than those of conventionally raised crops. But farmers offset this by not using expensive pesticides, herbicides, and synthetic fertilizers and by getting higher prices for or-

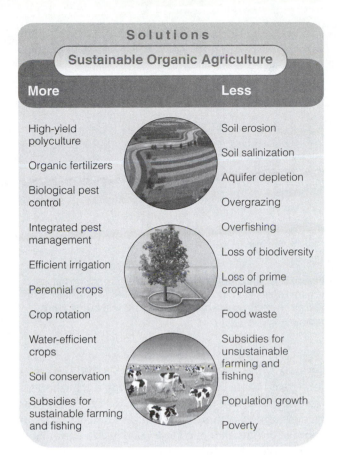

Solutions		
Sustainable Organic Agriculture		
More		**Less**
High-yield polyculture		Soil erosion
Organic fertilizers		Soil salinization
Biological pest control		Aquifer depletion
		Overgrazing
Integrated pest management		Overfishing
Efficient irrigation		Loss of biodiversity
Perennial crops		Loss of prime cropland
Crop rotation		Food waste
Water-efficient crops		Subsidies for unsustainable farming and fishing
Soil conservation		
Subsidies for sustainable farming and fishing		Population growth
		Poverty

Figure 8-18 Solutions: components of more sustainable, low-throughput agriculture based mostly on mimicking and working with nature. **QUESTION:** *Which two solutions do you think are the most important?*

ganic crops. Thus, the net economic return per unit of land for organic crop production is often equal to or higher than that from conventional crop production.

Most proponents of more sustainable agriculture are not opposed to high-yield agriculture. Instead, they see it as vital for protecting the earth's biodiversity by reducing the need to cultivate new and often marginal land. They call for using environmentally sustainable forms of both high-yield polyculture and high-yield monoculture for growing crops, with increasing emphasis on using organic methods for producing food.

Solutions: Making the Transition to More Sustainable Agriculture

More research, demonstration projects, government subsidies, and training can promote a shift to more sustainable organic agriculture.

Analysts suggest four major strategies to help farmers make the transition to more sustainable agriculture. *First,* greatly increase research on sustainable agriculture and human nutrition. *Second,* set up demonstration projects to show how sustainable organic agricultural

What Can You Do?

Sustainable Agriculture

- Waste less food
- Reduce or eliminate meat consumption
- Feed pets balanced grain foods instead of meat
- Use organic farming to grow some of your food
- Buy organic food
- Compost your food wastes

Figure 8-19 Individuals matter: ways to promote more sustainable agriculture. QUESTION: *Which three of these actions do you think are the most important? Which of these actions do you now take or plan to take?*

systems work. *Third,* provide subsidies and increased foreign aid to encourage its use. *Fourth,* establish training programs in sustainable organic agriculture for farmers and government agricultural officials, and encourage the creation of college curricula in sustainable organic agriculture and human nutrition.

Phasing in more sustainable agriculture involves *applying the four principles of sustainability* (Figure 1-11, p. 17) to producing food. This means relying more on solar energy and less on oil. It means sustaining nutrient cycling by practicing soil conservation and returning crop residues and animal wastes to the soil. It also means helping sustain natural and agricultural biodiversity by relying on a greater variety of crop and animal strains; by controlling pest populations through greater reliance on polyculture and integrated pest management; and by controlling human population growth and reducing food waste.

The goal is to feed the world's people while sustaining and restoring the earth's natural capital and living off the natural income it provides. This will not be easy, but it can be done. Figure 8-19 lists some ways in which you can promote more sustainable agriculture.

The sector of the economy that seems likely to unravel first is food. Eroding soils, deteriorating rangelands, collapsing fisheries, falling water tables, and rising temperatures are converging to make it difficult to expand food production fast enough to keep up with the demand.

LESTER R. BROWN

CRITICAL THINKING

1. Summarize the major economic and ecological advantages and limitations of each of the following proposals for increasing world food supplies and reducing hunger over the next 30 years: **(a)** cultivating more land by clearing tropical forests and irrigating arid lands, **(b)** catching more fish in the open sea, **(c)** producing more fish and shellfish with aquaculture, and **(d)** increasing the yield per area of cropland.

2. List three ways in which your lifestyle directly or indirectly contributes to soil erosion.

3. What can happen to energy-intensive agriculture in the United States and other industrialized countries when world oil prices rise sharply?

4. What are the three most important actions you would take to reduce hunger **(a)** in the country where you live and **(b)** in the world?

5. Should governments phase in agricultural tax breaks and subsidies to encourage farmers to switch to more sustainable farming? Explain your answer.

6. Explain why you support or oppose greatly increased use of **(a)** genetically modified food and **(b)** polyculture.

7. Explain how widespread use of a pesticide can **(a)** increase the damage done by a particular pest and **(b)** create new pest organisms.

8. Congratulations! You are in charge of the world. List the three most important features of your **(a)** agricultural policy, **(b)** policy to reduce soil erosion, and **(c)** policy for pest management.

9. List two questions that you would like to have answered as a result of reading this chapter.

LEARNING ONLINE

The website for this book contains helpful study aids and many ideas for further reading and research. They include a chapter summary, review questions for the entire chapter, flash cards for key terms and concepts, a multiple-choice practice quiz, interesting Internet sites, references, information about green careers, and a guide for accessing thousands of InfoTrac® College Edition articles. Log into

www.thomsonedu.com/biology/miller

Then choose Chapter 8, and select a learning resource. For access to animations, additional quizzes, chapter outlines and summaries, register and log into

Thomson™ NOW!

at **www.thomsonedu.com** using the access code card in the front of your book.

WATER RESOURCES AND WATER POLLUTION

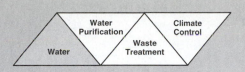

Our liquid planet glows like a soft blue sapphire in the hard-edged darkness of space. There is nothing else like it in the solar system. It is because of water.

JOHN TODD

WATER'S IMPORTANCE, AVAILABILITY, AND RENEWAL

Importance of Water (Science and Economics)

Water keeps us alive, moderates climate, sculpts the land, removes and dilutes wastes and pollutants, and moves continually through the hydrologic cycle.

We live on the water planet, with a precious film of water—most of it saltwater—covering about 71% of the earth's surface. Look in the mirror. What you see is about 60% water, most of it inside your cells.

You could survive for several weeks without food but only a few days without water. It takes huge amounts of water to supply you with food, provide shelter, and meet your other needs and wants. Water also plays a key role in sculpting the earth's surface, moderating climate, and removing and diluting water-soluble wastes and pollutants.

Despite its importance, water is one of our most poorly managed resources. We waste it and pollute it. We also charge too little for making it available. This encourages still greater waste and pollution of this resource, for which we have no substitute.

Only a tiny fraction—about 0.02%—of the planet's abundant water is readily available to us as freshwater. To illustrate how small a proportion this is, suppose the world's water supply were only 100 liters (26 gallons); then our usable supply of freshwater would be only about 0.02 liter, or 4 teaspoons!

Fortunately, the world's freshwater supply is continuously collected, purified, recycled, and distributed in the solar-powered *hydrologic cycle* (Figure 2-25, p. 45). This magnificent water recycling and purification system works well as long as we do not overload water systems with slowly degradable and nondegradable wastes or withdraw water from underground supplies faster than it is replenished. Unfortunately, in parts of the world, we are doing both of these things.

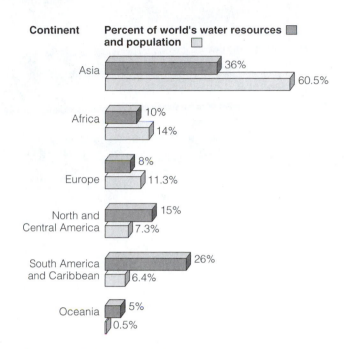

Figure 9-1 Natural capital shares: population and freshwater supplies on the world's continents. QUESTION: *What are two major conclusions that you can make from these data?* (Data from UNESCO, 2003)

Differences in average annual precipitation and economic resources divide the world's continents, countries, and people into water *haves* and *have-nots*. For example, Canada, with only 0.5% of the world's population, has 20% of the world's freshwater, whereas China, with 20% of the world's people, has only 7% of the supply. Figure 9-1 compares the population sizes and shares of the world's freshwater among the continents.

Groundwater and Surface Water (Science)

Some precipitation infiltrates the ground and is stored in spaces in soil and rock. Water that does not sink into the ground or evaporate into the air runs off into bodies of water.

Some precipitation infiltrates the ground and percolates downward through voids (pores, fractures, crevices, and other spaces) in soil, gravel, and rock (Figure 9-2, p. 172). The water in these spaces is called **groundwater**—one of our most important sources of freshwater.

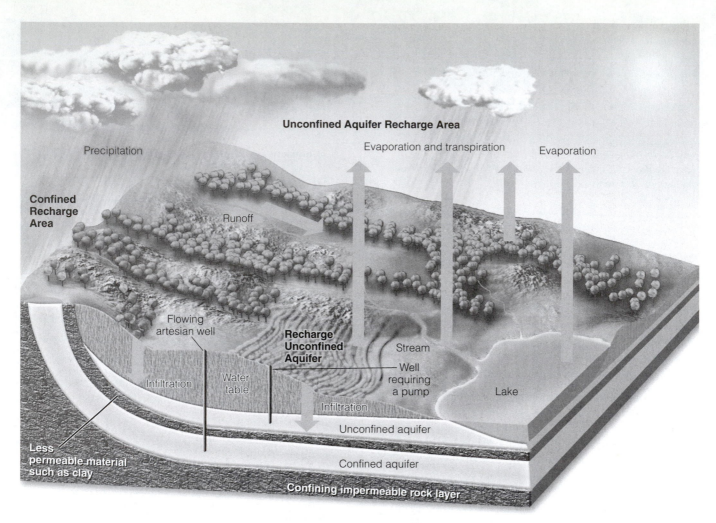

Confined Recharge Area

Unconfined Aquifer Recharge Area

Precipitation

Evaporation and transpiration

Evaporation

Runoff

Flowing artesian well

Recharge Unconfined Aquifer

Stream

Well requiring a pump

Lake

Infiltration

Water table

Infiltration

Unconfined aquifer

Less permeable material such as clay

Confined aquifer

Confining impermeable rock layer

Figure 9-2 Natural capital: groundwater system. An *unconfined aquifer* is an aquifer with a permeable water table. A *confined aquifer* is bounded above and below by less permeable beds of rock where the water is confined under pressure. Some aquifers are replenished by precipitation; others are not.

The spaces in soil, gravel, and rock hold little moisture. But below a certain depth, in the **zone of saturation,** these spaces are completely filled with water. The top of this zone is the **water table.** The water table falls in dry weather or when we remove groundwater faster than it is replenished and it rises in wet weather.

Deeper down are geological layers called **aquifers:** underground caverns and porous layers of sand, gravel, or bedrock through which groundwater flows. Groundwater normally moves from points of high elevation and pressure to points of lower elevation and pressure. Some underground caverns have rivers of groundwater flowing through them. In contrast, the porous layers of sand, gravel, or bedrock are like large elongated sponges through which groundwater seeps—typically moving only a meter or so (about 3 feet) per year and rarely more than 0.3 meter (1 foot) per day. Watertight layers of rock or clay below an aquifer keep the water from escaping deeper into the earth.

Most aquifers are replenished naturally by precipitation that percolates downward through soil and rock, a process called **natural recharge.** Others are recharged from the side by *lateral recharge* from nearby streams. Most aquifers recharge extremely slowly.

Some aquifers get very little, if any, recharge and on a human time scale are nonrenewable resources. They are found deep underground and were formed tens of thousands of years ago. Withdrawing water from them amounts to *mining* a nonrenewable resource. If kept up, such withdrawals will deplete these ancient deposits of liquid natural capital.

About one of every four people on the earth get their drinking water from aquifers, and aquifers are also a major source of water for irrigating cropland. Sustainable use of aquifer water means removing water no faster than it can be replenished and identifying and protecting aquifer recharge zones from development and pollution.

One of our most precious resources is **surface water,** the freshwater that flows across the earth's land surface and into rivers, streams, lakes, reservoirs, ponds, wetlands, and estuaries. Precipitation that does not infiltrate into the ground or return to the atmosphere by evaporation is called **surface runoff.** Surface water replenished by the *runoff* from precipitation and melting snow is classified as a renewable but finite resource. The land from which surface water drains into a river, lake, wetland, or other body of water is called its **watershed** or **drainage basin.**

Tapping the World's Reliable Surface Water Supply (Science)

We are using more than half the world's reliable runoff of surface water and could be using 70–90% by 2025.

Two-thirds of the annual surface runoff in rivers and streams is lost by seasonal floods and is not available for human use. The remaining one-third is **reliable runoff:** the amount of surface runoff that we can generally count on as a stable source of freshwater from year to year.

During the last century, the human population tripled, global water withdrawal increased sevenfold, and per capita withdrawal quadrupled. As a result, we now withdraw about 34% of the world's reliable runoff. We use another 20% of this runoff in streams to transport goods by boats, dilute pollution, and sustain fisheries and wildlife. Thus, *we directly or indirectly use about 54% of the world's reliable runoff of surface water.*

Because of increased population growth alone, global withdrawal rates of surface water could reach more than 70% of the reliable runoff by 2025 and 90% if per capita withdrawal of water continues increasing at the current rate. This is a global average. Withdrawal rates already exceed the reliable runoff in some areas.

Global Outlook: Uses of the World's Freshwater (Science and Economics)

Irrigation is the biggest user of water (70%) followed by industries (20%), and cities and residences (10%).

Worldwide, we use about 70% of the water we withdraw each year from rivers, lakes, and aquifers to irrigate one-fifth of the world's cropland that produces about 40% of the world's food. Industry uses about 20% of the water withdrawn each year, and cities and residences use the remaining 10%.

Affluent lifestyles require large amounts of water. For example, it takes about 400,000 liters (106,000 gallons) of water to produce an automobile and up to 100,000 liters (26,500 gallons) to produce 1 kilogram (2.2 pounds) of grain-fed beef. You can save more water by reducing your annual consumption of grain-fed beef by 1 kilogram (2.2 pounds) than by not taking a daily shower for almost 2 years.

Case Study: Freshwater Resources in the United States (Science and Economics)

The United States has plenty of freshwater, but supplies vary in different areas depending on climate.

The United States has more than enough renewable freshwater. But much of it is in the wrong place at the wrong time or is contaminated by agricultural and industrial practices. The eastern states usually have ample precipitation, whereas many western states have too little (Figure 9-3, top).

In the East, most water is used for energy production, cooling, and manufacturing. The largest use by far in the West (85%) is for irrigation.

In many parts of the eastern United States, the most serious water problems are flooding, occasional urban shortages, and pollution. The major water problem in the arid and semiarid areas of the western half of the country (Figure 9-3, bottom) is a shortage of runoff,

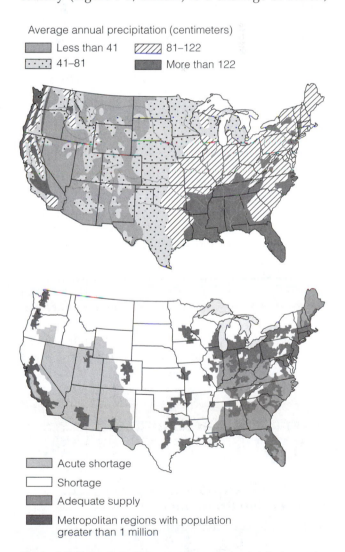

Figure 9-3 Natural capital: average annual precipitation and major rivers (top) and water-deficit regions in the continental United States and their proximity to metropolitan areas having populations greater than 1 million (bottom). (Data from U.S. Water Resources Council and U.S. Geological Survey)

Figure 9-4 Water *hot spot* areas in 17 western states that by 2025 could face intense conflicts over scarce water needed for urban growth, irrigation, recreation, and wildlife. Some analysts suggest that this is a map of places not to live over the next 25 years. (Data from U.S. Department of the Interior)

Legend:
- Highly likely conflict potential
- Substantial conflict potential
- Moderate conflict potential
- Unmet rural water needs

caused by low precipitation (Figure 9-3, top), high evaporation, and recurring prolonged drought.

Almost half (47%) of the water used in the United States comes from groundwater sources with the rest coming from surface waters (rivers, lakes, and reservoirs). Water tables in many areas are dropping quickly as farmers and rapidly growing cities deplete aquifers faster than they are recharged.

In 2003, the U.S. Department of the Interior mapped out *water hot spots* in 17 western states (Figure 9-4). In these areas, competition for scarce water to support growing urban areas, irrigation, recreation, and wildlife could trigger intense political and legal conflicts between states and between farmers and cities in the next 20 years.

WATER RESOURCE PROBLEMS: TOO LITTLE WATER AND TOO MUCH WATER

Global Outlook: Freshwater Shortages (Science and Economics)

About 41% of the world's population lives in river basins that do not have enough freshwater because of too many people relying on limited runoff.

The main factors causing water scarcity are a dry climate; drought (a prolonged period in which precipita-

tion is at least 70% lower and evaporation is higher than normal); too many people using the reliable supply of water; and lack of money to drill deep wells and to build dams, storage reservoirs, and water purification and distribution systems.

Figure 9-5 shows the degree of stress on the world's major river systems, based on a comparison of the amount of freshwater available with the amount used by humans. More than 30 countries—most of them in the Middle East and Africa—now face water scarcity. By 2050, some 60 countries are likely to be suffering from water stress or water scarcity.

In 2005, the United Nations reported that 1.2 billion people—one of every six people—do not have regular access to enough clean water for drinking, cooking, washing, and sanitation. According to the U.N., between 2 billion and 7 billion will face water shortages by 2050.

More than two-thirds of the world's households live in *hydrological poverty* and have to fetch water from outside their homes, typically from rivers, lakes, and village wells. In water-short rural areas in developing countries, many women and children must walk long distances each day, carrying heavy jars or cans, to get a meager and often contaminated supply of water.

Case Study: Water Conflicts in the Middle East—A Growing Problem

Some 40% of the world's population, especially in the Middle East, clash over water supplies from shared river basins and the situation is expected to get worse.

Many countries in the Middle East face water shortages and rising tensions over water sources they must share. Most water in this dry region comes from three shared river basins: the Nile, the Jordan, and the Tigris-Euphrates (Figure 9-6).

Three countries—Ethiopia, Sudan, and Egypt—use most of the water that flows in Africa's Nile River. Egypt, which gets more than 97% of its freshwater from the Nile, is last in line. To meet the water needs of its rapidly growing population, Ethiopia plans to divert more water from the Nile. So does Sudan. Such upstream diversions would reduce the amount of water available to Egypt, which cannot exist without irrigation water from the Nile.

Egypt has several options: it can work out water-sharing agreements with other countries or go to war with Sudan and Ethiopia for more water; cut population growth; waste less irrigation water; import more grain to reduce the need for irrigation water; or suffer the harsh human and economic consequences of *hydrological poverty.*

The Jordan Basin is by far the most water-short region, with fierce competition for its water among

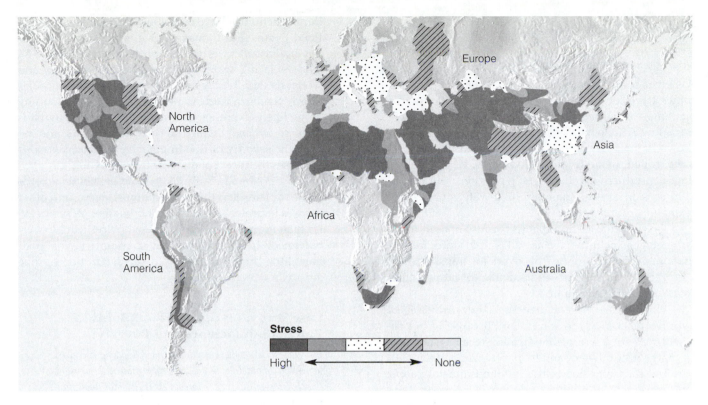

Figure 9-5 Natural capital degradation: stress on the world's major river basins, based on a comparison of the amount of water available with the amount used by humans. (Data from World Commission on Water Use in the 21st Century)

Jordan, Syria, Palestine (Gaza and the West Bank), and Israel. Syria plans to build dams and withdraw more water from the Jordan River, decreasing the downstream water supply for Jordan and Israel. Israel warns that it may destroy the largest dam that Syria plans to build. However, Israel has cooperated with Jordan and Palestine over their shared water resources.

Turkey, located at the headwaters of the Tigris and Euphrates Rivers, controls water flowing downstream to Syria and Iraq to where it empties into the Persian Gulf. Turkey is building 24 dams along the upper Tigris and Euphrates Rivers to generate electricity and irrigate a large area of land.

If completed, these dams will reduce the flow of water downstream to Syria and Iraq by up to 35% in normal years and by much more in dry years. Syria also plans to build a large dam along the Euphrates to divert water arriving from Turkey. This will leave little water for Iraq and could lead to a water war between Iraq and Syria.

Resolving these water distribution problems will require regional cooperation to allocate water supplies, slow population growth, improve efficiency in water use, raise water prices to help improve irrigation efficiency, and increase grain imports to reduce water needs. This will not be easy.

Figure 9-6 Threatened natural capital: many countries in the Middle East, with some of the world's highest population growth, face water shortages and conflicts over access to water because they share water from three major river basins.

Flooding (Science)

Heavy rainfall, rapid snowmelt, removal of vegetation, and destruction of wetlands cause flooding.

Whereas some areas have too little water, others sometimes have too much because of natural flooding by streams, caused mostly by heavy rain or rapidly melting snow. A flood happens when water in a stream overflows its normal channel and spills into the adjacent area, called a **floodplain.** Floodplains, which include highly productive wetlands, help provide natural flood and erosion control, maintain high water quality, and recharge groundwater.

People settle on floodplains because of their many advantages, including fertile soil, ample water for irrigation, availability of nearby rivers for transportation and recreation, and flat land suitable for crops, buildings, highways, and railroads.

Floods have several benefits. They provide the world's most productive farmland because of the nutrient-rich silt left after floodwaters recede. They also recharge groundwater and help refill wetlands.

But each year floods also kill thousands of people and cause tens of billions of dollars in property damage. Floods, like droughts, usually are considered natural disasters. But since the 1960s, several types of human activities have contributed to the sharp rise in flood deaths and damages. One is *removing water-absorbing vegetation*, especially on hillsides (Figure 9-7). And replacing water-absorbing vegetation, soil, and wetlands with highways, parking lots, and buildings, which cannot absorb rainwater, also makes flooding more likely. Another detrimental human activity is *draining wetlands*, which absorb floodwaters and reduce the severity of flooding. *Living on floodplains* also increases the threat of damage from flooding.

In developed countries, people deliberately settle on floodplains and then expect dams, levees, and other devices to protect them from floodwaters. When heavier than normal rains occur, these devices can be overwhelmed. In many developing countries, the poor have little choice but to try to survive in flood-prone areas (Case Study, below).

Case Study: Living on Floodplains in Bangladesh (Science and Poverty)

Bangladesh has experienced increased flooding because of upstream deforestation of Himalayan mountain slopes and the clearing of mangrove forests on its coastal floodplains.

Bangladesh is one of the world's most densely populated countries, with 144 million people (projected to

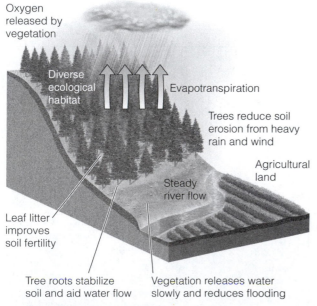

Forested Hillside

Oxygen released by vegetation

Diverse ecological habitat

Evapotranspiration

Trees reduce soil erosion from heavy rain and wind

Agricultural land

Steady river flow

Leaf litter improves soil fertility

Tree roots stabilize soil and aid water flow

Vegetation releases water slowly and reduces flooding

After Deforestation

Tree plantation

Roads destabilize hillsides

Evapotranspiration decreases

Ranching accelerates soil erosion by water and wind

Winds remove fragile topsoil

Agricultural land is flooded and silted up

Gullies and landslides

Heavy rain leaches nutrients from soil and erodes topsoil

Silt from erosion blocks rivers and reservoirs and causes flooding downstream

Rapid runoff causes flooding

ThomsonNOW™ Active Figure 9-7 Natural capital degradation: hillside before and after deforestation. Once a hillside has been deforested for timber and fuel wood, livestock grazing, or unsustainable farming, water from precipitation rushes down the denuded slopes, erodes precious topsoil, and can increase flooding in local streams. Such deforestation can also increase landslides and mudflows. A 3,000-year-old Chinese proverb says, "To protect your rivers, protect your mountains." *See an animation based on this figure and take a short quiz on the concept.*

reach 231 million by 2050) packed into an area roughly the size of the U.S. state of Wisconsin. It is a very flat country only slightly above sea level. It is also one of the world's poorest countries.

The people of Bangladesh depend on moderate annual flooding during the summer monsoon season to grow rice and help maintain soil fertility in the delta basin. The annual floods deposit eroded Himalayan soil on the country's crop fields.

In the past, great floods occurred every 50 years or so. But since the 1970s, they have come about every 4 years. Bangladesh's flooding problems begin in the Himalayan watershed. There, several factors—rapid population growth, deforestation, overgrazing, and unsustainable farming on steep and easily erodible slopes—have increased flows of monsoon rain water that runs more quickly off the denuded Himalayan foothills, carrying vital topsoil with it (Figure 9-7, right).

This increased runoff of soil, combined with heavier-than-normal monsoon rains, has increased the severity of flooding along Himalayan rivers and downstream in Bangladesh. A disastrous flood in 1998 covered two-thirds of Bangladesh's land area for 9 months, leveled 2 million homes, drowned at least 2,000 people, and left 30 million people homeless. It also destroyed more than one-fourth of the country's crops, which caused thousands of people to die of starvation. In 2002, another flood left 5 million people homeless and flooded large areas of rice fields. And yet another major flood occurred in 2004.

Living on Bangladesh's coastal floodplain also carries dangers from storm surges, cyclones, and tsunamis. In 1970, as many as 1 million people drowned in one storm. Another tropical cyclone in 2003 killed more than 1 million people and left tens of millions homeless.

In their struggle to survive, the poor in Bangladesh have cleared many of the country's coastal mangrove forests for fuelwood, farming, and aquaculture ponds for raising shrimp. This has led to more severe flooding because these coastal wetlands shelter Bangladesh's low-lying coastal areas from storm surges, cyclones, and tsunamis. Damages and deaths from cyclones and tsunamis in areas of Bangladesh still protected by mangrove forests have been much lower than in areas where the forests have been cleared.

Solutions: Reducing Flood Risks (Science, Economics, and Politics)

We can reduce flooding risks by controlling river water flows, protecting mountainside forests, preserving and restoring wetlands, identifying and managing flood-prone areas, and if possible, choosing not to live in such areas.

Figure 9-8 lists some ways to reduce flooding risks. Straightening and deepening, a process called *channel-*

Figure 9-8 Solutions: methods for reducing the harmful effects of flooding. QUESTION: *Which two of these solutions do you think are the most important?*

ization, reduces upstream flooding. However, it eliminates habitats for fish and other aquatic organisms and reduces groundwater discharge. It results in a faster flow, which can increase downstream flooding and sediment deposition. And it encourages human settlement in floodplains, increasing the risk of flood damage.

Levees or floodwalls along the sides of streams also contain and speed up stream flow, but they also increase the water's capacity for doing damage downstream. They do not protect against unusually high and powerful floodwaters, such as those occurring in 1993 when two-thirds of the levees built along the Mississippi River in the United States were damaged or destroyed. Dams can reduce the threat of flooding by storing water in a reservoir and releasing it gradually, but they also have a number of disadvantages (Figure 9-9, p. 178). Another way to reduce flooding is to *preserve existing wetlands* and *restore degraded wetlands* to take advantage of the natural flood control provided by floodplains.

The prevention or precautionary methods shown on the left side of Figure 9-8 are based on thousands of years of experience that can be summed up very simply: *Sooner or later the river (or the ocean) always wins.*

On a personal level, we can use the precautionary approach to *think carefully about where we live.* Many of the poor live in flood-prone areas because they have nowhere else to go. But most people can choose not to live in areas especially subject to flooding.

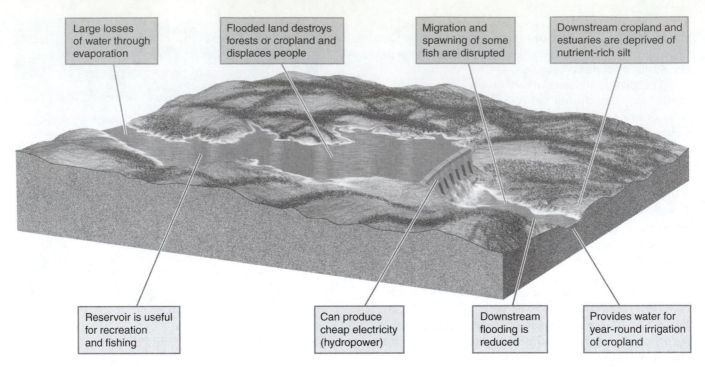

Figure 9-9 Trade-offs: advantages (lower boxes) and disadvantages (upper boxes) of large dams and reservoirs. The world's 45,000 large dams (higher than 15 meters or 50 feet) capture and store 14% of the world's runoff, provide water for almost half of all irrigated cropland, and supply more than half the electricity used by 65 countries. The United States has more than 70,000 large and small dams, capable of capturing and storing half of the country's entire river flow. QUESTION: *What are the single advantage and the single disadvantage that you think are the most important?*

The image contains the following labels:

Top boxes (disadvantages):
- Large losses of water through evaporation
- Flooded land destroys forests or cropland and displaces people
- Migration and spawning of some fish are disrupted
- Downstream cropland and estuaries are deprived of nutrient-rich silt

Bottom boxes (advantages):
- Reservoir is useful for recreation and fishing
- Can produce cheap electricity (hydropower)
- Downstream flooding is reduced
- Provides water for year-round irrigation of cropland

SUPPLYING MORE WATER

Increasing Freshwater Supplies

We can increase water supplies by withdrawing groundwater, building dams, bringing in water from somewhere else, converting saltwater to freshwater, wasting less water, and importing food.

There are several ways to increase the supply of freshwater in a particular area. One is to withdraw more groundwater. Another is to build dams and reservoirs to store runoff for release as needed. We can also bring in surface water from another area and convert saltwater to freshwater (desalination). Other strategies are to reduce water waste and import food to reduce the amount of water used in growing crops and raising livestock.

In *developed countries,* people tend to live where the climate is favorable and bring in water from another watershed. In *developing countries,* most people (especially the rural poor) must settle where the water is and try to capture and use the precipitation they need.

Trade-offs: Advantages and Disadvantages of Withdrawing Groundwater (Science and Economics)

Most aquifers are renewable sources unless water is removed faster than it is replenished or becomes contaminated.

Aquifers provide almost one-fourth of the water used throughout the world. In the United States, water pumped from aquifers supplies almost all of the drinking water in rural areas, one-fifth of that in urban areas, and 43% of irrigation water.

Relying more on groundwater has advantages and disadvantages (Figure 9-10). Aquifers are widely available and are renewable sources of water as long as the water is not withdrawn faster than it is replaced and as long as the aquifers do not become contaminated.

Unfortunately, water tables are falling in many areas of the world as the rate of pumping out water (mostly to irrigate crops) exceeds the rate of natural recharge from precipitation. The world's three largest grain-producing countries—China, India, and the United States—are overpumping many of their aquifers. Aquifer depletion also is a problem in Saudi Arabia, northern Africa (especially Libya and Tunisia), southern Europe, the Middle East, and parts of Mexico, Thailand, and Pakistan. Currently, more than half a billion people are being fed by grain produced through the unsustainable use of groundwater.

In the United States, groundwater is being withdrawn at a rate four times faster than the rate at which it is replenished. The most serious overdrafts are occurring in parts of the huge Ogallala Aquifer, extending from southern South Dakota to central Texas, and in parts of the arid southwestern United States

Trade-Offs

Withdrawing Groundwater

Advantages	Disadvantages

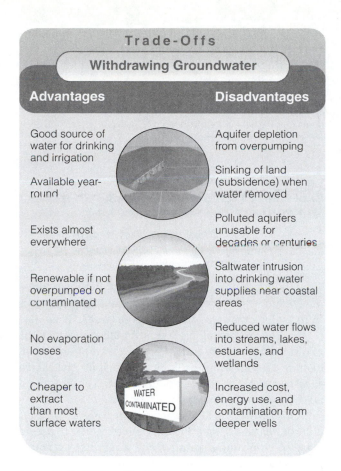

Good source of water for drinking and irrigation	Aquifer depletion from overpumping
Available year-round	Sinking of land (subsidence) when water removed
Exists almost everywhere	Polluted aquifers unusable for decades or centuries
Renewable if not overpumped or contaminated	Saltwater intrusion into drinking water supplies near coastal areas
No evaporation losses	Reduced water flows into streams, lakes, estuaries, and wetlands
Cheaper to extract than most surface waters	Increased cost, energy use, and contamination from deeper wells

Figure 9-10 Trade-offs: advantages and disadvantages of withdrawing groundwater. QUESTION: *What are the single advantage and the single disadvantage that you think are the most important?*

(Figure 9-11). Although it is gigantic, the Ogallala is essentially a one-time deposit of liquid natural capital with a very slow recharge rate. Serious groundwater depletion is also taking place in California's water-short Central Valley, which supplies half of the country's fruits and vegetables (long darker-shaded area in the California portion of Figure 9-11).

Groundwater overdrafts near coastal areas can contaminate groundwater supplies by causing the intrusion of saltwater into freshwater aquifers, making such water undrinkable and unusable for irrigation (Figure 9-12). This is an especially serious problem in many coastal areas in Florida, California, South Carolina, and Texas.

Figure 9-13 (p. 180) lists ways to prevent or slow the effects of groundwater depletion.

With global water shortages looming, scientists are evaluating deep aquifers—found at depths of 0.8 kilometer (0.5 mile) or more—as future water sources. Preliminary results suggest that some of these aquifers hold enough water to support billions of people for centuries. The quality of water in these aquifers may also be much higher than the water in most of the world's rivers and lakes.

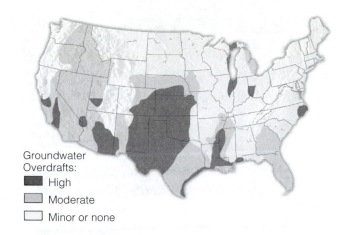

Groundwater Overdrafts:
- ■ High
- ▨ Moderate
- ☐ Minor or none

ThomsonNOW™ Active Figure 9-11 Natural capital degradation: areas of greatest aquifer depletion from groundwater overdraft in the continental United States. Aquifer depletion is also high in Hawaii and Puerto Rico (not shown on map). *See an animation based on this figure and take a short quiz on the concept.* (Data from U.S. Water Resources Council and U.S. Geological Survey)

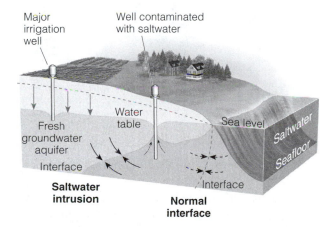

Figure 9-12 Natural capital degradation: *saltwater intrusion* along a coastal region. When the water table is lowered, the normal interface (dashed line) between fresh and saline groundwater moves inland, making groundwater drinking supplies unusable.

There are two major concerns about tapping these mostly one-time deposits of water. *First*, little is known about the geological and ecological impacts of pumping water from deep aquifers. *Second*, there are no international water treaties governing rights to, and ownership of, water found under several different countries. Without such treaties, legal and physical conflicts could ensue over who has the right to tap into and use these valuable resources.

Trade-offs: Advantages and Disadvantages of Large Dams and Reservoirs (Science and Economics)

Large dams and reservoirs can produce cheap electricity, reduce downstream flooding, and provide year-round water

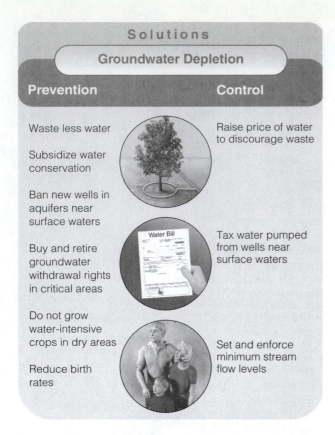

Solutions

Groundwater Depletion

Prevention	Control
Waste less water	Raise price of water to discourage waste
Subsidize water conservation	
Ban new wells in aquifers near surface waters	Tax water pumped from wells near surface waters
Buy and retire groundwater withdrawal rights in critical areas	
Do not grow water-intensive crops in dry areas	Set and enforce minimum stream flow levels
Reduce birth rates	

Figure 9-13 Solutions: ways to prevent or slow groundwater depletion by using water more sustainably. QUESTION: *Which two of these solutions do you think are the most important?*

for irrigating cropland, but they also displace people and disrupt aquatic systems.

Large dams and reservoirs have benefits and drawbacks (Figure 9-9). Their main purposes are to capture and store runoff and release it as needed to control floods, to generate electricity, and to supply water for irrigation and for towns and cities. Reservoirs also provide recreational activities such as swimming, fishing, and boating.

The more than 45,000 large dams built on the world's 227 largest rivers have increased the annual reliable runoff available for human use by nearly one-third. But a series of dams on a river in arid areas and withdrawals of river water for agricultural and urban uses can reduce downstream flow to a trickle and prevent it from reaching the sea as a part of the hydrologic cycle. According to the World Commission on Water in the 21st Century, more than half the world's major rivers go dry part of the year because of flow reduction by dams, especially in drought years.

This engineering approach to river management has displaced between 40 and 80 million people from their homes and has flooded an area of mostly productive land roughly equal to the area of the U.S. state of California. Such river management often impairs some of important ecological services rivers provide (Fig-

ure 9-14). In 2003, the World Resources Institute estimated that dams and reservoirs have strongly or moderately fragmented and disturbed 60% of the world's major river basins.

✗ *HOW WOULD YOU VOTE?* Do the advantages of large dams outweigh their disadvantages? Cast your vote online at www.thomsonedu.com/biology/miller.

Case Study: The California Experience (Science, Economics, and Politics)

The massive transfer of water from water-rich northern California to water-poor southern California has brought many benefits, but remains controversial.

Tunnels, aqueducts, and underground pipes can transfer stream runoff collected by dams and reservoirs from water-rich areas to water-poor areas. They also create environmental problems. Indeed, most of the world's dam projects and large-scale water transfers illustrate an important ecological principle: *You cannot do just one thing.* A number of unintended environmental consequences almost always occur.

One of the world's largest water transfer projects is the *California Water Project* (Figure 9-15). It uses a maze of giant dams, pumps, and aqueducts to transport water from water-rich northern California to southern California's heavily populated arid and semiarid agricultural regions and cities. In effect, this project supplies massive amounts of water to areas that without such water would be mostly desert.

Natural Capital

Ecological Services of Rivers

- Deliver nutrients to sea to help sustain coastal fisheries
- Deposit silt that maintains deltas
- Purify water
- Renew and renourish wetlands
- Provide habitats for wildlife

Figure 9-14 Natural capital: important ecological services provided by rivers. Currently, the services are given little or no monetary value when the costs and benefits of dam and reservoir projects are assessed. According to environmental economists, attaching even crudely estimated monetary values to these ecosystem services would help sustain them. QUESTION: *Which two of these services do you think are the most important?*

Figure 9-15 Solutions: California Water Project and the Central Arizona Project. These projects involve large-scale water transfers from one watershed to another. Arrows show the general direction of water flow.

For decades, northern and southern Californians have feuded over how the state's water should be allocated under this project. Southern Californians want more water from the north to grow more crops and to support Los Angeles, San Diego, and other growing urban areas. Agriculture uses three-fourths of the water withdrawn in California, much of it used inefficiently for water-thirsty crops growing in desert-like conditions.

Northern Californians say that sending more water south degrades the Sacramento River, threaten fisheries, and reduces the flushing action that helps clean San Francisco Bay of pollutants. They also argue that much of the water sent south is wasted. They point to studies showing that making irrigation just 10% more efficient would provide enough water for domestic and industrial uses in southern California.

According to a 2002 joint study by a group of scientists and engineers, projected global warming will sharply reduce water availability in California (especially southern California) and other water-short areas in the western United States (Figure 9-4) even under the study's best-case scenario. Some analysts project that sometime during this century, many of the people living in arid southern California cities (such as Los Angeles and San Diego) and farmers in this area, will have to move somewhere else because of a lack of water.

Pumping out more groundwater is not the answer because groundwater is already being withdrawn faster than it is replenished throughout much of California. Quicker and cheaper solutions would be to improve irrigation efficiency, stop growing water-thirsty crops in a desert climate, increase the price of water now sold by the government to farmers at low prices under long-term contracts, and allow farmers to sell

cities their legal rights to withdraw certain amounts of water from rivers.

Case Study: The Aral Sea Disaster (Science and Economics)

Diverting water from the Aral Sea and its two feeder rivers mostly for irrigation has created a major ecological, economic, and health disaster.

The shrinking of the Aral Sea is a result of a large-scale water transfer project in an area of the former Soviet Union with the driest climate in central Asia. Since 1960, enormous amounts of irrigation water have been diverted from the inland Aral Sea and its two feeder rivers to create one of the world's largest irrigated areas, mostly for raising cotton and rice. The irrigation canal, the world's longest, stretches more than 1,300 kilometers (800 miles).

This large-scale water diversion project, coupled with droughts and high evaporation rates due to the area's hot and dry climate, has caused a regional ecological, economic, and health disaster. Since 1961, the sea's salinity has tripled and the average level of its water has dropped by 22 meters (72 feet). It has lost 90% of its volume of water and has split into two parts. Water withdrawal for agriculture has reduced the sea's two supply rivers to mere trickles.

About 85% of the area's wetlands have been eliminated and roughly half the area's bird and mammal species have disappeared. In addition, a huge area of former lake bottom has been converted to a human-made desert covered with glistening white salt. The increased salt concentration of the water—three times saltier than ocean water—caused the presumed extinction of 20 of the area's 24 native fish species. This has devastated the area's fishing industry, which once provided work for more than 60,000 people. Fishing villages and boats once on the sea's coastline now are abandoned in the middle of a salt desert.

Winds pick up the salty dust that encrusts the lake's now-exposed seabed and blow it onto fields as far as 300 kilometers (190 miles) away. As the salt spreads, it chokes people, pollutes water, and kills wildlife, crops, and other vegetation. Aral Sea dust settling on glaciers in the Himalayas is causing them to melt at a faster than normal rate—another example of connections and unintended consequences.

To raise yields, farmers have increased inputs of herbicides, insecticides, fertilizers, and irrigation water on some crops. Many of these chemicals have percolated downward and accumulated to dangerous levels in the groundwater—the source of most of the region's drinking water.

Shrinkage of the Aral Sea has altered the area's climate. The once-huge sea acted as a thermal buffer that moderated the heat of summer and the extreme cold of winter. Now there is less rain, summers are hotter and drier, winters are colder, and the growing season is shorter. The combination of such climate change and severe salinization has reduced crop yields by 20–50% on almost a third of the area's cropland.

Finally, many of the 45 million people living in the Aral Sea's watershed have experienced increasing health problems from a combination of toxic dust, salt, and contaminated water.

Can the Aral Sea be saved, and can the area's serious ecological and human health problems be reduced? Since 1999, the United Nations and the World Bank have spent about $600 million to purify drinking water and upgrade irrigation and drainage systems, which improves irrigation efficiency and flushes salts from croplands. In 2005, a new dike was completed that by 2008 should raise the average level of the small Aral by 3 meters (10 feet). In addition, some artificial wetlands and lakes have been constructed to help restore aquatic vegetation, wildlife, and fisheries.

The five countries surrounding the lake and its two feeder rivers have worked to improve irrigation efficiency and to partially replace water-thirsty crops with others requiring less irrigation water. As a result, the total annual volume of water in the Aral Sea basin has stabilized. But experts expect the largest portion of the Aral Sea to continue shrinking.

Removing Salt from Seawater (Science and Economics)

Removing salt from seawater will probably not be done widely because of high costs and the question of what to do with the resulting salt.

Desalination involves removing dissolved salts from ocean water or from brackish (slightly salty) water in aquifers or lakes. It is another way to increase supplies of freshwater.

One method for desalinating water is *distillation*—heating saltwater until it evaporates, leaves behind salts in solid form, and condenses as freshwater. Another method is *reverse osmosis*, which uses high pressure to force saltwater through a membrane filter with pores small enough to remove the salt. In effect, high pressure is used to push freshwater out of saltwater.

There are about 13,500 desalination plants in 120 countries, especially in the arid, desert nations of the Middle East, North Africa, the Caribbean, and the Mediterranean. Saudi Arabia is the largest producer and accounts for more than a third of the world's output. These plants meet less than 0.3% of the world's water needs.

There are two major problems with the widespread use of desalination. One is the high cost—it takes a lot of energy to desalinate water. Currently, desalinating water by reverse osmosis costs two to three times as much as the conventional purification of freshwater, although recent advances in reverse osmosis have brought the energy costs down somewhat. Desalination by distillation requires even more energy and costs about 2–4 times more than reverse osmosis.

The second problem is that desalination produces large quantities of briny wastewater that contains a large amount of salt and other minerals. Dumping concentrated brine into a nearby ocean increases the salinity of the ocean water, threatening food resources and aquatic life in the vicinity. Dumping it on land could contaminate groundwater and surface water.

Bottom line: Currently, significant desalination is practical only for water-short wealthy countries and cities that can afford its high cost.

Scientists are working to develop new more effective membranes for reverse osmosis as well as on other desalinization technologies. Such technologies might bring down the cost of using desalination to produce drinking water. Even so, desalinated water probably will not be cheap enough to irrigate conventional crops or to meet much of the world's demand for freshwater. This may change if scientists can figure out how to use solar energy or other means to desalinate seawater cheaply and how to safely dispose of the salt.

INCREASING WATER SUPPLIES BY REDUCING WATER WASTE

Benefits of Reducing Water Waste (Green Economics)

We waste about two-thirds of the water we use, but we could reduce such waste to about 15% and meet projected global water needs.

Mohamed El-Ashry of the World Resources Institute estimates that *65–70% of the water people use throughout the world is lost through evaporation, leaks, and other losses.* The United States, the world's largest user of water, does slightly better but still loses about half the water it withdraws. El-Ashry believes it is economically and technically feasible to reduce such water losses to 15%, thereby meeting most of the world's water needs for the foreseeable future.

This win–win solution would decrease the burden on wastewater plants and reduce the need for expensive dams and water transfer projects that destroy wildlife habitats and displace people. It will also slow depletion of groundwater aquifers and save energy and money.

According to water resource experts, the main cause of water waste is that *we charge too little for water.* Such *underpricing* is mostly the result of government subsidies that provide irrigation water, electricity, and diesel fuel for farmers to pump water from rivers and aquifers at below-market prices.

Subsidies keep the price of water so low that users have little or no financial incentive to invest in water-saving technologies. According to water resource expert Sandra Postel, "By heavily subsidizing water, governments give out the false message that it is abundant and can afford to be wasted—even as rivers are drying up, aquifers are being depleted, fisheries are collapsing, and species are going extinct."

But farmers, industries, and others benefiting from government water subsidies argue that they promote settlement and agricultural production in arid and semiarid areas, and that they stimulate local economies and help keep the prices of food, manufactured goods, and electricity low.

Most water resource experts believe that when water scarcity afflicts many areas in this century, governments will have to make the unpopular decision to raise water prices. China did so in 2002 because it faced water shortages in most of its major cities, rivers were running dry, and water tables were falling in key agricultural areas.

Higher water prices encourage water conservation, but they also make it harder for low-income farmers and city dwellers to buy enough water to meet their needs. On the other hand, when South Africa raised water prices, it established *lifeline* rates that give each household a set amount of water at a low price to meet basic needs. When users exceed this amount the price rises.

The second major cause of water waste is *lack of government subsidies for improving the efficiency of water use*. A basic rule of economics is that you get more of the behavior you reward. Withdrawing subsidies that encourage water waste and providing subsidies for efficient water use would sharply reduce water waste.

X *HOW WOULD YOU VOTE?* Should water prices be raised sharply to help reduce water waste? Cast your vote online at www.thomsonedu.com/biology/miller.

Solutions: Wasting Less Water in Irrigation

Although 60% of the world's irrigation water is wasted, improved irrigation techniques could reduce this proportion to 5–20%.

About 60% of the irrigation water applied throughout the world does not reach the targeted crops. Most irrigation systems obtain water from a groundwater well or a surface water source. The water then flows by gravity through unlined ditches in crop fields so the crops can absorb it (Figure 9-16, p. 184, left). This *flood irrigation* method delivers far more water than needed for crop growth and typically loses 40% of the water through evaporation, seepage, and runoff.

More efficient and environmentally sound irrigation technologies can greatly reduce water demands and waste on farms by delivering water more precisely to crops. For example, the *center-pivot, low-pressure sprinkler* (Figure 9-16, right) uses pumps to spray water on a crop. Typically, it allows 80% of the water to reach crops. *Low-energy precision application (LEPA) sprinklers,* another form of center-pivot irrigation, put 90–95% of the water where crops need it by spraying the water closer to the ground and in larger droplets than the center-pivot system. Another method is to use *soil moisture detectors* to water crops only when they need it.

Drip irrigation or *microirrigation systems* (shown in Figure 9-16, center) are the most efficient ways to deliver small amounts of water precisely to crops. These systems consist of a network of perforated plastic tubes installed at or below the ground level. Small holes or emitters in the tubing deliver drops of water at a slow and steady rate, close to the plant roots.

Drip irrigation is costly but it drastically reduces water waste, with 90–95% of the water input reaching the crops. It also reduces soil salinization and waterlogging. The flexible and lightweight tubing system can easily be fitted to match the patterns of crops in a field, and it can be left in place or moved around.

Drip irrigation is used on just over 1% of the world's irrigated crop fields and 4% of those in the United States. This percentage rises to 90% in Cyprus, 66% in Israel, and 13% in the U.S. state of California. The capital cost of conventional drip irrigation systems remains too high for most poor farmers and for use on low-value row crops. But if water were priced closer to the value of the ecological services it provides and if government subsidies that encourage waste were reduced or eliminated, drip irrigation would quickly be adopted to irrigate most of the world's crops. *Good news.* The capital cost of a new type of drip irrigation system is one-tenth as much per hectare as conventional drip systems.

According to the United Nations, reducing current global withdrawal of water for irrigation by just 10% would save enough water to grow crops and meet the estimated additional water demands of cities and industries through 2025. Figure 9-17 (p. 184) lists other ways to reduce water waste in irrigating crops.

Since 1950, water-short Israel has used many of these techniques to slash irrigation water waste by about 84% while irrigating 44% more land. Israel now treats and reuses 30% of its municipal sewage water for crop production and plans to increase this to 80% by 2025. The government also gradually removed most water subsidies to raise the price of irrigation water to one of the highest in the world. Israelis also import most of their wheat and meat and concentrate on growing fruits, vegetables, and flowers that need less water.

Many of the world's poor farmers cannot afford to use most of the modern technological methods for

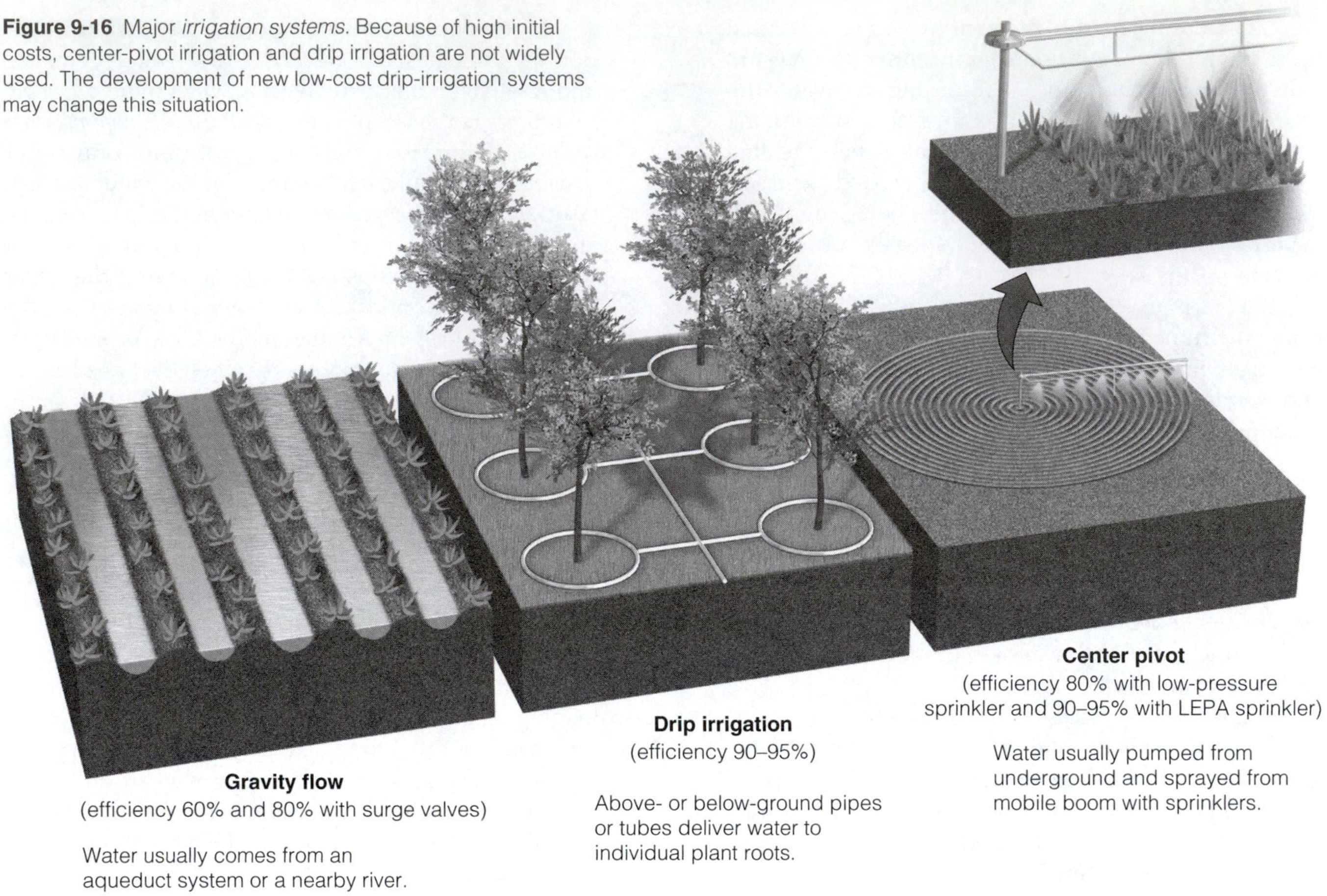

Figure 9-16 Major *irrigation systems.* Because of high initial costs, center-pivot irrigation and drip irrigation are not widely used. The development of new low-cost drip-irrigation systems may change this situation.

Gravity flow
(efficiency 60% and 80% with surge valves)

Water usually comes from an aqueduct system or a nearby river.

Drip irrigation
(efficiency 90–95%)

Above- or below-ground pipes or tubes deliver water to individual plant roots.

Center pivot
(efficiency 80% with low-pressure sprinkler and 90–95% with LEPA sprinkler)

Water usually pumped from underground and sprayed from mobile boom with sprinklers.

Solutions

Reducing Irrigation Water Waste

- Line canals bringing water to irrigation ditches

- Level fields with lasers

- Irrigate at night to reduce evaporation

- Monitor soil moisture to add water only when necessary

- Polyculture

- Organic farming

- Don't grow water-thirsty crops in dry areas

- Grow water-efficient crops using drought-resistant and salt-tolerant crop varieties

- Irrigate with treated urban waste water

- Import water intensive crops and meat

Figure 9-17 Solutions: methods for reducing water waste in irrigation. QUESTION: *Which two of these solutions do you think are the most important?*

increasing irrigation and its efficiency. Instead, they use small-scale and low-cost traditional technologies.

Millions of farmers in countries such as Bangladesh use human-powered treadle pumps to pump groundwater close to the earth's surface through irrigation ditches. Others use buckets, small tanks with holes, or cheap plastic tubing systems for drip irrigation.

Solutions: Wasting Less Water in Industry, Homes, and Businesses

We can save water by changing to yard plants that need little water, using drip irrigation, raising water prices, fixing leaks, and using water-saving toilets and other appliances.

Figure 9-18 lists ways to use water more efficiently in industries, homes, and businesses. Many homeowners and businesses in water-short areas are copying nature by replacing green lawns with native vegetation or aesthetically appealing stones. This win–win approach, called *Xeriscaping* (pronounced "ZER-i-scaping"), reduces water use by 30–85% and sharply reduces inputs of labor, fertilizer, and fuel. It also reduces polluted runoff, air pollution, and yard wastes.

About one-fifth of all U.S. public water systems do not have water meters and charge a single low rate for

Solutions

Reducing Water Waste

- Redesign manufacturing processes

- Repair leaking underground pipes

- Retire grass by landscaping yards with plants that require little water

- Use drip irrigation

- Fix water leaks

- Use water meters and charge for all municipal water use

- Raise water prices

- Use waterless composting toilets

- Require water conservation in water-short cities

- Use water-saving toilets, showerheads, and front-loading clothes washers

- Collect and reuse household water to irrigate lawns and nonedible plants

- Purify and reuse water for houses, apartments, and office buildings

- Don't waste energy

Figure 9-18 Solutions: methods of reducing water waste in industries, homes, and businesses. QUESTION: *Which two of these solutions do you think are the most important?*

almost unlimited use of high-quality water. Many apartment dwellers have little incentive to conserve water because water use is included in their rent. In Boulder, Colorado, introducing water meters reduced water use by more than one-third.

The FAO estimates that if current trends continue, within 40 years we will need the world's entire reliable flow of river water just to dilute and transport the wastes we produce. We can save much of this water by replacing waste treatment systems—in which we use large amounts of water good enough to drink to dilute and flush away industrial, animal, and household wastes—with systems that mimic the way nature deals with wastes.

One way to do this would be to return the nutrient-rich sludge produced by conventional waste treatment plants to the soil as a fertilizer, instead of using freshwater to transport it. Banning the discharge of industrial toxic wastes into municipal sewer systems would make this feasible. Another way is to rely more on waterless composting toilets that convert human fecal matter to a small amount of dry and odorless soil-like

humus material that can be removed from a composting chamber every year or so and returned to the soil as fertilizer. They work. I used one for 15 years without any problems in my environmental experimental office and house in the woods of North Carolina.

Solutions: Using Water More Sustainably

We can use water more sustainably by cutting waste, raising water prices, preserving forests in water basins, and slowing population growth.

A scarcity of freshwater is becoming one of the world's most serious environmental problems, highlighting the urgent need for all nations to develop and implement policies for using water resources more sustainably. Sustainable water use is based on the commonsense principle stated in an old Inca proverb: "The frog does not drink up the pond in which it lives." Figure 9-19 lists ways to implement this principle.

Solutions

Sustainable Water Use

- Not depleting aquifers

- Preserving ecological health of aquatic systems

- Preserving water quality

- Integrated watershed management

- Agreements among regions and countries sharing surface water resources

- Outside party mediation of water disputes between nations

- Marketing of water rights

- Raising water prices

- Wasting less water especially for agriculture which uses 10% of the world's water

- Decreasing government subsidies for supplying water

- Increasing government subsidies for reducing water waste

- Protecting forests, wetlands, and other natural ecosystems that store water

- Slowing population growth

Figure 9-19 Solutions: methods for achieving more sustainable use of the earth's water resources. QUESTION: *Which two of these solutions do you think are the most important?*

The challenge in encouraging such a *blue revolution*—in which we use the earth's water more sustainably—is to implement a mix of strategies. First, political leaders need to make protection of the health and functioning of freshwater ecosystems the top priority. In other words, begin by protecting the earth's freshwater natural capital. One approach involves using technologies described above to irrigate crops more efficiently and to save water in industries and homes.

What Can You Do?

Water Use and Waste

- Use water-saving toilets, showerheads, and faucet aerators.
- Shower instead of taking baths, and take short showers.
- Stop water leaks.
- Turn off sink faucets while brushing teeth, shaving, or washing.
- Flush toilet only when necessary.
- Wash only full loads of clothes or use the lowest possible water-level setting for smaller loads.
- Use recycled (gray) water for lawn, gardens, house plants, car washing.
- Wash a car from a bucket of soapy water, and use the hose for rinsing only.
- If you use a commercial car wash, try to find one that recycles its water.
- Replace your lawn with native plants that need little if any watering and decorative gravel or rocks.
- Water lawns and gardens in the early morning or evening.
- Sweep or blow off driveways instead of hosing off with water.
- Use drip irrigation and mulch for gardens and flowerbeds.

Figure 9-20 Individuals matter: ways you can reduce your use and waste of water. Visit **www.h2ouse.org** for an array of water-saving tips from the EPA and the California Urban Water Conservation Council that can be used anywhere. QUESTION: *Which four of these actions do you think are the most important? Which actions on this list do you already take or plan to take?*

A second approach uses economic and political decisions to remove government subsidies that cause water to be underpriced and thus wasted, while guaranteeing affordable prices for low-income consumers and adding subsidies that reward reduced water waste.

Third, we must leave enough water in rivers to protect wildlife, ecological processes, and the natural ecological services provided by rivers.

Fourth, we need to get very serious about slowing global warming, which can alter the natural distribution of water on the planet.

Finally, we ought to protect the forests, wetlands, mountain glaciers, and other natural ecosystems that store and release water, control erosion, facilitate groundwater recharge, provide flood control, and help maintain water quality.

Each of us can help bring about this blue revolution by using and wasting less water. Figure 9-20 lists ways you can reduce your water use and waste.

The four principles of sustainability can guide us in using water more sustainably. We can rely more on solar energy to desalinate water, recycle wastewater, preserve biodiversity by not disrupting aquatic systems, maintain interactions among species that help control population growth, and reduce human population growth and water waste.

WATER POLLUTION: SOURCES, TYPES, AND EFFECTS

Water Pollution and Its Sources (Science)

Water pollution degrades water quality and can come from single sources and larger dispersed sources.

Water pollution is any chemical, biological, or physical change in water quality that has a harmful effect on living organisms or makes water unsuitable for desired uses.

Water pollution can come from single (point) sources or from larger and dispersed (nonpoint) sources. **Point sources** discharge pollutants at specific locations through drain pipes, ditches, or sewer lines into bodies of surface water. Examples include factories, sewage treatment plants (which remove some but not all pollutants), underground mines, and oil tankers.

Because point sources are at specific places, they are fairly easy to identify, monitor, and regulate. Most developed countries have laws that help control point-source discharges of harmful chemicals into aquatic systems. But there is little control of such discharges in most developing countries.

Nonpoint sources are scattered and diffuse and cannot be traced to any single site of discharge. Examples include deposition from the atmosphere and runoff of chemicals and sediments into surface water from croplands, livestock feedlots, logged forests, surface mines, and urban streets, lawns, golf courses, and parking lots. We have made little progress in controlling nonpoint source water pollution because of the difficulty and expense of identifying and controlling discharges from so many diffuse sources.

The major sources of water pollution are agriculture, industries, and mining. *Agricultural activities* are by far the leading cause of water pollution. Sediment eroded from agricultural lands is the largest source. Other major agricultural pollutants include fertilizers and pesticides, bacteria from livestock and food processing wastes, and excess salt from soils of irrigated cropland.

Industrial facilities are another large source of water pollution from a variety of harmful inorganic and organic chemicals. *Mining* is a third source. Surface mining disturbs the earth's surface, creating a major source of eroded sediments and runoff of toxic chemicals.

Climate change from global warming can also affect water pollution. In a warmer world, some areas will get more precipitation and other areas will get less. Intense downpours can flush more harmful chemicals, plant nutrients, and microorganisms into waterways. Prolonged drought can reduce river flows, which dilute wastes, and spread infectious diseases more rapidly among people who lack enough water to stay clean.

Major Water Pollutants and Their Effects (Science)

Water is polluted by disease-causing agents, oxygen-demanding wastes, plant nutrients, organic and inorganic chemicals, sediment, and excess heat.

Table 9-1 lists the major classes and examples of water pollutants along with their human sources and their harmful effects.

Two large concerns about the effects of pollutants have to do with sanitation and the safety of drinking water. According to the United Nations, at least 2.4 billion people in developing countries (almost two-thirds of them in rural areas in India and China) have no access to basic sewer or sanitation facilities. And 1.1 billion people in these countries (42% of them in sub-Saharan Africa) lack access to clean drinking water. The World Health Organization (WHO) estimates that 5 million people—most of them children younger than age 5—die prematurely every year from infectious diseases spread by contaminated water or lack of water for adequate hygiene. Each year, diarrhea alone kills about 1.9 million people—about 90% of them children under age 5 in developing countries. This means that diarrhea kills a child every 17 seconds.

The United Nations estimates it would cost about $23 billion a year over 8–10 years to bring low-cost safe water and sanitation to the world's people who do not have it.

Table 9-1 Major Water Pollutants and Their Sources

Type/Effects	Examples	Major Sources
Infectious agents *Cause diseases*	Bacteria, viruses, parasites	Human and animal wastes
Oxygen-demanding wastes *Deplete dissolved oxygen needed by aquatic species*	Biodegradable animal wastes and plant debris	Sewage, animal feedlots, food processing facilities, pulp mills
Plant nutrients *Cause excessive growth of algae and other species*	Nitrates (NO_3^-) and phosphates (PO_4^{3-})	Sewage, animal wastes, inorganic fertilizers
Organic chemicals *Add toxins to aquatic systems*	Oil, gasoline, plastics, pesticides, cleaning solvents	Industry, farms, households
Inorganic chemicals *Add toxins to aquatic systems*	Acids, salts, metal compounds	Industry, households, surface runoff
Sediments *Disrupt photosynthesis, food webs, other processes*	Soil, silt	Land erosion
Thermal *Make some species vulnerable to disease*	Heat	Electric power and industrial plants

POLLUTION OF FRESHWATER STREAMS, LAKES, AND AQUIFERS

Learn more about how pollution affects the water in a stream and the creatures living there at ThomsonNOW.

Water Pollution Problems of Streams (Science)

Flowing streams can recover from a moderate level of degradable water pollutants if they are not overloaded and their flows are not reduced.

Rivers and other flowing streams can recover rapidly from moderate levels of degradable, oxygen-demanding wastes and excess heat through a combination of dilution and biodegradation of such wastes by bacteria. But this natural recovery process does not work if streams are overloaded with pollutants or when drought, damming, or water diversion for agriculture and industry reduce their flows. Likewise, these natural dilution and biodegradation processes do not eliminate slowly degradable and nondegradable pollutants.

In a flowing stream, the breakdown of degradable wastes by bacteria depletes dissolved oxygen and creates an *oxygen sag curve* (Figure 9-21). This reduces or eliminates populations of organisms with high oxygen requirements until the stream is cleansed of wastes. Similar oxygen sag curves can be plotted when heated water from industrial and power plants is discharged into streams.

Stream Pollution in Developed Countries (Science and Politics)

Most developed countries have sharply reduced point-source pollution, but toxic chemicals and pollution from nonpoint sources are still problems.

Water pollution control laws enacted in the 1970s have greatly increased the number and quality of wastewater treatment plants in the United States and most other developed countries. Such laws also require industries to reduce or eliminate point-source discharges of harmful chemicals into surface waters.

These efforts have enabled the United States to hold the line against increased pollution by disease-causing agents and oxygen-demanding wastes in most of its streams. This is an impressive accomplishment given the country's increased economic activity, resource consumption, and population since passage of these laws.

One success story is the cleanup of Ohio's Cuyahoga River. It was so polluted that in 1959 and again in 1969, it caught fire and burned for several

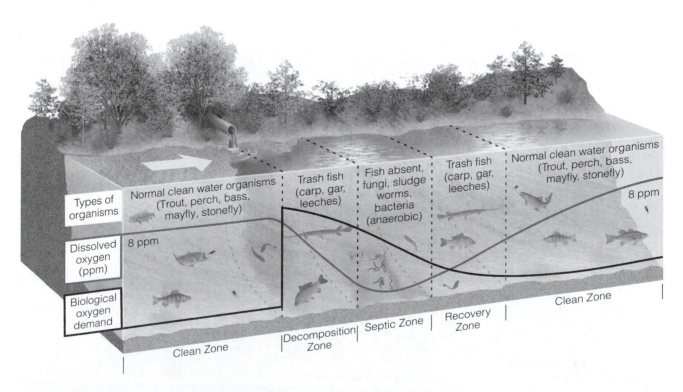

ThomsonNOW™ Active Figure 9-21 Natural capital: dilution and decay of degradable, oxygen-demanding wastes and heat in a stream, showing the oxygen sag curve and the oxygen demand curve. Depending on flow rates and the amount of pollutants, streams recover from oxygen-demanding wastes and heat if they are given enough time and are not overloaded. *See an animation based on this figure and take a short quiz on the concept.*

days as it flowed through Cleveland. The highly publicized image of this burning river prompted elected officials to enact laws limiting the discharge of industrial wastes into the river and sewage systems, and to provide funds to upgrade sewage treatment facilities. Today the river is cleaner and is widely used by boaters and anglers. This accomplishment illustrates the power of bottom-up pressure by citizens who prodded elected officials to change a severely polluted river into an economically and ecologically valuable public resource. Individuals matter!

But large fish kills and drinking water contamination still occur in parts of developed countries. Two causes of these problems are accidental or deliberate releases of toxic inorganic and organic chemicals by industries or mines and malfunctioning sewage treatment plants. A third cause is nonpoint runoff of pesticides and excess plant nutrients from cropland and animal feedlots.

Global Outlook: Stream Pollution in Developing Countries

Stream pollution in most developing countries is a serious and growing problem.

Stream pollution from discharges of untreated sewage and industrial wastes is a serious and growing problem in most developing countries. According to a 2003 report by the World Commission on Water in the 21st Century, half of the world's 500 rivers are heavily polluted, most of them running through developing countries. Most of these countries cannot afford to build waste treatment plants and do not have, or do not enforce, laws for controlling water pollution.

According to the Global Water Policy Project, most cities in developing countries discharge 80–90% of their untreated sewage directly into rivers, streams, and lakes whose waters are then used for drinking water, bathing, and washing clothes. Industrial wastes and sewage pollute more than two-thirds of India's water resources and 54 of the 78 streams monitored in China. Only about 10% of the sewage produced in Chinese cities is treated. In Latin America and Africa, most streams passing through urban or industrial areas suffer from severe pollution.

Pollution Problems of Lakes (Science)

Dilution of pollutants in lakes is less effective than in most streams because most lake water is not mixed well and has little flow.

In lakes and reservoirs, dilution of pollutants often is less effective than in streams for two reasons. *First,* lakes and reservoirs often contain stratified layers that undergo little vertical mixing. *Second,* they have little flow. The flushing and changing of water in lakes and large artificial reservoirs can take from 1 to 100 years, compared with several days to several weeks for streams.

This means that lakes and reservoirs are more vulnerable than streams to contamination by runoff or discharge of plant nutrients, oil, pesticides, and toxic substances such as lead, mercury, and selenium. These contaminants can kill bottom life and fish and birds that feed on contaminated aquatic organisms. Many toxic chemicals and acids also enter lakes and reservoirs from the atmosphere.

Cultural Eutrophication (Science)

Various human activities can overload lakes with plant nutrients, which decrease dissolved oxygen and kill some aquatic species.

Eutrophication is the name given to the natural nutrient enrichment of a shallow lake, estuary, or slow-moving stream, mostly from runoff of plant nutrients such as nitrates and phosphates from surrounding land. Over time, some lakes become more eutrophic, but others do not because of differences in the surrounding drainage basin.

Near urban or agricultural areas, human activities can greatly accelerate the input of plant nutrients to a lake—a process is called cultural eutrophication. It is mostly nitrate- and phosphate-containing effluents from various sources that cause such a change. These sources include runoff from farmland, animal feedlots, urban areas, and mining sites, and from the discharge of treated and untreated municipal sewage. Some nitrogen also reaches lakes by deposition from the atmosphere.

During hot weather or drought, this nutrient overload produces dense growths or "blooms" of organisms such as algae and cyanobacteria and thick growths of water hyacinths and other aquatic plants. These dense colonies of plant life can reduce lake productivity and fish growth by decreasing the input of solar energy needed for photosynthesis by the phytoplankton that support fish. If excess nutrients continue to flow into a lake, anaerobic bacteria take over and produce gaseous products such as smelly, highly toxic hydrogen sulfide and flammable methane.

According to the EPA, about one-third of the 100,000 medium to large lakes and about 85% of the large lakes near major population centers in the United States have some degree of cultural eutrophication. One-fourth of the lakes in China also suffer from cultural eutrophication.

There are several ways to *prevent* or *reduce* cultural eutrophication. We can use advanced (but expensive) waste treatment to remove nitrates and phosphates before wastewater enters lakes, ban or limit the use of phosphates in household detergents and other cleaning agents, and employ soil conservation and land-use control to reduce nutrient runoff.

There are also several ways to *clean up* lakes suffering from cultural eutrophication. We can mechanically remove excess weeds, control undesirable plant growth with herbicides and algicides, and pump air through lakes and reservoirs to prevent oxygen depletion (an expensive and energy-intensive method).

As usual, pollution prevention is more effective and usually cheaper in the long run than cleanup. The good news is that if excessive inputs of plant nutrients stop, a lake usually can recover from cultural eutrophication (Case Study, below).

Case Study: Pollution in the Great Lakes (Science, Economics, and Politics)

Pollution of the Great Lakes has decreased significantly but there is a long way to go.

The five connected Great Lakes contain at least 95% of the fresh surface water in the United States and one-fifth of the world's fresh surface water. At least 38 million people in the United States and Canada obtain their drinking water from these lakes.

Despite their enormous size, these lakes are vulnerable to pollution from point and nonpoint sources. One reason is that less than 1% of the water entering the Great Lakes flows out to the St. Lawrence River each year. Another reason is that in addition to land runoff, the lakes get atmospheric deposition of large quantities of acids, pesticides, and other toxic chemicals, often blown in from hundreds or thousands of kilometers away.

By the 1960s, many areas of the Great Lakes were suffering from severe cultural eutrophication, huge fish kills, and contamination from bacteria and a variety of toxic industrial wastes. The impact on Lake Erie was particularly intense because it is the shallowest of the Great Lakes and has the highest concentrations of people and industrial activity along its shores. Many bathing beaches had to be closed, and by 1970 the lake had lost most of its native fish.

Since 1972, Canada and the United States have joined forces and spent more than $20 billion on a Great Lakes pollution control program. This program has decreased algal blooms, increased dissolved oxygen levels and sport and commercial fishing catches in Lake Erie, and allowed most swimming beaches to reopen.

These improvements occurred mainly because of new or upgraded sewage treatment plants, better treatment of industrial wastes, and bans on the use of detergents, household cleaners, and water conditioners that contained phosphates.

Despite this important progress many problems remain. Each August, a large zone that is severely depleted of dissolved oxygen forms in Lake Erie's central area off Cleveland, Ohio. And according to a 2000 survey by the EPA, more than three-fourths of the shoreline of the Great Lakes is not clean enough for swimming or for supplying drinking water. Nonpoint land runoff of pesticides and fertilizers from urban sprawl now surpasses industrial pollution as the greatest threat to the lakes. Sediments in 26 toxic hot spots remain heavily polluted. And *biological pollution* from invasions by zebra mussels and more than 150 other alien species threaten some native species and cause billions of dollars in damages.

About half the toxic compounds entering the lakes come from atmospheric deposition of pesticides, mercury from coal-burning plants, and other toxic chemicals from as far away as Mexico and Russia. A recent survey by Wisconsin biologists found that one fish in four taken from the Great Lakes is unsafe for human consumption. Another problem has been an 80% drop in EPA funding for cleanup of the Great Lakes since 1992.

Some environmentalists call for banning the use of toxic chlorine compounds such as bleach in the pulp and paper industry around the Great Lakes. They would also ban new incinerators (which can release toxic chemicals into the atmosphere) in the area, and they would stop the discharge into the lakes of 70 toxic chemicals that threaten human health and wildlife. So far, officials of the industries involved have successfully opposed such bans.

The Threat from Groundwater Pollution (Science)

Groundwater can become contaminated with a variety of chemicals because it cannot effectively cleanse itself and dilute and disperse pollutants.

The drinking water for about half of the U.S. population and 95% of those in rural areas comes from groundwater. According to many scientists, groundwater pollution is a serious threat to human health.

Common pollutants such as fertilizers, pesticides, gasoline, and organic solvents can seep into groundwater from numerous sources (Figure 9-22). People who dump or spill gasoline, oil, and paint thinners and other organic solvents onto the ground also contaminate groundwater.

When groundwater becomes contaminated, it cannot cleanse itself of *degradable wastes* as flowing surface water does (Figure 9-21). Groundwater flows so slowly—usually less than 0.3 meter (1 foot) per day—that contaminants are not diluted and dispersed effectively. In addition, groundwater usually has much lower concentrations of dissolved oxygen (which helps decompose many contaminants) and smaller populations of decomposing bacteria. Also, the usually cold temperatures of groundwater slow down chemical reactions that decompose wastes.

Thus, it can take hundreds to thousands of years for contaminated groundwater to cleanse itself of

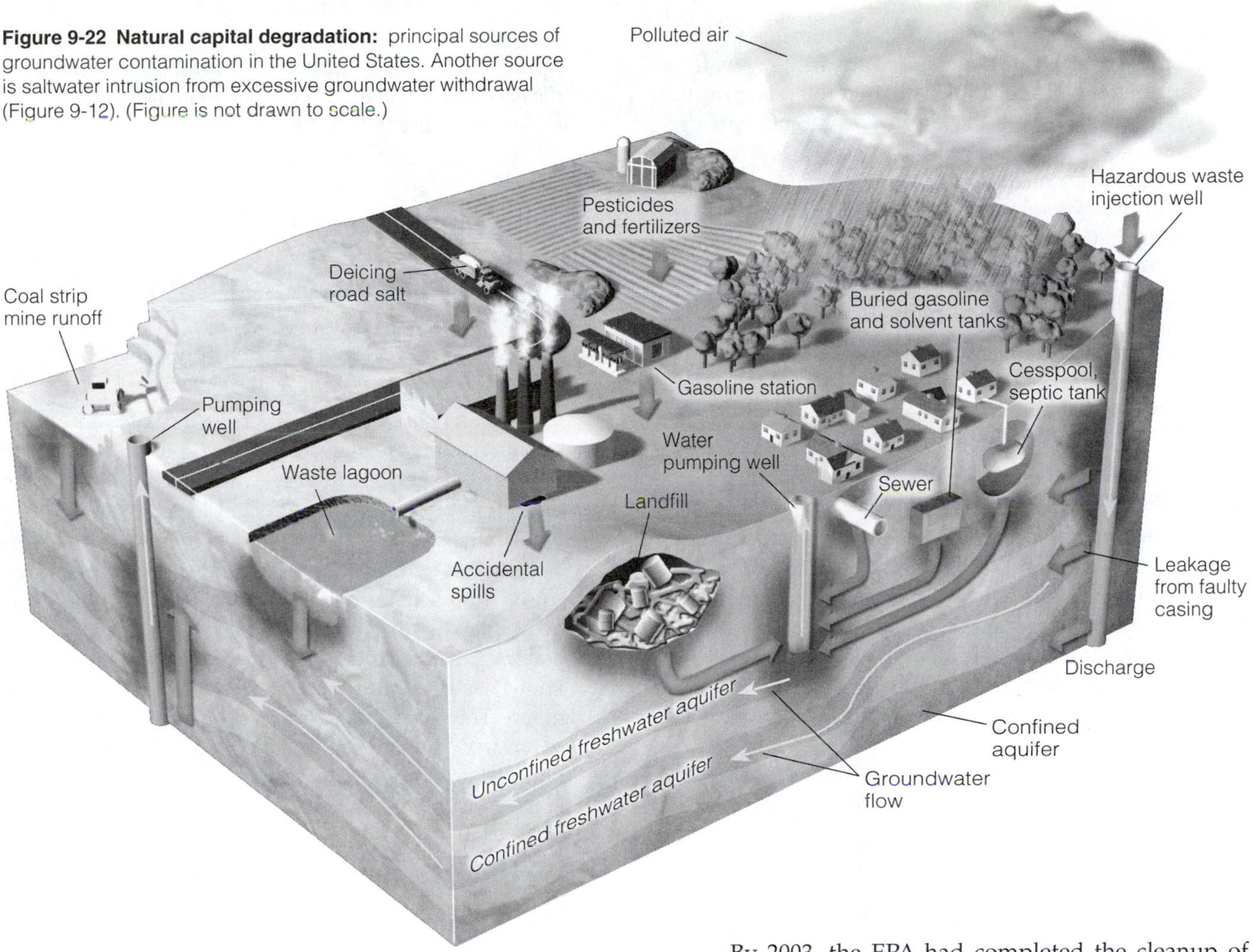

Figure 9-22 Natural capital degradation: principal sources of groundwater contamination in the United States. Another source is saltwater intrusion from excessive groundwater withdrawal (Figure 9-12). (Figure is not drawn to scale.)

Labels in figure: Polluted air; Hazardous waste injection well; Pesticides and fertilizers; Deicing road salt; Buried gasoline and solvent tanks; Cesspool, septic tank; Coal strip mine runoff; Gasoline station; Pumping well; Water pumping well; Sewer; Waste lagoon; Landfill; Leakage from faulty casing; Accidental spills; Discharge; Confined aquifer; Unconfined freshwater aquifer; Groundwater flow; Confined freshwater aquifer

degradable wastes. On a human time scale, *nondegradable wastes* (such as toxic lead, arsenic, and fluoride) are there permanently. And *slowly degradable wastes* (such as DDT) are there for decades.

Extent of Groundwater Pollution (Science)

Leaks from a number of sources have contaminated groundwater in parts of the world.

On a global scale, we do not know much about groundwater pollution because few countries go to the great expense of locating, tracking, and testing aquifers. But scientific studies in scattered parts of the world provide some bad news.

According to the EPA and the U.S. Geological Survey, one or more organic chemicals contaminate about 45% of municipal groundwater supplies in the United States. An EPA survey of 26,000 industrial waste ponds and lagoons in the United States found that one-third of them had no liners to prevent toxic liquid wastes from seeping into aquifers. One-third of these sites are within 1.6 kilometers (1 mile) of a drinking water well.

By 2003, the EPA had completed the cleanup of 297,000 of 436,000 underground tanks that were found to be leaking gasoline, diesel fuel, home heating oil, or toxic solvents into groundwater in the United States. During this century, scientists expect many of the millions of such tanks installed around the world in recent decades to corrode, leak, contaminate groundwater, and become a major global health problem.

Determining the extent of a leak from a single underground tank can cost $25,000–250,000, and cleanup costs range from $10,000 to more than $250,000. If the chemical reaches an aquifer, effective cleanup is often not possible or is too costly. *Bottom line:* wastes we think we have thrown away or stored safely can escape and come back to haunt us.

In the United States, groundwater pollution by MTBE (methyl tertiary butyl ether)—a gasoline additive used since 1979—is a serious problem. MTBE is a suspected carcinogen. But by the time this was discovered in the 1990s, about 250,000 leaking gasoline tanks had contaminated aquifers in many parts of the country. Use of MTBE is being phased out, but plumes of contaminated groundwater will move through aquifers for decades.

Toxic *arsenic* (As) contaminates drinking water when a well is drilled into an aquifer where soils and rock are naturally rich in arsenic. According to the WHO, more than 112 million people are drinking water with arsenic levels 5–100 times the WHO standard of 10 ppb, mostly in Bangladesh, China, and India's state of West Bengal.

There is also concern over arsenic levels in drinking water in parts of the United States, mostly in the western half of the country, where drinking water contains 3–10 ppb of arsenic. According to the WHO and other scientists, even the 10-ppb standard is not safe. Many scientists call for lowering the standard to 3–5 ppb, but this would be very expensive.

Solutions: Protecting Groundwater

Prevention is the most effective and affordable way to protect groundwater from pollutants.

Treating a contaminated aquifer involves eliminating the source of pollution and drilling monitoring wells to determine how far, in what direction, and how fast the contaminated plume is moving. Then a computer model is used to project future dispersion of the contaminant in the aquifer. The final step is to develop and implement a strategy to clean up the contamination (Figure 9-23, right).

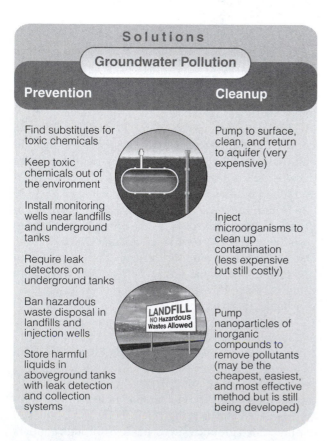

Figure 9-23 Solutions: methods for preventing and cleaning up contamination of groundwater. QUESTION: *Which two of these solutions do you think are the most important?*

Because of the difficulty and expense of cleaning up a contaminated aquifer, *preventing contamination is the most effective and cheapest way to protect groundwater resources* (Figure 9-23, left).

OCEAN POLLUTION

How Much Pollution Can the Oceans Tolerate? (Science)

Oceans, if they are not overloaded, can disperse and break down large quantities of degradable pollutants.

The oceans can dilute, disperse, and degrade large amounts of raw sewage and other types of degradable pollutants, especially in deep waters. Also, some forms of marine life have been affected less by some pollutants than expected.

Some scientists suggest it is safer to dump sewage sludge and most other harmful wastes into the deep ocean than to bury them on land or burn them in incinerators. Other scientists disagree, pointing out that we know less about the deep ocean than we do about the moon. They add that dumping harmful wastes in the ocean would delay urgently needed pollution prevention and promote further degradation of this vital part of the earth's life-support system.

Pollution of Coastal Waters (Science)

Pollution of coastal waters near heavily populated areas is a serious problem.

Coastal areas—especially wetlands and estuaries, coral reefs, and mangrove swamps—bear the brunt of our enormous inputs of pollutants and wastes into the ocean (Figure 9-24). This is not surprising because about 40% of the world's population lives on or near the coast and 14 of the world's 15 largest metropolitan areas (each with 10 million people or more) are near coastal waters (Figure 5-11, p. 97).

In most coastal developing countries and in some coastal developed countries, municipal sewage and industrial wastes are dumped into the sea without treatment. For example, about 85% of the sewage from large cities along the Mediterranean Sea (with a coastal population of 200 million people during tourist season) is discharged into the sea untreated. This causes widespread beach pollution and shellfish contamination.

Recent studies of some U.S. coastal waters have found vast colonies of human viruses from raw sewage, effluents from sewage treatment plants (which do not remove viruses), and leaking septic tanks. According to one study, about one-fourth of the people using coastal beaches in the United States develop ear infections, sore throats, eye irritations, respiratory disease, or gastrointestinal disease.

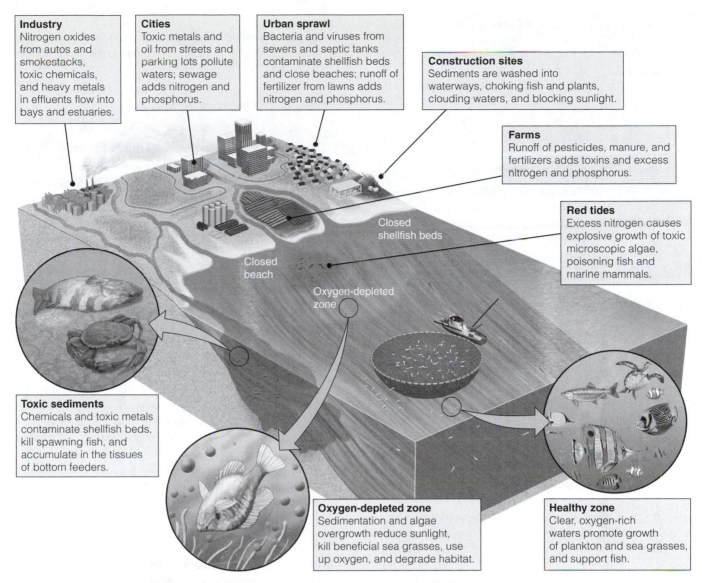

Industry
Nitrogen oxides from autos and smokestacks, toxic chemicals, and heavy metals in effluents flow into bays and estuaries.

Cities
Toxic metals and oil from streets and parking lots pollute waters; sewage adds nitrogen and phosphorus.

Urban sprawl
Bacteria and viruses from sewers and septic tanks contaminate shellfish beds and close beaches; runoff of fertilizer from lawns adds nitrogen and phosphorus.

Construction sites
Sediments are washed into waterways, choking fish and plants, clouding waters, and blocking sunlight.

Farms
Runoff of pesticides, manure, and fertilizers adds toxins and excess nitrogen and phosphorus.

Red tides
Excess nitrogen causes explosive growth of toxic microscopic algae, poisoning fish and marine mammals.

Closed shellfish beds

Closed beach

Oxygen-depleted zone

Toxic sediments
Chemicals and toxic metals contaminate shellfish beds, kill spawning fish, and accumulate in the tissues of bottom feeders.

Oxygen-depleted zone
Sedimentation and algae overgrowth reduce sunlight, kill beneficial sea grasses, use up oxygen, and degrade habitat.

Healthy zone
Clear, oxygen-rich waters promote growth of plankton and sea grasses, and support fish.

Figure 9-24 Natural capital degradation: residential areas, factories, and farms all contribute to the pollution of coastal waters and bays. According to the UN Environment Programme, coastal water pollution costs the world $16 billion annually—$731,000 a minute—due to ill health and premature death.

Runoffs of sewage and agricultural wastes into coastal waters introduce large quantities of nitrate and phosphate plant nutrients, which can cause explosive growth of harmful algae. These *harmful algal blooms* (HABs) are called red, brown, or green toxic tides. They can release waterborne and airborne toxins that damage fisheries, kill some fish-eating birds, reduce tourism, and poison seafood.

Case Study: The Chesapeake Bay (Science, Economics, and Politics)

Pollutants from six states contaminate the shallow Chesapeake Bay estuary, but cooperative efforts have reduced some of the pollution inputs.

Since 1960, the Chesapeake Bay—America's largest estuary—has been in serious trouble from water pol-

lution, mostly because of human activities. Between 1940 and 2004, the number of people living in the Chesapeake Bay area grew from 3.7 million to 17 million, and may soon reach 18 million.

The estuary receives wastes from point and nonpoint sources scattered throughout a huge drainage basin that includes 9 large rivers and 141 smaller streams and creeks in parts of six states (Figure 9-25, p. 194). The bay has become a huge pollution sink because only 1% of the waste entering it is flushed into the Atlantic Ocean.

Phosphate and nitrate levels have risen sharply in many parts of the bay, causing algal blooms and oxygen depletion. Commercial harvests of its once-abundant oysters, crabs, and several important fish have fallen sharply since 1960 because of a combination of pollution, overfishing, and disease.

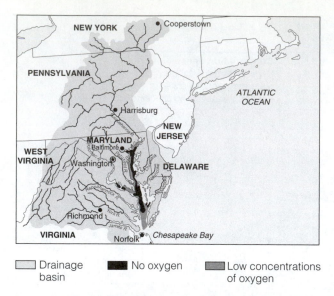

Drainage basin · No oxygen · Low concentrations of oxygen

Figure 9-25 Natural capital degradation: *Chesapeake Bay*, the largest estuary in the United States, is severely degraded as a result of water pollution from point and nonpoint sources in six states and from the atmospheric deposition of air pollutants.

Point sources, primarily sewage treatment plants and industrial plants (often in violation of their discharge permits), account for 60% by weight of the phosphates. Nonpoint sources—mostly runoff of fertilizer and animal wastes from urban, suburban, and agricultural land and deposition from the atmosphere—account for about 60% by weight of the nitrates.

In 1983, the United States implemented the Chesapeake Bay Program. In this ambitious attempt at *integrated coastal management*, citizens' groups, communities, state legislatures, and the federal government are working together to reduce pollution inputs into the bay.

Strategies include establishing land-use regulations in the bay's six watershed states to reduce agricultural and urban runoff, banning phosphate detergents, upgrading sewage treatment plants, and better monitoring of industrial discharges. In addition, wetlands are being restored and large areas of the bay are being replanted with sea grasses to help filter out nutrients and other pollutants. Some have proposed rebuilding the bay's once huge oyster population as a way to help clean up the water.

This hard work has paid off. Between 1985 and 2000, phosphorus levels declined 27%, nitrogen levels dropped 16%, and grasses growing on the bay's floor have made a comeback. This is a significant achievement given the increasing population in the watershed and the fact that nearly 40% of the nitrogen inputs come from the atmosphere.

However, there is still a long way to go, and a sharp drop in state and federal funding has slowed

progress. Despite some setbacks, the Chesapeake Bay Program shows what can be done when diverse groups work together to achieve goals that benefit both wildlife and people.

Ocean Oil Pollution (Science)

Most oil pollution comes from human activities on land.

Crude petroleum (oil as it comes out of the ground) and *refined petroleum* (fuel oil, gasoline, and other processed petroleum products) reach the ocean from a number of sources.

Tanker accidents and blowouts at offshore drilling rigs (when oil escapes under high pressure from a borehole in the ocean floor) get most of the publicity because of their high visibility. But *studies show that most ocean oil pollution comes from activities on land.* According to a 2004 study by the Pew Oceans Commission, every 8 months an amount of oil equal to that spilled by the *Exxon Valdez* tanker into Alaska's Prince William Sound in 1989 drains from the land into the oceans. At least 37% and perhaps half of the oil reaching the oceans is waste oil dumped, spilled, or leaked onto the land or into sewers by cities, industries, and people changing their own motor oil.

Volatile organic hydrocarbons in oil immediately kill a number of aquatic organisms, especially in their vulnerable larval forms. Other chemicals in oil form tar-like globs that float on the surface and coat the feathers of birds (especially diving birds) and the fur of marine mammals. This oil coating destroys their natural insulation and buoyancy, causing many of them to drown or die of exposure from loss of body heat.

Heavy oil components that sink to the ocean floor or wash into estuaries can smother bottom-dwelling organisms such as crabs, oysters, mussels, and clams or make them unfit for human consumption. Some oil spills have killed reef corals.

Research shows that populations of many forms of marine life recover from exposure to large amounts of *crude oil* within about 3 years. But recovery from exposure to *refined oil*, especially in estuaries and salt marshes, can take 10–15 years. Oil slicks that wash onto beaches can have a serious economic impact on coastal residents, who lose income from fishing and tourist activities.

If they are not too large, oil spills can be partially cleaned up by mechanical means including floating booms, skimmer boats, and absorbent devices such as large pillows filled with feathers or hair. Chemical, fire, and natural methods such as using cocktails of bacteria to speed up oil decomposition are also employed.

But scientists estimate that current methods can recover no more than 15% of the oil from a major spill. This explains why preventing oil pollution is the most effective and, in the long run, the least costly approach. One of the best ways to reduce oil pollution is to use oil tankers with double hulls that help prevent the release of oil from collisions. However, by 2005—16 years after the *Exxon Valdez* spill—about half of the world's 10,000 oil tankers still had the older and more vulnerable single hulls.

Solutions: How Can We Protect Coastal Waters? (Science and Politics)

Preventing or reducing the flow of pollution from the land and from streams emptying into the ocean is the key to protecting the oceans.

Figure 9-26 lists ways of preventing and reducing excessive pollution of coastal waters. The key to protecting oceans is to reduce the flow of pollution from the land and from streams emptying into the ocean. Thus, ocean pollution control must be linked with land-use and air pollution policies.

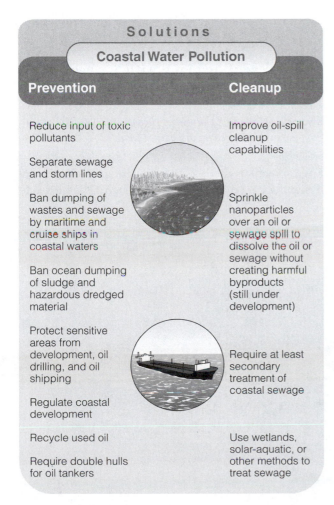

Figure 9-26 Solutions: methods for preventing and cleaning up excessive pollution of coastal waters. QUESTION: *Which two of these solutions do you think are the most important?*

PREVENTING AND REDUCING SURFACE WATER POLLUTION

Solutions: Reducing Surface Water Pollution from Nonpoint Sources

The key to reducing nonpoint pollution, most of it from agriculture, is to prevent it from reaching bodies of water.

There are a number of ways to reduce nonpoint water pollution, most of which comes from agriculture. Farmers can reduce soil erosion by keeping cropland covered with vegetation. They can also reduce the amount of fertilizer that runs off into surface waters and leaches into aquifers by using slow-release fertilizers, using no fertilizer at all on steeply sloped land, and planting buffer zones of vegetation between cultivated fields and nearby surface water.

Applying pesticides only when needed and relying more on integrated pest management (p. 168) can reduce pesticide runoff. Farmers can control runoff and infiltration of manure from animal feedlots by planting buffers and locating feedlots and animal waste sites away from steeply sloped land, surface water, and flood zones.

X *HOW WOULD YOU VOTE?* Should we greatly increase efforts to reduce water pollution from nonpoint sources? Cast your vote online at www.thomsonedu.com/biology/miller.

Laws for Reducing Water Pollution from Point Sources (Politics and Economics)

Most developed countries use laws to set water pollution standards, but such laws rarely exist or are poorly enforced in most developing countries.

The Federal Water Pollution Control Act of 1972 (renamed the Clean Water Act when it was amended in 1977) and the 1987 Water Quality Act form the basis of U.S. efforts to control pollution of the country's surface waters. The Clean Water Act sets standards for allowable levels of key water pollutants and requires polluters to get permits limiting how much of various pollutants they can discharge into aquatic systems.

The EPA is also experimenting with a *discharge trading policy* that uses market forces to reduce water pollution in the United States. Under this program, a water pollution source is allowed to pollute at levels higher than allowed in its permits by buying credits from permit holders with pollution levels below their allowed levels.

Environmental scientists warn that such a system is no better than the caps set for total pollution levels in various areas. They call for careful scrutiny of the cap levels and for gradually lowering the caps to encourage prevention of water pollution and development of better technology for controlling water pollution. Neither scrutiny of the cap levels or the gradual lowering of caps is a part of the current EPA water pollution discharge trading system. Scientists also warn that discharge trading could allow pollutants to build up to dangerous levels in areas where credits are bought.

Reducing Water Pollution through Sewage Treatment (Science)

Septic tanks and various levels of sewage treatment can reduce point-source water pollution.

In rural and suburban areas with suitable soils, sewage from each house usually is discharged into a **septic tank** (Figure 9-27). About one-fourth of all homes in the United States are served by septic tanks.

In urban areas of the United States and most developed countries, most waterborne wastes from homes, businesses, factories, and storm runoff flow through a network of sewer pipes to *wastewater* or *sewage treatment plants*. Raw sewage reaching a treatment plant typically undergoes one or both of two levels of wastewater treatment. The first is **primary sewage treatment**—a *physical* process that uses screens and a grit tank to remove large floating objects and allow solids such as sand and rock to settle out. Then the waste stream flows into a primary settling tank where suspended, organic solids settle out as sludge (Figure 9-28). The second level is called **secondary sewage** treatment—a *biological* process in which aerobic bacteria remove up to 90% of dissolved and biodegradable, oxygen-demanding organic wastes.

Before discharge, water from primary, secondary, or more advanced treatment undergoes *bleaching* to remove water coloration and *disinfection* to kill disease-carrying bacteria and some (but not all) viruses. The usual method for doing this is *chlorination*. But chlorine can react with organic materials in water to form small amounts of chlorinated hydrocarbons. Some of these chemicals cause cancers in test animals, can increase the risk of miscarriages, and may damage the human nervous, immune, and endocrine systems. Use of other disinfectants, such as ozone and ultraviolet light, is increasing but they cost more and their effects do not last as long as chlorination.

U.S. federal law requires primary and secondary treatment for all municipal sewage treatment plants but exemptions from secondary treatment are possible when the cost of installing such treatment poses an excessive financial burden. And according to the EPA, at least two-thirds of these plants have at times violated water pollution regulations. Also, 500 cities have failed to meet federal standards for sewage treatment plants, and 34 East Coast cities simply screen out large floating objects from their sewage before discharging it into coastal waters.

Using Prevention along with Sewage Treatment (Science and Politics)

Preventing toxic chemicals from reaching sewage treatment plants would eliminate such chemicals from the sludge and water discharged from such plants.

Environmental scientist Peter Montague calls for redesigning the sewage treatment system. The idea is to

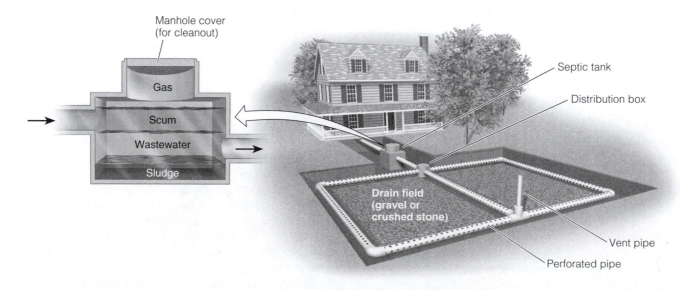

Figure 9-27 Solutions: *septic tank system* used for disposal of domestic sewage and wastewater in rural and suburban areas.

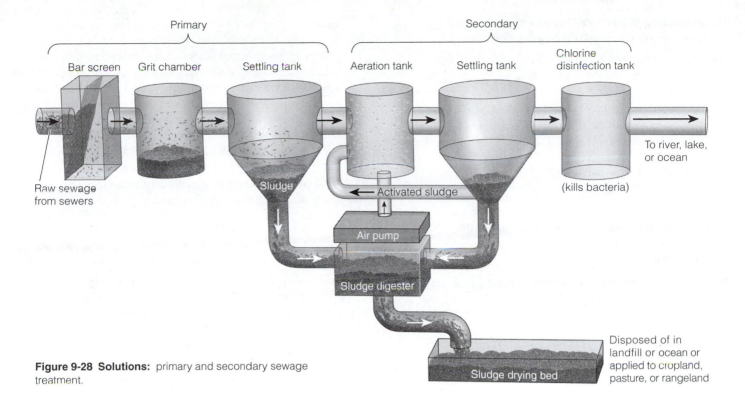

Figure 9-28 Solutions: primary and secondary sewage treatment.

prevent toxic and hazardous chemicals from reaching sewage treatment plants and thus from getting into sludge and the water discharged from such plants.

Montague suggests several ways to do this. One is to require industries and businesses to remove toxic and hazardous wastes from water sent to municipal sewage treatment plants. Another is to encourage industries to reduce or eliminate the use and waste of toxic chemicals.

Another suggestion is to have more households, apartment buildings, and offices eliminate sewage outputs by switching to waterless *composting toilet systems* that are installed, maintained, and managed by professionals. Such systems would be cheaper to install and maintain than current sewage systems because they do not require vast systems of underground pipes connected to centralized sewage treatment plants. They also save large amounts of water.

X *HOW WOULD YOU VOTE?* Should we ban the discharge of toxic chemicals into pipes leading to sewage treatment plants? Cast your vote online at **www.thomsonedu.com/biology/miller**.

Solutions: Treating Sewage by Working with Nature

Natural and artificial wetlands and other ecological systems can be used to treat sewage.

Biologist John Todd has developed an ecological approach to treating sewage, which he calls *living ma-* *chines.* This purification process begins when sewage flows into a passive solar greenhouse or outdoor sites containing rows of large open tanks populated by an increasingly complex series of organisms. In the first set of tanks, algae and microorganisms decompose organic wastes, with sunlight speeding up the process. Water hyacinths, cattails, bulrushes, and other aquatic plants growing in the tanks take up the resulting nutrients.

After flowing though several of these natural purification tanks, the water passes through an artificial marsh of sand, gravel, and bulrush plants, which filter out algae and remaining organic waste. Some of the plants also absorb (sequester) toxic metals such as lead and mercury and secrete natural antibiotic compounds that kill pathogens.

Next, the water flows into aquarium tanks. Snails and zooplankton in these tanks consume microorganisms and are in turn consumed by crayfish, tilapia, and other fish that can be eaten or sold as bait. After ten days, the clear water flows into a second artificial marsh for final filtering and cleansing.

The water can be made pure enough to drink by exposing it to ultraviolet light or by passing the water through an ozone generator, usually immersed out of sight in an attractive pond or wetland habitat. Operating costs are about the same as for a conventional sewage treatment plant. Some communities also work with nature by using nearby natural wetlands to treat sewage, and others create artificial wetlands for such purposes.

Reducing Water Pollution from Point Sources in the United States (Politics)

Water pollution laws have significantly improved water quality in many U.S. streams and lakes, but there is a long way to go.

Great news. According to the EPA, the Clean Water Act of 1972 led to a number of improvements in U.S. water quality. Between 1992 and 2002:

- The number of Americans served by community water systems that met federal health standards increased from 79% to 94%.

- The percentage of U.S. stream lengths found to be fishable and swimmable increased from 36% to 60% of those tested.

- The amount of topsoil lost through agricultural runoff was cut by about 1.1 billion metric tons (1 billion tons) annually.

- The proportion of the U.S. population served by sewage treatment plants increased from 32% to 74%.

- Annual wetland losses decreased by 80%.

These are impressive achievements given the increases in the U.S. population and per capita consumption of water and other resources since 1972.

Bad news. In 2000, the EPA found that 45% of the country's lakes and 40% of the streams surveyed were too polluted for swimming or fishing. Runoff of animal wastes from hog, poultry, and cattle feedlots and meat processing facilities pollutes 7 of every 10 U.S. rivers. Most livestock wastes are not treated and are stored in lagoons that sometimes leak. They can also overflow or rupture as a result of excessive rainfall and spill their contents into nearby streams and rivers and sometimes into residential areas.

Fish caught in more than 1,400 different waterways and more than a fourth of the nation's lakes are unsafe to eat because of high levels of pesticides, mercury, and other toxic substances. And a 2003 internal study by the EPA found that at least half of the country's 6,600 largest industrial facilities and municipal wastewater treatment plants have illegally discharged toxic or biological wastes into waterways for years without government enforcement actions or fines.

Should the U.S. Clean Water Act Be Strengthened or Weakened? (Science and Politics)

Some want to strengthen the Clean Water Act while others want to weaken it.

Some environmental scientists and a 2001 report by the EPA's inspector general call for the Clean Water Act to be strengthened. Suggested improvements include shifting the emphasis to preventing water pollution instead of focusing mostly on end-of-pipe removal of specific pollutants. The report also calls for increased funding and authority to control nonpoint sources of pollution, greatly increased monitoring for compliance with the law, much larger mandatory fines for violators, and stronger programs to prevent and control toxic water pollution.

Other suggestions include providing more funding and authority for integrated watershed and airshed planning to protect groundwater and surface water from contamination, and expanding the rights of citizens to bring lawsuits to ensure that water pollution laws are enforced. The National Academy of Sciences also calls for halting the loss of wetlands, higher standards for wetland restoration, and creating and evaluating new wetlands before filling any natural wetlands.

Many people oppose these proposals, contending that the Clean Water Act's regulations and government wetlands regulations are already too restrictive and costly. Farmers and developers see the law as limiting their rights as property owners to fill in wetlands. They also believe they should be compensated for any property value losses because of federal wetland protection.

State and local officials want more discretion in testing for and meeting water quality standards. They argue that in many communities it is unnecessary and too expensive to do all the water pollutant testing required by federal law.

X *HOW WOULD YOU VOTE?* Should the U.S. Clean Water Act be strengthened? Cast your vote online at www.thomsonedu.com/biology/miller.

DRINKING WATER QUALITY

Purifying Urban Drinking Water (Science)

Centralized water treatment plants and watershed protection can provide safe drinking water for city dwellers in developed countries. Simpler and cheaper ways can be used to purify drinking water in developing countries.

Most developed countries have laws establishing drinking water standards, but most developing countries do not have such laws or do not enforce them.

Areas in developed countries that depend on surface water usually store it in a reservoir for several days. This improves clarity and taste by increasing dissolved oxygen content and allowing suspended matter to settle. Next, the water is pumped to a purification plant and treated to meet government drinking water standards. In areas with very pure groundwater or surface water sources, little treatment except disinfection is necessary. Some cities (including New York City) have found that protecting watersheds that supply their drinking water is a lot cheaper than building water purification plants.

Simple measures can also be used to purify drinking water. In tropical countries without centralized water treatment systems, the WHO urges people to purify drinking water by exposing a clear plastic bottle filled with contaminated water to intense sunlight. Heat and the sun's UV rays can kill infectious microbes in as little as 3 hours. Painting one side of the bottle black can improve heat absorption in this simple solar disinfection method. Where this measure has been used, incidences of dangerous childhood diarrhea have decreased by 30–40%.

In Bangladesh, households receive strips of cloth for filtering cholera-producing bacteria from drinking water. Villages using this approach have cut the number of cholera cases in half.

Is Bottled Water the Answer? (Science and Economics)

Some bottled water is not as pure as tap water and costs much more.

Despite some problems, experts say the United States has some of the world's cleanest drinking water. Municipal water systems in the United States are required to test the water regularly for a number of pollutants and to make such information available to citizens. Yet about half of all Americans worry about getting sick from tap water contaminants, and many drink bottled water or install expensive water purification systems. Other countries must rely on bottled water wherever their tap water is too polluted to drink.

Studies reveal that in the United States bottled water is 240 times to 10,000 times more expensive than tap water. Yet, about one-fourth of it is tap water, and bacteria contaminate about 40% of bottled water.

Before drinking expensive bottled water and buying costly home water purifiers, health officials suggest that consumers have their water tested by local health authorities or private labs (not companies trying to sell water purification equipment). The goals are to identify what contaminants (if any) must be removed and to determine the type of purification needed to remove such contaminants. Independent experts contend that unless tests show otherwise, for most urban and suburban Americans (or people in other countries) served by large municipal drinking water systems, home water treatment systems are not worth the expense and maintenance hassles.

Buyers should check out companies selling water purification equipment and be wary of claims that the EPA has approved a treatment device. Although the EPA does *register* such devices, it neither tests nor approves them.

X HOW WOULD YOU VOTE? Should pollution standards be established for bottled water? Cast your vote online at www.thomsonedu.com/biology/miller.

Solutions: Reducing Water Pollution— The Road Ahead

Shifting our priorities from cleaning up water pollution to preventing and reducing it will require bottom-up political action by individuals and groups.

It is encouraging that since 1970 most developed countries have enacted laws and regulations that have significantly reduced point-source water pollution. These improvements were largely the result of *bottom-up* political pressure on elected officials by individuals and organized groups. However, little has been done to reduce water pollution in most developing countries.

To environmental and health scientists the next step is to increase efforts to reduce and prevent water pollution in developed and developing countries by asking the question: *"How can we avoid producing water pollutants in the first place?"* Figure 9-29 lists ways to do this over the next several decades.

This shift to *preventing water pollution* will not take place anywhere without *bottom-up* political pressure on elected officials. Developing countries will also

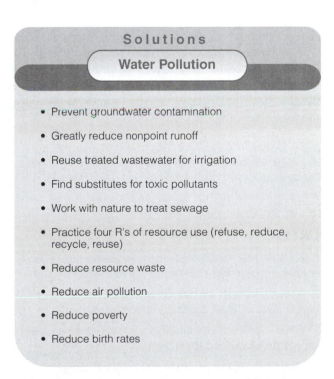

Solutions

Water Pollution

- Prevent groundwater contamination
- Greatly reduce nonpoint runoff
- Reuse treated wastewater for irrigation
- Find substitutes for toxic pollutants
- Work with nature to treat sewage
- Practice four R's of resource use (refuse, reduce, recycle, reuse)
- Reduce resource waste
- Reduce air pollution
- Reduce poverty
- Reduce birth rates

Figure 9-29 Solutions: methods for preventing and reducing water pollution. QUESTION: *Which two of these solutions do you think are the most important?*

What Can You Do?

Water Pollution

- Fertilize your garden and yard plants with manure or compost instead of commercial inorganic fertilizer.

- Minimize your use of pesticides.

- Never apply fertilizer or pesticides near a body of water.

- Grow or buy organic foods.

- Compost your food wastes.

- Do not use water fresheners in toilets.

- Do not flush unwanted medicines down the toilet.

- Do not pour pesticides, paints, solvents, oil, antifreeze, or other products containing harmful chemicals down the drain or onto the ground.

Figure 9-30 Individuals matter: ways to help reduce water pollution. QUESTION: *Which three of these actions do you think are the most important? Which of these actions do you already take or hope to take?*

need financial and technical aid from developed countries. Figure 9-30 lists some actions you can take to help reduce water pollution.

The four principles of sustainability can be used to prevent and reduce water pollution. For example solar energy, natural recycling and purifying processes, and a diversity of organisms can be used to purify water in living machines (p. 197) and in natural and artificial wetlands. And of course, we can reduce water pollution by slowing population growth and by reducing the unnecessary waste of resources.

It is a hard truth to swallow, but nature does not care if we live or die. We cannot survive without the oceans, for example, but they can do just fine without us.

ROGER ROSENBLATT

CRITICAL THINKING

1. How do human activities increase the harmful effects of prolonged drought? How can we reduce these effects?

2. How do human activities contribute to flooding and flood damage? How can these effects be reduced?

3. What role does population growth play in **(a)** water supply problems and **(b)** water pollution problems?

4. Should the prices of water for all uses be raised sharply to include more of its environmental costs and to encourage water conservation? Explain. What harmful and beneficial effects might this have on **(a)** business and jobs, **(b)** your lifestyle and the lifestyles of any children or grandchildren you might have, **(c)** the poor, and **(d)** the environment?

5. Explain why dilution is not always the solution to water pollution. Give examples and conditions for which this solution is or is not applicable.

6. Is each of the seven categories of pollutants listed in Table 9-1 most likely to originate from **(a)** point sources or **(b)** nonpoint sources?

7. When you flush a toilet, where does the wastewater go? Trace the actual flow of this wastewater in your community from your toilet through sewers to a wastewater treatment plant and from there to the environment. Try to visit a local sewage treatment plant to see what it does with your wastewater. Compare the processes it uses with those shown in Figure 9-28. What happens to the sludge produced by this plant? What improvements, if any, would you suggest for this plant?

8. Find out the price of tap water where you live. Then go to a grocery or other store and get prices per liter (or other volume unit) on all the available types of bottled water. Use these data to compare the price per liter of various brands of bottled water with the price of tap water.

9. Congratulations! You are in charge of the world. What are the three most important actions you will take to **(a)** manage the world's water resources, **(b)** sharply reduce point-source water pollution in developed countries, **(c)** sharply reduce nonpoint-source water pollution throughout the world, **(d)** sharply reduce groundwater pollution throughout the world, and **(e)** provide safe drinking water for the poor and other people in developing countries?

10. List two questions that you would like to have answered as a result of reading this chapter.

LEARNING ONLINE

The website for this book contains helpful study aids and many ideas for further reading and research. They include a chapter summary, review questions for the entire chapter, flash cards for key terms and concepts, a multiple-choice practice quiz, interesting Internet sites, references, information about green careers, and a guide for accessing thousands of InfoTrac® College Edition articles. Log into

www.thomsonedu.com/biology/miller

Then choose Chapter 9, and select a learning resource. For access to animations, additional quizzes, chapter outlines and summaries, register and log into

at **www.thomsonedu.com** using the access code card in the front of your book.

10 ENERGY

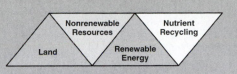

Land | Nonrenewable Resources | Renewable Energy | Nutrient Recycling

Typical citizens of advanced industrialized nations each consume as much energy in six months as typical citizens in developing countries consume during their entire life.

MAURICE STRONG

EVALUATING ENERGY RESOURCES

Solar and Commercial Energy (Science)

About 99% of the energy that heats the earth and our buildings comes from the sun, and the remaining 1% comes mostly from burning fossil fuels.

Almost all the energy that heats the earth and our buildings comes from the sun at no cost to us. Without this essentially inexhaustible solar energy (*solar capital*), the earth's average temperature would be −240°C (−400°F), and life as we know it would not exist.

This direct input of solar energy also produces several other *indirect forms of renewable solar energy*. Examples are *wind, hydropower* (falling and flowing water), and *biomass* (solar energy converted to chemical energy and stored in the chemical bonds of organic compounds in trees and other plants).

Commercial energy sold in the marketplace makes up the remaining 1% of the energy we use to supplement the earth's direct input of solar energy. Most commercial energy comes from extracting and burning *nonrenewable mineral resources* obtained from the earth's crust, primarily carbon-containing fossil fuels—oil, natural gas, and coal—as shown in Figure 10-1.

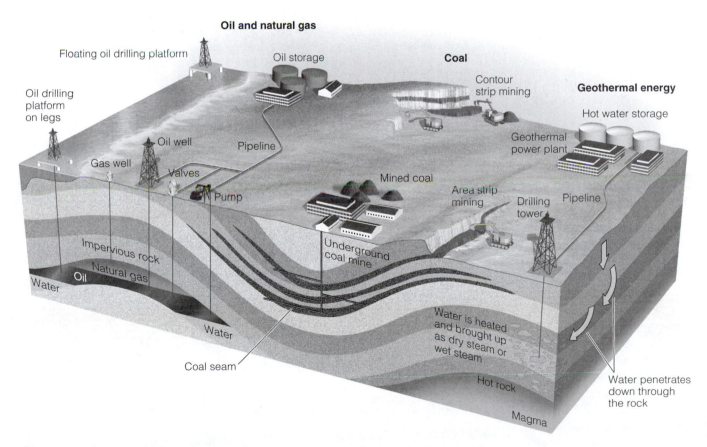

Figure 10-1 Natural capital: important nonrenewable energy resources that can be removed from the earth's crust are coal, oil, natural gas, and some forms of geothermal energy. Nonrenewable uranium ore is also extracted from the earth's crust and processed to increase its concentration of uranium-235, which can serve as a fuel in nuclear reactors to produce electricity.

Types of Commercial Energy

About 76% of the commercial energy we use comes from nonrenewable fossil fuels with the remainder coming from renewable sources.

About 82% of the commercial energy consumed in the world comes from *nonrenewable* energy resources—76% from fossil fuels and 6% from nuclear power (Figure 10-2, left). The remaining 18% comes from *renewable* energy resources—biomass (11%), hydropower (4.5%), and a combination of geothermal, wind, and solar energy (2.5%).

Roughly half the world's people in developing countries burn wood and charcoal to heat their dwellings and cook their food. This *biomass energy* is renewable as long as wood supplies are not harvested faster than they are replenished. But many people in developing countries face a *fuelwood shortage* that is expected to get worse because of unsustainable harvesting of fuelwood.

 Examine and compare energy sources used in developing and developed countries at ThomsonNOW.

Net Energy (Science)

Net energy is the amount of high-quality, usable energy available from a resource after subtracting the energy needed to make it available for use.

It takes energy to get energy. For example, before oil is useful to us it must be found, pumped up from beneath the ground or ocean floor, transferred to a refinery and converted to useful fuels, transported to users, and burned in furnaces and cars. Each of these steps uses high-quality energy. The second law of thermodynamics tells us that some of the high-quality energy we used in each step is wasted and degraded to lower-quality energy.

The usable amount of *high-quality energy* available from a given quantity of an energy resource is its **net energy.** It is the total amount of energy available from an energy resource minus the energy needed to find, extract, process, and get that energy to consumers. It is calculated by estimating the total energy available from the resource over its lifetime minus the amount of energy *used* (the first law of thermodynamics), *automatically wasted* (the second law of thermodynamics), and *unnecessarily wasted* in finding, processing, concentrating, and transporting the useful energy to users.

Net energy is like your net spendable income—your wages minus taxes and job-related expenses. For example, suppose for every 10 units of energy in oil in the ground we have to use and waste 8 units of energy to find, extract, process, and transport that oil to users. Then we have only 2 units of *useful energy* available from every 10 units of energy in the oil.

We can express net energy as the ratio of useful energy produced to the energy used to produce it. In

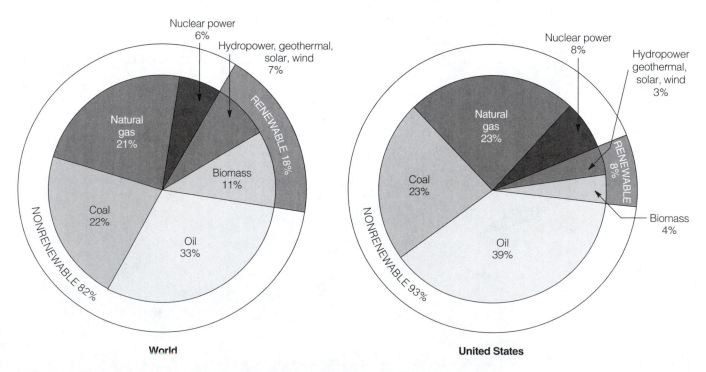

Figure 10-2 Natural capital: commercial energy use by source for the world (left) and the United States (right) in 2004. Commercial energy amounts to only 1% of the energy used in the world; the other 99% is direct solar energy received from the sun and is not sold in the marketplace. (Data from U.S. Department of Energy, British Petroleum, Worldwatch Institute, and International Energy Agency)

Space Heating

Passive solar — 5.8
Natural gas — 4.9
Oil — 4.5
Active solar — 1.9
Coal gasification — 1.5
Electric resistance heating (coal-fired plant) — 0.4
Electric resistance heating (natural-gas-fired plant) — 0.4
Electric resistance heating (nuclear plant) — 0.3

High-Temperature Industrial Heat

Surface-mined coal — 28.2
Underground-mined coal — 25.8
Natural gas — 4.9
Oil — 4.7
Coal gasification — 1.5
Direct solar (highly concentrated by mirrors, heliostats, or other devices) — 0.9

Transportation

Natural gas — 4.9
Gasoline (refined crude oil) — 4.1
Biofuel (ethyl alcohol) — 1.9
Coal liquefaction — 1.4
Oil shale — 1.2

Figure 10-3 Science: *Net energy ratios* for various energy systems over their estimated lifetimes: the higher the net energy ratio, the greater the net energy available. (Data from U.S. Department of Energy and Colorado Energy Research Institute, *Net Energy Analysis*, 1976; and Howard T. Odum and Elisabeth C. Odum, *Energy Basis for Man and Nature*, 3rd ed., New York: McGraw-Hill, 1981)

the example just given, the *net energy ratio* would be 10/8, or approximately 1.25. The higher the ratio, the greater the net energy. When the ratio is less than 1, there is a net energy loss.

Figure 10-3 shows estimated net energy ratios for various types of space heating, high-temperature heat for industrial processes, and transportation.

Currently, oil has a high net energy ratio because much of it comes from large, accessible, and cheap-to-extract deposits such as those in the Middle East. When those sources are depleted, the net energy ratio of oil will decline and prices will rise.

Electricity produced at a nuclear plant has a low net energy ratio because large amounts of energy are needed to extract and process uranium ore, convert it into nuclear fuel, build and operate nuclear power plants, dismantle the highly radioactive plants after their 15–60 years of useful life, and store the resulting highly radioactive wastes safely for 10,000–240,000 years. Each of these steps in the *nuclear power fuel cycle* uses energy and costs money. Some analysts estimate that ultimately the conventional nuclear fuel cycle will lead to a net energy loss because we will have to put more energy into it than we will ever get out of it.

NONRENEWABLE FOSSIL FUELS

Crude Oil (Science)

Crude oil is a thick liquid containing hydrocarbons that we extract from underground deposits and separate into products such as gasoline, heating oil, and asphalt.

Petroleum, or **crude oil** (oil as it comes out of the ground), is a thick and gooey liquid consisting of hundreds of combustible hydrocarbons along with small amounts of sulfur, oxygen, and nitrogen impurities. It is also known as *conventional oil* or *light oil*.

Deposits of crude oil and natural gas often are trapped together under a dome deep within the earth's crust on land or under the seafloor (Figure 10-1). The crude oil is dispersed in pores and cracks in underground rock formations, somewhat like water saturating a sponge. To extract the oil, a well is drilled into the deposit. Then oil, drawn by gravity out of the rock pores and into the bottom of the well, is pumped to the surface.

On average, producers get only about 35–50% of the oil out of an oil deposit. The remaining *heavy crude oil* is too difficult or expensive to recover. Some believe

Figure 10-4 Refining crude oil. Based on their boiling points, components are removed at various levels in a giant distillation column. The most volatile components with the lowest boiling points are removed at the top of the column.

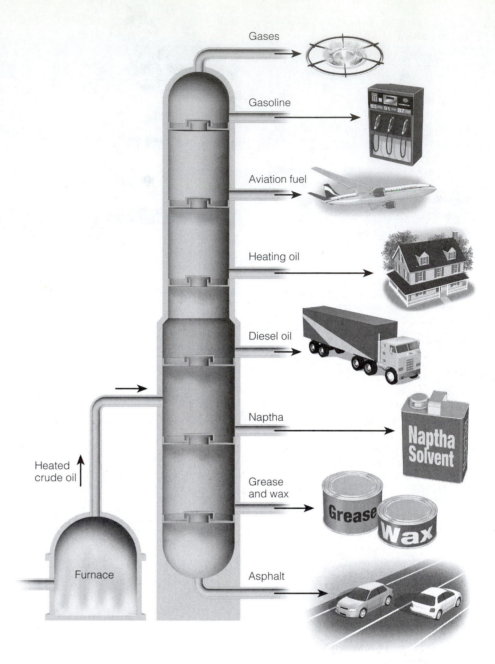

Gases

Gasoline

Aviation fuel

Heating oil

Diesel oil

Naptha

Grease and wax

Asphalt

Heated crude oil

Furnace

that improved drilling technology may increase the oil recovery rate to 75%.

After it is extracted, crude oil is transported to a *refinery* by pipeline, truck, or ship (oil tanker). There it is heated and distilled in gigantic columns to separate it into components with different boiling points (Figure 10-4)—a technological marvel based on complex chemistry and engineering. Some of the products of oil distillation, called **petrochemicals,** are used as raw materials in industrial organic chemicals, pesticides, plastics, synthetic fibers, paints, medicines, and many other products.

Global Oil Supplies—OPEC Rules (Geology and Economics)

Eleven OPEC countries—most of them in the Middle East—have 78% of the world's proven oil reserves and most of the world's unproven reserves.

The oil industry is the world's largest business. Control of the world's current and future oil reserves is the single greatest source of global economic and political power.

Oil *reserves* are identified deposits from which oil can be extracted profitably at current prices with current technology. The 11 countries that make up the Organization of Petroleum Exporting Countries (OPEC) have 78% of the world's crude oil reserves. Thus OPEC is expected to have long-term control over the supplies and prices of the world's conventional oil. Today OPEC's members are Algeria, Indonesia, Iran, Iraq, Kuwait, Libya, Nigeria, Qatar, Saudi Arabia, the United Arab Emirates, and Venezuela.

Saudi Arabia has by far the largest proportion of the world's crude oil reserves (25%). It is followed by

Canada (15%) whose huge supply of oil sand was recently classified as a conventional source of oil. Other countries with large proven reserves are Iraq (11%), the United Arab Emirates (9.3%), Kuwait (9.2%), and Iran (8.6%).

We are not yet running out of oil. But like all nonrenewable resources, the world's oil supplies are eventually expected to decline and lead to a sharp rise in its price. After we have depleted about 80% of the world's oil reserves it will cost too much to extract what is left.

According to geologists, known and projected global reserves of oil are expected to be 80% depleted sometime between 2050 and 2100, depending on how rapidly we use oil. If these estimates are correct, oil should be reaching its sunset years sometime during this century.

We have three options: look for more oil, use or waste less oil, or use something else. Some analysts—mostly economists—contend that rising oil prices will stimulate exploration and lead to enough new oil to keep the party going. Many oil geologists disagree.

Others argue that even if much more oil is somehow found, we are ignoring the consequences of the high (2–5% per year) exponential growth rate in global oil consumption. Suppose we continue to use oil reserves at the current rate of about 2.8% per year with the unlikely assumption that the rate won't increase. Here are some of the results of this scenario:

- Saudi Arabia, with the world's largest known crude oil reserves, could supply the world's entire oil needs for about 10 years.

- The estimated reserves under Alaska's North Slope—the largest ever found in North America—would meet current world demand for only 6 months or U.S. demand for 3 years.

- The estimated reserves in Alaska's Arctic National Wildlife Refuge would meet current world demand for only 1–5 months and U.S. demand for 7–24 months.

Thus, if the world were merely to keep using conventional oil at the projected rate of increase, we would have to discover global oil reserves equivalent to two new Saudi Arabian supplies every 10 years. According to most geologists, this is highly unlikely.

Sometime during this century, probably before 2050, we will shift from an abundant supply of fairly cheap oil to a dwindling supply of increasingly expensive oil that oil-using countries will compete to buy. As oil prices rise, the prices of food produced by industrialized agriculture and products such as plastics, pesticides, asphalt, and other widely used materials produced from petrochemicals will also rise.

Case Study: U.S. Oil Supplies (Science, Economics, and Politics)

The United States—the world's largest oil user—has only 2.9% of the world's proven oil reserves and only a small percentage of its unproven reserves.

The United States has only 2.9% of the world's proven oil reserves. It produces about 9% of the world's oil but it uses about 25% of global oil production (68% of it for transportation). According to geologists, U.S. oil production peaked in 1974 and has declined since then. Most geologists project that domestic oil production will reach the 80% depletion point by about 2055. Already, the United States produces most of its dwindling supply of oil at a high cost, about $7.50–10 per barrel compared to $1–2 per barrel in Saudi Arabia.

In 2005, the United States imported about 60% of the oil it used (up from 36% in 1973 when OPEC imposed an oil embargo against the United States and other nations). Reasons for this high dependence on imported oil are declining domestic oil reserves in the United States, higher production costs for domestic oil than for most sources of imported oil, and increased oil use.

According to the Department of Energy (DOE), the United States could be importing as much as 70% of the oil it uses by 2020. But it will be facing stiff competition for world oil supplies from rapidly industrializing China. By 2020, China expects to equal U.S. oil consumption, but because it has little oil, its imports will be twice those of the United States.

Bottom line: If you think of U.S. oil reserves as a six-pack of oil, four of the cans are empty. Geologists estimate that if the country opens up all of its public lands and coastal regions to oil exploration, it may find at best about half a can of new oil at a high cost (compared to much cheaper OPEC oil) and with harmful environmental effects. Thus, *opening all U.S. coastal waters, forests, and wild places to drilling would increase the profits of oil companies but would hardly put a dent in world oil prices or meet much of the U.S. demand for oil.*

Trade-Offs: Major Advantages and Disadvantages of Conventional Oil (Science and Economics)

Conventional oil is a versatile fuel that can last for at least 50 years, but burning it produces air pollution and releases the greenhouse gas carbon dioxide into the atmosphere.

Figure 10-5 (p. 206) lists the advantages and disadvantages of using conventional crude oil as an energy resource.

A serious problem is that burning oil or any carbon-containing fossil fuel releases CO_2 into the atmosphere

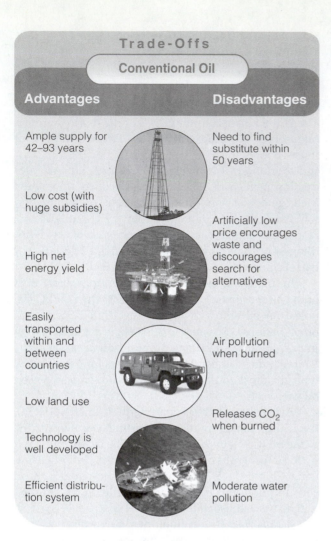

Trade-Offs

Conventional Oil

Advantages	Disadvantages
Ample supply for 42–93 years	Need to find substitute within 50 years
Low cost (with huge subsidies)	Artificially low price encourages waste and discourages search for alternatives
High net energy yield	
Easily transported within and between countries	Air pollution when burned
Low land use	
Technology is well developed	Releases CO_2 when burned
Efficient distribution system	Moderate water pollution

Figure 10-5 Trade-offs: advantages and disadvantages of using conventional crude oil as an energy resource. QUESTION: *What are the single advantage and the single disadvantage that you think are the most important?*

and thus can help promote climate change from global warming. Currently, burning oil mostly as gasoline and diesel fuel for transportation accounts for 43% of global CO_2 emissions. Figure 10-6 compares the relative amounts of CO_2 emitted per unit of energy by the major fossil fuels, nuclear power, and geothermal energy.

X HOW WOULD YOU VOTE? Do the advantages of relying on conventional oil as the world's major energy resource outweigh its disadvantages? Cast your vote online at www.thomsonedu.com/biology/miller.

Heavy Oils from Oil Sand and Oil Shale (Science and Economics)

Heavy oils from oil sand and oil shale could supplement conventional oil, but there are environmental problems.

Oil sand, or **tar sand,** is a mixture of clay, sand, water, and a combustible organic material called *bitumen*—a thick and sticky heavy oil with a high sulfur content. Oil sands nearest the earth's surface are dug up by gigantic electric shovels and mixed with hot water and steam to extract the bitumen, which is heated in huge cookers and converted to a low-sulfur synthetic crude oil suitable for refining.

Northeastern Alberta in Canada has about three-fourths of the world's oil sand resources, about a tenth of them close enough to the surface to be recovered by open pit and underground mining. Improved technology may allow extraction of twice that amount.

Currently, these deposits supply about a fifth of Canada's oil needs and this proportion is expected to increase. Because of the dramatic reductions in development and production costs, in 2003 the oil industry began counting Canada's oil sands as reserves of conventional oil. This means that Canada has 15% of the world's oil reserves, second only to Saudi Arabia. However, oil sand companies produce oil at a cost of about $15 per barrel, compared to $1–2 a barrel for conventional oil produced in Saudi Arabia. If a pipeline is built to transfer some of this synthetic crude oil from western Canada to the northwestern United States, Canada could greatly reduce future U.S. dependence on oil imports from the Middle East and add to Canadian income.

As is the case with most nonrenewable energy and mineral resources, extracting and processing oil sands

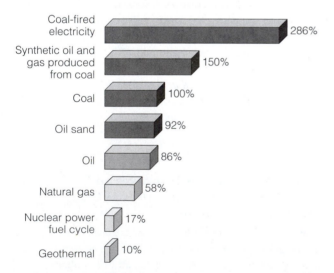

Coal-fired electricity	286%
Synthetic oil and gas produced from coal	150%
Coal	100%
Oil sand	92%
Oil	86%
Natural gas	58%
Nuclear power fuel cycle	17%
Geothermal	10%

Figure 10-6 Natural capital degradation: CO_2 emissions per unit of energy produced by using various energy resources to produce electricity, expressed as percentages of emissions produced by burning coal directly. These emissions can enhance the earth's natural greenhouse effect (Figure 2-11, p. 32) and lead to warming of the troposphere. (Data from U.S. Department of Energy)

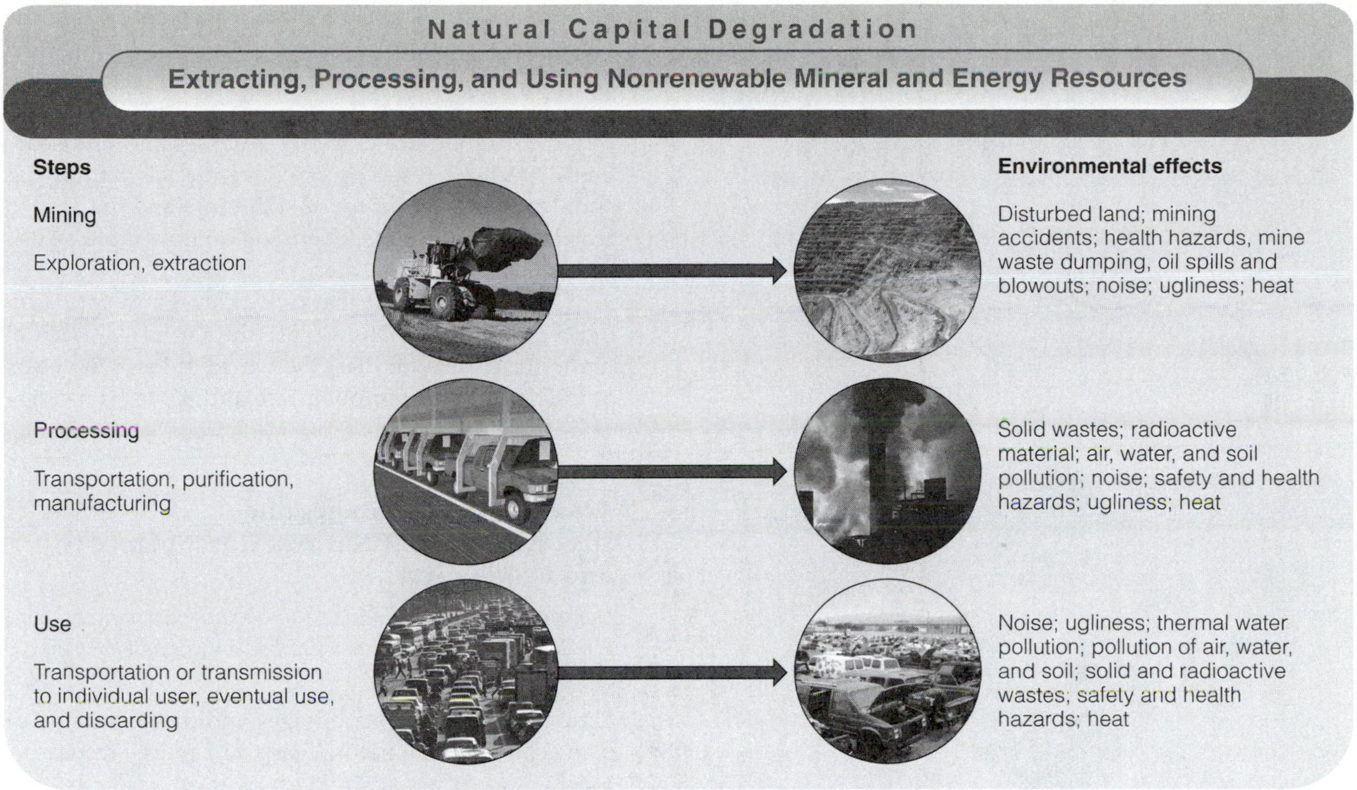

Steps			Environmental effects
Mining Exploration, extraction			Disturbed land; mining accidents; health hazards, mine waste dumping, oil spills and blowouts; noise; ugliness; heat
Processing Transportation, purification, manufacturing			Solid wastes; radioactive material; air, water, and soil pollution; noise; safety and health hazards; ugliness; heat
Use Transportation or transmission to individual user, eventual use, and discarding			Noise; ugliness; thermal water pollution; pollution of air, water, and soil; solid and radioactive wastes; safety and health hazards; heat

Figure 10-7 Natural capital degradation: some harmful environmental effects of extracting, processing, and using nonrenewable mineral and energy resources. The energy required to carry out each step causes additional pollution and environmental degradation.

has a severe impact (Figure 10-7). It yields huge amounts of toxic sludge and produces much more water pollution, much more air pollution (especially sulfur dioxide), and more CO_2 per unit of energy than conventional crude oil.

Oily rocks are another potential supply of heavy oil. Such rocks, called *oil shales,* contain a solid combustible mixture of hydrocarbons called *kerogen.* It can be extracted from crushed oil shales by heating them in a large container, a process which yields a distillate called **shale oil.** Before the thick shale oil can be sent by pipeline to a refinery, it must be heated to increase its flow rate and processed to remove sulfur, nitrogen, and other impurities.

Estimated potential global supplies of shale oil are about 240 times larger than estimated global supplies of conventional oil. But most deposits are locked up in rock and are of such a low grade that it takes a great deal of energy and money to mine and convert the kerogen to crude oil. Producing and using shale oil also has a much higher environmental impact than conventional oil.

Figure 10-8 (p. 208) lists the advantages and disadvantages of using oil sand and oil shales as energy resources.

Natural Gas (Science)

Natural gas, consisting mostly of methane, is often found above reservoirs of crude oil.

In its underground gaseous state, **natural gas** is a mixture of gases, of which 50–90% by volume is methane (CH_4), the simplest hydrocarbon. It contains smaller amounts of heavier gaseous hydrocarbons such as ethane (C_2H_6), propane (C_3H_8), and butane (C_4H_{10}), and small amounts of highly toxic hydrogen sulfide (H_2S).

Conventional natural gas lies above most reservoirs of crude oil (Figure 10-1). However, unless a natural gas pipeline has been built, deposits cannot be used. Indeed, the natural gas found above oil reservoirs in deep sea and remote land areas is often viewed as an unwanted byproduct and is burned off. This wastes a valuable energy resource and releases carbon dioxide into the atmosphere.

When a natural gas field is tapped, propane and butane gases are liquefied and removed as **liquefied petroleum gas (LPG).** LPG is stored in pressurized tanks for use mostly in rural areas not served by natural gas pipelines. The rest of the gas (mostly methane) is dried to remove water vapor, cleansed of poisonous

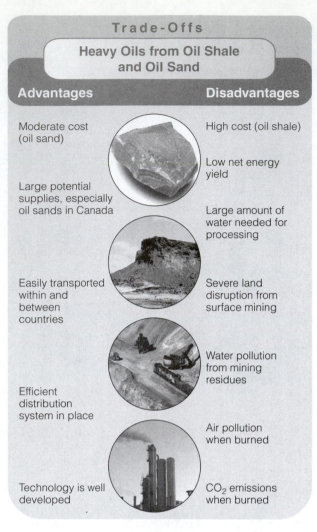

Trade-Offs

Heavy Oils from Oil Shale and Oil Sand

Advantages	Disadvantages
Moderate cost (oil sand)	High cost (oil shale)
Large potential supplies, especially oil sands in Canada	Low net energy yield
	Large amount of water needed for processing
Easily transported within and between countries	Severe land disruption from surface mining
	Water pollution from mining residues
Efficient distribution system in place	Air pollution when burned
Technology is well developed	CO_2 emissions when burned

Figure 10-8 Trade-offs: advantages and disadvantages of using heavy oils from oil sand and oil shale as energy resources. QUESTION: *What are the single advantage and the single disadvantage that you think are the most important?*

hydrogen sulfide and other impurities, and pumped into pressurized pipelines for distribution. At a very low temperature, natural gas can be converted to **liquefied natural gas (LNG).** This highly flammable liquid can then be shipped to other countries in refrigerated tanker ships.

Unconventional natural gas is also found in underground sources. One is coal bed methane gas found in coal beds across parts of the United States and Canada. Another is methane hydrate—methane trapped in cagelike structures of water molecules buried under Arctic permafrost and and deep beneath the ocean bottom. So far, it costs too much to get natural gas from such unconventional sources, and the environmental impacts are large, but the extraction technology is being developed rapidly.

Russia has about 31% of the world's proven conventional natural gas reserves, followed by Iran (15%)

and Qatar (9%). The United States has only 3% of the world's proven natural gas reserves.

The long-term global outlook for natural gas supplies is better than for conventional oil. At the current consumption rate, known reserves and undiscovered, potential reserves of conventional natural gas should last the world for 62–125 years and the United States for 55–80 years, depending on how rapidly they are used.

Geologists project that *conventional* and *unconventional* supplies of natural gas (the latter available at higher prices) should last the world at least 200 years at the current consumption rate and 80 years if usage rates rise 2% per year.

Trade-offs: Advantages and Disadvantages of Natural Gas (Science and Economics)

Natural gas is a versatile and clean-burning fuel, but using it releases the greenhouse gases carbon dioxide and methane into the atmosphere.

Figure 10-9 lists the advantages and disadvantages of using conventional natural gas as an energy resource.

Because of its advantages over oil, coal, and nuclear energy, some analysts see natural gas as the best fuel to help us make the transition to improved energy efficiency and greater use of renewable energy over the next 50 years. Converting natural gas to its liquid form and shipping it in special tankers could greatly increase the global distribution of natural gas supplies.

Coal: Origins and Uses (Science)

Coal is an abundant energy resource that is burned mostly to produce electricity and steel.

Coal is a solid fossil fuel formed in several stages as the buried remains of land plants that lived 300–400 million years ago were subjected to intense heat and pressure over many millions of years (Figure 10-10). Coal is mostly carbon and contains small amounts of sulfur, released into the atmosphere as sulfur dioxide when the coal is burned. Burning coal also releases trace amounts of toxic mercury and radioactive materials. Coal is burned in power plants (Figure 10-11, p. 210) to generate 62% of the world's electricity (53% in the United States). It is also burned to make three-fourths of the world's steel.

Coal is the world's most abundant fossil fuel. According to the U.S. Geological Survey, identified and unidentified supplies of coal could last for 214–1,125 years, depending on the rate of usage. The United States has one-fourth of the world's proven coal reserves. Russia has 16% and China 12%. In 2005, just over half of global coal consumption was split almost evenly between China and the United States.

China has enough proven coal reserves to last 300 years at its current rate of consumption. According to the U.S. Geological Survey, identified U.S. coal reserves should also last about 300 years at the current consumption rate, and unidentified U.S. coal resources could extend those supplies for perhaps another 100 years, at a higher cost. However, if U.S. coal use should increase by 4% a year—as the coal industry projects—the country's proven coal reserves would last only 64 years.

Trade-Offs: Advantages and Disadvantages of Coal (Science and Economics)

Coal is the most abundant fossil fuel, but compared to oil and natural gas it is not as versatile, has a much higher environmental impact, and releases much more carbon dioxide into the troposphere.

Figure 10-12 (p. 211) lists the advantages and disadvantages of using coal as an energy resource. *Bottom line:* Coal is the world's most abundant fossil fuel but mining and burning coal has a severe environmental impact on the earth's air, water, and land, and accounts for more than a third of the world's annual CO_2 emissions.

Each year in the United States alone, air pollutants from burning coal prematurely kill at least 23,500 people, and cause 38,200 nonfatal heart attacks, and 554,000 asthma attacks. They also result in several billion dollars of property damage each year. Many people are unaware that burning coal is responsible for one-fourth of atmospheric mercury pollution in the United States, and it releases far more radioactive particles into the air than normally operating nuclear power plants.

Many analysts project a decline in coal use over the next 40–50 years because of its high CO_2 emissions (Figure 10-6) and harmful health effects, and the availability of cleaner and cheaper ways to produce electricity

Trade-Offs

Conventional Natural Gas

Advantages	Disadvantages
Ample supplies (125 years)	Nonrenewable resource
High net energy yield	Releases CO_2 when burned
Low cost (with huge subsidies)	Methane (a greenhouse gas) can leak from pipelines
Less air pollution than other fossil fuels	
Lower CO_2 emissions than other fossil fuels	Difficult to transfer from one country to another
Moderate environmental impact	Shipped across ocean as highly explosive LNG
Easily transported by pipeline	
Low land use	Sometimes burned off and wasted at wells because of low price
Good fuel for fuel cells and gas turbines	Requires pipelines

Figure 10-9 Trade-offs: advantages and disadvantages of using conventional natural gas as an energy resource. QUESTION: *What are the single advantage and the single disadvantage that you think are the most important?*

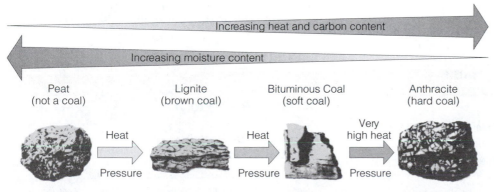

Peat (not a coal)	Lignite (brown coal)	Bituminous Coal (soft coal)	Anthracite (hard coal)
Partially decayed plant and animal matter in swamps and bogs; low heat content	Low heat content; low sulfur content; limited supplies in most areas	Extensively used as a fuel because of its high heat content and large supplies; normally has a high sulfur content	Highly desirable fuel because of its high heat content and low sulfur content; supplies are limited in most areas

Figure 10-10 Natural capital: stages in coal formation over millions of years. Peat is a soil material made of moist, partially decomposed organic matter. Lignite and bituminous coal are sedimentary rocks, whereas anthracite is a metamorphic rock.

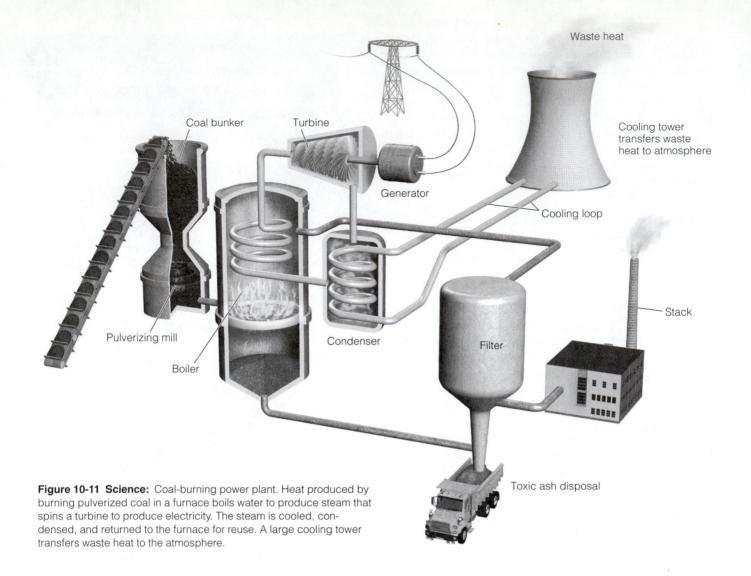

Waste heat

Coal bunker   Turbine

Generator

Cooling tower transfers waste heat to atmosphere

Cooling loop

Pulverizing mill

Boiler

Condenser

Filter

Stack

Toxic ash disposal

Figure 10-11 Science: Coal-burning power plant. Heat produced by burning pulverized coal in a furnace boils water to produce steam that spins a turbine to produce electricity. The steam is cooled, condensed, and returned to the furnace for reuse. A large cooling tower transfers waste heat to the atmosphere.

such as wind energy and burning natural gas in efficient gas turbines.

☒ *How Would You Vote?* Should coal use be phased out over the next 20 years? Cast your vote online at www.thomsonedu.com/biology/miller.

Trade-Offs: Advantages and Disadvantages of Converting Solid Coal into Gaseous and Liquid Fuels (Science and Economics)

Coal can be converted to gaseous and liquid fuels that burn cleaner than coal, but the costs are high, and producing and burning them add more carbon dioxide to the atmosphere than burning coal.

Solid coal can be converted into **synthetic natural gas (SNG or syngas)** by **coal gasification** or into a liquid fuel such as methanol or synthetic gasoline by **coal liquefaction.** Figure 10-13 lists the advantages and disadvantages of using these *synfuels*.

Without huge government subsidies, most analysts expect synfuels to play a minor role as an energy resource in the next 20–50 years. Compared to burning conventional coals, they require mining 50% more coal and producing and burning them could add 50% more carbon dioxide to the atmosphere. Also, they cost more to produce than coal.

NUCLEAR ENERGY

What Is Nuclear Fission? Splitting Nuclei

Neutrons can split apart or fission the nuclei of certain isotopes with large mass numbers and release a large amount of energy.

The source of energy for nuclear power is **nuclear fission,** a nuclear change in which nuclei of certain isotopes with large mass numbers (such as uranium-235)

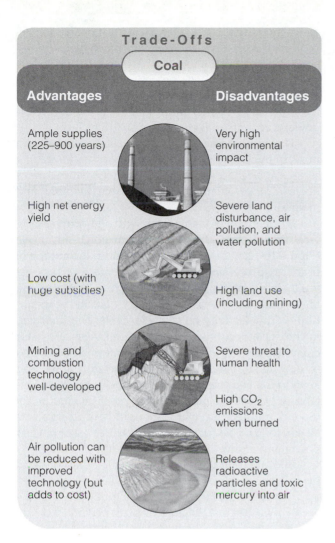

Trade-Offs

Coal

Advantages	Disadvantages
Ample supplies (225–900 years)	Very high environmental impact
High net energy yield	Severe land disturbance, air pollution, and water pollution
Low cost (with huge subsidies)	High land use (including mining)
Mining and combustion technology well-developed	Severe threat to human health
Air pollution can be reduced with improved technology (but adds to cost)	High CO_2 emissions when burned
	Releases radioactive particles and toxic mercury into air

Figure 10-12 Trade-offs: advantages and disadvantages of using coal as an energy resource. QUESTION: *What are the single advantage and the single disadvantage that you think are the most important?*

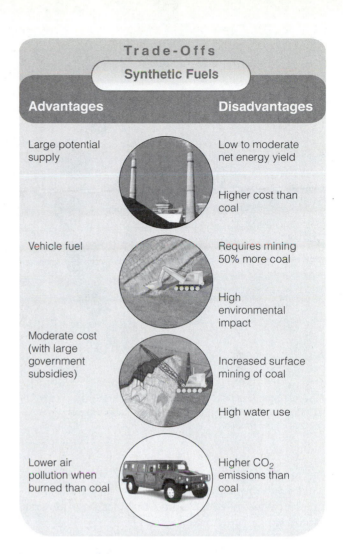

Trade-Offs

Synthetic Fuels

Advantages	Disadvantages
Large potential supply	Low to moderate net energy yield
Vehicle fuel	Higher cost than coal
Moderate cost (with large government subsidies)	Requires mining 50% more coal
	High environmental impact
	Increased surface mining of coal
	High water use
Lower air pollution when burned than coal	Higher CO_2 emissions than coal

Figure 10-13 Trade-offs: advantages and disadvantages of using synthetic natural gas (syngas) and liquid synfuels produced from coal. QUESTION: *What are the single advantage and the single disadvantage that you think are the most important?*

are split apart into lighter nuclei when struck by neutrons. Each fission releases two or three more neutrons plus energy. These multiple fissions within a critical mass of nuclear fuel form a **chain reaction,** which releases an enormous amount of energy (Figure 10-14, p. 212) that can be used to produce electricity at a nuclear power plant.

Nuclear fission produces radioactive fission fragments containing isotopes that spontaneously shoot out fast-moving particles (alpha and beta particles), gamma rays (a form of high-energy electromagnetic radiation; Figure 2-5, p. 26), or both. These unstable isotopes are called **radioactive isotopes,** or **radioisotopes.**

Exposure to alpha, beta, and gamma radiation and the high-speed neutrons emitted in a nuclear fission chain reaction and by the resulting radioactive wastes can harm human cells in two ways. *First,* harmful mutations of DNA molecules in genes and chromosomes can cause genetic defects in one or more generations of offspring. *Second,* tissue damage such as burns, miscarriages, eye cataracts, and cancers (bone, thyroid, breast, skin, and lung) can occur during the victim's lifetime.

How Does a Nuclear Fission Reactor Work? (Science)

In a conventional nuclear reactor, isotopes of uranium and plutonium undergo controlled nuclear fission. The resulting heat is used to produce steam that spins turbines to generate electricity.

To evaluate the advantages and disadvantages of nuclear power, we must know how a conventional nuclear power plant and its accompanying nuclear power fuel cycle work. In the reactor of a nuclear power plant,

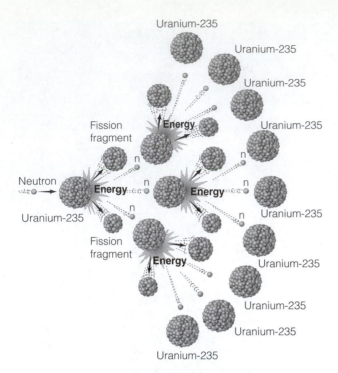

Figure 10-14 Fission of a uranium-235 nucleus by a neutron (*n*) releases more neutrons that can cause multiple fissions in a *nuclear chain reaction.*

the rate of fission in a nuclear chain reaction is controlled and the heat generated is used to produce high-pressure steam, which spins turbines that generate electricity.

Light-water reactors (LWRs) like the one diagrammed in Figure 10-15 produce about 85% of the world's nuclear-generated electricity (100% in the United States). *Control rods* are moved in and out of the reactor core to absorb neutrons, thereby regulating the rate of fission and amount of power produced. A *coolant*, usually water, circulates through the reactor's core to remove heat to keep fuel rods and other materials from melting and to produce steam for generating electricity. The greatest danger in water-cooled reactors is a loss of coolant that would allow the nuclear fuel to quickly overheat, melt down, and possibly release radioactive materials to the environment. An LWR reactor has an emergency core cooling system as a backup to help prevent such meltdowns.

As a further safety backup, a *containment vessel* with strong, thick walls surrounds the reactor core. It is designed to keep radioactive materials from escaping into the environment in case of an internal explosion or core meltdown within the reactor, and to protect the core from external threats such as a plane crash.

When reactors are refueled about once a year, intensely hot and radioactive fuel rod assemblies are removed and stored on site outside of the nuclear reactor building in *water-filled pools* or *dry casks* with thick steel walls. Spent-fuel pools or casks are not nearly as well protected as the reactor core and are much more vulnerable to accidental or intentional damage. The long-term goal is to transport spent fuel rods and other long-lived radioactive wastes to an underground facility for long-term storage. But after more than 50 years of nuclear power no country has developed such a facility.

The overlapping and multiple safety features of a modern nuclear reactor greatly reduce the chance of a serious nuclear accident. But these safety features make nuclear power plants very expensive to build and maintain.

Nuclear power plants, each with one or more reactors, are only one part of the nuclear power fuel cycle (Figure 10-16, p. 214). This cycle includes the mining of uranium, processing it to make a satisfactory fuel, using it in a reactor, and safely storing the resulting highly radioactive wastes for 10,000–240,000 years until their radioactivity falls to safe levels. The final step in the cycle occurs when after 15–60 years a reactor comes to the end of its useful life and must be retired or decommissioned. It cannot just be shut down and abandoned because it contains large quantities of intensely radioactive materials that must be kept out of the environment for many thousands of years. Each step in the nuclear fuel cycle adds to the cost of nuclear power and reduces its net energy yield.

In evaluating the safety, economic feasibility, and overall environmental impact of nuclear power, energy experts and economists caution us to look at the entire cycle, not just the nuclear plant itself.

What Happened to Nuclear Power?

After more than 50 years of development and enormous government subsidies, nuclear power has not lived up to its promise.

In the 1950s, researchers predicted that by the year 2000 at least 1,800 nuclear power plants would supply 21% of the world's commercial energy (25% in the United States) and most of the world's electricity.

After more than 50 years of development, enormous government subsidies, and an investment of $2 trillion, these goals have not been met. Instead, by 2005, 441 commercial nuclear reactors in 30 countries were producing only 6% of the world's commercial energy and 17% of its electricity.

Since 1989, electricity production from nuclear power has increased only slightly and is now the world's slowest-growing energy source. The U.S. Department of Energy projects that the percentage of the world's electricity produced by nuclear power will fall to 12% by 2025 because the retirement of aging plants is expected to exceed the construction of new ones.

No new nuclear power plants have been ordered in the United States since 1978, and all 120 plants or-

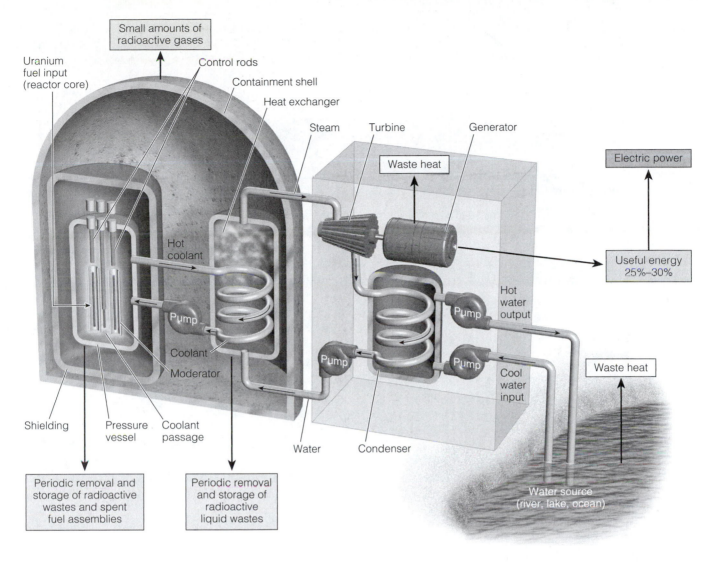

Figure 10-15 **Science:** light-water moderated and cooled nuclear power plant with a pressurized water reactor.

dered since 1973 have been canceled. In 2005, there were 103 licensed and operating commercial nuclear power reactors in 31 states—most of them in the eastern half of the country. These reactors generate about 20% of the country's electricity and 8% of its total energy. This percentage is expected to decline over the next two decades as existing plants wear out and are retired.

According to energy analysts and economists, several reasons explain the failure of nuclear power to grow as projected. They include multibillion-dollar construction cost overruns, higher operating costs and more malfunctions than expected, and poor management. Two other major obstacles have been public concerns about safety and stricter government safety regulations, especially after the accidents in 1979 at the Three Mile Island nuclear plant in the U.S. state of

Pennsylvania and in 1986 at the Chernobyl nuclear plant in Ukraine. (See Case Study, p. 214.)

Another problem is investor concerns about the economic feasibility of nuclear power, taking into account the entire nuclear fuel cycle. Even with massive government subsidies, using the nuclear fuel cycle to generate electricity costs more than burning coal or natural gas and using wind power to produce electricity. At Three Mile Island, investors lost over a billion dollars in one hour from damaged equipment and repair. Also, concern has risen about the vulnerability of nuclear power plants to terrorist attack after the destruction of the World Trade Center buildings in New York City on September 11, 2001. Experts are especially concerned about the vulnerability of poorly protected and intensely radioactive spent fuel rods stored in water pools or casks outside of reactor buildings.

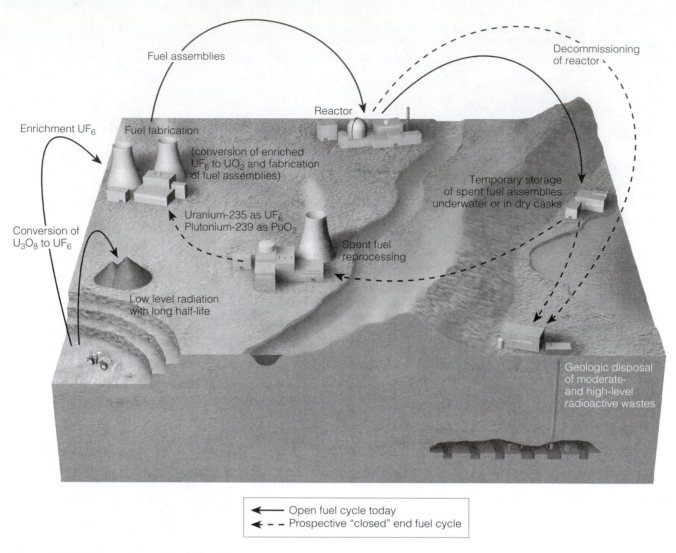

Fuel assemblies

Decommissioning of reactor

Enrichment UF$_6$

Fuel fabrication

Reactor

(conversion of enriched UF$_6$ to UO$_2$ and fabrication of fuel assemblies)

Temporary storage of spent fuel assemblies underwater or in dry casks

Conversion of U$_3$O$_8$ to UF$_6$

Uranium-235 as UF$_6$
Plutonium-239 as PuO$_2$

Spent fuel reprocessing

Low level radiation with long half-life

Geologic disposal of moderate- and high-level radioactive wastes

⟵——— Open fuel cycle today
⟵- - - Prospective "closed" end fuel cycle

Figure 10-16 Science: the nuclear fuel cycle.

Case Study: The Chernobyl Nuclear Power Plant Accident

The accident at Chernobyl was due to poor design and human error and likely caused thousands of premature deaths.

Chernobyl is known around the globe as the site of the world's most serious nuclear power plant accident. On April 26, 1986, a series of explosions in one of the reactors in this nuclear power plant in Ukraine (then part of the Soviet Union) blew the massive roof off a reactor building. The reactor melted down and its graphite moderator caught fire and burned for 10 days. The initial explosion and the prolonged graphite fires released huge radioactive clouds that spread in an unpredictable manner over much of Ukraine, Belarus, Russia, and other parts of eastern Europe, and eventually encircled the planet.

According to various U.N. studies, the disaster, which was caused by poor reactor design and human error, had serious consequences. By 2005, 56 people had died from radiation released by the accident, mostly reactor staff and rescue workers and 9 children who died from thyroid cancer. According to a 2005 U.N. study, at least 4,000 more people will eventually die from cancers caused by radiation exposure from the accident. Some researchers say the toll may be higher.

Some 286,000 people had to abandon their homes. Most were not evacuated until at least 10 days after the accident. In 2003, Ukraine officials downgraded the 27-kilometer (17-mile) area surrounding the reactor to a "zone with high risk" to allow those willing to accept the health risk to return to their homes.

The total cost of the accident is at least $140 billion, according to the U.S. Department of Energy, and could eventually reach at least $358 billion according to Ukrainian officials—many times more than the value of all nuclear electricity ever generated in the former Soviet Union.

Chernobyl taught us that a major nuclear accident anywhere has effects that reverberate throughout much of the world.

Trade-Offs: Advantages and Disadvantages of the Conventional Nuclear Power Fuel Cycle (Science and Economics)

The nuclear power fuel cycle has a fairly low environmental impact, an ample supply of fuel, and a very low risk of an accident, but costs are high, radioactive wastes must be stored safely for thousands of years, facilities are vulnerable to terrorist attacks, and the spread of nuclear reactors gives more countries the knowledge to build nuclear weapons.

Figure 10-17 lists the major advantages and disadvantages of the conventional nuclear fuel cycle. Using nuclear power to produce electricity has some important advantages over coal-burning power plants (Figure 10-18, p. 216).

However, the net energy yield for the nuclear fuel cycle is low, and, according to some calculations, there is a net energy loss. Its low net-energy yield is a major reason why nuclear power cannot compete in the marketplace with many other energy alternatives without huge government subsidies.

In 1995, the World Bank said nuclear power is too costly and risky. *Forbes* business magazine has called the failure of the U.S. nuclear power program "the largest managerial disaster in U.S. business history, involving $1 trillion in wasted investment and $10 billion in direct losses to stockholders."

Nuclear power plants also spread the knowledge needed to make nuclear weapons. Currently, 60 countries—1 of every 3 in the world—have nuclear weapons or the knowledge and ability to build them. Information and fuel needed to build these nuclear weapons have come mostly from the research and commercial nuclear reactors that the United States and 14 other countries have been giving away and selling in the international marketplace for decades. Experts warn that a significant increase in the number of nuclear power plants in the world could greatly increase the threat of nuclear war by rogue nations and nuclear terrorism by organized groups.

Because of the built-in safety features, the risk of exposure to radioactivity from nuclear power plants in the United States and most other developed countries is extremely low. However, a partial or complete meltdown or explosion is possible, as accidents at the Chernobyl nuclear power plant in Ukraine (Case Study, p. 214) and the Three Mile Island plant in Pennsylvania taught us.

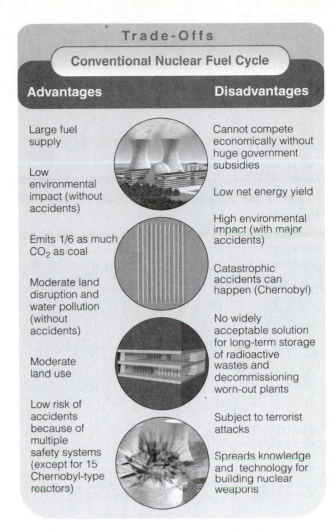

Figure 10-17 Trade-offs: advantages and disadvantages of using the conventional nuclear fuel cycle (Figure 10-16) to produce electricity. QUESTION: *What are the single advantage and the single disadvantage that you think are the most important?*

The U.S. Nuclear Regulatory Commission (NRC) estimates there is a 15–45% chance of a complete core meltdown at a U.S. reactor during the next 20 years. The NRC also found that 39 U.S. reactors have an 80% chance of containment shell failure from a meltdown or an explosion of gases inside the containment structures. In the United States, there is considerable public distrust of government agencies' abilities to enforce nuclear safety in commercial (NRC) and military (DOE) nuclear facilities.

A 2005 study by the U.S. National Academy of Sciences warned that deep water pools and dry concrete casks used to store spent fuel rods at 68 nuclear power plants in 31 U.S. states are vulnerable to sabotage or terrorist attack. A spent-fuel pool typically holds 5–10 times more long-lived radioactivity than the radioactive core inside a plant's reactor. Unlike the reactor core with its thick concrete protective dome,

Trade-Offs

Coal vs. Nuclear

Coal	Nuclear
Ample supply	Ample supply of uranium
High net energy yield	Low net energy yield
Very high air pollution	Low air pollution (mostly from fuel reprocessing)
High CO_2 emissions	Low CO_2 emissions (mostly from fuel reprocessing)
High land disruption from surface mining	Much lower land disruption from surface mining
High land use	Moderate land use
Low cost (with huge subsidies)	High cost (with huge subsidies)

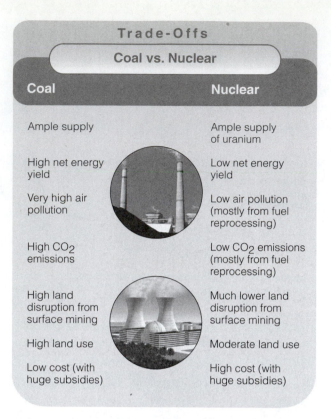

Figure 10-18 Trade-offs: comparison of the risks of using nuclear power and coal-burning plants to produce electricity. A 1,000-megawatt nuclear plant is refueled once a year, whereas a coal plant of the same size requires 80 rail cars of coal a day. QUESTION: *If you had to, would you rather live next door to a coal-fired power plant or a nuclear power plant? Explain.*

spent-fuel pools and dry storage casks have little protective cover. According to this study, an earthquake, a deliberate crash by a small airplane, or an attack by a group of suicidal terrorists could release significant amounts of radioactive materials into the troposphere, contaminating large areas for decades and creating economic and psychological havoc.

According to critics, the problem is that the NRC is reluctant to require utilities to significantly upgrade plant security because this would increase the costs of nuclear power and make it less competitive in the marketplace.

High-Level Radioactive Waste

Scientists disagree about the best methods for the long-term storage of high-level radioactive waste.

Each part of the nuclear power fuel cycle produces low-level and high-level solid, liquid, and gaseous radioactive wastes. *High-level radioactive wastes* must be stored safely for at least 10,000 years, or 240,000 years if plutonium-239 is not removed by reprocessing. These wastes consist mainly of spent fuel rods from commercial nuclear power plants and assorted wastes from the production of nuclear weapons.

After more than 50 years of research, scientists still don't agree on whether there is a safe way of storing high-level radioactive waste. Some believe the long-term safe storage or disposal of high-level radioactive wastes is technically possible. Others disagree, pointing out that it is impossible to demonstrate that any method will work for 10,000–240,000 years. The favored strategy is to bury it deep underground—an option being pursued by the United States (Case Study, at right) and several other countries.

X *HOW WOULD YOU VOTE?* Should highly radioactive spent fuel be stored in casks at high-security sites for 100 years instead of shipping them to a single site for underground burial? Cast your vote online at www.thomsonedu.com/biology/miller.

Retiring Worn-Out Nuclear Power Plants (Science)

When a nuclear reactor reaches the end of its useful life we must keep its highly radioactive materials from reaching the environment for thousands of years.

When a nuclear power plant comes to the end of its useful life, mostly because of corrosion and radiation damage to metal parts, it must be *decommissioned*, or retired—the last step in the nuclear power fuel cycle. Scientists have proposed three ways to do this.

One strategy is to dismantle the plant after it closes and store its large volume of highly radioactive materials in a high-level, nuclear waste storage facility. A second approach is to put up a physical barrier around the plant and set up full-time security for 30–100 years before the plant is dismantled after its radioactive materials have reached safer levels. A third option is to enclose the entire plant in a tomb that must last and be monitored for several thousand years. Regardless of the method chosen, decommissioning adds to the total costs of nuclear power and reduces its net energy. Experience indicates that dismantling a plant and storing the resulting radioactive wastes costs 2–10 times more than building the plant in the first place.

At least 228 large commercial reactors worldwide (20 in the United States) are scheduled for retirement by 2012. However, operators of half the current reactors in the United States have applied to the NRC to extend their soon-to-expire 40-year licenses to 60 years. Opponents contend this could increase the risk of nuclear accidents in aging reactors.

Can Nuclear Power Reduce Dependence on Imported Oil and Help Reduce Global Warming? (Science and Politics)

Building more nuclear power plants is neither a way to lessen dependence on imported oil nor a major way to reduce carbon dioxide emissions compared to other quicker and cheaper alternatives.

High-Level Radioactive Wastes in the United States (Science, Economics, and Politics)

CASE STUDY

In 1985, the U.S. DOE announced plans to build a repository for underground storage of high-level radioactive wastes from commercial nuclear reactors on federal land in the Yucca Mountain desert region, 160 kilometers (100 miles) northwest of Las Vegas, Nevada.

The proposed facility is expected to cost at least $58 billion to build (financed jointly by a tax on nuclear power and taxpayers) and could reach $100 billion. The original target opening date was 2015, but now there is no target date because of a number of legal battles and scientific problems with the site.

Some scientists argue that it should never be allowed to open, mostly because rock fractures and tiny cracks may allow water to leak into the site and eventually corrode radioactive waste storage casks. Heat from the fuel rods would accelerate such corrosion. Rain that percolates into the mountain each year could carry radioactive wastes leaking from corroded containers into groundwater, irrigation systems, and drinking-water wells for hundreds of generations. In 1998, Jerry Szymanski, formerly the DOE's top geologist at Yucca Mountain and now an outspoken opponent of the site, said that if water flooded the site it could cause an explosion so large that "Chernobyl would be small potatoes."

In 2002, the U.S. National Academy of Sciences, in collaboration with Harvard University and University of Tokyo scientists, urged the U.S. government to slow down and rethink the nuclear waste storage process. These scientists contend that storing spent fuel rods in dry-storage casks in well-protected buildings at nuclear plant sites is an adequate solution for at least 100 years in terms of safety and national security. This would buy time to carry out more research on this complex problem and to evaluate other sites and storage methods that might be more acceptable scientifically and politically.

Despite objections from scientists and citizens, during the summer of 2002, Congress approved Yucca Mountain as the official site for storing the country's commercial nuclear wastes. In 2004, a federal appeals court ruled that the radiation standard set by the EPA to protect the public at Yucca Mountain is inadequate and illegal. In 2005, Congress and the DOE temporarily halted efforts to get final approval for the Yucca Mountain project after hearing allegations that some of the scientific data used to justify the project were made up.

Critical Thinking

If you lived (or if you do live) in Las Vegas, would you (or do you) favor the development of the nuclear waste storage facility at Yucca Mountain? Explain.

Some proponents of nuclear power in the United States claim it will help reduce dependence on imported oil. Other analysts point out that use of nuclear power has little effect on U.S. oil use because burning oil typically produces only 2–3% of the electricity in the United States (and in most other countries). And the major use of oil is to produce gasoline and diesel fuel for motor vehicles, which do not run on electricity produced by nuclear or other power plants.

Nuclear power advocates also contend that increased use of nuclear power would reduce the threat of global warming by reducing emissions of CO_2. Scientists point out this argument is only partially correct. Nuclear plants themselves do not produce CO_2, but the nuclear fuel cycle does—a fact that is rarely mentioned by the nuclear industry or by the press.

Such emissions are much less than those produced by burning coal or natural gas to produce electricity (Figure 10-6). However, according to a 2004 study by German scientists, considering the entire nuclear fuel cycle, carbon dioxide emissions per kilowatt-hour of electricity are much higher than the numbers in Figure 10-6 indicate. Many energy experts point out that reducing energy waste and increasing the use of wind turbines, solar cells, and hydrogen to produce electricity are much better ways to reduce CO_2 emissions.

In 2005, the U.S. nuclear industry persuaded the U.S. Congress to provide $150 billion more in government (taxpayer) subsidies to help build hundreds of smaller, second-generation plants using standardized designs. The industry claims these plants are safer and can be built quickly (in 3–6 years).

However, according to *Nucleonics Week,* an important nuclear industry publication, "Experts are flatly unconvinced that safety has been achieved—or even substantially increased—by the new designs." In addition, these new designs do not eliminate the threats and the expense and hazards of long-term radioactive waste storage, power plant decommissioning, and the spread of knowledge about the production of nuclear weapons.

Nuclear Fusion (Science)

Nuclear fusion has advantages, but after more than five decades of research and billions of dollars in government subsidies, it is still at the laboratory stage.

Nuclear fusion is a nuclear change in which two isotopes of light elements, such as hydrogen, are forced together at extremely high temperatures until they fuse to form a heavier nucleus, releasing energy in the process. Scientists hope that controlled nuclear fusion will provide an almost limitless source of high-temperature heat and electricity. Research has focused on the D–T nuclear fusion reaction, in which two isotopes of hydrogen—deuterium (D) and tritium (T)—fuse at about 100 million degrees (Figure 10-19).

With nuclear fusion, there would be no risk of meltdown or release of large amounts of radioactive materials from a terrorist attack and little risk from additional proliferation of nuclear weapons because bomb-grade materials (such as enriched uranium-235 and plutonium-239) are not required for fusion energy. Fusion power might also be used to destroy toxic wastes, supply electricity for ordinary use, and decompose water to produce the hydrogen gas needed to run a hydrogen economy by the end of this century.

This sounds great. So what is holding up fusion energy? In the United States, after more than 50 years of research and a $25 billion investment of mostly government funds, controlled nuclear fusion is still in the laboratory stage. None of the approaches tested so far have produced more energy than it used.

If researchers can eventually get more energy out of nuclear fusion than they put in, the next step would be to build a small fusion reactor and then scale it up to commercial size. This is an extremely difficult engineering problem. Also, the estimated cost of building and operating a commercial fusion reactor (even with huge government subsidies) is several times that of a comparable conventional fission reactor.

Proponents contend that with greatly increased federal funding, a commercial nuclear fusion power plant might be built by 2030. However, many energy experts do not expect nuclear fusion to be a significant energy source until 2100, if then.

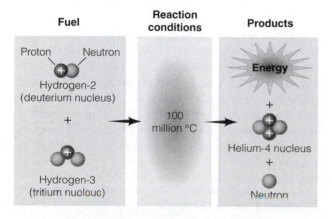

Fuel	Reaction conditions	Products

Proton   Neutron

Hydrogen-2
(deuterium nucleus)

+

Hydrogen-3
(tritium nucleus)

100 million °C

Energy

+

Helium-4 nucleus

+

Neutron

Figure 10-19 The deuterium–tritium (D–T) *nuclear fusion* reaction takes place at extremely high temperatures.

Nuclear Power's Future (Economics and Politics)

There is disagreement over whether the United States and other countries should phase out nuclear power or keep this option open in case other alternatives do not pan out.

Since 1948, nuclear energy (fission and fusion) has received about 58% of all federal energy research and development funds in the United States—compared to 22% for fossil fuels, 11% for renewable energy, and 8% for energy efficiency and conservation. Because the results of this huge investment of taxpayer dollars have been largely disappointing, some analysts call for phasing out all or most government subsides and tax breaks for nuclear power and using the money to accelerate the development of other more promising energy technologies.

To these analysts, conventional nuclear power is a complex, expensive, inflexible, and centralized way to produce electricity that is too vulnerable to terrorist attack. They believe it is a technology whose time has passed in a world where electricity will increasingly be provided by small, decentralized, easily expandable power plants such as natural gas turbines, farms of wind turbines, arrays of solar cells, and hydrogen-powered fuel cells. According to investors and World Bank economists, conventional nuclear power simply cannot compete in today's increasingly open, decentralized, and unregulated energy market unless it is artificially shielded from free-market competition by huge government subsidies.

Proponents of nuclear power argue that governments should continue funding research and development and pilot plant testing of potentially safer and cheaper conventional fission reactor designs along with breeder fission (a process that creates more nuclear fuel as it operates) and nuclear fusion. They say we need to keep these nuclear options available for use in the future if various renewable energy options fail to keep up with electricity demands and help reduce CO_2 emissions to acceptable levels.

Nuclear fission breeder reactors generate more nuclear fuel than they consume. However, because of very high costs and bad safety experiences with several of these reactors, the United States and most countries have abandoned their plans to build full-size commercial breeder reactors.

In 2005, the U.S. Congress provided $2 billion in government (taxpayer) subsidies to encourage utility companies to build new nuclear reactors in the United States. Opponents say it makes better sense to invest this money in spurring the more rapid development of energy conservation and renewable energy resources that are much safer and can be developed more quickly.

A serious long-term problem is that use of nonrenewable fossil and nuclear fuels violates the four principles of sustainability. It does not depend on solar energy and it involves disrupting chemical cycles and energy flow by emitting large quantities of greenhouse gases. It also degrades land and causes pollution, thereby destroying and degrading biodiversity and interfering with natural population control within ecosystems.

REDUCING ENERGY WASTE AND IMPROVING ENERGY EFFICIENCY

Wasting Energy (Science)

About 43% of the energy used in the United States is wasted unnecessarily.

Energy conservation involves reducing or eliminating the unnecessary waste of energy. You may be surprised to learn that 84% of all commercial energy used in the United States is wasted (Figure 10-20). About 41% of this energy is wasted automatically because of the degradation of energy quality imposed by the sec-

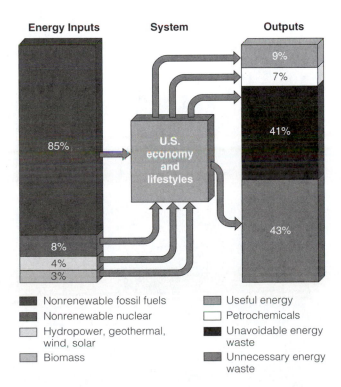

Figure 10-20 Flow of commercial energy through the U.S. economy. Only 16% of all commercial energy used in the United States ends up performing useful tasks or being converted to petrochemicals; the rest is unavoidably wasted because of the second law of thermodynamics (41%) or is wasted unnecessarily (43%). (Data from U.S. Department of Energy)

Solutions

Reducing Energy Waste

Prolongs fossil fuel supplies

Reduces oil imports

Very high net energy

Low cost

Reduces pollution and environmental degradation

Buys time to phase in renewable energy

Less need for military protection of Middle East oil resources

Improves local economy by reducing flow of money out to pay for energy

Creates local jobs

Figure 10-21 Solutions: advantages of reducing unnecessary energy waste and improving energy efficiency. Global improvements in energy efficiency could save the world about $1 trillion per year—an average of $114 million per hour! QUESTION: *Which two of these advantages do you believe are the most important?*

ond law of thermodynamics (p. 27). The other 43% is wasted unnecessarily, mostly by inefficient motor vehicles, power plants, furnaces, industrial motors and other devices, and leaky, poorly insulated, poorly designed buildings. See the Guest Essay on this topic by Amory Lovins on the website for this chapter. According to Lovins, "If the United States wants to save a lot of oil and money and increase national security, there are two simple ways to do it: Stop driving petropigs and stop living in energy sieves."

One way to reduce unnecessary energy waste is to improve the *energy efficiency* of devices by using less energy to accomplish a particular task. It involves getting more work from each unit of energy we use.

Energy efficiency in the United States has improved since the oil price shock in 1979. But unnecessary energy waste still costs the United States about $300 billion per year—an average of $570,000 per minute—and the world more than $1 trillion a year.

Reducing such energy waste has a number of economic and environmental advantages (Figure 10-21). Indeed, to most energy analysts, *reducing energy waste is the quickest, cheapest, and cleanest way to provide more energy, reduce pollution and environmental degradation, and slow global warming.*

Four widely used devices waste large amounts of energy:

- An *incandescent light bulb* uses only 5% of the electricity it draws to produce light; the other 95% is wasted as heat. It is really a *heat bulb*.

- A *motor vehicle* with an *internal combustion engine* wastes 94% of the energy in its fuel.

- A *nuclear power plant* producing electricity for space heating or water heating wastes about 86% of the energy in its nuclear fuel and probably 92% when we include the energy needed to deal with its radioactive wastes and to retire the plant.

- In a *coal-burning power plant*, two-thirds of the energy released by burning coal ends up as waste heat in the environment.

Energy experts call for us to replace these devices or greatly improve their energy efficiency over the next few decades.

On an individual level, we can save energy and money by buying more energy-efficient cars, lighting, heating systems, water heaters, air conditioners, and appliances. Some energy-efficient models may cost more initially, but in the long run they usually save money by having a lower **life cycle cost**: initial cost plus lifetime operating costs.

Saving Energy and Money in Industry (Science and Economics)

Industries can save energy and money by producing both heat and electricity from an energy source and by using more energy-efficient electric motors and lighting.

Some industries save energy and money by using **cogeneration,** or **combined heat and power (CHP)** systems. In such systems, two useful forms of energy (such as steam and electricity) are produced from the same fuel source. The energy efficiency of these systems is as high as 80% (compared to about 30–40% for coal-fired boilers and nuclear power plants) and such systems emit two-thirds less CO_2 per unit of energy produced than do conventional coal-fired boilers.

Cogeneration has been widely used in western Europe for years. Its use in the United States (where it now produces 9% of the country's electricity) and China is growing.

Another way to save energy and money in industry is to *replace energy-wasting electric motors*, which consume about one-fourth of the electricity produced in the United States. Most of these motors are inefficient because they run only at full speed with their output throttled to match the task—somewhat like driving a car very fast with your foot on the brake pedal. Each year, a heavily used electric motor consumes 10 times its purchase cost in electricity—equivalent to using $200,000 worth of gasoline each year to fuel a $20,000

car. The costs of replacing such motors with new adjustable-speed drive motors would be paid back in about 1 year and save an amount of energy equal to that generated by 150 large (1,000-megawatt) power plants.

A third way to save energy is to *switch from low-efficiency incandescent lighting to higher-efficiency fluorescent lighting.*

Saving Energy in Transportation (Science, Economics, and Politics)

The best way to save energy in transportation is to increase the fuel efficiency of motor vehicles.

Good news: Between 1973 and 1985, the average fuel efficiency rose sharply for motor vehicles sold in the United States primarily because of government-mandated *Corporate Average Fuel Economy* (CAFE) standards. *Bad news:* Between 1985 and 2006, the average fuel efficiency of new vehicles sold in the United States leveled off or declined slightly mostly because there has been no increase in the CAFE standards.

Fuel-efficient cars are available but account for less than 1% of all car sales. One reason is that the inflation-adjusted price of gasoline today in the United States is low (Figure 10-22).

But gasoline costs U.S. consumers much more then the price they pay at the pump. Indeed, if the hid-

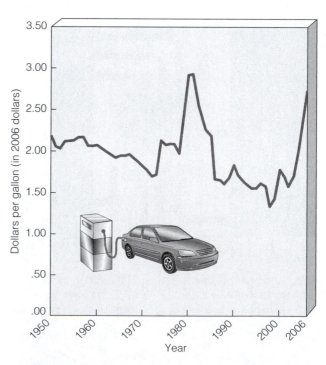

Figure 10-22 Economics: inflation-adjusted price of gasoline (in 1993 dollars) in the United States, 1920–2006. Motor vehicles in the United States use 40% of the world's gasoline. Gasoline is one of the cheapest items American consumers buy—costing less per liter than bottled water. (Data from U.S. Department of Energy)

den costs of gasoline were included as taxes in the market price of gasoline, its *true cost* for U.S. consumers would be about $2.90 per liter ($11 per gallon). Filling a 20-gallon tank would then cost about $220 and consumers would demand much more energy-efficient and less polluting cars.

These hidden costs include government subsidies and tax breaks for oil companies and road builders, pollution control and cleanup, military protection of oil supplies in the Middle East, increased medical bills and insurance premiums, time wasted in traffic jams, and increased deaths from air and water pollution. Politically powerful oil and car companies benefit financially by being able to pass these hidden costs on to consumers, future generations, and the environment.

A second reason for low fuel efficiency is that 54% of U.S. consumers own SUVs, pickup trucks, minivans, and other large, inefficient vehicles compared to 5% in 1990. A third reason is that the government has not provided large enough tax breaks, rebates, and low-interest, long-term loans to encourage people to buy more fuel-efficient cars.

In 1991, energy expert Amory Lovins proposed a *feebate* program in which fuel-inefficient vehicles would be taxed heavily with the revenue returned directly as rebates (not tax deductions) to buyers of efficient models. For example, the tax on a $50,000 Hummer H2 that averages about 5 kilometers per liter (12 miles per gallon) might be $7,000. The revenue from such taxes would go to buyers of fuel-efficient vehicles. For example, the rebate on a $22,000 hybrid car that averages 21 kilometers per liter (50 miles per gallon) might be $7,000. Within a short time, such a program—endorsed in 2001 by the U.S. National Academy of Sciences—would greatly increase sales of gas-sipping vehicles. It would also focus car makers on producing and making their profits from such vehicles, and it would cost the government (taxpayers) nothing.

Hybrid Vehicles and Sustainable Wind Power (Science)

Fuel-efficient vehicles powered by a hybrid gasoline-electric engine are expected to dominate the automotive industry between 2020 and 2030. With an extra plug-in battery, these cars could be powered mostly by electricity produced by wind.

There is growing interest in developing *superefficient cars* that could eventually get 34–128 kilometers per liter (80–300 miles per gallon). One of these vehicles is the energy-efficient *hybrid-electric car* (Figure 10-23). It has a small traditional gasoline-powered motor and an electric motor used for acceleration and hill climbing. Current models of these cars get up to 21 kilometers per liter (50 miles per gallon).

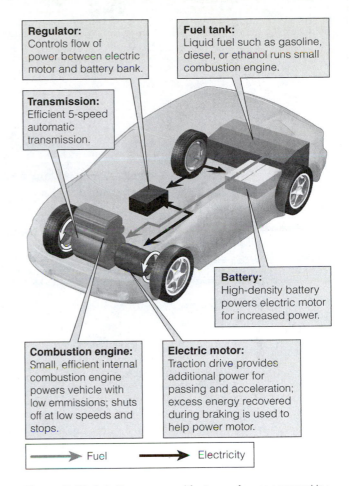

Regulator: Controls flow of power between electric motor and battery bank.

Fuel tank: Liquid fuel such as gasoline, diesel, or ethanol runs small combustion engine.

Transmission: Efficient 5-speed automatic transmission.

Battery: High-density battery powers electric motor for increased power.

Combustion engine: Small, efficient internal combustion engine powers vehicle with low emmissions; shuts off at low speeds and stops.

Electric motor: Traction drive provides additional power for passing and acceleration; excess energy recovered during braking is used to help power motor.

→ Fuel   → Electricity

Figure 10-23 Solutions: general features of a car powered by a *hybrid gasoline–electric engine.* (Concept information from DaimlerChrysler, Ford, Honda, and Toyota)

Sales of hybrid vehicles are projected to grow rapidly and could dominate motor vehicle sales by 2025. However, gas mileage for most newer models of hybrids has stayed the same or decreased because automakers are adding more horsepower and weight to satisfy the desires of many buyers for speed and power over fuel efficiency.

The next step is *plug-in hybrids* with a second battery that can be plugged in and recharged at night like laptops and cell phones. Such cars could easily get the equivalent of 32 kilometers per liter (75 miles per gallon). Most of the electricity used to recharge the extra battery would be generated by coal and nuclear power plants that have a number of disadvantages (Figure 10-12 and Figure 10-17). But according to DOE, building a network of wind energy farms in just four wind-rich states—North Dakota, Texas, Kansas, and South Dakota—could meet all U.S. electricity needs, and drastically cut CO_2 emissions. And millions of plug-in batteries in hybrid cars would act as storage systems for energy produced by wind.

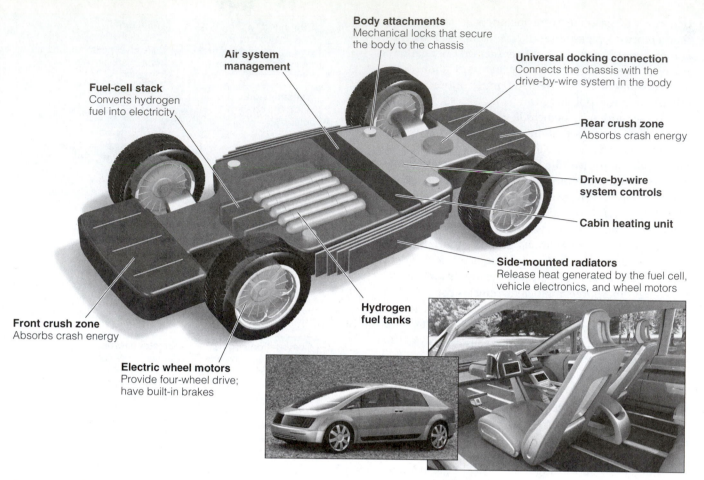

Fuel-cell stack
Converts hydrogen fuel into electricity

Air system management

Body attachments
Mechanical locks that secure the body to the chassis

Universal docking connection
Connects the chassis with the drive-by-wire system in the body

Rear crush zone
Absorbs crash energy

Drive-by-wire system controls

Cabin heating unit

Side-mounted radiators
Release heat generated by the fuel cell, vehicle electronics, and wheel motors

Hydrogen fuel tanks

Front crush zone
Absorbs crash energy

Electric wheel motors
Provide four-wheel drive; have built-in brakes

Figure 10-24 Solutions: prototype hydrogen fuel cell car developed by General Motors. This ultralight and ultrastrong car consists of a skateboard-like chassis and a variety of snap-on fiberglass bodies. It handles like a high-speed sports car, zips along with no engine noise, and emits only wisps of warm water vapor and heat—no smelly exhaust, no smog, no greenhouse gases. General Motors claims the car could be on the road within a decade, but some analysts believe that it will be 2020 before this and fuel cell cars from other manufacturers will be mass produced. (Basic information from General Motors)

The fuel efficiency for hybrid and all types of cars could nearly double if car bodies are made out of *ultralight* and *ultrastrong* composite materials such as fiberglass and the carbon-fiber composites used in bicycle helmets and in some racing cars. Such materials also provide much better crash protection.

Another type of superefficient car is an electric vehicle that uses a *fuel cell*—a device that combines hydrogen gas (H_2) and oxygen gas (O_2) in the air to produce electrical energy to power the car and water vapor (H_2O), which is emitted into the troposphere.

Fuel cells are at least twice as efficient as internal combustion engines, have no moving parts, require little maintenance, and emit no air pollutants or CO_2 if the hydrogen is produced from renewable energy resources. Most major automobile companies have developed prototype fuel-cell cars and plan to market a variety of such vehicles by 2020 (with a few models available by 2010; Figure 10-24) and to greatly increase their use by 2050. Energy experts see the following progression: today's gasoline hybrids, followed by plug-in gasoline hybrids, ultralight hybrids, and ultimately fuel-cell ultralight hybrids.

Designing Buildings to Save Energy and Other Resources (Science and Economics)

We can save energy in buildings by getting heat from the sun, superinsulating them, using plant-covered ecoroofs, and using recycled and more sustainable building materials.

Atlanta's 13-story Georgia Power Company building uses 60% less energy than conventional office buildings of the same size. The largest surface of the building faces south to capture solar energy. Each floor extends out over the one below it. This blocks out the higher summer sun to reduce air conditioning costs but allows warming by the lower winter sun. Energy-efficient lights focus on desks rather than illuminating entire rooms. Thermostats are programmed to trigger warming and cooling as needed during the workday. Such "green architecture," based on energy-efficient and money-saving designs and use of recycled materials, is beginning to catch on in Europe, the United States, and Japan.

Another energy-efficient design is a *superinsulated house*. Such houses typically cost 5% more to build than conventional houses of the same size. But this extra cost is paid back by energy savings within about 5 years and can save a homeowner $50,000–100,000 over a 40-year period. Superinsulated houses in Sweden use 90% less energy for heating and cooling that the typical American home uses.

Living roofs or *green roofs* covered with plants have been used for decades in Germany, in other parts of Europe, and in Iceland. With proper design, aesthetically pleasing and low-maintenance plant-covered roof gardens provide insulation that cuts heating and cooling costs, reduces noise, and outlasts conventional roofs.

Saving Energy in Existing Buildings (Science and Economics)

We can save energy in existing buildings by insulating them, plugging leaks, and using energy-efficient heating and cooling systems, appliances, and lighting.

Here are some ways to save energy in existing buildings.

Insulate and plug leaks. About one-third of heated air in U.S. homes and buildings escapes through closed windows and holes and cracks—roughly equal to the energy in all the oil flowing through the Alaska pipeline every year. During hot weather, these windows and cracks also let heat in, increasing the use of air conditioning. Adding insulation and plugging leaks in a house are two of the quickest, cheapest, and best ways to save energy and money.

Use energy-efficient windows. Replacing all windows in the United States with energy-efficient windows would cut expensive heat losses from houses by two-thirds, lessen cooling costs in the summer, and reduce CO_2 emissions. Widely available superinsulating windows insulate as well as 8–12 panes of glass. Although they cost 10–15% more than double-glazed windows, this cost is paid back rapidly by the energy they save. Even better windows will reach the market soon.

Heat houses more efficiently. In order, the most energy-efficient ways to heat space are superinsulation, a geothermal heat pump that transfers heat stored in the earth to a home, passive solar heating, a conventional heat pump (in warm climates only), small cogenerating microturbines, and a high-efficiency (85–98%) natural gas furnace. The most wasteful and expensive way is to use electric resistance heating with the electricity produced by a coal-fired or nuclear power plant.

Heat water more efficiently. One approach is to use a *tankless instant water heater* (about the size of a small suitcase) fired by natural gas or LPG but not by electricity. These devices, widely used in many parts of Europe, heat water instantly as it flows through a small burner chamber, providing hot water only when it is needed and using less energy than traditional water heaters. They work. I used them in a passive solar office and living space for 15 years. For information visit **http://foreverhotwater.com**.

Use energy-efficient appliances and energy-efficient lighting. If all households in the United States used the most efficient frost-free refrigerator now available, 18 large (1,000-megawatt) power plants could close. Microwave ovens can cut electricity use for cooking by 25–50% (but not if used for defrosting food). Clothes dryers with moisture sensors cut energy use by 15%, and front-loading washers use 50% less energy than top-loading models but cost about the same. A compact fluorescent bulb uses 75–80% less electricity than does an incandescent lightbulb, lasts 10–13 times longer, and saves about $70 over its lifetime. Replacing 20 incandescent bulbs with compact fluorescent bulbs could save a homeowner about $1,400.

Why Are We Still Wasting So Much Energy? (Economics and Politics)

Low-priced fossil fuels and lack of government tax breaks and other financial incentives for saving energy promote energy waste.

With such an impressive array of benefits to be gained from reducing energy waste (Figure 10-21), why is there so little emphasis on improving energy efficiency? One reason is a glut of low-cost gasoline (Figure 10-22) and other fossil fuels. As long as energy remains artificially cheap, with its market price not including its harmful environmental and health costs, people are more likely to waste it and are not likely to make investments in improving energy efficiency.

Another reason is a lack of large enough tax breaks, rebates, low-interest, long-term loans, and other economic incentives for consumers and businesses to invest in improving energy efficiency.

Would you like to earn about 20% a year on your money, tax-free and risk free? Invest it in improving the energy efficiency of your home, lights, and appliances. You will get your investment back in a few years and then make about 20% per year by having lower heating, cooling, and electricity bills. This is a win–win deal for you and the earth.

X *How Would You Vote?* Should the United States or the country where you live greatly increase its emphasis on improving energy efficiency? Cast your vote online at www.thomsonedu.com/biology/miller.

USING RENEWABLE ENERGY TO PROVIDE HEAT AND ELECTRICITY

Why Don't We Use More Renewable Energy? (Economics and Politics)

A variety of renewable energy resources are available, but their use has been hindered by a lack of government support compared to the support for nonrenewable fossil fuels and nuclear power.

One of nature's four principles of sustainability (Figure 1-11, p. 17) is to *rely mostly on renewable solar energy.* We can get renewable solar energy directly from the sun or indirectly from moving water, wind, and biomass. Another form of renewable energy is geothermal energy from the earth's interior. Studies show that with increased and consistent government backing, renewable energy could provide 20% of the world's electricity by 2020 and 50% by 2050.

If renewable energy is so great, why does it now provide only 18% of the world's energy and only 6% of the energy in the United States? One reason is that renewable energy resources have received and continue to receive much lower government tax breaks, subsidies, and research and development funding than fossil fuels and nuclear power have received for decades. Another reason is that the prices we pay for fossil fuels and nuclear power do not include the costs of their harm to the environment and to human health.

In other words, the economic dice have been loaded against solar, wind, and other forms of renewable energy. If these two economic handicaps—inequitable subsidies and inaccurate pricing—were eliminated, energy analysts say that many forms of renewable energy would be cheaper than fossil fuels and nuclear energy.

Heating Buildings and Water with Solar Energy (Science)

We can heat buildings by orienting them toward the sun (passive solar heating) or by pumping a liquid such as water through rooftop collectors (active solar heating).

Buildings and water can be heated by solar energy using two methods: passive and active (Figure 10-25).

A **passive solar heating system** absorbs and stores heat from the sun directly within a structure without the need for pumps or fans to distribute the heat (Figure 10-25, left). Energy-efficient windows and attached greenhouses face the sun to collect solar energy directly. Walls and floors of concrete, adobe, brick, stone, salt-treated timber, and water in metal or plastic containers store much of the collected solar energy as heat and release it slowly throughout the day and night. A small backup heating system such as a vented natural gas or propane heater may be used but is not necessary in many climates. See the Guest Essay by Nancy Wicks on this topic on the website for this chapter.

On a life cycle cost basis, good passive solar and superinsulated design is the cheapest way to heat a home or small building in sunny areas. The typical payback time for passive solar features is 3–7 years.

An **active solar heating system** absorbs energy from the sun by pumping a heat-absorbing fluid (such as water or an antifreeze solution) through special collectors usually mounted on a roof or on special racks to face the sun (Figure 10-25, right). Some of the collected heat can be used directly. The rest can be stored in a large insulated container filled with gravel, water,

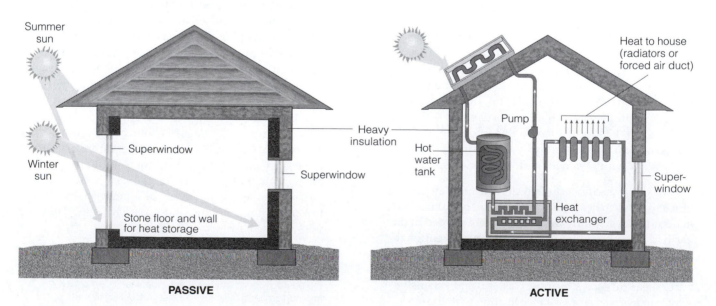

PASSIVE

ACTIVE

Figure 10-25 Solutions: passive and active solar heating for a home.

Trade-Offs

Passive or Active Solar Heating

Advantages	Disadvantages
Energy is free	Need access to sun 60% of time
Net energy is moderate (active) to high (passive)	Sun blocked by other structures
Quick installation	Need heat storage system
No CO_2 emissions	High cost (active)
Very low air and water pollution	Active system needs maintenance and repair
Very low land disturbance (built into roof or window)	Active collectors unattractive
Moderate cost (passive)	

Figure 10-26 Trade-offs: advantages and disadvantages of heating a house with passive or active solar energy. QUESTION: *What are the single advantage and the single disadvantage that you think are the most important?*

clay, or a heat-absorbing chemical for release as needed. Active solar collectors also supply hot water for more than 32 million households.

Figure 10-26 lists the major advantages and disadvantages of using passive or active solar energy for heating buildings. Passive and active solar energy can be used to heat new homes in areas with adequate sunlight. But solar energy cannot be used to heat existing homes and buildings not oriented to receive sunlight or blocked from sunlight by other buildings or trees. Most analysts do not expect widespread use of active solar collectors for heating houses because of high costs, maintenance requirements, and unappealing appearance.

Using Solar Energy to Generate High-Temperature Heat and Electricity (Science)

Large arrays of solar collectors in sunny deserts can produce high-temperature heat to spin turbines and produce electricity, but costs are high.

Solar thermal systems can collect and transform energy from the sun into high-temperature thermal energy (heat), which can be used directly or converted to electricity. These systems are used mostly in desert areas

with ample sunlight. Figure 10-27 lists the advantages and disadvantages of concentrating solar energy to produce high-temperature heat or electricity.

One method uses a *central receiver system*, called a *power tower*. Huge arrays of computer-controlled mirrors called *heliostats* track the sun and focus sunlight on a central heat collection tower (top drawing in Figure 10-27).

Another approach is a *solar thermal plant* in which sunlight is collected and focused on oil-filled pipes running through the middle of a large area of curved solar collectors (bottom drawing in Figure 10-27). This concentrated sunlight can generate temperatures high enough to produce steam to run turbines and generate electricity. At night or on cloudy days, high-efficiency, natural gas turbines can supply backup electricity as needed.

Most analysts do not expect widespread use of such technologies over the next few decades because of high costs, limited suitable sites, and availability of much cheaper ways to produce electricity such as *combined-cycle* natural gas turbines and wind turbines.

On an individual scale, inexpensive *solar cookers* can focus and concentrate sunlight to cook food, especially in rural villages in sunny areas. They can be made by fitting an insulated box big enough to hold three or four pots with a transparent, removable top. Solar cookers can reduce deforestation from fuelwood harvesting and save the time and labor needed to collect firewood. They also reduce indoor air pollution from smoky fires.

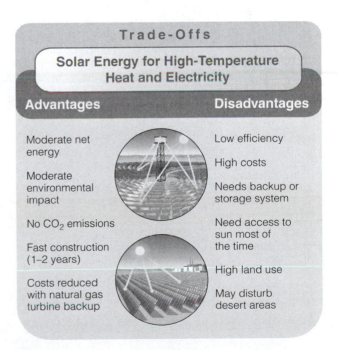

Trade-Offs

Solar Energy for High-Temperature Heat and Electricity

Advantages	Disadvantages
Moderate net energy	Low efficiency
Moderate environmental impact	High costs
No CO_2 emissions	Needs backup or storage system
Fast construction (1–2 years)	Need access to sun most of the time
Costs reduced with natural gas turbine backup	High land use
	May disturb desert areas

Figure 10-27 Trade-offs: advantages and disadvantages of using solar energy to generate high-temperature heat and electricity. QUESTION: *What are the single advantage and the single disadvantage that you think are the most important?*

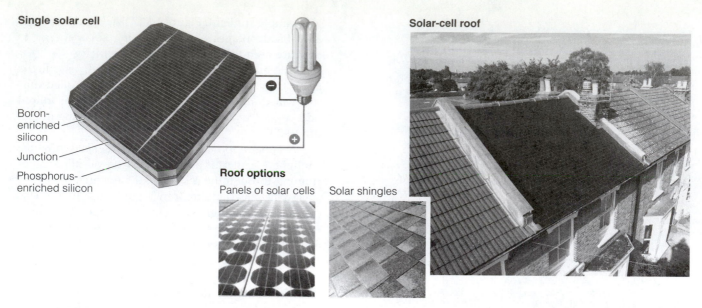

Single solar cell

Boron-enriched silicon

Junction

Phosphorus-enriched silicon

Solar-cell roof

Roof options

Panels of solar cells   Solar shingles

Figure 10-28 Solutions: Photovoltaic (PV) or solar cells can provide electricity for a house or building using solar-cell roof shingles, as shown in this house in Richmond, Surrey, England. Solar cell roof systems that look like a metal roof are also available. In addition, new thin-film solar cells can be applied to windows and outside glass walls.

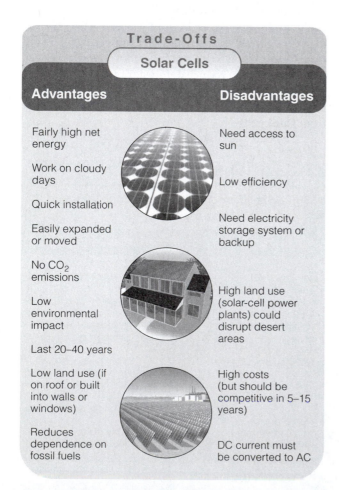

Trade-Offs

Solar Cells

Advantages	Disadvantages
Fairly high net energy	Need access to sun
Work on cloudy days	Low efficiency
Quick installation	
Easily expanded or moved	Need electricity storage system or backup
No CO_2 emissions	
Low environmental impact	High land use (solar-cell power plants) could disrupt desert areas
Last 20–40 years	
Low land use (if on roof or built into walls or windows)	High costs (but should be competitive in 5–15 years)
Reduces dependence on fossil fuels	DC current must be converted to AC

Figure 10-29 Trade-offs: advantages and disadvantages of using solar cells to produce electricity. QUESTION: *What are the single advantage and the single disadvantage that you think are the most important?*

Producing Electricity with Solar Cells—A Step on the Path to Sustainability (Science)

Solar cells that convert sunlight to electricity can be incorporated into roofing materials or windows and soon on almost any surface including clothing, and the high costs are expected to fall.

Solar energy can be converted directly into electrical energy by **photovoltaic (PV) cells,** commonly called **solar cells** (Figure 10-28). A typical solar cell is a transparent wafer that contains a semiconductor with a thickness ranging from less than that of a human hair to a sheet of paper. Sunlight energizes electrons in the semiconductor, causing them to flow, creating an electrical current.

These devices have no moving parts, are safe and quiet, require little maintenance, produce no pollution during operation, and last as long as a conventional fossil fuel or nuclear power plant. The cells can be connected to batteries to store the electrical energy until needed or to existing electrical grid systems. The semiconductor material used in solar cells can be made into paper-thin rigid or flexible sheets that can be incorporated into traditional-looking roofing materials (Figure 10-28, right) and building walls or windows.

Easily expandable banks of solar cells can be used in developing countries to provide electricity for the 1.7 billion people in rural villages with ample sunlight who are not connected to an electrical grid. With financing from the World Bank, India is installing solar-cell systems in 38,000 villages. The United Nations is mapping the potential solar and wind energy resources of 13 developing countries in Africa, Asia, and South and Central America.

Figure 10-29 lists the advantages and disadvantages of solar cells. Current costs of producing electricity from solar cells are high but dropped 10-fold between 1975 and 2004, and they are expected to drop much more because of savings from mass production and new designs.

Currently, solar cells supply less than 1% of the world's electricity. But with increased government and private research and development—and much greater and consistent government tax breaks and other subsidies—they could provide a quarter of the world's electricity by 2040. If such projections are correct, the production, sale, and installation of solar cells could become one of the world's largest and fastest-growing businesses. Japan is now the world's leader in solar cell technology and sales and Germany is number two.

Using Flowing Water to Produce Electricity (Science)

Water flowing in rivers and streams can be trapped in reservoirs behind dams and released as needed to spin turbines and produce electricity.

Water flowing from higher to lower elevations in rivers and streams can be controlled by dams and reservoirs and used to produce electricity (Figure 9-9, p. 178). This indirect form of renewable solar energy is called *hydropower*.

The most widely used approach is to build a high dam across a large river to create a reservoir. Some of the water stored in the reservoir is allowed to flow through huge pipes at controlled rates, spinning turbines and producing electricity.

X *How Would You Vote?* Should the world greatly increase its dependence on large-scale dams for producing electricity? Cast your vote online at www.thomsonedu.com/biology/miller.

In 2004, hydropower supplied about one-fifth of the world's electricity, 99% of the electricity in Norway, 75% in New Zealand, 25% in China, and 7% in the United States (but about 50% on the West Coast).

Because of increasing concern about the harmful environmental and social consequences of large dams, the World Bank and other development agencies have been pressured to stop funding new large-scale hydropower projects. Some analysts expect the contribution of large-scale hydropower plants to fall slowly over the next several decades as many existing reservoir systems fill with silt and become useless faster than new ones are built.

Figure 10-30 lists the advantages and disadvantages of using large-scale hydropower plants to produce electricity.

Small-scale hydropower projects eliminate most of the harmful effects of large-scale projects. But their electrical output can vary with seasonal changes in stream flow.

Trade-Offs
Large-Scale Hydropower

Advantages	Disadvantages
Moderate to high net energy	High construction costs
High efficiency (80%)	High environmental impact from flooding land to form a reservoir
Large untapped potential	High CO_2 emissions from biomass decay in shallow tropical reservoirs
Low-cost electricity	
Long life span	Floods natural areas behind dam
No CO_2 emissions during operation in temperate areas	Converts land habitat to lake habitat
May provide flood control below dam	Danger of collapse
Provides water for year-round irrigation of cropland	Uproots people
	Decreases fish harvest below dam
Reservoir is useful for fishing and recreation	Decreases flow of natural fertilizer (silt) to land below dam

Figure 10-30 Trade-offs: advantages and disadvantages of using large dams and reservoirs to produce electricity. QUESTION: *What are the single advantage and the single disadvantage that you think are the most important?*

We can also produce electricity from water flows by tapping into the energy from tides and waves. Most analysts expect these sources to make little contribution to world electricity production because of high costs and few areas that have the right conditions.

Wind Power—An Astounding Energy Alternative (Science and Economics)

Wind power is the world's most promising energy resource because it is abundant, inexhaustible, widely distributed, cheap, clean, and emits no greenhouse gases.

The differences in solar heating of the earth between the equator and the poles, together with the earth's rotation set up flows of air called *wind*. Wind turbines can capture this indirect form of solar energy and convert it to electrical energy (Figure 10-31, p. 228). Today's wind turbines can be as tall as 30-stories, which allows them to tap into stronger, more reliable, and less turbulent winds found at higher elevations.

Figure 10-31 Solutions:
Wind turbines can be used individually to produce electricity. But increasingly they are being used in interconnected arrays of ten to hundreds of turbines. These *wind farms* or *wind parks* can be located on land or offshore.

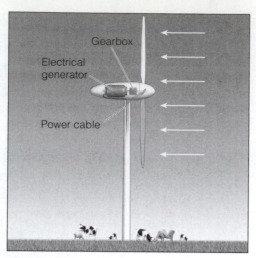

Gearbox
Electrical generator
Power cable

Wind turbine

Wind farm

The price of wind-generated electricity has fallen from 38 cents per kilowatt-hour in the early 1980s to 3–7 cents per kilowatt hour in 2005. And the price is expected to reach 2.5 cents per kilowatt hour within a few years. Table 10-1 compares the total costs of producing electricity from a number of sources by adding their generating costs and their estimated harmful environmental (external) costs. The prices of all sources listed are expected to stay about the same, except for wind and solar cells, which are expected to drop because of improved designs and mass production.

Much of the world's potential for wind power remains untapped. According to recent studies, wind farms on only one-tenth of the earth's land and 13% of the land area in developing countries could produce twice the world's projected demand for electricity by 2020. And a 2004 study found that capturing only 20% of the wind energy at the world's best energy sites could meet all of the world's energy demand and more than seven times the amount of electricity currently used in the world.

Analysts expect to see increasing use of offshore wind farms because wind speeds over water are often greater than those over land. Any noise produced is muffled by surf sounds, and negotiations with multiple landowners are unnecessary. However, offshore installation costs are higher and some coastal towns and cities oppose them because of their visual pollution—just as they oppose offshore oil drilling rigs.

The DOE calls the Great Plains states of North Dakota, South Dakota, Nebraska, Kansas, Oklahoma, and Texas the "Saudi Arabia of wind power." In theory, these states have enough wind resources to more than meet all the nation's electricity needs. According to the American Wind Energy Association, with increased and consistent government subsidies and tax breaks, wind power could produce almost a fourth of the country's electricity by 2025. This electricity can also be passed through water to produce hydrogen gas, which can run fuel cells.

Some U.S. farmers and ranchers make much more money by using their land for wind power production than by growing crops or raising cattle. They typically receive $3,000–5,000 a year in royalties for each turbine they allow on a small plot of land that can still be used to grow corn worth about $120 and graze cattle to provide about $10 worth of beef.

Figure 10-32 lists the advantages and disadvantages of using wind to produce electricity. According to energy analysts, wind power has more advantages and fewer serious disadvantages than any other energy resource. Many governments and corporations are recognizing that wind is a vast, climate-benign, renewable energy resource that can supply both electricity and hydrogen fuel for running fuel cells at an affordable cost. They understand that *there is money in wind* and that our energy future may be *blowing in the wind.*

X How Would You Vote? Should we greatly increase our dependence on wind power? Cast your vote online at www.thomsonedu.com/biology/miller.

Table 17-1 Total Costs of Electricity from Different Sources in 2004 (In U.S. cents per kilowatt hour)*			
Electricity Source	Generating Costs	Environmental Costs	Total Costs
Wind	3–7	0.1–0.3	3.1–7.3
Hydropower	3–8	0–1.1	3–9.1
Natural gas	4–7.0	1.1–4.5	5.5–11.5
Coal	3–4	2.3–17.0	5.3–21.0
Geothermal	5–8	1?	6–9
Biomass	7–9	1–3.4	8–12.4
Nuclear	10–14	0.2–0.7	10.2–14.7
Solar cells	24–28	0.7	24.7–28.7

*Data from U.S. Department of Energy and a variety of sources compiled by the Worldwatch Institute.

Trade-Offs

Wind Power

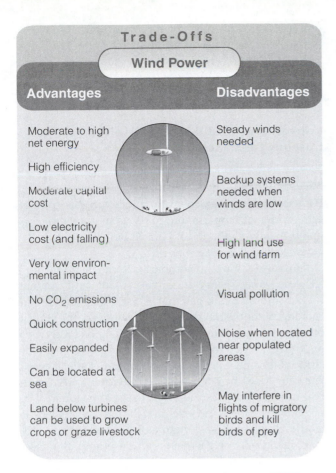

Advantages	Disadvantages
Moderate to high net energy	Steady winds needed
High efficiency	
Moderate capital cost	Backup systems needed when winds are low
Low electricity cost (and falling)	
Very low environmental impact	High land use for wind farm
No CO_2 emissions	Visual pollution
Quick construction	
Easily expanded	Noise when located near populated areas
Can be located at sea	
Land below turbines can be used to grow crops or graze livestock	May interfere in flights of migratory birds and kill birds of prey

Figure 10-32 Trade-offs: advantages and disadvantages of using wind to produce electricity. By 2020, wind power could supply more than 10% of the world's electricity and 10–25% of the electricity used in the United States. QUESTION: *What are the single advantage and the single disadvantage that you think are the most important?*

Burning Solid Biomass (Science)

Plant materials and animal wastes can be burned to provide heat or electricity or converted into gaseous or liquid biofuels.

Biomass consists of plant materials (such as wood and agricultural waste) and animal wastes that can be burned directly as a solid fuel or converted into gaseous or liquid **biofuels.** Biomass is an indirect form of solar energy because it consists of combustible organic compounds produced by photosynthesis.

Biomass is burned for heating, cooking, and industrial processes and to drive turbines and produce electricity. Wood, charcoal made from wood, animal manure, and other forms of biomass supply about 10% of the world's energy, 35% of the energy used in developing countries, and 95% of energy needs in the poorest countries.

Almost 70% of the people living in developing countries heat their homes and cook their food by burning wood or charcoal. But wood is a renewable fuel only as long as it is not harvested faster than it is re-

plenished. *Bad news:* About 2.7 billion people in 77 developing countries face a *fuelwood crisis* and are forced to meet their fuel needs by over-harvesting wood.

One way to produce biomass fuel is to plant, harvest, and burn large numbers of fast-growing trees (such as cottonwoods, poplars, and sycamores), shrubs, perennial grasses (such as switchgrass), and water hyacinths in *biomass plantations*. But repeated cycles of growing and harvesting biomass plantations can deplete the soil of key nutrients. And clearing forests to plant biomass plantations destroys and degrades biodiversity.

In agricultural areas, *crop residues* (such as sugarcane residues, rice husks, cotton stalks, and coconut shells) and *animal manure* can be collected and burned or converted into biofuels. Some ecologists argue that it makes more sense to use animal manure as a fertilizer and crop residues to feed livestock, retard soil erosion, and fertilize the soil.

Figure 10-33 lists the general advantages and disadvantages of burning solid biomass as a fuel. One

Trade-Offs

Solid Biomass

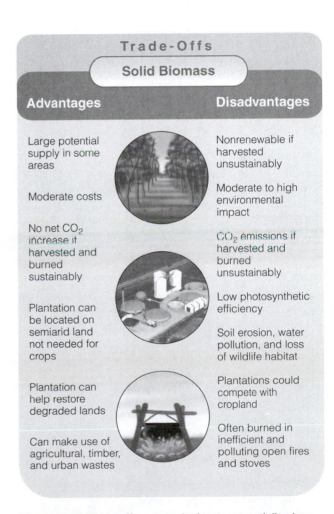

Advantages	Disadvantages
Large potential supply in some areas	Nonrenewable if harvested unsustainably
Moderate costs	Moderate to high environmental impact
No net CO_2 increase if harvested and burned sustainably	CO_2 emissions if harvested and burned unsustainably
	Low photosynthetic efficiency
Plantation can be located on semiarid land not needed for crops	Soil erosion, water pollution, and loss of wildlife habitat
Plantation can help restore degraded lands	Plantations could compete with cropland
Can make use of agricultural, timber, and urban wastes	Often burned in inefficient and polluting open fires and stoves

Figure 10-33 Trade-offs: general advantages and disadvantages of burning solid biomass as a fuel. QUESTION: *What are the single advantage and the single disadvantage that you think are the most important?*

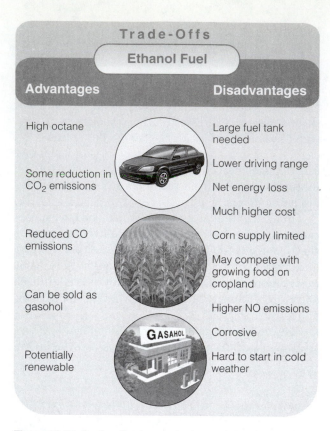

Trade-Offs

Ethanol Fuel

Advantages	Disadvantages
High octane	Large fuel tank needed
Some reduction in CO_2 emissions	Lower driving range
	Net energy loss
	Much higher cost
Reduced CO emissions	Corn supply limited
	May compete with growing food on cropland
Can be sold as gasohol	Higher NO emissions
	Corrosive
Potentially renewable	Hard to start in cold weather

Figure 10-34 Trade-offs: general advantages and disadvantages of using ethanol as a vehicle fuel compared to gasoline. QUESTION: *What are the single advantage and the single disadvantage that you think are the most important?*

Trade-Offs

Methanol Fuel

Advantages	Disadvantages
High octane	Large fuel tank needed
Some reduction in CO_2 emissions	Half the driving range of gasoline
Lower total air pollution (30–40%)	Corrodes metal, rubber, plastic
Can be made from natural gas, agricultural wastes, sewage sludge, garbage, and CO_2	High CO_2 emissions if made from coal
	Expensive to produce
Can be used to produce H_2 for fuel cells	Hard to start in cold weather

Figure 10-35 Trade-offs: general advantages and disadvantages of using methanol as a vehicle fuel compared to using gasoline. QUESTION: *What are the single advantage and the single disadvantage that you think are the most important?*

problem is that burning biomass produces CO_2. However, if the rate of use of biomass does not exceed the rate at which it is replenished by new plant growth (which takes up CO_2), there is no net increase in CO_2 emissions.

Converting Solid Biomass into Liquid and Gaseous Fuels (Science)

Some forms of biomass can be converted into gaseous and liquid biofuels.

Bacteria and various chemical processes can convert some forms of biomass into gaseous biofuel, or *biogas* that is mostly methane, and liquid biofuels such as *liquid ethanol,* and *liquid methanol.*

In rural China, anaerobic bacteria in more than 500,000 *biogas digesters* convert plant and animal wastes into methane gas that is used for heating and cooking. After the biogas has been removed, the almost odorless solid residue is used as fertilizer on food crops or on trees. When they work, biogas digesters are very efficient. But they are slow and unpredictable, a problem that could be corrected by developing more reliable models. They also add CO_2 to the atmosphere.

Some analysts believe that liquid fuels such as biodiesel, ethanol, and methanol produced from biomass could replace gasoline and diesel fuel. Biodiesel is a diesel fuel made from renewable resources such as vegetable oils (extracted from palm and soybean plants) and fats, including used vegetable oils from restaurants. Its use is increasing in Europe.

However, many of the palm and soybean plants used to produce biodiesel fuel in Europe are produced on plantations in tropical developing countries such as Malaysia, Indonesia, and Brazil. Most of these plantations are established by clearing tropical forests. Thus, increased use of biodiesel in Europe and elsewhere can lead to a major loss of biodiversity and increased CO_2 emissions from tropical deforestation—more than offsetting the environmental benefits of burning biodiesel.

Ethanol can be made from sugar and grain crops (sugarcane, sugar beets, sorghum, sunflowers, and corn) by fermentation and distillation. Gasoline mixed with 10–23% pure ethanol makes *gasohol,* which can be burned in conventional gasoline engines. Figure 10-34 lists the advantages and disadvantages of using ethanol as a vehicle fuel compared to gasoline.

Methanol is made mostly from natural gas, but it also can be produced at a higher cost from carbon dioxide, coal, and biomass such as wood, wood wastes, agricultural wastes, sewage sludge, and garbage. Figure 10-35 lists the advantages and disadvantages of using methanol as a vehicle fuel compared to using gasoline.

Chemist George A. Olah believes that establishing a *methanol economy* is preferable to the highly publicized hydrogen economy. He points out that methanol can be produced chemically from carbon dioxide in the atmosphere, which could also help slow projected global warming. In addition, methanol can be converted to other hydrocarbon compounds that can be used to produce a variety of useful chemicals like those made from petroleum and natural gas.

GEOTHERMAL ENERGY

Tapping the Earth's Internal Heat (Science)

We can use geothermal energy stored in the earth's mantle to heat and cool buildings and to produce electricity.

Geothermal energy consists of heat stored in soil, underground rocks, and fluids in the earth's mantle. Scientists have developed several ways to tap into this stored energy to heat and cool buildings and to produce electricity.

A *geothermal heat pump* (GHP) system can heat and cool a house by exploiting the difference between the earth's surface and underground temperatures. These devices extract heat from the earth in winter and in summer can store heat removed from a house in the earth. They are a very efficient and cost-effective way to heat or cool a space.

According to the EPA, a well-designed geothermal heat pump system is the most energy-efficient, reliable, cost-effective, and environmentally clean way to heat or cool a building. It produces no air pollutants and emits no CO_2. A typical system can cut home heating and cooling costs by about half.

We have also learned to tap into deeper and more concentrated underground *hydrothermal reservoirs* of geothermal energy. One type of reservoir contains *dry steam* with water vapor but no water droplets. Another consists of *wet steam*, a mixture of steam and water droplets. The third is *hot water* trapped in fractured or porous rock at various places in the earth's crust.

If such geothermal sites are close to the surface, wells can be drilled to extract the dry steam, wet steam, or hot water (Figure 10-1). It can be used to heat homes and buildings or to spin turbines and produce electricity.

Currently, about 22 countries (most of them in the developing world) are extracting energy from geothermal sites to produce about 1% of the world's electricity. Geothermal energy is used to heat 8 of every 10 buildings in Iceland, produce electricity, and heat greenhouses that produce most of the country's fruits and vegetables

Geothermal electricity meets the electricity needs of 6 million Americans and supplies 6% of the U.S. state of California's electricity. In 1999, Santa Monica, California, became the first city in the world to get all

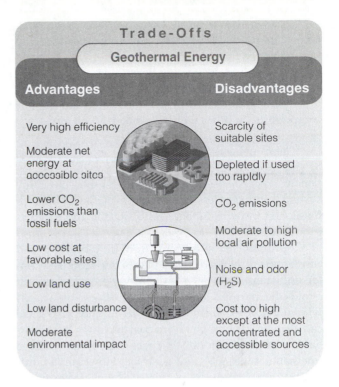

Figure 10-36 Trade-offs: advantages and disadvantages of using geothermal energy for space heating and to produce electricity or high-temperature heat for industrial processes. QUESTION: *What are the single advantage and the single disadvantage that you think are the most important?*

its electricity from geothermal energy. Figure 10-36 lists the advantages and disadvantages of using geothermal energy.

X *HOW WOULD YOU VOTE?* Should the United States, or the country where you live, greatly increase its dependence on geothermal energy to provide heat and to produce electricity? Cast your vote online at www.thomsonedu.com/biology/miller.

HYDROGEN

Can Hydrogen Replace Oil? (Science)

Some energy analysts view hydrogen gas as the best fuel to replace oil during the last half of this century, but there are big hurdles to overcome.

Many scientists and executives of major oil companies and automobile companies say the fuel of the future is hydrogen gas (H_2).

When hydrogen gas burns in air or in fuel cells, it combines with oxygen gas in the air and produces nonpolluting water vapor. Widespread use of hydrogen as a fuel would eliminate most of the air pollution problems we face today and greatly reduce the threats from global warming by emitting no CO_2—as long as the hydrogen is not produced from fossil fuels or other carbon-containing compounds.

So what is the catch? There are four major challenges in turning the vision of widespread use of hydrogen as a fuel into reality. *First,* hydrogen is chemically locked up in water and organic compounds such as methane and gasoline. *Second,* it takes energy and money to produce hydrogen from water and organic compounds. In other words, *hydrogen is not a source of energy. It is a fuel produced by using energy.*

Third, fuel cells are the best way to use hydrogen to produce electricity, but current versions are expensive. *Fourth,* whether a hydrogen-based energy system produces less air pollution and CO_2 than a fossil fuel system depends on how the hydrogen is produced.

We could use electricity from coal-burning and conventional nuclear power plants to decompose water and make hydrogen. But this approach does not avoid the harmful environmental effects associated with using these fuels (Figures 10-12 and 10-17). We can also make hydrogen from coal and strip it from organic compounds found in fuels such as natural gas (methane), methanol, and gasoline. However, according to a 2002 study by physicist Marin Hoffer and a team of other scientists, producing hydrogen from coal and organic compounds will add more CO_2 to the atmosphere per unit of heat generated than does burning these carbon-containing fuels directly.

Most proponents of hydrogen believe that if we are to receive its very low pollution and low CO_2 benefits, the energy used to produce hydrogen must come from low-polluting, renewable sources that also emit little or no CO_2. The most likely sources are electricity generated by wind farms, geothermal energy, solar cells, micro-hydropower plants, or biological processes in bacteria and algae.

In 1999, a a public–private partnership consisting of DaimlerChrysler, Royal Dutch Shell, Norsk Hydro, and Icelandic New Energy announced government-approved plans to turn the tiny country of Iceland into the world's first "hydrogen economy" by 2050. The plan is under way. In 2003, the world's first commercial hydrogen filling station opened in Reykjavik. It is being used to fuel several hydrogen buses. Norsk Hydro is providing the electricity and technology to produce the hydrogen. The next goal is to use hydrogen to power fuel cells in all the country's cars.

Once produced, hydrogen can be stored in a pressurized tank, as liquid hydrogen, and in solid metal hydride compounds, which when heated release hydrogen gas. Scientists are also evaluating ways to store hydrogen by absorbing it onto the surfaces of activated charcoal or carbon nanofibers, which when heated release hydrogen gas. Another possibility is to store it inside tiny glass microspheres. More research is needed to convert these possibilities into reality. However, use of ultralight car bodies made of composites and energy-efficient aerodynamic design would im-

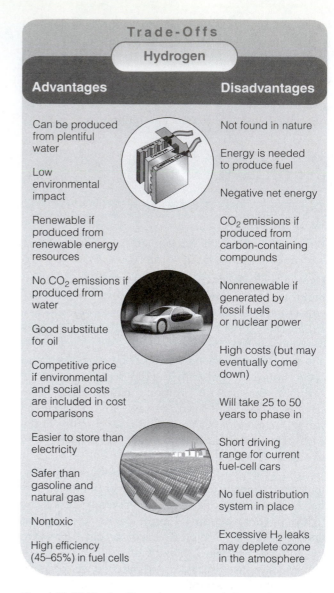

Figure 10-37 Trade-offs: advantages and disadvantages of using hydrogen as a fuel for vehicles and for providing heat and electricity. QUESTION: *What are the single advantage and the single disadvantage that you think are the most important?*

prove fuel efficiency so that large hydrogen fuel tanks would not be needed.

Some good news. Metal hydrides, sodium borohydride, charcoal powders, ammonia borane, carbon nanotubes, and glass microspheres containing hydrogen will not explode or burn if a vehicle's fuel tank or system is ruptured in an accident. This makes hydrogen a much safer fuel than gasoline, diesel fuel, methanol, and natural gas. Figure 10-37 lists the advantages and disadvantages of using hydrogen as an energy resource.

X *HOW WOULD YOU VOTE?* Do the advantages of producing and burning hydrogen as an energy resource outweigh the disadvantages? Cast your vote online at www.thomsonedu.com/biology/miller.

A SUSTAINABLE ENERGY STRATEGY

Choosing Energy Paths

We need to answer several questions in deciding which energy resources to promote.

Energy policies need to be developed with the future in mind. Experience shows that it usually takes at least 50 years and huge investments to phase in new energy alternatives to the point at which they provide 10–20% of total energy use. Figure 10-38 shows the shifts in use of commercial sources of energy in the United States since 1800, and one scenario projecting changes to a solar–hydrogen economy by 2100.

Making projections such as those in Figure 10-38 and then converting such projections to energy policy involves trying to answer the following questions for *each* energy alternative:

- How much of the energy resource is likely to be available in the near future (the next 25 years) and the long term (the next 50 years)?

- What is the net energy yield for the resource?

- How much will it cost to develop, phase in, and use the resource?

- What government research and development subsidies and tax breaks will be used to help develop the resource?

- How will dependence on the resource affect national and global economic and military security?

- How vulnerable is the resource to terrorism?

- How will extracting, transporting, and using the resource affect the environment, human health, and the earth's climate? Should these harmful costs be included in the market price of each energy resource

through mechanisms like taxing and reducing subsidies?

What Are the Best Energy Alternatives? A New Vision

A more sustainable energy policy would improve energy efficiency, rely more on renewable energy, and reduce the harmful effects of using fossil fuels and nuclear energy.

Many scientists and energy experts who have evaluated energy alternatives have come to these three general conclusions:

First, *there will be a gradual shift from large, centralized macropower systems to smaller, decentralized micropower systems* such as small natural gas turbines and stationary fuel cells for commercial buildings, wind turbines, fuel cells for cars, and household solar panels (Figure 10-39, p. 234). This shift from centralized *macropower* to dispersed *micropower* is analogous to the computer industry's shift from large centralized mainframes to increasingly smaller, widely dispersed PCs, laptops, and handheld computers. This shift also improves national and economic security by enabling a country to rely on a diversity of small and dispersed energy resources instead of on a small number of large and vulnerable coal and nuclear power plants.

Second, *the best alternatives combine improved energy efficiency and the use of natural gas as a fuel to make the transition to a diverse mix of locally available renewable energy resources and possibly nuclear fusion in the distant future (if it becomes feasible and affordable).*

Third, *over the next 50 years, the choice is not between using nonrenewable fossil fuels and using renewable energy sources.* Because of their supplies and low prices, fossil fuels will continue to be used in large quantities. The challenge is to find ways to reduce the harmful environmental impacts of widespread fossil fuel use, with special emphasis on reducing air pollution and emissions of greenhouse gases, as less harmful alternatives are phased in.

Figure 10-40 (p. 235) lists strategies for making the transition to a more sustainable energy future over the next 50 years.

Economics, Politics, Education, and Energy Resources

Governments can use a combination of subsidies, tax breaks, rebates, taxes, and public education to promote or discourage use of various energy alternatives.

To most analysts the key to making a shift to more sustainable energy resources and societies is economics and politics. Governments can use three strategies to help stimulate or dampen the short-term and long-term use of a particular energy resource.

First, they can *keep energy prices artificially low to encourage use of selected energy resources.* They can provide

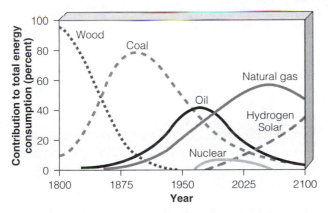

Figure 10-38 Science, economics, and politics: shifts in the use of commercial energy resources in the United States since 1800 and projected changes to 2100. Shifts from wood to coal, and then from coal to oil and natural gas, have each taken about 50–75 years. Note that since 1800, the United States has shifted from wood to coal to oil for its primary energy resource. A shift by 2100 to increased use of natural gas, hydrogen gas produced mostly by solar cells, and wind is one of many possible scenarios. (Data from U.S. Department of Energy)

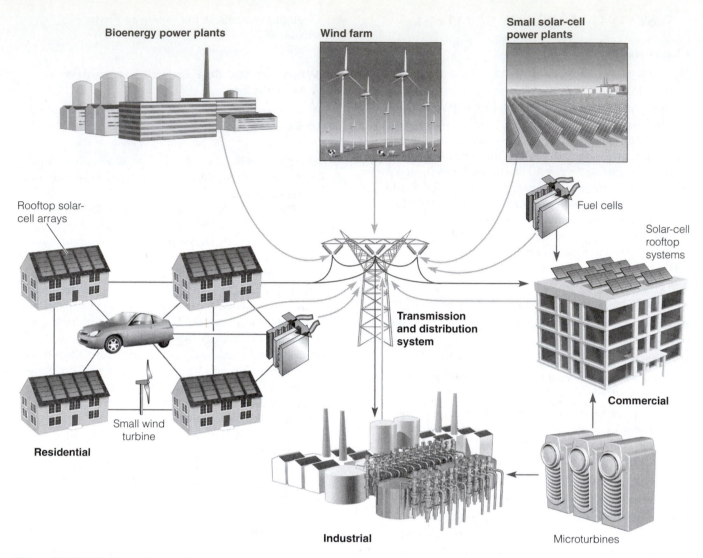

Figure 10-39 Solutions: *decentralized power system* in which electricity is produced by a large number of dispersed, small-scale *micropower systems.* Some would produce power on site; others would feed the power they produce into a conventional electrical distribution system. Over the next few decades, many energy and financial analysts expect a shift to this type of power system.

research and development subsidies and tax breaks and enact regulations that help stimulate the development and use of energy resources receiving such support. For decades, this approach has been used to help stimulate the development and use of fossil fuels and nuclear power in the United States and in most other developed countries. This approach has created an uneven economic playing field that encourages energy waste and rapid depletion of nonrenewable energy resources. It also discourages improvements in energy efficiency and the development of renewable energy.

Second, governments can *keep energy prices artificially high to discourage use of a resource.* They can raise the price of an energy resource by eliminating existing tax breaks and other subsidies, enacting restrictive regulations, or adding taxes on its use. This increases gov-

ernment revenues, encourages improvements in energy efficiency, reduces dependence on imported energy, and decreases use of an energy resource that has a limited future supply. To make this acceptable to the public, analysts call for the government to offset energy taxes by reducing income and payroll taxes and providing an energy safety net for the poor and lower middle class.

✗ HOW WOULD YOU VOTE? Should the government increase taxes on fossil fuels and offset this by reducing income and payroll taxes and providing an energy safety net for the poor and the lower middle class? Cast your vote online at www.thomsonedu.com/biology/miller.

Third, *emphasize consumer education.* Even if governments offer generous financial incentives for en-

Improve Energy Efficiency

Increase fuel-efficiency standards for vehicles, buildings, and appliances

Mandate government purchases of efficient vehicles and other devices

Provide large tax credits for buying efficient cars, houses, and appliances

Offer large tax credits for investments in energy efficiency

Reward utilities for reducing demand for electricity

Encourage independent power producers

Greatly increase energy efficiency research and development

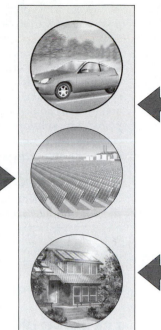

More Renewable Energy

Increase renewable energy to 20% by 2020 and 50% by 2050

Provide large subsidies and tax credits for renewable energy

Use full-cost accounting and life-cycle cost for comparing all energy alternatives

Encourage government purchase of renewable energy devices

Greatly increase renewable energy research and development

Reduce Pollution and Health Risk

Cut coal use 50% by 2020

Phase out coal subsidies

Levy taxes on coal and oil use

Phase out nuclear power or put it on hold until 2020

Phase out nuclear power subsidies

Figure 10-40 Solutions: suggestions of various energy analysts to help make the transition to a more sustainable energy future.

ergy efficiency and renewable energy, people will not make such investments if they are uninformed—or misinformed—about the availability and advantages and disadvantages of such options. For example, there are more solar water heaters in cloudy Germany than in sunny France and Spain mostly because the German government has made the public aware of the benefits of this technology.

We have the technology, creativity, and wealth to make the transition to a more sustainable energy future. Making this transition depends primarily on *politics,* which depends largely on the pressure individuals put on elected officials using their ballots. People can also vote with their wallets by refusing to buy inefficient and environmentally harmful products and by letting company executives know about it. Figure 10-41 lists some ways you can contribute to making this transition by reducing the amount of energy you use and waste.

We can make the transition to a more sustainable energy future by applying the four principles of sustainability. This means

■ Relying much more on direct and indirect forms of solar energy

■ Recycling and reusing materials and reducing wasteful and excessive consumption of energy and matter

■ Mimicking nature's reliance on biodiversity by relying mostly on a diverse mix of locally and regionally available renewable energy resources instead of on nonrenewable oil and coal

What Can You Do?

Energy Use and Waste

- Drive a car that gets at least 15 kilometers per liter (35 miles per gallon) and join a carpool.

- Use mass transit, walking, and bicycling.

- Superinsulate your house and plug all air leaks.

- Turn off lights, TV sets, computers, and other electronic equipment when they are not in use.

- Wash laundry in warm or cold water.

- Use passive solar heating.

- For cooling, open windows and use ceiling fans or whole-house attic or window fans.

- Turn thermostats down in winter and up in summer.

- Buy the most energy-efficient homes, lights, cars, and appliances available.

- Turn down the thermostat on water heaters to 43–49°C (110–120°F) and insulate hot water heaters and pipes.

Figure 10-41 Individuals matter: ways to reduce your use and waste of energy.

■ Reducing the number of people using and wasting energy and other resources by controlling population growth

A transition to renewable energy is inevitable, not because fossil fuel supplies will run out—large reserves of oil, coal, and gas remain in the world—but because the costs and risks of using these supplies will continue to increase relative to renewable energy.

Mohamed El-Ashry

CRITICAL THINKING

1. Just to continue using oil at the current rate (not the projected higher exponential increase in its annual use), we must discover and add to global oil reserves the equivalent of a new Saudi Arabian supply (the world's largest) *every 10 years.* Do you believe this is possible? If not, what effects might this have on your life and on the life of a child or grandchild you might have?

2. List three actions you can take to reduce your dependence on oil and the gasoline derived from it. Which of these actions do you actually plan to take?

3. Explain why you are for or against continuing to increase oil imports in the United States or in the country where you live. If you favor reducing dependence on oil imports, list the three best ways to do this.

4. Explain why you agree or disagree with the following proposals by various energy analysts to solve U.S. energy problems: **(a)** find and develop more domestic supplies of oil, **(b)** place a heavy federal tax on gasoline and imported oil to help reduce the waste of oil resources, **(c)** increase dependence on nuclear power, and **(d)** phase out all nuclear power plants by 2025.

5. Would you favor having high-level nuclear waste transported by truck or train through the area where you live to a centralized underground storage site? Explain. What are the options?

6. Someone tells you that we can save energy by recycling it. How would you respond?

7. Explain why you agree or disagree with the following proposals by various energy analysts: **(a)** Government subsidies for all energy alternatives should be eliminated so all energy choices can compete in a true free-market system. **(b)** All government tax breaks and other subsidies for conventional fuels (oil, natural gas, and coal), synthetic natural gas and oil, and nuclear power (fission and fusion) should be phased out. They should be replaced with subsidies and tax breaks for improving energy efficiency and developing solar, wind, geothermal, hydrogen, and biomass energy alternatives. **(c)** Development of solar, wind, and hydrogen energy should be left to private enterprise and receive little or no help from the federal government, but nuclear energy and fossil fuels should continue to receive large federal subsidies.

8. Congratulations! You are in charge of the world. List the five most important features of your energy policy.

9. List two questions that you would like to have answered as a result of reading this chapter.

LEARNING ONLINE

The website for this book contains helpful study aids and many ideas for further reading and research. They include a chapter summary, review questions for the entire chapter, flash cards for key terms and concepts, a multiple-choice practice quiz, interesting Internet sites, references, information about green careers, and a guide for accessing thousands of InfoTrac® College Edition articles. Log into

www.thomsonedu.com/biology/miller

Then choose Chapter 10, and select a learning resource. For access to animations, additional quizzes, chapter outlines and summaries, register and log into

at **www.thomsonedu.com** using the access code card in the front of your book.

11 ENVIRONMENTAL HAZARDS AND HUMAN HEALTH

The dose makes the poison.

PARACELSUS, 1540

RISKS AND HAZARDS

Risk and Risk Assessment (Science)

Risk is a measure of the likelihood that you will suffer harm from a hazard.

Risk is the *possibility* of suffering harm from a hazard that can cause injury, disease, death, economic loss, or environmental damage. It is usually expressed in terms of *probability*—a mathematical statement about how likely it is that harm will be suffered from a hazard. Scientists often state probability in terms such as "The lifetime probability of developing lung cancer from smoking a pack of cigarettes a day is 1 in 250." This means that 1 of every 250 people who smoke a pack of cigarettes every day will develop lung cancer over a typical lifetime (usually considered 70 years).

It is important to distinguish between *possibility* and *probability*. When we say that it is *possible* that a smoker can get lung cancer we are saying that this event could happen. *Probability* gives us an estimate of the likelihood of such an event.

Risk assessment is the scientific process of estimating how much harm a particular hazard can cause to human health or the environment. **Risk management** involves deciding whether or how to reduce a particular risk to a certain level and at what cost. Figure 11-1 summarizes how risks are assessed and managed.

Types of Hazards (Science)

We can suffer harm from cultural hazards, biological hazards, chemical hazards, and physical hazards, but determining the risks involved is difficult.

We can suffer harm from four major types of hazards:

- *Cultural hazards* such as smoking (Case Study, at right)

- *Biological hazards* from pathogens (bacteria, viruses, and parasites) that cause infectious disease

- *Chemical hazards* from harmful chemicals in air, water, soil, and food

- *Physical hazards* such as a fire, earthquake, volcanic eruption, flood, tornado, and hurricane

Figure 11-1 Science: *risk assessment* and *risk management.*

Case Study: Death from Smoking (Science)

Smoking is the number one preventable cause of premature death among adults, yet tobacco is subsidized by the U.S. government, which does not regulate nicotine as the highly addictive drug that it is.

What is roughly the diameter of a 30-caliber bullet, can be bought almost anywhere, is highly addictive, and kills about 13,700 people every day, or one every 6 seconds? It is a cigarette. *Cigarette smoking is the world's most preventable major cause of suffering and premature death among adults.*

According to the World Health Organization (WHO), tobacco helped kill about 85 million people between 1950 and 2005—almost three times the 30 million people killed in battle in all wars since 1900! The WHO estimates that each year tobacco contributes to the premature deaths of at least 5 million people (about half from developed countries and half from developing countries) from 34 illnesses including *heart disease, lung cancer, other cancers, bronchitis, emphysema,* and *stroke.* By 2030, the annual death toll from smoking-related diseases is projected to reach 10 million—an average of about 27,400 preventable deaths per day or 1 death every 3 seconds. About 70% of these deaths are expected to occur in developing countries.

According to a 2002 study by the Centers for Disease Control and Prevention, smoking kills about

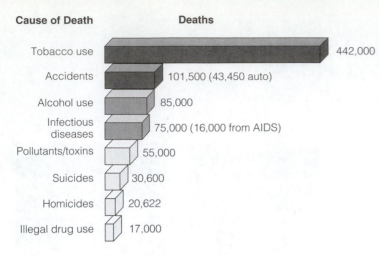

Cause of Death — **Deaths**

- Tobacco use — 442,000
- Accidents — 101,500 (43,450 auto)
- Alcohol use — 85,000
- Infectious diseases — 75,000 (16,000 from AIDS)
- Pollutants/toxins — 55,000
- Suicides — 30,600
- Homicides — 20,622
- Illegal drug use — 17,000

Figure 11-2 Annual deaths in the United States from tobacco use and other causes in 2003. Smoking is by far the nation's leading cause of preventable death, causing more premature deaths each year than all the other categories in this figure combined. (Data from U.S. National Center for Health Statistics and Centers for Disease Control and Prevention and U.S. Surgeon General)

442,000 Americans per year prematurely, an average of 1,210 deaths per day (Figure 11-2). This death toll is roughly equivalent to three fully loaded 400-passenger jumbo jets crashing *every day* with no survivors! Yet this ongoing major human tragedy rarely makes the news.

The overwhelming consensus in the scientific community is that the nicotine inhaled in tobacco smoke is highly addictive. Only 1 in 10 people who try to quit smoking succeed, about the same relapse rate as for recovering alcoholics and those addicted to heroin or crack cocaine. A British government study showed that adolescents who smoke more than one cigarette have an 85% chance of becoming smokers.

Passive smoking, or breathing secondhand smoke, also poses health hazards for children and adults. The EPA estimates that each year secondhand smoke causes an estimated 3,000 lung-cancer deaths and 37,000 deaths from heart disease in the United States.

A 50-year study published in 2004 found that cigarette smokers die on average 10 years earlier than non-smokers, but that kicking the habit—even at 50 years old—can cut a person's risk in half. If people quit smoking by the age of 30, they can avoid nearly all the risk of dying prematurely.

Many health experts urge that a $3–5 federal tax be added to the price of a pack of cigarettes in the United States. Then users of tobacco products—not the rest of society—would pay a much greater share of the $158 billion per year in health, economic, and social costs associated with smoking.

Other suggestions for reducing the death toll and the health effects of smoking in the United States (and in other countries) include banning all cigarette advertis-ing, prohibiting the sale of cigarettes and other tobacco products to anyone under 21 (with strict penalties for violators), and banning cigarette vending machines and candy-flavored cigarettes.

Analysts also call for classifying and regulating the use of nicotine as an addictive and dangerous drug, eliminating all federal subsidies and tax breaks to tobacco farmers and tobacco companies, and using cigarette tax income to finance an aggressive anti-tobacco advertising and education program. So far, the U.S. Congress has not enacted such reforms.

X *How Would You Vote?* Do you favor classifying and regulating nicotine as an addictive and dangerous drug? Cast your vote online at www.thomsonedu.com/biology/miller.

BIOLOGICAL HAZARDS: DISEASE IN DEVELOPED AND DEVELOPING COUNTRIES

Nontransmissible and Transmissible Diseases (Science)

Diseases not caused by living organisms do not spread from one person to another, while those caused by living organisms such as bacteria and viruses can spread from person to person.

A **nontransmissible disease** is not caused by living organisms and does not spread from one person to another. Such diseases tend to develop slowly and have multiple causes. Examples are cardiovascular (heart and blood vessel) disorders, asthma, emphysema, and malnutrition.

In an infection, a pathogen in the form of a bacterium, virus, protozoa, or parasite invades the body and multiplies in its cells and tissues. This can lead to an **infectious** or **transmissible disease** if the body cannot mobilize its defenses fast enough to keep the pathogen from interfering with bodily functions.

Figure 11-3 shows major pathways to infectious diseases in humans. Such diseases can then be spread from one person to another by air, water, food, and body fluids such as pathogen-loaded droplets present in the sneezes, coughs, feces, urine, or blood of infected people. There is also concern about *bioterrorism,* which involves the deliberate release of disease-causing bacteria or viruses into the air, water supply, or food supply of concentrated urban populations.

A large-scale outbreak of an infectious disease in an area or country is called an *epidemic* and a global epidemic is called a *pandemic.* Figure 11-4 shows the annual death toll from the world's seven deadliest infectious diseases. The deaths *each year* from these diseases are 58 times the 221,000 people killed by the December 2004 tsunamis.

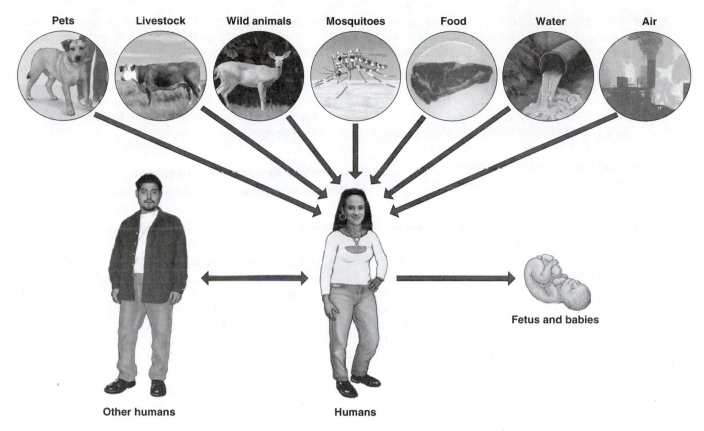

Figure 11-3 Science: pathways for infectious disease in humans.

Labels in figure: Pets, Livestock, Wild animals, Mosquitoes, Food, Water, Air, Other humans, Humans, Fetus and babies

Great news: Since 1900, and especially since 1950, the incidence of infectious diseases and the death rates from such diseases have been greatly reduced. This has been achieved mostly by a combination of better health care, using antibiotics to treat infectious disease caused by bacteria, and developing vaccines to prevent the spread of some infectious viral diseases.

Bad news: Many disease-carrying bacteria have developed genetic immunity to widely used antibiotics. Also, many disease-transmitting species of insects have become immune to widely used pesticides that once helped control their populations.

Growing Germ Resistance to Antibiotics (Science)

Rapidly producing infectious bacteria become genetically resistant to widely used antibiotics.

We may be falling behind in our efforts to prevent infectious bacterial diseases because of the astounding reproductive rate of bacteria, some of which can produce well over 16 million offspring in 24 hours. Their high reproductive rate allows them to become genetically resistant to an increasing number of antibiotics through natural selection. They can also transfer such resistance to nonresistant bacteria even more quickly by exchanging genetic material.

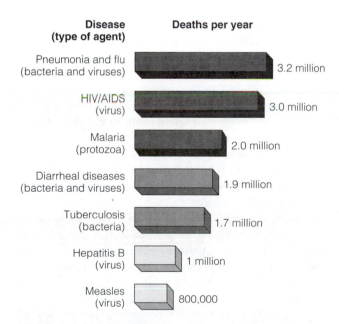

Disease (type of agent)	Deaths per year
Pneumonia and flu (bacteria and viruses)	3.2 million
HIV/AIDS (virus)	3.0 million
Malaria (protozoa)	2.0 million
Diarrheal diseases (bacteria and viruses)	1.9 million
Tuberculosis (bacteria)	1.7 million
Hepatitis B (virus)	1 million
Measles (virus)	800,000

Figure 11-4 Global outlook: the World Health Organization estimates that each year the world's seven deadliest infectious diseases kill 12.9 million people—most of them poor people in developing countries. This amounts to about 35,000 mostly preventable deaths every day. QUESTION: *What are three things that you would do to reduce this death toll?* (Data from the World Health Organization)

Other factors play a key role in fostering resistance. One is the spread of bacteria (some beneficial and some harmful) around the globe by human travel and international trade. Another is overuse of pesticides, which increase populations of pesticide-resistant insects and other carriers of bacterial diseases.

Yet another factor is overuse of antibiotics by doctors. According to a 2000 study by Richard Wenzel and Michael Edward, at least half of all antibiotics used to treat humans are prescribed unnecessarily. In many countries antibiotics are available without a prescription, which also promotes unnecessary use. Resistance to some antibiotics has also increased because of their widespread use in livestock and dairy animals to control disease and to promote growth.

The result of these factors acting together is that every major disease-causing bacterium now has strains that resist at least one of the roughly 160 antibiotics we use to treat bacterial infections, such as those from tuberculosis (Case Study, below). Each year genetic resistance to antibiotics prematurely kills at least 90,000 Americans who picked up mostly preventable infections while they were in the hospital.

Case Study: The Growing Global Threat from Tuberculosis (Science)

Tuberculosis (TB) kills about 1.7 million people a year and could kill 25 million people between 2006 and 2020.

Since 1990, one of the world's most underreported stories has been the rapid spread of *tuberculosis* (TB). According to the WHO, this highly infectious bacterial disease strikes 9 million people per year and kills 1.7 million people—about 84% of them in developing countries. The WHO projects that between 2006 and 2020, about 25 million people will die of the disease unless current efforts and funding to control TB are greatly strengthened and expanded.

Many infected people do not appear to be sick, and about half of them do not know they are infected. Left untreated, each person with active TB typically infects 10–15 other people.

Several factors account for the recent increase in TB. One is the lack of TB screening and control programs, especially in developing countries, where about 95% of the new cases occur. A second problem is that most strains of the TB bacterium have developed genetic resistance to most of the effective antibiotics.

Population growth, urbanization, and air travel have increased contacts between people and spread TB, especially in areas where large numbers of the poor are crowded together. In addition, AIDS greatly weakens the immune system and allows TB bacteria to multiply in AIDS victims.

Slowing the spread of the disease involves early identification and treatment of people with active TB,

especially those with a chronic cough. Treatment with a combination of four inexpensive drugs can cure 90% of those with active TB. To be effective, the drugs must be taken every day for 6–8 months. Because the symptoms disappear after a few weeks, many patients think they are cured and stop taking the drugs, allowing the disease to recur in a drug-resistant form and spread to other people.

Viral Diseases (Science)

Flu, HIV, and hepatitis B viruses infect and kill many more people each year than the highly publicized West Nile and SARS viruses.

What are the world's three most widespread and dangerous viruses? The biggest killer is the *influenza* or *flu* virus, which is transmitted by the body fluids or airborne emissions of an infected person. Easily transmitted and especially potent flu viruses could spread around the world in a pandemic that could kill tens to hundreds of millions of people in only a few months.

The second biggest killer is the *human immunodeficiency virus* (HIV), which is transmitted by unsafe sex, sharing of needles by drug users, infected mothers who pass the virus to offspring before or during birth, and exposure to infected blood. On a global scale, HIV infects at least 5 million new people each year, and the resulting complications from AIDS kill about 3 million people annually.

The third largest killer is the *hepatitis B virus* (HBV), which damages the liver and kills about 1 million people a year. Like HIV, it is transmitted by unsafe sex, sharing of needles by drug users, infected mothers who pass the virus to offspring before or during birth, and exposure to infected blood.

 Examine the HIV virus and how it replicates by using a host cell at ThomsonNOW.

In recent years, several other viruses have received widespread coverage in the media. One is the *West Nile virus,* which is transmitted to humans by the bite of a common mosquito that becomes infected by feeding on birds that carry the virus. Fortunately, the chance of being infected and killed by this disease is low (about 1 in 2,500).

A second is the *severe acute respiratory syndrome (SARS) virus,* which first appeared in humans in China in 2002. With flu-like symptoms, SARS can quickly turn into life-threatening pneumonia. Like flu, it can easily spread from person to person. During six months in 2003, the disease began spreading beyond China, infecting at least 8,500 people and causing 812 deaths. Swift local action by the WHO and other health agencies helped contain the spread of this disease by July 2003. But without careful vigilance, it might break out again.

Health officials are concerned about the emergence and spread of West Nile virus, SARS, and other *emerging viral diseases*, and they are working hard to control their spread. But in terms of annual infection rates and deaths, the three most dangerous viruses by far are flu, HIV, and HBV. In 2004, for example, flu killed about 36,000 Americans and West Nile virus killed 100.

You can greatly reduce your chances of getting infectious diseases by practicing good old-fashioned hygiene. Wash your hands frequently and thoroughly, avoid touching your face, and stay away from people who have flu or other viral diseases.

Case Study: The Global HIV/AIDS Epidemic

The spread of acquired immune deficiency syndrome (AIDS), caused by HIV, is one of the world's most serious and rapidly growing health threats.

The global spread of *acquired immune deficiency syndrome* (AIDS), caused by the human immunodeficiency virus (HIV), is a serious and rapidly growing health threat. The virus itself is not deadly, but it kills immune cells and leaves the body defenseless against infectious bacteria and other viruses.

According to the WHO, in 2005, about 40 million people worldwide were infected with HIV. About 96% of them lived in developing countries, especially African countries south of the Sahara Desert. Every day 12,000 to 16,000 more people—most of them between the ages of 15 and 24—become infected with HIV.

Within 7–10 years, at least half of all HIV-infected people develop AIDS. This long incubation period means that infected people often spread the virus for several years without knowing they are infected. *There is no vaccine to prevent HIV and no cure for AIDS. If you get AIDS, you will eventually die from it.* Drugs help some infected people live longer, but only a tiny fraction of those suffering from AIDS can afford to use these costly drugs.

Between 1980 and 2005, about 34 million people (506,000 in the United States) died of AIDS-related diseases. Each year AIDS claims about 3 million more lives.

AIDS has reduced the life expectancy of 700 million people living in sub-Saharan Africa from 62 to 47 years—40 years in some countries. The premature deaths of teachers, health care workers, and other young productive adults in such countries leads to diminished education and health care, decreased food production and economic development, and disintegrating families. Such deaths drastically alter a country's age structure diagram (Figure 11-5). AIDS has also left 15 million children orphaned—roughly equal to the number of children under age 5 in the United States.

Between 2005 and 2020, the WHO estimates 57 million more deaths from AIDS and a death toll reaching as high as 5 million a year.

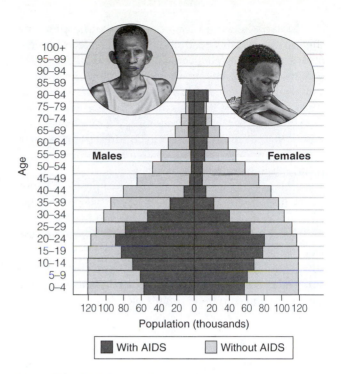

Figure 11-5 **Global outlook:** worldwide, AIDS is the leading cause of death for people of ages 15–49. This loss of productive working adults can affect the age structure of a population. In Botswana, more than 39% of this age group is infected with HIV. This figure shows the projected age structure of Botswana's population in 2020 with and without AIDS. (Data from the U.S. Census Bureau)

Case Study: Malaria (Science)

Malaria kills about 2 million people a year and has probably killed more people than all of the wars ever fought.

About one of every five people in the world—most of them living in poor African countries—is at risk from malaria (Figure 11-6). Malaria should also concern anyone traveling to malaria-prone areas because there is no vaccine for this disease.

Figure 11-6 **Global outlook:** distribution of malaria. About 40% of the world's population live in areas in which malaria is prevalent. Malaria kills about 2 million people a year. (Data from the World Health Organization and U.S. Centers for Disease Control and Prevention).

Malaria is caused by a parasite that is spread by the bites of certain mosquito species. It infects and destroys red blood cells, causing fever, chills, drenching sweats, anemia, severe abdominal pain, headaches, vomiting, extreme weakness, and greater susceptibility to other diseases. It kills about 2 million to 3 million people each year—an average of 5,500 deaths per day. About 90% of these victims are children younger than age 5. Many of those who survive have brain damage or impaired learning ability.

Malaria is caused by four species of protozoan parasites in the genus *Plasmodium*. Most cases of the disease are transmitted when an uninfected female of any of about 60 *Anopheles* mosquito species bites a person infected with *Plasmodium,* ingests blood that contains the parasite, and later bites an uninfected person (Figure 11-7). *Plasmodium* parasites then move out of the mosquito and into the human's bloodstream, multiply in the liver, and enter blood cells to continue multiplying. Malaria can also be transmitted by blood transfusions or by sharing needles.

The malaria cycle repeats itself until immunity develops, treatment is given, or the victim dies. *Over the course of human history, malarial protozoa probably have killed more people than all the wars ever fought.*

During the 1950s and 1960s, the spread of malaria was sharply curtailed by draining swamplands and marshes, spraying breeding areas with insecticides, and using drugs to kill the parasites in the bloodstream. Since 1970, malaria has come roaring back. Most species of the malaria-carrying *Anopheles* mosquito have become genetically resistant to most insecticides. Worse, the *Plasmodium* parasites have become genetically resistant to common antimalarial drugs.

Researchers are working to develop new antimalarial drugs, (such as *artemisinins* derived from the Chinese herbal remedy qinghaosu), vaccines, and biological controls for *Anopheles* mosquitoes. But such approaches receive too little funding and have proved more difficult than originally thought.

Other approaches include providing poor people in malarial regions with window screens for their dwellings and insecticide-treated bed nets, cultivating fish that feed on mosquito larvae (biological control), clearing vegetation around houses, planting trees that soak up water in low-lying marsh areas where mosquitoes thrive (a method that can degrade or destroy ecologically important wetlands), and using zinc and vitamin A supplements to boost resistance to malaria in children.

Spraying the inside of homes with low concentrations of DDT twice a year greatly reduces the number of malaria cases. But under an international treaty enacted in 2002, DDT and five similar pesticides are being phased out in developing countries. However, the treaty allows 25 countries to continue using DDT for malaria control until other alternatives are available.

The cost of lifesaving malaria treatment for one person is 25 cents to $2.40. Columbia University economist Jeffrey Sachs estimates that spending $2–3 billion on malaria might save more than 1 million lives a year. To him, "This is probably the best bargain on the planet."

Thomson NOW! Watch through a microscope what happens when a mosquito infects a human with malaria at ThomsonNOW.

Solutions: Reducing the Incidence of Infectious Diseases

There are a number of ways to reduce the incidence of infectious diseases if the world is willing to provide the necessary funds and assistance.

Good news: According to the WHO, the global death rate from infectious diseases decreased by about two-thirds between 1970 and 2000 and is projected to continue dropping. Also, between 1971 and 2000, the percentage of children in developing countries immunized with vaccines to prevent tetanus, measles, diphtheria, typhoid fever, and polio increased from 10% to 84%—saving about 10 million lives a year.

Figure 11-8 lists measures health scientists and public health officials suggest to help prevent or reduce the incidence of infectious diseases—especially in developing countries. An important breakthrough has been the development of simple *oral rehydration therapy* to help prevent death from dehydration for victims of diarrheal diseases, which cause about one-fourth of all deaths of children under age 5. It involves

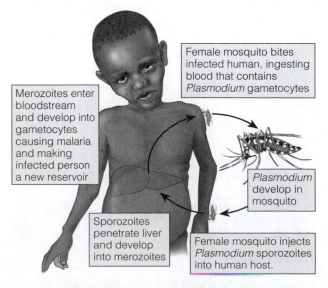

Figure 11-7 Science: the life cycle of malaria. *Plasmodium* parasites circulate from mosquito to human and back to mosquito.

Solutions

Infectious Diseases

- Increase research on tropical diseases and vaccines
- Reduce poverty
- Decrease malnutrition
- Improve drinking water quality
- Reduce unnecessary use of antibiotics
- Educate people to take all of an antibiotic prescription
- Reduce antibiotic use to promote livestock growth
- Careful hand washing by all medical personnel
- Immunize children against major viral diseases
- Oral rehydration for diarrhea victims
- Global campaign to reduce HIV/AIDS

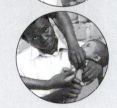

Figure 11-8 Solutions: ways to prevent or reduce the incidence of infectious diseases, especially in developing countries. QUESTION: *Which three of these approaches do you think are the most important?*

administering a simple solution of boiled water, salt, and sugar or rice, at a cost of only a few cents per person. It has been the major factor in reducing the annual number of deaths from diarrhea from 4.6 million in 1980 to 1.9 million in 2004. Few investments have saved so many lives at such a low cost.

Bad news: The WHO estimates that only about 10% of global medical research and development money is spent on preventing infectious diseases in developing countries, even though more people worldwide suffer and die from these diseases than from all other diseases combined.

CHEMICAL HAZARDS

Toxic and Hazardous Chemicals (Science)

Toxic chemicals can kill and hazardous chemicals can cause various types of harm.

A **toxic chemical** can cause temporary or permanent harm or death to humans or animals. A **hazardous chemical** can harm humans or other animals because it is flammable or explosive or because it can irritate or damage the skin or lungs, interfere with oxygen uptake, or induce allergic reactions.

There are three major types of potentially toxic agents. **Mutagens** are chemicals or forms of radiation that cause or increase the frequency of *mutations*, or changes in the DNA molecules found in cells. Most mutations cause no harm but some can lead to cancers and other disorders. For example, nitrous acid formed by the digestion of nitrite preservatives in foods can cause mutations linked to increases in stomach cancer in people who consume large amounts of processed foods and wine that contain nitrate preservatives.

Harmful mutations occurring in reproductive cells can be passed on to offspring and to future generations. There is no safe threshold for exposure to mutagens .

Teratogens are chemicals that cause harm or birth defects to a fetus or embryo. Ethyl alcohol is a teratogen. Drinking during pregnancy can lead to offspring with low birth weight and a number of physical, developmental, behavioral, and mental problems. Thalidomide is also a potent teratogen.

Carcinogens are chemicals or types of radiation that can cause or promote *cancer*— a disease in which malignant cells multiply uncontrollably and create tumors that can damage the body and often lead to death. An example is benzene, a widely used chemical solvent. Many cancerous tumors spread by **metastasis,** when malignant cells break off from tumors and travel in body fluids to other parts of the body. There they start new tumors, making treatment much more difficult. Typically, 10–40 years may elapse between the initial exposure to a carcinogen and the appearance of detectable symptoms. Partly because of this time lag, many healthy teenagers and young adults have trouble believing that their smoking, drinking, eating, and other lifestyle habits today could lead to some form of cancer before they reach age 50.

Effects of Chemicals on the Immune, Nervous, and Endocrine Systems (Science)

Long-term exposure to some chemicals at low doses may disrupt the body's immune, nervous, and endocrine systems.

Since the 1970s, a growing body of research on wildlife and laboratory animals, along with some studies of humans, suggest that long-term exposure to some chemicals in the environment can disrupt the body's immune, nervous, and endocrine systems.

The *immune system* consists of specialized cells and tissues that protect the body against disease and harmful substances by forming antibodies that make invading agents harmless. Some chemicals such as arsenic and dioxins can weaken the human immune system and can leave the body vulnerable to attacks by allergens, infectious bacteria, viruses, and protozoans.

Some natural and synthetic chemicals in the environment, called *neurotoxins,* can harm the human *nervous system* (brain, spinal cord, and peripheral nerves). They inhibit, damage, or destroy nerve cells (neurons) that transmit electrochemical messages throughout the body. Effects can include behavioral changes, learning and other disabilities, attention deficit disorder, paralysis, and death. Examples of neurotoxins are PCBs, methyl mercury, and certain pesticides.

The *endocrine system* is a complex network of glands that releases very small amounts of *hormones* into the bloodstreams of humans and other vertebrate animals. Low levels of these chemical messengers turn on and off bodily systems that control sexual reproduction, growth, development, learning ability, and behavior. Exposure to low levels of certain synthetic chemicals may disrupt the effects of natural hormones in animals, but more research is needed to verify the effects of these chemicals on humans.

TOXICOLOGY: ASSESSING CHEMICAL HAZARDS

What Determines Whether a Chemical Is Harmful? (Science)

The harm caused by exposure to a chemical depends on the amount of exposure (dose), frequency of exposure, who is exposed, how well the body's detoxification systems work, and one's genetic makeup.

Toxicology is the science that examines the effects of harmful chemicals on humans, wildlife, and ecosystems. **Toxicity** is a measure of how harmful a substance is in causing injury, illness, or death to a living organism. This depends on several factors. One is the **dose,** the amount of a substance a person has ingested, inhaled, or absorbed through the skin. Other factors are frequency of exposure, who is exposed (adult or child, for example), and how well the body's detoxification systems (such as the liver, lungs, and kidneys) work, and the genetic makeup of individuals.

The type and amount of health damage that result from exposure to a chemical or other agent are called the **response.** An *acute effect* is an immediate or rapid harmful reaction to an exposure—ranging from dizziness to death. A *chronic effect* is a permanent or long-lasting consequence (kidney or liver damage, for example) from exposure to a single dose or to repeated lower doses of a harmful substance.

A basic concept of toxicology is that *any synthetic or natural chemical can be harmful if ingested in a large enough quantity.* For example, drinking 100 cups of strong coffee one after another would expose most people to a lethal dosage of caffeine. Similarly, downing 100 tablets of aspirin or 1 liter (1.1 quarts) of pure alcohol (ethanol) would kill most people.

The critical question is *how much exposure to a particular toxic chemical causes a harmful response?* This is the meaning of the chapter-opening quote by the German scientist Paracelsus about the dose making the poison.

Effects of Trace Levels of Toxic Chemicals (Science)

Trace amounts of chemicals in the environment or your body may or may not be harmful.

Should we be concerned about trace amounts of various chemicals in air, water, food, and our bodies? The honest answer is that, in most cases, we do not know because of a lack of data and the difficulty of determining the effects of exposures to low levels of chemicals.

Some scientists view trace amounts of synthetic pesticides in our food as a minor risk compared to the risks from trace levels of harmful natural chemicals in what we eat. Others argue that naturally harmful chemicals in food are usually more readily broken down and excreted by the body while some harmful synthetic chemicals in food persist and accumulate in fatty tissues.

Chemists are able to detect increasingly small amounts of potentially toxic chemicals in air, water, and food. This is good news, but it can give the false impression that dangers from toxic chemicals are increasing. In reality, we may simply be uncovering levels of chemicals that have been around for a long time.

Some people also have the mistaken idea that natural chemicals are safe and synthetic chemicals are harmful. In fact, many synthetic chemicals are quite safe if used as intended, and many natural chemicals are deadly.

Estimating the Toxicity of a Chemical (Science)

Chemicals vary widely in their toxicity to humans and other animals.

A **poison** or **toxin** is a chemical that adversely affects the health of a human or animal by causing injury, illness, or death. One way to determine the relative toxicity of various chemicals is to measure their effects on test animals. For example, we can determine a chemical's *lethal dose* (LD)—that which can kill. A chemical's *median lethal dose* (LD50) is that dose that kills 50% of the animals (usually rats and mice) in a test population within an 18-day period.

Chemicals vary widely in their toxicity (Table 11-1). Some poisons can cause serious harm or death after a single acute exposure at very low dosages. Others cause

Table 11-1 Toxicity Ratings and Average Lethal Doses for Humans

Toxicity Rating	LD50 (milligrams per kilogram of body weight)*	Average Lethal Dose†	Examples
Supertoxic	Less than 0.01	Less than 1 drop	Nerve gases, botulism toxin, mushroom toxins, dioxin (TCDD)
Extremely toxic	Less than 5	Less than 7 drops	Potassium cyanide, heroin, atropine, parathion, nicotine
Very toxic	5–50	7 drops to 1 teaspoon	Mercury salts, morphine, codeine
Toxic	50–500	1 teaspoon to 1 ounce	Lead salts, DDT, sodium hydroxide, sodium fluoride, sulfuric acid, caffeine, carbon tetrachloride
Moderately toxic	500–5,000	1 ounce to 1 pint	Methyl (wood) alcohol, ether, phenobarbital, amphetamines (speed), kerosene, aspirin
Slightly toxic	5,000–15,000	1 pint to 1 quart	Ethyl alcohol, Lysol, soaps
Essentially nontoxic	15,000 or greater	More than 1 quart	Water, glycerin, table sugar

*Dosage that kills 50% of individuals exposed
†Amounts of substances in liquid form at room temperature that are lethal when given to a 70.4-kilogram (155-pound) human

such harm only at dosages so huge that it is nearly impossible to get enough into the body to cause injury or death. Most chemicals fall between these two extremes.

Using Case Reports, Wildlife Studies, and Epidemiological Studies to Estimate Toxicity (Science)

We can estimate toxicity by using case reports about the harmful effects of chemicals on human health, studying the effects of various chemicals on wild species, and comparing the health of a group of people exposed to a chemical with the health of a similar group not exposed to the chemical.

Scientists use various methods to get information about the harmful effects of chemicals on human health. For example, *case reports,* usually made by physicians, provide information about people suffering some adverse health effect or death after exposure to a chemical. Such information often involves poisonings, drug overdoses, homicides, or suicide attempts.

Most case reports are not reliable sources for estimating toxicity because the actual dosage and the exposed person's health status are often not known. But such reports can provide clues about environmental hazards and suggest the need for laboratory investigations.

Toxicological studies of the effects of various chemicals on wildlife can provide clues about possible harmful effects of such chemicals on humans. Still another source of information is *epidemiological studies,* which compare the health of people exposed to a particular chemical (the *experimental group*) with the health of a similar group of people not exposed to the agent (the *control group*). The goal is to determine whether the statistical association between exposure to a toxic chemical and a health problem is strong, moderate, weak, or undetectable.

Four factors can limit the usefulness of epidemiological studies. *First,* in many cases, too few people have been exposed to high enough levels of a toxic agent to detect statistically significant differences. *Second,* they usually take a long time. *Third,* conclusively linking an observed effect with exposure to a particular chemical is difficult because people are exposed to many different toxic agents throughout their lives and can vary in their sensitivity to such chemicals. *Fourth,* we cannot use epidemiological studies to evaluate hazards from new technologies or chemicals to which people have not yet been exposed.

Using Laboratory Experiments to Estimate Toxicity (Science, Ethics, and Economics)

Exposing a population of live laboratory animals (especially mice and rats) to known amounts of a chemical is the most widely used method for determining its toxicity.

The most widely used method for determining toxicity is to expose a population of live laboratory animals to measured doses of a specific substance under controlled conditions. Animal tests take 2–5 years, involve hundreds to thousands of test animals, and cost as much as $2,000,000 per substance tested. Such tests can be painful to the test animals and can kill or harm them. The goal is to develop data on the responses of the test animals to various doses of a chemical, but estimating the effects of low doses is difficult.

Animal welfare groups want to limit or ban the use of test animals or ensure that they are treated in the most humane manner possible. More humane methods for carrying out toxicity tests are available. They include computer simulations and using tissue cultures of cells and bacteria, chicken egg membranes, and measurements of changes in the electrical properties of individual animal cells.

These alternatives can greatly decrease the use of animals for testing toxicity. But scientists point out that some animal testing is needed because the alternative methods cannot adequately mimic the complex biochemical interactions of a live animal.

Experimenters run acute toxicity tests to develop a **dose-response curve,** which shows the responses of a group of test animals to various dosages of a toxic agent. In *controlled experiments,* the responses of a *test group* are compared with the responses of a *control group* of organisms not exposed to the chemical. Care is taken that organisms in each group are as identical as possible in age, health status, and genetic makeup, and that all are exposed to the same environmental conditions.

There are two general types of dose-response curves (Figure 11-9). With the *nonthreshold dose-response model* (Figure 11-9, left), any dosage of a toxic chemical causes harm that increases with the dosage. With the *threshold dose-response model* (Figure 11-9, right), a threshold dosage must be reached before any detectable harmful effects occur, presumably because the body can repair the damage caused by low dosages of some substances.

Establishing which of these models applies at low dosages is extremely difficult and controversial. To be on the safe side, the nonthreshold dose-response model often is assumed. Fairly high dosages are used to reduce the number of test animals needed, obtain results quickly, and lower costs. Otherwise, tests would have to be run on millions of laboratory animals for many years, and manufacturers could not afford to test most chemicals.

For the same reasons, scientists usually use mathematical models to extrapolate the results of high-dose exposures to low-dose exposures. Then they extrapolate the low-dose results from the test organisms to humans to estimate LD50 values for acute toxicity (Table 11-1).

Some scientists challenge the validity of extrapolating data from test animals to humans because human physiology and metabolism differ from those of the test animals. Other scientists say that such tests and models work fairly well (especially for revealing cancer risks) when the correct experimental animal is chosen or a when a chemical is toxic or harmful to several different test-animal species.

The problem of estimating toxicities gets worse. In real life, each of us is exposed to a variety of chemicals, some of which can interact in ways that decrease or enhance their individual effects over the short and long term. Toxicologists already have great difficulty estimating the toxicity of a single substance. But adding the problem of evaluating *mixtures of potentially toxic substances,* separating out which ones are the culprits, and determining how they can interact with one another is overwhelming from a scientific and economic standpoint. For example, just studying the interactions of three of the 500 most widely used industrial chemicals would take 20.7 million experiments—a physical and financial impossibility.

The effects of a particular chemical can also depend upon when in a subject's life exposure occurs. For example, toxic chemicals usually have a greater effect on children than on adults.

Why Do We Know So Little about the Harmful Effects of Chemicals? (Science, Politics, and Economics)

Under existing laws, most chemicals are considered innocent until proven guilty, and estimating their toxicity to establish guilt is difficult, uncertain, and expensive.

As we have seen, all methods for estimating toxicity levels and risks have serious limitations. But they are all we have. To take this uncertainty into account and minimize harm, scientists and regulators typically set allowed exposure levels to toxic substances and ionizing radiation at 1/100 or even 1/1,000 of the estimated harmful levels.

According to risk assessment expert Joseph V. Rodricks, "Toxicologists know a great deal about a few chemicals, a little about many, and next to nothing about most." The U.S. National Academy of Sciences estimates that only 10% of 85,000 registered synthetic chemicals in commercial use have been thoroughly screened for toxicity, and only 2% have been ade-

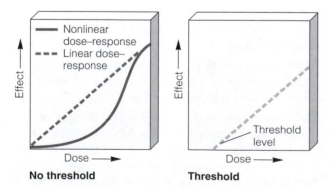

Figure 11-9 Science: two types of *dose-response curves.* The linear and nonlinear curves in the left graph apply if even the smallest dosage of a chemical has a harmful effect that increases with the dosage. The curve on the right applies if a harmful effect occurs only when the dosage exceeds a certain *threshold level.* Which model is better for a specific harmful agent is uncertain and controversial because of the difficulty in estimating the response to very low dosages.

quately tested to determine whether they are carcinogens, teratogens, or mutagens. Hardly any of the chemicals in commercial use have been screened for possible damage to the human nervous, endocrine, and immune systems. As a result, federal and state governments do not regulate about 99.5% of the commercially used chemicals in the United States.

Pollution Prevention and the Precautionary Principle (Science and Economics)

Some scientists and health officials say that preliminary but not conclusive evidence that a chemical causes significant harm should spur preventive action, but others disagree.

So where does this leave us? We do not know a lot about the potentially toxic chemicals around us and inside us, and estimating their effects is very difficult, time consuming, and expensive. Is there a way to deal with this problem?

Some scientists and health officials, especially those in European Union countries, are pushing for much greater emphasis on *pollution prevention.* They say we should not release into the environment chemicals that we know or suspect can cause significant harm. This means looking for harmless or less harmful substitutes for toxic and hazardous chemicals or recycling them within production processes so they do not reach the environment.

This prevention approach is based on the *precautionary principle.* Recall that this means when there is reasonable but incomplete scientific evidence (frontier science evidence) of significant harm to humans or the environment from a proposed or existing chemical or technology, we should take action to prevent or reduce the risk instead of waiting for more conclusive (sound or consensus science) evidence.

Under this approach, those proposing to introduce a new chemical or technology would bear the burden of establishing its safety. This means two major changes in the way we evaluate risks. *First,* new chemicals and technologies would be assumed harmful until scientific studies can show otherwise. *Second,* existing chemicals and technologies that appear to have a strong chance of causing significant harm would be removed from the market until their safety can be established.

Some movement is being made in this direction, especially in the European Union. In 2000, negotiators agreed to a global treaty that would ban or phase out use of 12 of the most notorious *persistent organic pollutants* (POPs), also called the *dirty dozen.* The list included DDT and eight other persistent pesticides, PCBs, and dioxins and furans. New chemicals would be added to the list when the harm they could cause is seen as outweighing their usefulness. This treaty went into effect in 2004.

Manufacturers and businesses contend that widespread application of the precautionary principle would make it too expensive and almost impossible to introduce any new chemical or technology. They argue that we can never have a risk-free society. Proponents of increased reliance on the precautionary principle agree, but argue we have an ethical responsibility to make greater use of the precautionary principle to reduce known or potentially serious risks. They also point out that pollution prevention and waste reduction has increased profits for some and promoted economic development because it stimulates the development and sale of less harmful and wasteful products and technologies in the global marketplace.

X *How Would You Vote?* Should we rely more on the precautionary principle as a way to reduce the risks from chemicals and various technologies? Cast your vote online at www.thomsonedu.com/biology/miller.

RISK ANALYSIS

Estimating Risks (Science, Economics, and Ethics)

Scientists have developed ways to evaluate and compare risks, decide how much risk is acceptable, and find affordable ways to reduce it.

Risk analysis involves identifying hazards and evaluating their associated risks (*risk assessment;* Figure 11-1, left), ranking risks (*comparative risk analysis*), determining options and making decisions about reducing or eliminating risks (*risk management;* Figure 11-1, right), and informing decision makers and the public about risks (*risk communication*).

Statistical probabilities based on past experience, animal testing, and other tests, and epidemiological studies are used to estimate risks from older technologies and chemicals. To evaluate new technologies and products, risk evaluators use more uncertain statistical probabilities, based on models rather than actual experience and testing.

Figure 11-10 (p. 248) lists the results of a *comparative risk analysis,* summarizing the greatest ecological and health risks identified by a panel of scientists acting as advisers to the EPA.

The greatest risks many people face today are rarely dramatic enough to make the daily news. In terms of the number of premature deaths per year (Figure 11-11, p. 248) and reduced life span (Figure 11-12, p. 249), *the greatest risk by far is poverty.* Its high death toll is a result of malnutrition, increased susceptibility to normally nonfatal infectious diseases, and often fatal infectious diseases from lack of access to a safe water supply.

Comparative Risk Analysis

Most Serious Ecological and Health Problems

High-Risk Health Problems
- Indoor air pollution
- Outdoor air pollution
- Worker exposure to industrial or farm chemicals
- Pollutants in drinking water
- Pesticide residues on food
- Toxic chemicals in consumer products

High-Risk Ecological Problems
- Global climate change
- Stratospheric ozone depletion
- Wildlife habitat alteration and destruction
- Species extinction and loss of biodiversity

Medium-Risk Ecological Problems
- Acid deposition
- Pesticides
- Airborne toxic chemicals
- Toxic chemicals, nutrients, and sediment in surface waters

Low-Risk Ecological Problems
- Oil spills
- Groundwater pollution
- Radioactive isotopes
- Acid runoff to surface waters
- Thermal pollution

Figure 11-10 Science: *comparative risk analysis* of the most serious ecological and health problems according to scientists acting as advisers to the EPA. Risks under each category are not listed in rank order. QUESTION: *Which two risks in each of the high-risk problem areas do you think are the most serious?* (Data from Science Advisory Board, *Reducing Risks*, Washington, D.C.: Environmental Protection Agency, 1990)

A sharp reduction in or elimination of poverty would do far more to improve longevity and human health than any other measure. It would also greatly improve human rights, provide more people with income to stimulate economic development, and reduce environmental degradation and the threat of terrorism.

After the health risks associated with poverty and gender, the greatest risks of premature death mostly result from voluntary choices people make about their lifestyles (Figure 11-12). By far the best ways to reduce one's risk of premature death and serious health risks are to avoid smoking and exposure to smoke, lose excess weight, reduce consumption of foods containing cholesterol and saturated fats, eat a variety of fruits and vegetables, exercise regularly, drink little or no alcohol (no more than two drinks in a single day), avoid excess sunlight (which ages skin and can cause skin cancer), and practice safe sex. A 2005 study by Majjid Ezzati with participation by 100 scientists around the world estimated that one-third of the 7 million annual deaths from cancer could be prevented if people were to follow these guidelines.

Cause of death   **Annual deaths**

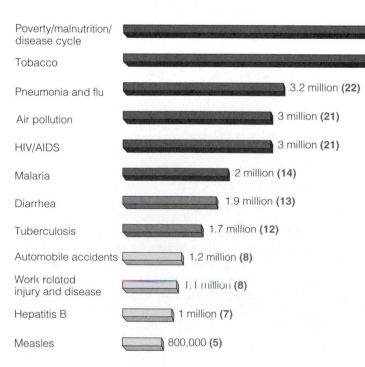

Poverty/malnutrition/disease cycle — 11 million **(75)**

Tobacco — 5 million **(34)**

Pneumonia and flu — 3.2 million **(22)**

Air pollution — 3 million **(21)**

HIV/AIDS — 3 million **(21)**

Malaria — 2 million **(14)**

Diarrhea — 1.9 million **(13)**

Tuberculosis — 1.7 million **(12)**

Automobile accidents — 1.2 million **(8)**

Work related injury and disease — 1.1 million **(8)**

Hepatitis B — 1 million **(7)**

Measles — 800,000 **(5)**

Figure 11-11 Global outlook: number of deaths per year in the world from various causes. Numbers in parentheses give these deaths in terms of the number of fully loaded 400-passenger jumbo jets crashing *every day of the year* with no survivors. Because of sensational media coverage, most people have a distorted view of the largest annual causes of death. QUESTION: *Which three of these items are most likely to shorten your life span?* (Data from World Health Organization)

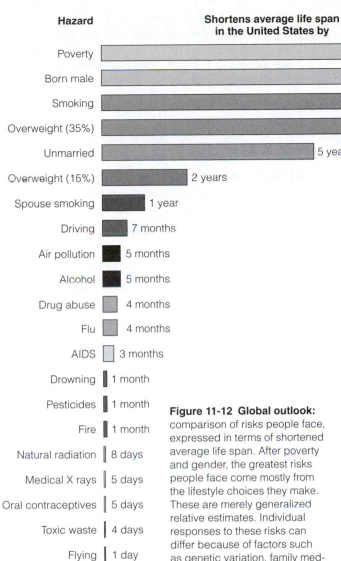

Hazard	Shortens average life span in the United States by
Poverty	7–10 years
Born male	7.5 years
Smoking	6–10 years
Overweight (35%)	6 years
Unmarried	5 years
Overweight (15%)	2 years
Spouse smoking	1 year
Driving	7 months
Air pollution	5 months
Alcohol	5 months
Drug abuse	4 months
Flu	4 months
AIDS	3 months
Drowning	1 month
Pesticides	1 month
Fire	1 month
Natural radiation	8 days
Medical X rays	5 days
Oral contraceptives	5 days
Toxic waste	4 days
Flying	1 day
Hurricanes, tornadoes	1 day
Living lifetime near nuclear plant	10 hours

Figure 11-12 Global outlook: comparison of risks people face, expressed in terms of shortened average life span. After poverty and gender, the greatest risks people face come mostly from the lifestyle choices they make. These are merely generalized relative estimates. Individual responses to these risks can differ because of factors such as genetic variation, family medical history, emotional makeup, stress, and social ties and support. (Data from Bernard L. Cohen)

Estimating Risks from Technologies (Science)

Estimating risks of using certain technologies is difficult because of unpredictable human behavior, chance, and sabotage.

The more complex a technological system and the more people needed to design and run it, the more difficult it is to estimate the risks. The overall *reliability* or the probability (expressed as a percentage) that a person or device will complete a task without failing is the product of two factors:

$$\text{System reliability (\%)} = \frac{\text{Technological}}{\text{reliability}} \times \frac{\text{Human}}{\text{reliability}}$$

With careful design, quality control, maintenance, and monitoring, a highly complex system such as a nu-

clear power plant or space shuttle can achieve a high degree of technological reliability. But human reliability usually is much lower than technological reliability and almost impossible to predict: To err is human.

Suppose the technological reliability of a nuclear power plant is 95% (0.95) and human reliability is 75% (0.75). Then the overall system reliability is 71% (0.95 × 0.75 × 100 = 71%). Even if we could make the technology 100% reliable (1.0), the overall system reliability would still be only 75% (1.0 × 0.75 × 100 = 75%). The crucial dependence of even the most carefully designed systems on unpredictable human reliability helps explain allegedly "almost impossible" tragedies such as the Chernobyl nuclear power plant accident and the *Challenger* and *Columbia* space shuttle accidents.

One way to make a system more foolproof or failsafe is to move more of the potentially fallible elements from the human side to the technological side. However, chance events such as a lightning bolt can knock out an automatic control system, and no machine or computer program can completely replace human judgment. Also, the parts in any automated control system are manufactured, assembled, tested, certified, and maintained by fallible human beings. In addition, computer software programs used to monitor and control complex systems can be flawed because of human error or can be deliberately modified by computer viruses to malfunction.

Perceiving Risks

Most individuals are poor at evaluating the relative risks they face, mostly because of misleading information, denial, and irrational fears.

Most of us are not good at assessing the relative risks from the hazards that surround us. Also, many people deny or shrug off the high-risk chances of death (or injury) from voluntary activities they enjoy, such as *motorcycling* (1 death in 50 participants), *smoking* (1 in 250 by age 70 for a pack-a-day smoker), *hang gliding* (1 in 1,250), and *driving* (1 in 3,300 without a seatbelt and 1 in 6,070 with a seatbelt). Indeed, the most dangerous thing most people in many countries do each day is drive or ride in a car.

Yet some of these same people may be terrified about the possibility of being killed by a *gun* (1 in 28,000 in the United States), *flu* (1 in 130,000), *nuclear power plant accident* (1 in 200,000), *West Nile virus* (1 in 1 million), *lightning* (1 in 3 million), *commercial airplane crash* (1 in 9 million), *snakebite* (1 in 36 million), or *shark attack* (1 in 281 million).

Four factors can cause people to see a technology or a product as being riskier than experts judge it to be. First is the *degree of control* we have. Most of us have a greater fear of things over which we have no personal control. For example, some individuals feel safer driving their own car for long distances through bad traffic than traveling the same distance on a plane. But look at the math. The risk of dying in a car accident while using your seatbelt is 1 in 6,070 whereas the risk of dying in a commercial airliner crash is 1 in 9 million.

Second is the *fear of the unknown*. Most people have greater fear of a new, unknown product or technology than they do of an older, more familiar one. Examples include a greater fear of genetically modified food than of food produced by traditional plant-breeding techniques, and a greater fear of nuclear power plants than of more familiar coal-fired power plants.

Third is *whether or not we voluntarily take the risk.* For example, we might perceive that the risk from driving, which is largely *voluntary*, is less than that from a nuclear power plant, which is mostly imposed on us whether we like it or not.

Fourth is *whether a risk is catastrophic,* not chronic. We usually have a much greater fear of a well-publicized death toll from a single catastrophic accident rather than the same or an even larger death toll spread out over a longer time. Examples include a severe nuclear power plant accident, an industrial explosion, or an accidental plane crash, as opposed to coal-burning power plants, automobiles, or smoking.

There is also concern over the *unfair distribution of risks* from the use of a technology or chemical. Citizens are outraged when government officials decide to put a hazardous waste landfill or incinerator in or near their neighborhood. Even when the decision is based on careful risk analysis, it is usually seen as politics, not science. Residents will not be satisfied by estimates that the lifetime risks of cancer death from the facility are not greater than, say, 1 in 100,000. Instead, they point out that living near the facility means that they will have a much higher risk of dying from cancer than would people living farther away.

Becoming Better at Risk Analysis

To become better at risk analysis, evaluate the bad news covered in the media, compare risks, concentrate on reducing risks over which you have some control, and help to reduce the five most serious global risks.

You can do four things to become better at estimating risks. *First,* recognize that everything is risky. The question is, how risky. *Second,* while carefully evaluating news reports, recognize that the media often give an exaggerated view of risks to capture our interest and thus sell newspapers or gain TV viewers.

Third, compare risks. Do you risk getting cancer by eating a charcoal-broiled steak once or twice a week? Yes, because in theory anything can harm you. The question is whether this danger is great enough for you to worry about. In evaluating a risk the question is not, "Is it safe?" but rather, *"How risky is it compared to other risks?"*

Fourth, concentrate on the most serious risks to your life and health that you have some control over and stop worrying about smaller risks and those over which you have no control. When you worry about something, the most important question to ask is, "Do I have any control over this?"

You have control over major ways to reduce your risks from heart attack, stroke, and many forms of cancer. You decide whether you smoke, what you eat, and how much alcohol you drink—the three biggest factors within your control. Other factors under your control are whether or not you practice safe sex, how much exercise you get, how safely you drive, and how often you expose yourself to the ultraviolet rays from the sun or in tanning booths. Concentrate on evaluating these important choices, and you will have a much greater chance of living a longer, healthier, happier, and less fearful life.

We can use the four principles of sustainability to help us reduce these major risks to human health. This involves shifting from polluting fossil fuels to solar energy, reducing pollution and the waste of energy and matter resources by reusing and recycling them, emphasizing the use of diverse strategies for solving environmental and health problems and reducing poverty, and controlling population growth.

The burden of proof imposed on individuals, companies, and institutions should be to show that pollution prevention options have been thoroughly examined, evaluated, and used before lesser options are chosen.

JOEL HIRSCHORN

CRITICAL THINKING

1. Explain why you agree or disagree with the proposals for reducing the death toll and other harmful effects of smoking listed on pp. 237–238. Do you think there should be a ban on smoking indoors in all public places? Explain.

2. How can changes in the age structure of a human population increase the spread of infectious diseases? How can the spread of infectious diseases affect the age structure of human populations?

3. Should we have zero pollution levels for all toxic and hazardous chemicals? Explain. What are the alternatives?

4. Evaluate the following statements:
   a. We should not get worked up about exposure to toxic chemicals because almost any chemical can cause some harm at a large enough dosage.
   b. We should not worry so much about exposure to toxic chemicals because, through genetic adaptation, we can develop immunity to such chemicals.

c. We should not worry so much about exposure to toxic chemicals because we can use genetic engineering to reduce or eliminate the harmful effects of toxic chemicals on humans.

5. Workers in a number of industries are exposed to higher levels of various toxic substances than is the general public. Should allowable workplace levels for such chemicals be reduced? What economic effects might this have?

6. Should laboratory-bred animals be used in laboratory experiments in toxicology? Explain. What are the alternatives?

7. What are the three major risks you face from **(a)** your lifestyle, **(b)** where you live, and **(c)** what you do for a living? Which of these risks are voluntary and which are involuntary? List the five most important things you can do to reduce these risks. Which of these things do you actually plan to do?

8. Congratulations! You are in charge of the world. List the three most important features of your program to reduce the risk from exposure to **(a)** toxic and hazardous chemicals, and **(b)** infectious disease organisms.

9. List two questions that you would like to have answered as a result of reading this chapter.

LEARNING ONLINE

The website for this book contains helpful study aids and many ideas for further reading and research. They include a chapter summary, review questions for the entire chapter, flash cards for key terms and concepts, a multiple-choice practice quiz, interesting Internet sites, references, information about green careers, and a guide for accessing thousands of InfoTrac® College Edition articles. Log into

www.thomsonedu.com/biology/miller

Then choose Chapter 11, and select a learning resource. For access to animations, additional quizzes, chapter outlines and summaries, register and log into

at **www.thomsonedu.com** using the access code card in the front of your book.

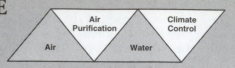

12 AIR POLLUTION, CLIMATE CHANGE, AND OZONE DEPLETION

We are embarked on the most colossal ecological experiment of all time—doubling the concentration in the atmosphere of an entire planet of one of its most important gases—and we really have little idea of what might happen.

PAUL A. COLINVAUX

ATMOSPHERE, WEATHER, AND CLIMATE

The Troposphere (Science)

The atmosphere's innermost layer is made up mostly of nitrogen and oxygen, with smaller amounts of water vapor and carbon dioxide.

We live at the bottom of a thin layer of gases surrounding the earth, called the **atmosphere.** It is divided into several spherical layers (Figure 12-1), each characterized by abrupt changes in temperature because of differences in the absorption of incoming solar energy.

About 75–80% of the mass of the earth's air is found in the **troposphere,** the atmospheric layer closest to the earth's surface. This layer extends only about 17 kilometers (11 miles) above sea level at the equator and about 8 kilometers (5 miles) over the poles. If the earth were the size of an apple, this lower layer containing the air we breathe would be no thicker than the apple's skin.

Take a deep breath. About 99% of the volume of the air you just inhaled consists of two gases: nitrogen (78%) and oxygen (21%). The remainder consists of water vapor (varying from 0.01% at the frigid poles to 4% in the humid tropics), 0.93% argon (Ar), 0.038% carbon dioxide (CO_2), and trace amounts of dust and soot and several other gases including methane (CH_4), ozone (O_3), and nitrous oxide (N_2O).

The troposphere is a dynamic system involved in the chemical cycling of many of the earth's vital nutrients. This thin and turbulent layer of rising and falling air currents and winds is also largely responsible for the planet's short-term *weather* and long-term *climate.*

The Stratosphere (Science)

Ozone in the atmosphere's second layer filters out most of the sun's ultraviolet radiation that is harmful to us and most other species.

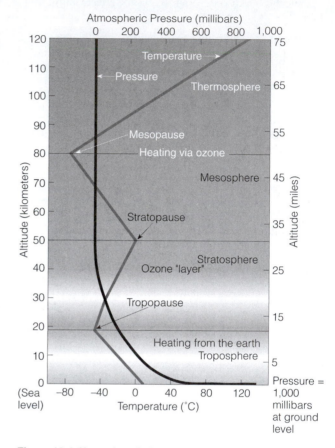

Figure 12-1 Natural capital: the earth's atmosphere is a dynamic system that consists of four layers. The average temperature of the atmosphere varies with altitude (red line). Most UV radiation from the sun is absorbed by ozone (O_3), found primarily in the stratosphere in the *ozone layer* 17–26 kilometers (10–16 miles) above sea level.

The atmosphere's second layer is the **stratosphere,** which extends from about 17 to 48 kilometers (11–30 miles) above the earth's surface (Figure 12-1). Although the stratosphere contains less matter than the troposphere, its composition is similar, with two notable exceptions: its volume of water vapor is about 1/1,000 as much and its concentration of ozone (O_3) is much higher.

Stratospheric ozone is produced when some of the oxygen molecules there interact with ultraviolet (UV) radiation emitted by the sun. This "global sunscreen" of ozone in the stratosphere—the *ozone layer*—keeps about 95% of the sun's harmful UV radiation from reaching the earth's surface.

This UV filter of "good" ozone in the lower stratosphere allows us and other forms of life to exist on land and helps protect us from sunburn, skin and eye cancers, cataracts, and damage to our immune systems. It also prevents much of the oxygen in the troposphere from being converted to photochemical ozone, a harmful air pollutant.

Much evidence indicates that some human activities are *decreasing* the amount of beneficial or "good" ozone in the stratosphere and *increasing* the amount of harmful or "bad" ozone (a pollutant that makes breathing difficult for some people) in the troposphere—especially in some urban areas.

Weather and Climate

Weather is a local area's short-term physical conditions such as temperature and precipitation, and climate is a region's average weather conditions over a long time.

Weather includes an area's short-term temperature, precipitation, humidity, wind speed, cloud cover, and other physical conditions of the lower atmosphere over a short period of time—the atmospheric conditions we experience day by day.

Climate is a region's general pattern of atmospheric or weather conditions over a long time—years, decades, and centuries. *Average temperature* and *average precipitation* are the two main factors determining climate, along with the closely related factors of *latitude* (distance from the equator) and *altitude*. The earth's major climate zones are shown in Figure 12-2.

Many factors contribute to a local climate, including the amount of solar radiation reaching the area, the earth's daily rotation and annual path around the sun, air circulation over the earth's surface, the global distribution of landmasses and seas, the circulation of ocean currents, and the elevation of landmasses.

Four major factors determine global air circulation patterns. One is the *uneven heating of the earth's surface.* Air is heated much more at the equator where the sun's rays strike directly than at the poles where sunlight strikes at an angle and thus is spread out over a much greater area. These differences in the amount of incoming solar energy help explain why tropical re-

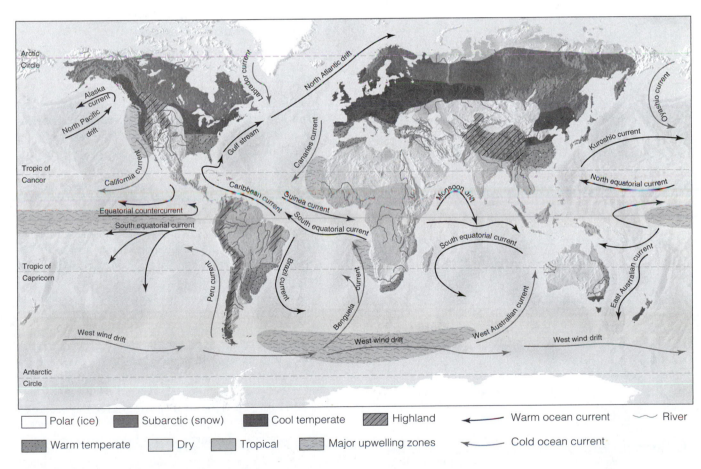

ThomsonNOW™ Active Figure 12-2 Natural capital: generalized map of the earth's current climate zones, showing the major contributing ocean currents and drifts. *See an animation based on this figure and take a short quiz on the concept.*

gions near the equator are hot, polar regions are cold, and temperate regions in between generally have intermediate average temperatures.

A second factor is *seasonal changes in temperature and precipitation*. The earth's axis—an imaginary line connecting the north and south poles—is tilted with respect to the sun's rays. As a result, various regions are tipped toward or away from the sun during most of the earth's year-long revolution around the sun. This creates opposite seasons in the northern and southern hemispheres.

A third factor is *rotation of the earth on its axis*, which causes warm air masses moving north and south from the equator to be deflected to the west or east over different parts of the planet's surface. The direction of air movement in the resulting huge atmospheric regions called *cells* sets up belts of *prevailing winds*—major surface winds that blow almost continuously and distribute air and moisture over the earth's surface (Figure 12-3).

The fourth factor affecting global air circulation is *properties of air, water, and land*. Heat from the sun evap-

orates ocean water and transfers heat from the oceans to the atmosphere, especially near the hot equator. This evaporation of water creates cyclical convection cells that circulate air, heat, and moisture both vertically and from place to place in the troposphere.

Also, land masses give up heat faster than the oceans. The earth's air circulation patterns, prevailing winds, and mixture of continents and oceans result in the six giant convection cells in which warm, moist air rises and cools, and the cool, dry air sinks. This leads to an irregular distribution of climates and patterns of vegetation, as shown in Figure 12-3.

Ocean currents also affect the climates of regions. The oceans absorb heat from the air circulation patterns just described, with the bulk of this heat absorbed near the warm tropical areas. This heat plus differences in water density create warm and cold ocean currents (Figure 12-2). Driven by winds and the earth's rotation, these currents redistribute heat from the sun and thus influence climate and vegetation, especially near coastal areas. They also help mix ocean waters and distribute nutrients and dissolved oxygen needed by aquatic organisms.

Atmospheric Gases and Climate: The Natural Greenhouse Effect

Water vapor, carbon dioxide, and other gases influence climate by warming the lower troposphere and the earth's surface.

Small amounts of certain gases play a key role in determining the earth's average temperatures and thus its climates. These gases include water vapor (H_2O), carbon dioxide (CO_2), methane (CH_4), and nitrous oxide (N_2O).

Together, these gases, known as **greenhouse gases,** allow mostly visible light and some infrared radiation and UV radiation from the sun to pass through the troposphere. The earth's surface absorbs much of this solar energy and transforms it to longer-wavelength, infrared radiation (heat), which rises into the troposphere.

Some of this heat escapes into space, and some is absorbed by molecules of greenhouse gases and emitted into the troposphere as even longer-wavelength infrared radiation. Some of this released energy is radiated into space and some warms the troposphere and the earth's surface. This natural warming effect in the troposphere is called the **greenhouse effect** (Figure 2-11, p. 32). Without its current greenhouse gases (especially water vapor, which is found in the largest concentration), the earth would be a cold and mostly lifeless planet.

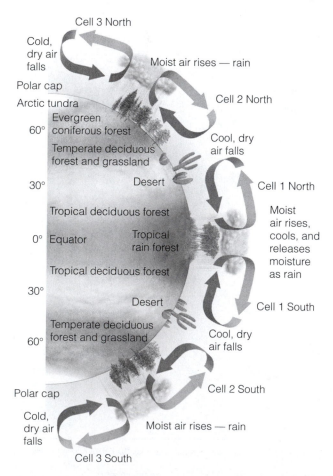

Figure 12-3 **Natural capital:** global air circulation and biomes. Heat and moisture are distributed over the earth's surface by vertical currents, which form six giant convection cells at different latitudes. The resulting uneven distribution of heat and moisture over the planet's surface leads to the forests, grasslands, and deserts that make up the earth's biomes.

Cell 3 North
Cold, dry air falls
Moist air rises — rain
Polar cap
Arctic tundra
Cell 2 North
Evergreen coniferous forest
60°
Temperate deciduous forest and grassland
Cool, dry air falls
30°
Desert
Cell 1 North
Tropical deciduous forest
Moist air rises, cools, and releases moisture as rain
0° Equator
Tropical rain forest
Tropical deciduous forest
30°
Desert
Cell 1 South
Temperate deciduous forest and grassland
60°
Cool, dry air falls
Cell 2 South
Polar cap
Cold, dry air falls
Moist air rises — rain
Cell 3 South

See how greenhouse gases trap heat in the atmosphere and raise the earth's temperature at ThomsonNOW.

AIR POLLUTION

Types and Sources of Air Pollution (Science)

Air pollutants come mostly from natural sources and from burning fossil fuels; indoor air pollutants come from outside but include chemicals used and generated inside and are a bigger health threat.

Air pollution is the presence of chemicals in the troposphere in concentrations high enough to harm humans, other organisms, ecosystems, or materials, and to change climate. Table 12-1 lists the major classes and sources of air pollutants. Most of these air pollutants are gases or volatile liquids that evaporate into the air. Some, called *suspended particulate matter* (SPM), or *aerosols,* consist of tiny particles of solids or liquid droplets suspended in the air.

Table 12-1 Major Air Pollutants and Their Sources	
Pollutant	**Sources**
Carbon oxides Carbon monoxide (CO) Carbon dioxide (CO_2)	Industries, motor vehicles, forest and grassland fires, open fires and poorly designed stoves for indoor cooking (CO), faulty furnaces (CO)
Nitrogen oxides Nitric oxide (NO) Nitrogen dioxide (NO_2) Nitrous oxide (N_2O)	Industries, motor vehicles, fires, volcanoes, fertilized cropland (N_2O), woodstoves, unvented gas stoves, kerosene heaters
Sulfur oxides Sulfur dioxide (SO_2) Sulfur trioxide (SO_3)	Coal-burning electric power plants, industries, volcanoes, formations in the atmosphere
Suspended particulate matter (SPM) Soot (C) Lead (Pb) Sulfuric acid (H_2SO_4) Nitric acid (HNO_3)	Industries, motor vehicles, windstorms, fires, volcanoes, reactions in the atmosphere (H_2SO_4 and HNO_3), open fires and poorly designed stoves for indoor cooking (soot), tobacco smoke, pollen, pet dander, dust mites
Volatile organic compounds (VOCs) Methane (CH_4) Isoprene (C_5H_8) Benzene (C_6H_6)	Industries, green plants (isoprene), natural gas wells (CH_4) and facilities, aerosol sprays, paint strippers and thinners
Ozone (O_3)	Reactions in the atmosphere resulting from a combination of nitrogen oxides, VOCs, and sunlight
Radioactive radon (Rn)	Release into homes from certain types of rock formations
Toxics Chlorine (Cl) Hydrogen sulfide (H_2S) Formaldehyde (CH_2O_2) Carbon tetrachloride (CCl_4)	Industries, businesses such as dry cleaning (CCl_4), decomposing plants (H_2S), volcanoes (H_2S), building materials (CH_2O_2), carpets, furniture stuffing, paneling, particleboard, foam insulation, formaldehyde

Globally, nature produces the largest quantities of air pollutants. Natural pollutants include dust from windstorms; CO, CO_2, nitrogen oxides (NO_x), and particulates from wildfires; and SO_2, NO_x, particulates, and H_2S from volcanoes.

Good news. Chemicals from natural sources rarely reach harmful levels. Most natural air pollutants are spread out over the globe or removed by chemical cycles, precipitation, and gravity. However, chemicals emitted by volcanic eruptions and some natural forest fires can reach harmful levels. Most outdoor pollutants from human activities enter the atmosphere from the burning of fossil fuels in power plants and factories (*stationary sources*) and in motor vehicles (*mobile sources*).

Scientists classify outdoor air pollutants into two categories. **Primary pollutants** are harmful substances emitted directly into the air. While in the troposphere, some of these primary pollutants may react with one another or with the basic components of air to form new harmful pollutants, called **secondary pollutants** (Figure 12-4, p. 256).

With their high concentration of cars and factories, cities normally have higher outdoor air pollution levels than rural areas have. But prevailing winds can spread long-lived primary and secondary air pollutants from urban and industrial areas to the countryside and to other urban areas. Indoor pollutants come from infiltration of polluted outside air and various chemicals used or produced inside buildings.

Good news. Over the past 30 years the quality of outdoor air in most developed countries has greatly improved. This occurred mostly because grassroots pressure from citizens led governments to pass and enforce air pollution control laws.

Bad news. According to the WHO, one of every six people on the earth or more than 1.1 billion people live in urban areas where the outdoor air is unhealthy to breathe. Most of them live in densely populated cities in developing countries where air pollution control laws do not exist or are poorly enforced. The biggest health threat comes from *indoor air pollution,* which occurs when the poor must burn wood, charcoal, coal, or dung in open fires or poorly designed stoves to heat their dwellings and cook their food.

In the United States and most other developed countries, government-mandated standards set maximum allowable atmospheric concentrations, or criteria, for six *criteria* or *conventional air pollutants* commonly found in outdoor air. Table 12-1 lists these eight types of pollutants and their sources.

Carbon dioxide is not regulated as a pollutant under the U.S. Clean Air Act. But most environmental scientists and many business leaders believe it should be so regulated because increasing levels of this greenhouse gas can warm the troposphere, contributing to global climate change with harmful environmental

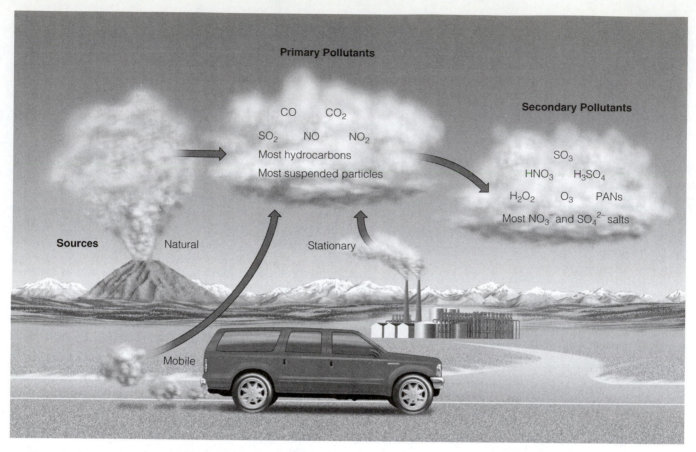

Primary Pollutants

CO CO_2

SO_2 NO NO_2

Most hydrocarbons

Most suspended particles

Secondary Pollutants

SO_3

HNO_3 H_3SO_4

H_2O_2 O_3 PANs

Most NO_3^- and SO_4^{2-} salts

Sources   Natural   Stationary

Mobile

Figure 12-4 Natural capital degradation: sources and types of air pollutants. Human inputs of air pollutants may come from *mobile sources* (such as cars) and *stationary sources* (such as industrial and power plants). Some *primary air pollutants* may react with one another or with other chemicals in the air to form *secondary air pollutants*.

and economic effects. Oil, automobile, electric utility, and coal companies oppose this because it would cost them money, and they have successfully lobbied the U.S. Congress to keep CO_2 off the list of regulated pollutants.

X *How Would You Vote?* Should carbon dioxide be classified as an air pollutant? Cast your vote online at www.thomsonedu.com/biology/miller.

Industrial Smog (Science)

Industrial smog is a mixture of sulfur dioxide, droplets of sulfuric acid, and a variety of suspended solid particles emitted by burning coal and oil.

Fifty years ago, cities such as London, England, and Chicago and Pittsburgh in the United States burned large amounts of heavy oil and coal (both of which contain sulfur impurities) in power plants and factories and for heating homes and cooking food. During the winter, people in such cities were exposed to **in-**dustrial smog, consisting mostly of sulfur dioxide, aerosols containing suspended droplets of sulfuric acid, and a variety of suspended solid particles that give the resulting smog a gray color, explaining why it is sometimes called *gray-air smog*.

Today, urban industrial smog is rarely a problem in most developed countries where coal and heavy oil are burned in large boilers with reasonably good pollution controls or with tall smokestacks that transfer the pollutants to downwind rural areas. However, industrial smog remains a problem in industrialized urban areas of China, India, Ukraine, and some eastern European countries, where large quantities of coal are burned in factories with inadequate pollution controls.

Photochemical Smog (Science)

Photochemical smog is a mixture of air pollutants formed by the reaction of nitrogen oxides and volatile organic hydrocarbons (VOCs) under the influence of sunlight.

A *photochemical reaction* is any chemical reaction activated by light. Air pollution known as **photochemical**

smog is a mixture of primary and secondary pollutants formed under the influence of sunlight.

In greatly simplified terms,

VOCs + NO$_x$ + heat + sunlight → ground level ozone (O$_3$) + other photochemical oxidants + aldehydes + other secondary air pollutants

The formation of photochemical smog begins when exhaust from morning commuter traffic releases large amounts of NO and VOCs into the air over a city. The NO is converted to reddish-brown NO$_2$, explaining why photochemical smog is sometimes called *brown-air smog.* When exposed to UV radiation from the sun, the NO$_2$ engages in a complex series of reactions with hydrocarbons that produce the mixture of chemicals called photochemical smog. Hotter days lead to higher levels of ozone and other components of smog. As traffic increases on a sunny day, photochemical smog (dominated by ozone) usually builds up to peak levels by late morning, irritating people's eyes and respiratory tracts.

All modern cities have some photochemical smog, but it is much more common in cities with sunny, warm, dry climates and lots of motor vehicles. Examples are Los Angeles, Denver, and Salt Lake City in the United States; Sydney, Australia; Mexico City, Mexico; São Paulo, Brazil; and Buenos Aires, Argentina. According to a 1999 study, if 400 million Chinese drive gasoline-powered cars by 2050 as projected, the resulting photochemical smog could cover the entire western Pacific with ozone, extending to the United States.

Thomson NOW! See how photochemical smog forms and how it affects us at ThomsonNOW.

Factors Influencing Levels of Outdoor Air Pollution (Science)

Outdoor air pollution can be reduced by gravitational settling out, precipitation, sea spray, winds, and chemical reactions, and increased by urban buildings, mountains, and high temperatures.

Four natural factors help *reduce* air pollution. First, particles heavier than air settle out as a result of gravitational attraction to the earth. Second, *rain and snow* help cleanse the air of pollutants. Third, *salty sea spray from the oceans* can wash out much of the particulates and other water-soluble pollutants from air that flows from land over the oceans. Fourth, *winds,* sweep pollutants away, dilute them by mixing them with cleaner air, and bring in fresh air.

Five other factors can *increase* air pollution. First, *urban buildings* can slow wind speed and reduce dilution and removal of pollutants. Second, *hills and mountains* can reduce the flow of air in valleys below them and allow pollutant levels to build up at ground level. Third, *higher temperatures* promote the chemical reactions leading to smog formation. Fourth, *VOC emissions from certain species of trees and plants* in heavily wooded urban areas can play a large role in the formation of photochemical smog.

Fifth, *temperature inversions* can cause buildup of high levels of air pollutants. During daylight, the sun warms the air near the earth's surface. Normally, this warm air and most of the pollutants it contains rise to mix and disperse the pollutants with the cooler air above it. Under certain atmospheric conditions, however, a layer of warm air can lie atop a layer of cooler air nearer the ground, a situation known as a **temperature inversion.** Because the cooler air is denser than the warmer air above it, the air near the surface does not rise and mix with the air above it. Pollutants can concentrate in this stagnant layer of cool air near the ground. Under certain conditions, temperature inversions can last for several days and allow pollutants to build up to dangerous concentrations.

Thomson NOW! Learn more about thermal inversions and what they can mean for people in some cities at ThomsonNOW.

What Is Acid Deposition (Science)

Sulfur dioxide, nitrogen oxides, and particulates can react in the atmosphere to produce acidic chemicals that can travel long distances before returning to the earth's surface.

Most coal-burning power plants, ore smelters, and other industrial plants in developed countries use tall smokestacks to emit sulfur dioxide, suspended particles, and nitrogen oxides high into the troposphere where wind can mix, dilute, and disperse them.

These tall smokestacks reduce *local* air pollution, but can increase *regional* air pollution downwind. Sulfur dioxide and nitrogen oxides, the primary pollutants, are emitted high into the atmosphere and may be transported as far as 1,000 kilometers (600 miles) by prevailing winds. During their trip, they form secondary pollutants such as nitric acid vapor, droplets of sulfuric acid, and particles of acid-forming sulfate and nitrate salts.

These acidic substances remain in the atmosphere for 2–14 days, depending mostly on prevailing winds, precipitation, and other weather patterns. During this period, they descend to the earth's surface in two forms: *wet deposition,* or acidic rain, snow, fog, and cloud vapor, and *dry deposition* consisting of acidic particles. The resulting mixture is called **acid deposition** (Figure 12-5, p. 258), sometimes called *acid rain*—with a pH below 5.6 (Figure 2-24, p. 44). Most dry deposition

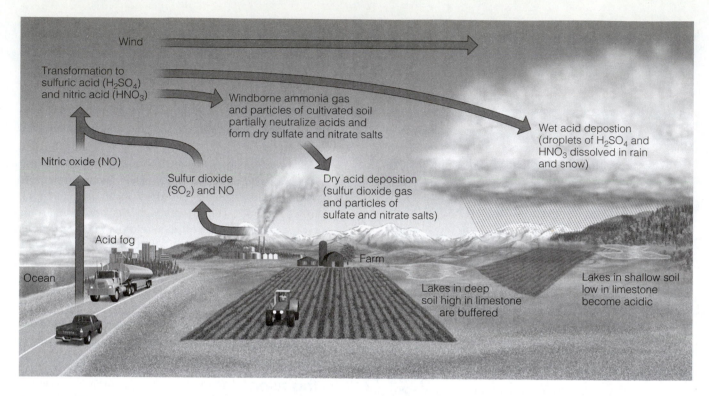

ThomsonNOW Active Figure 12-5 Natural capital degradation: *acid deposition,* which consists of rain, snow, dust, or gas with a pH lower than 5.6, is commonly called *acid rain.* Soils and lakes vary in their ability to buffer or remove excess acidity. *See an animation based on this figure and take a short quiz on the concept.*

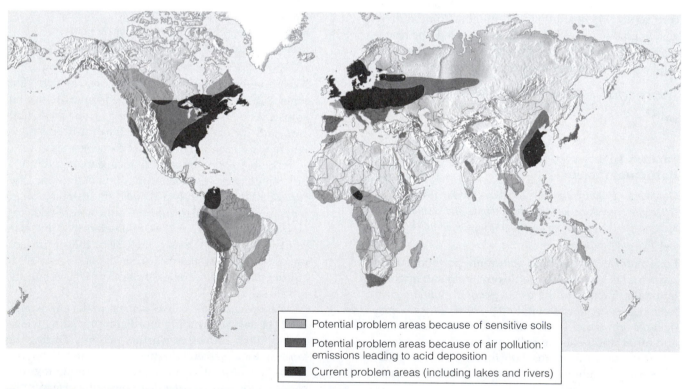

Potential problem areas because of sensitive soils

Potential problem areas because of air pollution: emissions leading to acid deposition

Current problem areas (including lakes and rivers)

Figure 12-6 Natural capital degradation: regions where acid deposition is now a problem and regions with the potential to develop this problem. Such regions have large inputs of air pollution (mostly from power plants, industrial plants, and ore smelters) or are sensitive areas with soils and bedrock that cannot neutralize (buffer) inputs of acidic compounds. (Data from World Resources Institute and U.S. Environmental Protection Agency)

occurs within about 2–3 days, fairly near the emission sources, whereas most wet deposition takes place within 4–14 days in more distant downwind areas.

Acid deposition is a *regional* air pollution problem in areas that lie downwind from coal-burning facilities and from urban areas with large numbers of cars. Such areas include the eastern United States and other parts of the world (Figure 12-6).

In some areas, soils contain basic compounds that can react with and neutralize, or *buffer*, some input of acids. The areas most sensitive to acid deposition are those with thin, acidic soils without such natural buffering (Figure 12-6, some darker shaded areas) and those where the buffering capacity of soils has been depleted by decades of acid deposition.

Many acid-producing chemicals generated in one country are exported to other countries by prevailing winds. Acidic emissions from industrialized areas of western Europe (especially the United Kingdom and Germany) and eastern Europe blow into Norway, Switzerland, Austria, Sweden, the Netherlands, and Finland. Some SO_2 and other emissions from coal-burning power and industrial plants in the Ohio Valley of the United States end up in Southeastern Canada. Some of China's acidic emissions end up in Japan and North and South Korea.

The worst acid deposition area is in Asia, especially China, which gets about 59% of its energy from burning coal. Scientists estimate that by 2025, China will emit more sulfur dioxide than the United States, Canada, and Japan combined.

Learn more about the sources of acid deposition, how it forms, and what it can do to lakes and soils at ThomsonNOW.

Harmful Effects of Acid Deposition (Science)

Acid deposition can cause or worsen respiratory disease, damage metallic and stone objects, decrease atmospheric visibility, kill fish, deplete soils of vital plant nutrients, and harm crops and plants.

Acid deposition has a number of harmful effects. It contributes to chronic respiratory diseases such as bronchitis and asthma, and can leach toxic metals (such as lead and mercury) from soils and rocks into acidic lakes used as sources of drinking water. These toxic metals can accumulate in the tissues of fish eaten by people, mammals, and birds. Acid deposition also damages statues, national monuments, buildings, metals, and car finishes. Acidic particles in the air, especially sulfates, decrease atmospheric visibility.

Acid deposition also harms aquatic systems. Because of excess acidity, several thousand lakes in Norway and Sweden contain no fish, and many more

lakes there have lost most of their acid-neutralizing capacity. In Ontario, Canada, at least 1,200 acidified lakes contain few if any fish, and some fish populations in many thousands of other lakes are declining because of increased acidity. In the United States, several hundred lakes (most in the Northeast) are threatened with excess acidity.

Acid deposition (often along with other air pollutants such as ozone) can harm forests and crops by leaching essential plant nutrients such as calcium and magnesium salts from soils. This reduces plant productivity, tree growth, and the ability of the soils to buffer or neutralize acidic inputs. Acid deposition rarely kills trees directly, but can weaken them and leave them vulnerable to other stresses such as severe cold, diseases, insect attacks, drought, and harmful mosses.

Mountaintop forests are the terrestrial areas hardest hit by acid deposition. These areas tend to have thin soils without much buffering capacity. And trees on mountaintops (especially conifers such as red spruce and balsam fir, which keep their leaves year-round) are bathed almost continuously in very acidic fog and clouds.

Most of the world's lakes and forests are not being destroyed or seriously harmed by acid deposition. Rather, this regional problem is harming forests and lakes downwind from coal-burning facilities and from large car-dominated cities without adequate controls.

Examine how acid deposition can harm a pine forest and what it means to surrounding land and waters at ThomsonNOW.

Figure 12-7 (p. 260) summarizes ways to reduce acid deposition. According to most scientists studying the problem, the best solutions are *prevention approaches* that reduce or eliminate emissions of sulfur dioxide, nitrogen oxides, and particulates.

Controlling acid deposition is a political hot potato. One problem is that the people and ecosystems it affects often are quite distant from those who cause the problem. Also, countries with large supplies of coal (such as China, India, Russia, and the United States) have a strong incentive to use it as a major energy resource. Owners of coal-burning power plants say the costs of adding pollution control equipment, using low-sulfur coal, or removing sulfur from coal are too high and would increase the cost of electricity for consumers.

Environmental scientists respond that affordable and much cleaner resources are available to produce electricity, including wind and natural gas. They also point out that the largely hidden health and environmental costs of burning coal are 2 to 3 times its market cost. Including these costs would spur prevention of acid deposition.

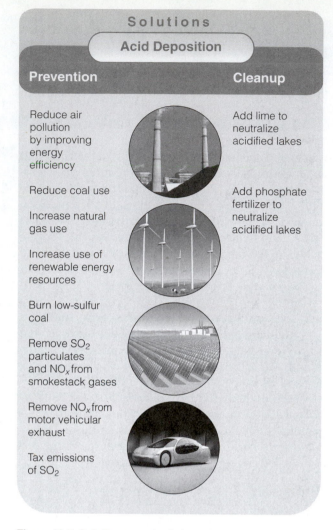

Solutions

Acid Deposition

Prevention	Cleanup
Reduce air pollution by improving energy efficiency	Add lime to neutralize acidified lakes
Reduce coal use	Add phosphate fertilizer to neutralize acidified lakes
Increase natural gas use	
Increase use of renewable energy resources	
Burn low-sulfur coal	
Remove SO$_2$ particulates and NO$_x$ from smokestack gases	
Remove NO$_x$ from motor vehicular exhaust	
Tax emissions of SO$_2$	

Figure 12-7 Solutions: methods for reducing acid deposition and its damage. QUESTION: *Which two of these solutions do you think are the most important?*

Indoor Air Pollution (Science)

Indoor air pollution usually is a much greater threat to human health than outdoor air pollution.

If you are reading this book indoors, you may be inhaling more air pollutants with each breath than you would if you were outside. Figure 12-8 shows some typical sources of indoor air pollution.

EPA studies have revealed some alarming facts about indoor air pollution in the United States and other developed countries. *First,* levels of 11 common pollutants generally are 2 to 5 times higher inside homes and commercial buildings than outdoors, and as much as 100 times higher in some cases. *Second,* pollution levels inside cars in traffic-clogged urban areas can be up to 18 times higher than outside. *Third,* the health risks from exposure to such chemicals are magnified because most people in developed countries typically spend 70–98% of their time indoors or inside vehicles.

In 1990, the EPA placed indoor air pollution at the top of the list of 18 sources of cancer risk—causing as many as 6,000 premature cancer deaths per year in the U.S. At greatest risk are smokers, infants and children under age 5, the old, the sick, pregnant women, people with respiratory or heart problems, and factory workers.

Danish and U.S. EPA studies have linked pollutants found in buildings to a number of health effects, a phenomenon known as the *sick-building syndrome* (SBS). Such effects include dizziness, headaches, coughing, sneezing, shortness of breath, nausea, burning eyes, sore throats, chronic fatigue, irritability, skin dryness and irritation, respiratory infections, flu-like symptoms, and depression. EPA and Labor Department studies indicate that almost one in five commercial buildings in the United States expose employees to health risks from indoor air pollution.

According to the EPA and public health officials, the four most dangerous indoor air pollutants in developed countries are *tobacco smoke, formaldehyde, radioactive radon-222 gas* (see Case Study, below), and *very small fine and ultrafine particles.*

In developing countries, the indoor burning of wood, charcoal, dung, crop residues, and coal in open fires or in unvented or poorly vented stoves for cooking and heating exposes inhabitants to dangerous levels of particulate air pollution. According to the WHO and the World Bank, *indoor air pollution for the poor is by far the world's most serious air pollution problem*—a glaring example of the relationship between poverty and environmental quality.

The chemical that causes most people in developed countries difficulty is *formaldehyde,* a colorless, extremely irritating chemical. According to the EPA and the American Lung Association, 20–40 million Americans suffer from chronic breathing problems, dizziness, rash, headaches, sore throat, sinus and eye irritation, skin irritation, wheezing, and nausea caused by daily exposure to low levels of formaldehyde emitted from common household materials.

These materials include building materials (such as plywood, particleboard, paneling, and high-gloss wood used in floors and cabinets), furniture, drapes, upholstery, adhesives in carpeting and wallpaper, urethane-formaldehyde foam insulation, fingernail hardener, and wrinkle-free coating on permanent-press clothing (Figure 12-8). The EPA estimates that as many as 1 of every 5,000 people who live in manufactured homes for more than 10 years will develop cancer from formaldehyde exposure.

Case Study: Radioactive Radon Gas (Science)

Radon-222, a radioactive gas found in some soil and rocks, can seep into some houses and increase the risk of lung cancer.

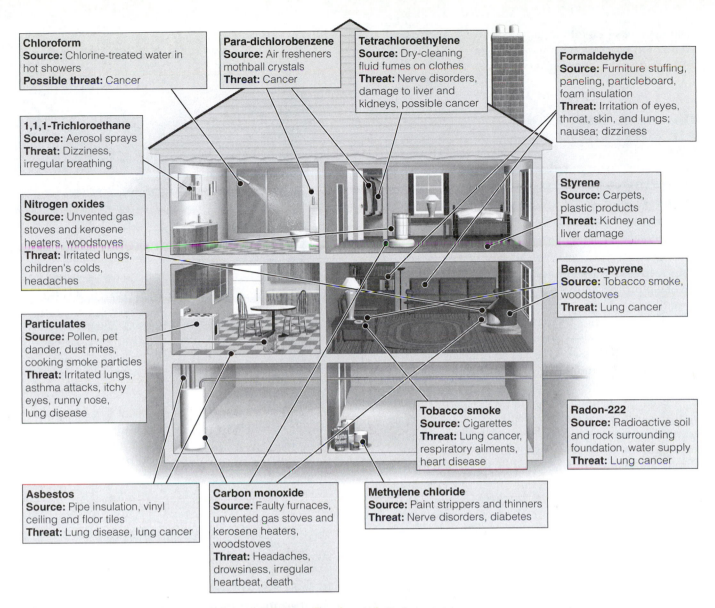

Chloroform
Source: Chlorine-treated water in hot showers
Possible threat: Cancer

1,1,1-Trichloroethane
Source: Aerosol sprays
Threat: Dizziness, irregular breathing

Nitrogen oxides
Source: Unvented gas stoves and kerosene heaters, woodstoves
Threat: Irritated lungs, children's colds, headaches

Particulates
Source: Pollen, pet dander, dust mites, cooking smoke particles
Threat: Irritated lungs, asthma attacks, itchy eyes, runny nose, lung disease

Para-dichlorobenzene
Source: Air fresheners mothball crystals
Threat: Cancer

Tetrachloroethylene
Source: Dry-cleaning fluid fumes on clothes
Threat: Nerve disorders, damage to liver and kidneys, possible cancer

Formaldehyde
Source: Furniture stuffing, paneling, particleboard, foam insulation
Threat: Irritation of eyes, throat, skin, and lungs; nausea; dizziness

Styrene
Source: Carpets, plastic products
Threat: Kidney and liver damage

Benzo-α-pyrene
Source: Tobacco smoke, woodstoves
Threat: Lung cancer

Tobacco smoke
Source: Cigarettes
Threat: Lung cancer, respiratory ailments, heart disease

Radon-222
Source: Radioactive soil and rock surrounding foundation, water supply
Threat: Lung cancer

Asbestos
Source: Pipe insulation, vinyl ceiling and floor tiles
Threat: Lung disease, lung cancer

Carbon monoxide
Source: Faulty furnaces, unvented gas stoves and kerosene heaters, woodstoves
Threat: Headaches, drowsiness, irregular heartbeat, death

Methylene chloride
Source: Paint strippers and thinners
Threat: Nerve disorders, diabetes

Figure 12-8 Science: some important indoor air pollutants. (Data from U.S. Environmental Protection Agency)

Radon-222—a naturally occurring radioactive gas that you cannot see, taste, or smell—is produced by the radioactive decay of uranium-238. Most soil and rock contain small amounts of uranium-238. But this isotope is much more concentrated in underground deposits of minerals such as uranium, phosphate, granite, and shale.

When radon gas from such deposits seeps upward through the soil and is released outdoors, it disperses quickly in the atmosphere and decays to harmless levels. However, in buildings above such deposits radon gas can enter through cracks in foundations and walls, openings around sump pumps and drains, and hollow concrete blocks. Once inside, it can build up to high levels, especially in unventilated lower levels of homes and buildings.

Radon-222 gas quickly decays into solid particles of other radioactive elements that, if inhaled, expose lung tissue to a large amount of ionizing radiation from alpha particles. This exposure can damage lung tissue and lead to lung cancer over the course of a 70-year lifetime. Your chances of getting lung cancer from radon depend mostly on how much radon is in your home, how much time you spend in your home, and whether you are a smoker or have ever smoked.

Ideally, radon levels should be monitored continuously in the main living areas (not basements or crawl spaces) for 2 months to a year. By 2003, only about 6% of U.S. households had followed the EPA's recommendation to conduct radon tests (most lasting only 2–7 days and costing $20–100 per home).

For information about radon testing visit the EPA website at **http://www.epa.gov/iaq/radon**. According to the EPA, radon control could add $350–500 to the cost of a new home, and correcting a radon problem in an existing house could run $800–2,500. Remedies include sealing cracks in the foundation and walls, increasing ventilation by cracking a window or installing vents, and using a fan to create cross ventilation.

HARMFUL EFFECTS OF AIR POLLUTION

How Does Your Respiratory System Help Protect You from Air Pollution? (Science)

Your respiratory system can help protect you from air pollution, but some air pollutants can overcome these defenses.

Your respiratory system has a number of mechanisms that help protect you from air pollution. Hairs in your nose filter out large particles. Sticky mucus in the lining of your upper respiratory tract captures smaller (but not the smallest) particles and dissolves some gaseous pollutants. Sneezing and coughing expel contaminated air and mucus when pollutants irritate your respiratory system.

In addition, hundreds of thousands of tiny mucus-coated hair-like structures called *cilia* line your upper respiratory tract. They continually wave back and forth and transport mucus and the pollutants to your throat where they are swallowed or expelled.

But prolonged or acute exposure to air pollutants including tobacco smoke can overload or break down these natural defenses. This can cause or contribute to various respiratory diseases such as asthma, bronchitis, and emphysema. Years of smoking and breathing air pollutants can lead to *lung cancer* and *chronic bronchitis*. Damage deeper in the lung can cause *emphysema,* which is irreversible damage to air sacs, or alveoli, leading to loss of lung elasticity, and acute shortness of breath.

Premature Deaths from Air Pollution (Science)

Each year, air pollution prematurely kills about 3 million people, mostly from indoor air pollution in developing countries.

According to the WHO, at least 3 million people worldwide (most of them in Asia) die prematurely each year from the effects of air pollution—an average of 8,200 deaths per day. About 2.8 million of these deaths (93%) are from *indoor* air pollution, typically from heart attacks, respiratory diseases, and lung cancer related to daily breathing of polluted air.

In the United States, the EPA estimates that annual deaths related to indoor and outdoor air pollution range from 150,000 to 350,000 people—equivalent to one to two fully loaded 400-passenger jumbo jets crashing *each day* with no survivors. Millions more become ill and lose work time. Most of these deaths are related to inhalation of fine and ultrafine particulates in indoor air and in outdoor air from coal-burning power plants, mostly in the eastern half of the United States (Figure 12-9).

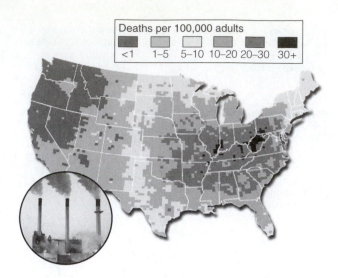

Figure 12-9 Premature deaths from air pollution in the United States, mostly from fine and ultrafine particles added to the troposphere by coal-burning power plants. QUESTION: *What is the risk where you live?* (Data from U.S. Environmental Protection Agency)

PREVENTING AND REDUCING AIR POLLUTION

U.S. Air Pollution Laws (Science, Economics, and Politics)

The Clean Air Acts in the United States have greatly reduced outdoor air pollution from six major pollutants.

The U.S. Congress passed Clean Air Acts in 1970, 1977, and 1990. With these acts, the federal government established air pollution regulations for key pollutants that are enforced by states and by major cities.

Great news: According to a 2005 EPA report, combined emissions of the six criteria air pollutants decreased by 54% between 1970 and 2004, even with significant increases in gross domestic product, vehicle miles traveled, energy consumption, and population.

Bad news: After dropping in the 1980s, photochemical smog levels did not drop between 1993 and 2004, mostly because reducing smog requires much bigger cuts in emissions of nitrogen oxides from power and industrial plants and motor vehicles. Also, according to the EPA, in 2003 more than 170 million people lived in areas where air is unhealthy to breathe during part of the year because of high levels of air pollutants—primarily ozone and fine particles.

Improving U.S. Air Pollution Laws (Science, Economics, and Politics)

Environmental scientists applaud the success of U.S. air pollution control laws, but suggest several ways to make them more effective.

The reduction of outdoor air pollution in the United States since 1970 has been a remarkable success story. This occurred because of two factors. *First*, U.S. citizens insisted that laws be passed and enforced to improve air quality. *Second*, the country was affluent enough to afford such controls and improvements.

But more can be done. Environmental scientists point to several deficiencies in the Clean Air Acts. First, *we continue to rely mostly on pollution cleanup rather than prevention.* The power of prevention is clear: In the United States, the air pollutant with the largest drop in its atmospheric level was lead (98% between 1970 and 2004), which was largely banned in gasoline. This is viewed as one the greatest environmental success stories in the country's history.

Second, *Congress has failed to increase fuel-efficiency standards for cars, sport utility vehicles (SUVs), and light trucks.* Many scientists and economists believe that a *feebate* program (p. 221) is the best way to increase the fuel efficiency of motor vehicles. According to environmental scientists, increased fuel efficiency would reduce air pollution from motor vehicles more quickly and effectively than any other method. It would also reduce CO_2 emissions, reduce dependence on imported oil, save energy, and save consumers enormous amounts of money.

Third, *regulation of emissions from motorcycles and two-cycle gasoline engines remains inadequate.* A 2005 Swiss study indicated that motorcycles collectively emit 16 times more hydrocarbons and three times more carbon monoxide that do cars. And two-cycle engines used in lawn mowers, leaf blowers, chain saws, jet skis, outboard motors, and snowmobiles emit high levels of pollutants. According to the California Air Resources Board, a 1-hour ride on a typical jet ski creates more air pollution than the average U.S. car does in a year, and operating a 100-horsepower boat engine for 7 hours emits more air pollutants than driving a new car driven 160,000 kilometers (100,000 miles). In 2001, the EPA announced plans to reduce emissions from most of these sources by 2007. Manufacturers push for extending these deadlines and weakening the standards.

Fourth, *the acts do not regulate emissions of the greenhouse gas CO_2.* Fifth, *urban ozone levels are still too high in many areas.* Sixth, *the acts have failed to deal seriously with indoor air pollution,* even though it is by far the most serious air pollution problem in terms of poorer health, premature death, and economic losses from lost work time and increased health costs.

Finally, there is a need for *better enforcement of the Clean Air Acts.* According to a 2002 government study, more rigorous enforcement would save about 6,000 lives and prevent 140,000 asthma attacks each year in the United States.

Executives of companies affected by implementing such policies claim that correcting these problems would cost too much and harm economic growth. Proponents contend that history has shown that most industry cost estimates for implementing various air pollution control standards in the United States were many times the actual cost. In addition, implementing such standards has boosted economic growth and created jobs by stimulating companies to develop new technologies for reducing air pollution emissions.

X *HOW WOULD YOU VOTE?* Should the 1990 U.S. Clean Air Act be strengthened? Cast your vote online at www.thomsonedu .com/biology/miller.

Using the Marketplace to Reduce Outdoor Air Pollution (Economics)

Allowing producers of air pollutants to buy and sell government air pollution allotments in the marketplace can help reduce emissions.

To help reduce SO_2 emissions, the Clean Air Act of 1990 authorizes an *emissions trading,* or *cap-and-trade, policy,* which enables the 110 most polluting power plants in 21 states (primarily in the Midwest and East) to buy and sell SO_2 pollution rights.

Each year, a coal-burning power plant is given a certain number of pollution credits, or rights, that allow it to emit a certain amount of SO_2. A utility that emits less SO_2 than it is allowed to emit has a surplus of pollution credits. It can use these credits to avoid reductions in SO_2 emissions at another of its plants, keep them for future plant expansions, or sell them to other utilities, private citizens, or environmental groups.

Proponents argue that this approach is cheaper and more efficient than having the government dictate how to control air pollution. Critics contend that it allows utilities with older, dirtier power plants to buy their way out of their environmental responsibilities and continue emitting unacceptable levels of SO_2. This approach also creates incentives to cheat because air quality regulation is based largely on self-reporting of emissions.

Ultimately, the success of any emissions trading approach depends on how low the initial cap is set and then on how much it is reduced annually to promote continuing innovation in air pollution prevention and control. Without these elements, emissions trading programs mostly move air pollutants from one area to another without achieving an overall reduction in air quality.

Between 1990 and 2002, the emissions trading system helped reduce SO_2 emissions from electric power

Solutions

Stationary Source Air Pollution

Prevention	Dispersion or Cleanup
Burn low-sulfur coal	Disperse emissions above thermal inversion layer with tall smokestacks
Remove sulfur from coal	
Convert coal to a liquid or gaseous fuel	Remove pollutants after combustion
Shift to less polluting fuels	Tax each unit of pollution produced

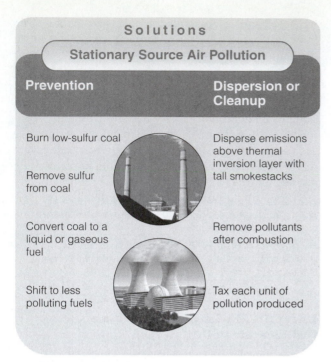

Figure 12-10 Solutions: methods for reducing emissions of sulfur oxides, nitrogen oxides, and particulate matter from stationary sources such as coal-burning electric power plants and industrial plants. QUESTION: *Which two of these solutions do you think are the most important?*

Solutions

Motor Vehicle Air Pollution

Prevention	Cleanup
Mass transit	Emission control devices
Bicycles and walking	
Less polluting engines	
Less polluting fuels	
Improve fuel efficiency	Car exhaust inspections twice a year
Get older, polluting cars off the road	
Give buyers large tax write-offs for buying low-polluting, energy-efficient vehicles	
Restrict driving in polluted areas	Stricter emission standards

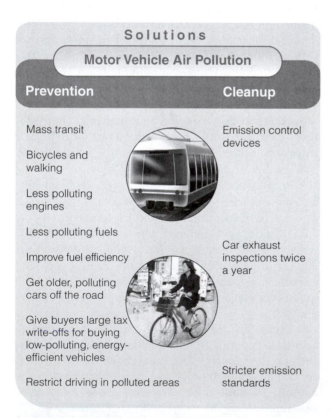

Figure 12-11 Solutions: methods for reducing emissions from motor vehicles. QUESTION: *Which two of these solutions do you think are the most important?*

plants in the United States by 35% at a cost of less than one-tenth the cost projected by industry. Emissions trading may also be implemented for other air pollutants.

However, environmental and health scientists strongly oppose using a cap-and-trade program to control emissions of mercury by coal-burning power plants and industries because mercury is highly toxic and does not break down in the environment. Coal-burning plants choosing to buy permits instead of sharply reducing their mercury emissions would create toxic hot spots with unacceptably high levels of mercury.

X *HOW WOULD YOU VOTE?* Should emissions trading be used to help control emissions of all major air pollutants? Cast your vote online at www.thomsonedu.com/biology/miller.

Reducing Outdoor Air Pollution (Science and Economics)

There are a number of ways to prevent and control air pollution from coal-burning facilities and motor vehicles.

Figure 12-10 summarizes ways to reduce emissions of sulfur oxides, nitrogen oxides, and particulate matter from stationary sources such as electric power plants and industrial plants that burn coal.

About 20,000 older coal burning plants, industrial plants, and oil refineries in the United States have not been required to meet the air pollution standards required for new facilities under the Clean Air Acts. Officials of states subject to pollution from such plants have been trying to get Congress to correct this problem since 1970. But they have been not been successful because of strong lobbying efforts by U.S. coal and electric power industries.

X *HOW WOULD YOU VOTE?* Should older coal-burning power and industrial plants have to meet the same air pollution standards as new facilities? Cast your vote online at www.thomsonedu.com/biology/miller.

Figure 12-11 lists ways to reduce emissions from motor vehicles, the primary culprits in producing photochemical smog.

Good news: Over the next 10–20 years, air pollution from motor vehicles should decrease from increased use of *partial zero-emission vehicles* (PZEVs) that emit almost no air pollutants because of improved engine and emission systems. *Hybrid-electric vehicles* and vehicles powered by *fuel cells running on hydrogen* will also be used more.

Bad news: The growing number of motor vehicles in urban areas of many developing countries is worsening the already poor air quality there. Many of these vehicles are 10 or more years old, have no pollution control devices, and burn leaded gasoline.

Solutions: Reducing Indoor Air Pollution

Little effort has been devoted to reducing indoor air pollution even though it is a much greater threat to human health than outdoor air pollution.

Reducing indoor air pollution does not require setting indoor air quality standards and monitoring the more than 100 million homes and buildings in the United States (or the buildings in any country). Instead, air pollution experts suggest several ways to prevent or reduce indoor air pollution (Figure 12-12).

In developing countries, indoor air pollution from open fires and leaky and inefficient stoves that burn wood, charcoal, or coal could be reduced. Governments could give people inexpensive clay or metal stoves that burn biofuels more efficiently while venting their exhausts to the outside, or stoves that use solar energy to cook food (solar cookers). These improvements would also reduce deforestation by using less fuelwood and charcoal.

Figure 12-13 lists some ways that you can reduce your exposure to indoor air pollution.

Solutions

Indoor Air Pollution

Prevention	Cleanup or Dilution
Cover ceiling tiles and lining of AC ducts to prevent release of mineral fibers	Use adjustable fresh air vents for work spaces
Ban smoking or limit it to well-ventilated areas	Increase intake of outside air
Set stricter formaldehyde emissions standards for carpet, furniture, and building materials	Change air more frequently
Prevent radon infiltration	Circulate a building's air through rooftop greenhouses
Use office machines in well-ventilated areas	Use exhaust hoods for stoves and appliances burning natural gas
Use less polluting substitutes for harmful cleaning agents, paints, and other products	Install efficient chimneys for wood-burning stoves

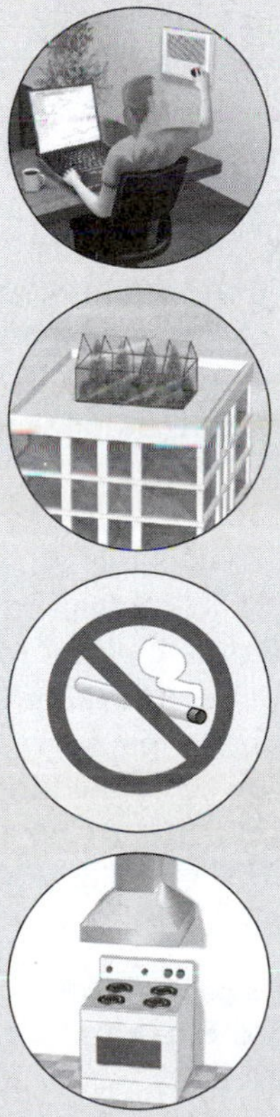

Figure 12-12 Solutions: ways to prevent and reduce indoor air pollution. QUESTION: *Which two of these solutions do you think are the most important?*

What Can You Do?

Indoor Air Pollution

- Test for radon and formaldehyde inside your home and take corrective measures as needed.
- Do not buy furniture and other products containing formaldehyde.
- Remove your shoes before entering your house to reduce inputs of dust, lead, and pesticides.
- Test your house or workplace for asbestos fiber levels and for any crumbling asbestos materials if it was built before 1980.
- Don't live in a pre-1980 house without having its indoor air tested for asbestos and lead.
- Do not store gasoline, solvents, or other volatile hazardous chemicals inside a home or attached garage.
- If you smoke, do it outside or in a closed room vented to the outside.
- Make sure that wood-burning stoves, fireplaces, and kerosene- and gas-burning heaters are properly installed, vented, and maintained.
- Install carbon monoxide detectors in all sleeping areas.

Figure 12-13 Individuals matter: ways to reduce your exposure to indoor air pollution. QUESTION: *Which three of these actions do you think are the most important?*

Solutions: The Next Step

Environmental and health scientists call for us to focus on preventing air pollution, with emphasis on sharply reducing indoor air pollution in developing countries.

Encouraging news. Since 1970, most of the world's developed countries have enacted laws and regulations that have significantly reduced outdoor air pollution. Without grassroots pressure on elected officials in the 1970s and 1980s, these laws and regulations would not have been enacted, funded, and implemented. In turn, these legal requirements spurred companies, scientists, and engineers to come up with better ways to control outdoor pollution.

The current laws represent a useful *output approach* to controlling pollution. To environmental and health scientists, however, the next step is to shift to *preventing air pollution*. With this approach, the question is not *"What can we do about the air pollutants we produce?"* but *"How can we avoid producing pollutants in the first place?"*

Figure 12-14 shows ways to prevent outdoor and indoor air pollution over the next 30–40 years. Like the shift to *controlling* air pollution between 1970 and 2000, this new shift to *preventing* air pollution will not take place without political pressure on elected officials and economic pressure on companies by individual citizens and groups.

Many of these solutions involve applying the four principles of sustainability—shifting from use of polluting fossil fuels to solar energy; reusing and recycling materials to reduce waste of matter and energy and the resulting air pollution; using diverse strategies to reduce air pollution with increased emphasis on prevention; and reducing population growth to decrease the number of people and devices that emit air pollutants.

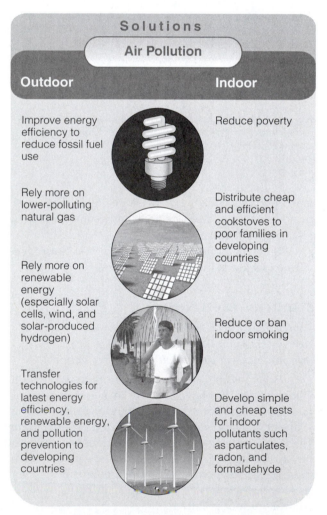

Figure 12-14 Solutions: ways to prevent outdoor and indoor air pollution over the next 30–40 years. QUESTION: *Which two of these solutions do you think are the most important?*

CLIMATE CHANGE AND HUMAN ACTIVITIES

Historic Changes in the Earth's Temperature (Science)

The earth has experienced prolonged periods of global warming and global cooling.

Climate change is neither new nor unusual. Over the past 4.7 billion years, the planet's climate has been altered by volcanic emissions, changes in solar input, continents moving as a result of shifting tectonic plates, strikes by large meteors, and other factors.

Over the past 900,000 years, the troposphere has undergone prolonged periods of *global cooling* and *global warming* (Figure 12-15, top left). These alternating cycles of freezing and thawing are known as *glacial* and *interglacial (between ice ages) periods.*

During each cold period, thick glacial ice covered much of the earth's surface for about 100,000 years. Most of it melted during a *warmer interglacial period* lasting 10,000–12,500 years that followed each glacial period. For roughly 12,000 years, we have had the good fortune to live in an interglacial period with a fairly stable climate and average global surface temperature (Figure 12-15, top right, and bottom left).

According to measurements of CO_2 in bubbles at various depths in glacial ice, estimated changes in tropospheric CO_2 levels correlate fairly closely with estimated variations in the average global temperature near the earth's surface during the past 160,000 years (Figure 12-16).

For the past 275 years since the beginning of the Industrial Revolution, human actions have lead to significant increases in the tropospheric concentration of three greenhouse gases—CO_2, CH_4, and N_2O—mostly from agriculture, deforestation, and burning fossil fuels (Figure 12-17, p. 268).

The United States, which uses 26% of the world's oil and 26% of its coal, is the largest emitter of greenhouse gases, with 21% of global emissions. Motor vehicles, which are essentially global warming factories on wheels, account for about 28% of U.S. greenhouse gas emissions. The second biggest greenhouse gas emitter is China (15%) followed by the European Union (14%), Russia (6%), and India (6%). China's emissions are expected to surpass U.S. emissions by 2025.

Signs That the Troposphere Is Warming: The Human Connection (Science)

There is considerable evidence that the earth's troposphere is warming, partly because of human activities.

In 1988, the United Nations and the World Meteorological Organization established the Intergovernmental Panel on Climate Change (IPCC) to document past

Average temperature over past 900,000 years

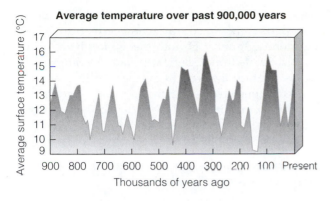

Temperature change over past 22,000 years

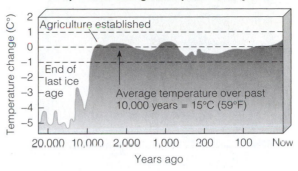

Temperature change over past 1,000 years

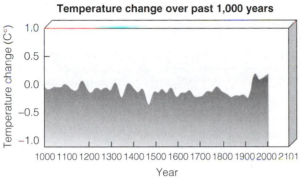

Average temperature over past 130 years

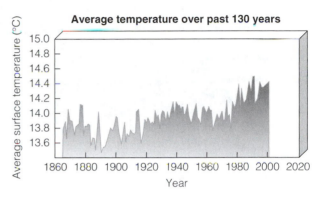

Figure 12-15 Science: estimated changes in the average global temperature of the atmosphere near the earth's surface over different periods of time. Although a particular place might have much lower or much higher readings than the troposphere's average temperature, such averages provide a valuable way to measure long-term trends. (Data from Goddard Institute for Space Studies, Intergovernmental Panel on Climate Change, National Academy of Sciences, National Aeronautics and Space Agency, National Center for Atmospheric Research, and National Oceanic and Atmospheric Administration)

climate change and project future changes. The IPCC network includes more than 2,000 climate experts from 70 nations.

In its 2001 report, the IPCC listed a number of findings indicating that *very likely* (90–99% probability) that the troposphere is getting warmer (Figure 12-15, bottom right). *First,* the 20th century was the hottest century in the past 1,000 years (Figure 12-15, bottom left). *Second,* since 1861 the average global temperature of the troposphere near the earth's surface has risen 0.6 C° (1.1 F°) over the entire globe and about 0.8 C°(1.4 F°) over the continents (Figure 12-15, bottom right). Most of this increase has taken place since 1980.

Third, the 10 warmest years since 1861 have occurred since 1990. In order, the five hottest years have been 1998, 2005, 2002, 2004, and 2001. *Fourth,* over the past 50 years Arctic temperatures have risen almost twice as fast as those in the rest of the world, with parts of Alaska and Siberia warming even faster.

Fifth, glaciers and floating sea ice in some parts of the world are melting and shrinking at increasing

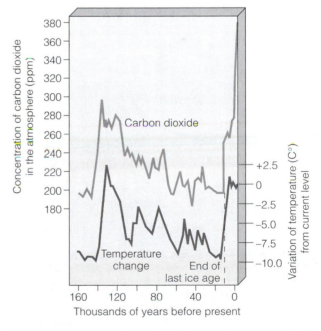

Figure 12-16 Science: atmospheric carbon dioxide levels and global temperature. Estimated long-term variations in average global temperature of the atmosphere near the earth's surface are graphed along with average tropospheric CO_2 levels over the past 160,000 years. The rough correlation between CO_2 levels in the troposphere and temperature shown in these estimates based on ice core data suggests a connection between the two variables. In 2005, an ice core showed that CO_2 levels in the troposphere are the highest they have been in 650,000 years. (Data from Intergovernmental Panel on Climate Change and National Center for Atmospheric Research)

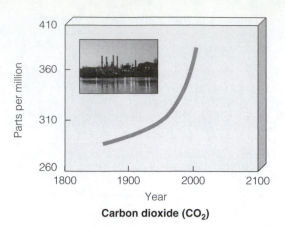

Carbon dioxide (CO₂)

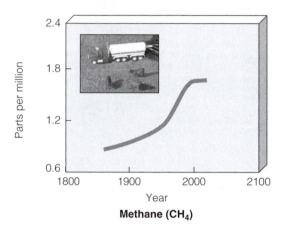

Methane (CH₄)

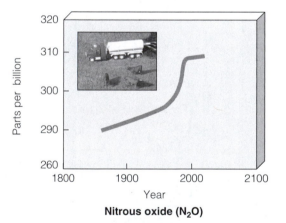

Nitrous oxide (N₂O)

ThomsonNOW™ Active Figure 12-17 Science: increases in average concentrations of three greenhouse gases—carbon dioxide, methane, and nitrous oxide—in the troposphere between 1860 and 2004, mostly because of fossil fuel burning, deforestation, and agriculture. The fluctuations in the CO_2 curve (top) reflect seasonal changes in photosynthetic activity, which cause small differences between summer and winter concentrations of CO_2. *See an animation based on this figure and take a short quiz on the concept.* (Data from Intergovernmental Panel on Climate Change, National Center for Atmospheric Research, and World Resources Institute)

rates. Sixth, warmer temperatures in Alaska and in other parts of the Arctic are melting permafrost, releasing more CO_2 and CH_4 into the troposphere, which may accelerate tropospheric warming. *Seventh*, during the last century the world's average sea level rose by 10–20 centimeters (4–8 inches), mostly because of runoff from melting land-based ice and the expansion of ocean water as its temperature increases.

The Scientific Consensus about Future Climate Change (Science)

There is strong evidence that human activities will play an important role in changing the earth's climate during this century.

To project the effects of increases in greenhouse gases on average global temperature, scientists develop complex *mathematical models* of the earth's climate system that simulate interactions among the earth's sunlight, clouds, landmasses, oceans and ocean currents, and concentrations of greenhouse gases and pollutants. Then they run the models on supercomputers and use the results to project future changes in the earth's average temperature. Figure 12-18 gives a greatly simplified summary of some interactions in the global climate system.

Such models provide projections of what is *very likely* (90–99% level of confidence) or *likely* (66–90% level of confidence) to happen to the average temperature of the troposphere. How well the results correspond to the real world depends on the assumptions and variables built into the model and the accuracy of the data used.

In 1990, 1995, and 2001, the IPCC published reports that evaluated how global temperatures have changed and are likely to change during this century. According to the 2001 report, "There is new and stronger evidence that most of the warming observed over the last 50 years is attributable to human activities." The 2001 report and more recent runs of various climate models suggest that it is *very likely* (90–99% level of confidence) that the earth's mean surface temperature will increase by 2.4–5.4C° (4.5°–9.7F°) between 2000 and 2100 (Figure 12-19, p. 270), with a most probable increase of 3.2C° (5.8F°)—a major increase in such a short period.

Global warming refers to temperature increases in the troposphere, which in turn can cause climate change. Natural changes, human activities, or both can cause such warming. **Global climate change** is a broader term that refers to changes in any aspects of the earth's climate, including temperature, precipitation, and storm intensity and patterns.

There is overwhelming consensus among the world's climate scientists that global warming is oc-

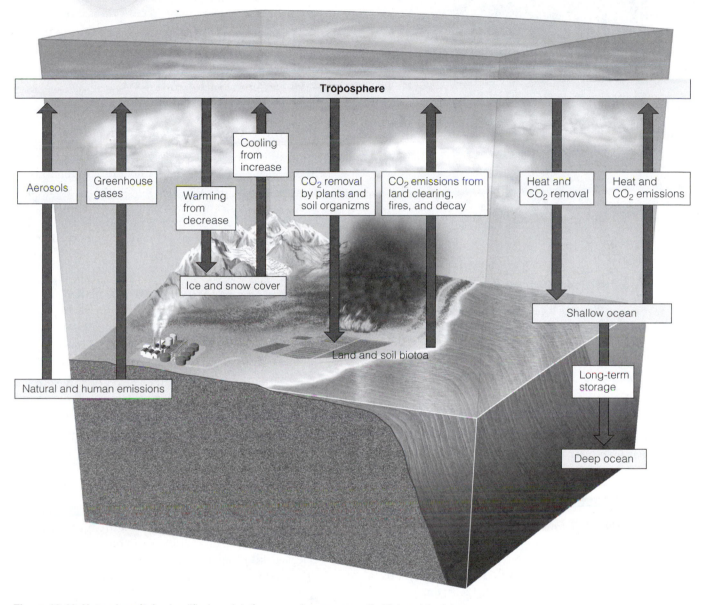

Figure 12-18 **Natural capital:** simplified model of some major processes that interact to determine the average temperature and greenhouse gas content of the troposphere and thus the earth's climate.

curring and that human activities are a major factor in this temperature increase. But as always in science, there is some uncertainty and a few scientists disagree with this consensus view.

In 2005, however, environmental writer Bill McKibben wrote: "It's about time for denial to come to an end. We're no longer talking about theory, about

computer models of what might happen. We're talking about what is happening, all around the world, with almost unimaginable speed."

✗ HOW WOULD YOU VOTE? Do you believe that we will experience significant global warming during this century? Cast your vote online at **www.thomsonedu.com/biology/miller**

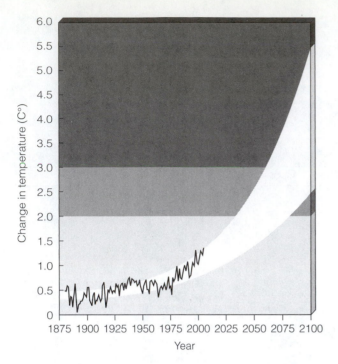

Figure 12-19 Natural capital degradation: comparison of measured changes in the average temperature of the atmosphere at the earth's surface between 1860 and 2005 and the projected range of temperature increase during the rest of this century. (Data from U.S. National Academy of Sciences, National Center for Atmospheric Research, and Intergovernmental Panel on Climate Change, Hadley Center for Climate Prediction and Research)

Why Should We Be Concerned about a Warmer Earth? (Science, Economics, and Ethics)

A rapid increase in the temperature of the troposphere during this century would give us little time to deal with its effects.

Climate scientists warn that the concern is not only about how much the temperature changes but about how rapidly it occurs. Most past changes in the temperature of the troposphere took place over thousands to a hundred thousand years (Figure 12-15, top left).

The problem we face is a significant increase in the temperature of the troposphere during this century (Figure 12-19). Such rapid change could drastically affect life on earth. A 2003 U.S. National Academy of Sciences report laid out a nightmarish *worst-case scenario* in which human activities or a combination of human activities and natural factors trigger a new and abrupt climate change. It describes ecosystems suddenly collapsing, low-lying cities being flooded, forests being consumed in vast fires, grasslands dying out and turning into dust bowls, much wildlife disappearing, more frequent and intensified coastal storms and hurricanes, and tropical waterborne and insect-transmitted infectious diseases spreading rapidly beyond their current ranges.

These possibilities were supported by a 2004 analysis carried out by Peter Schwartz and Doug Randall for the U.S. Department of Defense. They concluded that global warming "must be viewed as a serious threat to global stability and should be elevated beyond a scientific debate to a U.S. national security concern."

FACTORS AFFECTING THE EARTH'S TEMPERATURE

Possible Wild Cards (Science)

A number of factors could affect climate change.

Scientists have identified a number of natural and human-influenced factors that might *amplify* or *dampen* projected changes in the average temperature of the troposphere. Let us examine some possible wild cards that could slow or hasten climate change during this century.

Can the Oceans Store More CO₂ and Heat? (Science)

There is uncertainty about how much CO_2 and heat the oceans can remove from the troposphere and how long the heat and CO_2 might remain in the oceans.

The oceans help moderate the earth's average surface temperature by removing almost half of the excess CO_2 we pump into the atmosphere from fossil fuel combustion. The oceans also absorb heat from the atmosphere and slowly transfer some of it to the deep ocean, where it is removed from the climate system for long, but unknown, periods of time (Figure 12-18).

We do not know whether the oceans can absorb more CO_2. But the solubility of CO_2 in ocean water decreases with increasing temperature. Thus, if the oceans heat up, some of their dissolved CO_2 could be released into the atmosphere—like CO_2 bubbling out of a warm carbonated soft drink. This could amplify global warming. Scientific measurements show that the upper portion of the ocean has warmed by an average of 0.037 C° (0.067 F°) since 1955—an astounding increase considering the huge volume involved. Hence, we can expect global warming to increase the world's rainfall and humidity but to decrease snowfall and to melt some of the world's ice and snow, all of which are happening in parts of the world.

Effects of Cloud Cover (Science)

Warmer temperatures create more clouds that could warm or cool the troposphere.

A major unknown in global climate models is the effect of changes in the global distribution of clouds on the temperature of the troposphere. Warmer temperatures increase evaporation of surface water and create more clouds. These additional clouds can have a *warming effect*, by absorbing and releasing heat into the troposphere, or a *cooling effect* by reflecting more sunlight back into space.

The net result of these two opposing effects depends mostly on the type (thin or thick), coverage (continuous or discontinuous), and altitude of clouds. An increase in thick and continuous clouds at low altitudes can *decrease* surface warming by reflecting and blocking more sunlight. However, an increase in thin and discontinuous cirrus clouds at high altitudes can warm the lower troposphere and *increase* surface warming.

In addition, infrared satellite images indicate that the wispy condensation trails (contrails) left behind by jet planes might have a greater impact on the temperature of the troposphere than scientists once thought. NASA scientists found that jet contrails expand and turn into large cirrus clouds that tend to release heat into the upper troposphere. If these preliminary results are confirmed, emissions from jet planes could be responsible for as much as half of the tropospheric warming in the northern hemisphere.

What climate scientists know about the effects of clouds has been included in the latest climate models, but much uncertainty remains.

Effects of Outdoor Air Pollution (Science)

Aerosol pollutants and soot produced by human activities can warm or cool the atmosphere, but such effects will decrease with any decline in outdoor air pollution.

Aerosols (microscopic droplets and solid particles) of various air pollutants are released or formed in the troposphere by volcanic eruptions and human activities. They can either warm or cool the air depending on factors such as their size and the reflectivity of the underlying surface.

Most tropospheric aerosols, such as sulfate particles produced by fossil fuel combustion, tend to cool the atmosphere and thus can temporarily slow global warming. But a recent study by Mark Jacobson of Stanford University indicated that tiny particles of *soot* or *black carbon aerosols*—produced mainly from incomplete combustion in coal burning, diesel engines, and open fires—may be the second biggest contributor to global warming after the greenhouse gas CO_2.

Climate scientists do not expect aerosol pollutants to counteract or enhance projected global warming very much in the next 50 years for two reasons. *First*, aerosols and soot fall back to the earth or are washed out of the lower atmosphere within weeks or months, whereas CO_2 and other greenhouse gases remain in the atmosphere for decades to several hundred years. *Second*, aerosol inputs into the atmosphere are being reduced—especially in developed countries.

Effects of Higher CO_2 Levels on Photosynthesis (Science)

Increased CO_2 in the troposphere could increase plant photosynthesis, but several factors can limit or counter this effect.

Some studies suggest that more CO_2 in the atmosphere could increase the rate of photosynthesis in some areas with adequate water and soil nutrients. This would remove more CO_2 from the troposphere and help slow global warming.

However, recent studies indicate that this effect would be temporary for three reasons. *First*, the increase in photosynthesis would slow as the plants reach maturity and take up less CO_2 from the troposphere. *Second*, carbon stored by plants would be returned to the atmosphere as CO_2 when the plants die and decompose or burn. *Third*, a 2005 study found that increased photosynthesis decreased the amount of carbon stored in the soil.

Effects of a Warmer Troposphere on Methane Emissions (Science)

Warmer air can release methane gas stored in bogs, wetlands, and tundra soils and accelerate global warming.

Methane, the second most important greenhouse gas, after CO_2, is produced when organic matter decays in the absence of oxygen. Global warming could be accelerated by an increased release of methane from two major sources: freshwater wetlands and ice-like compounds called *methane hydrates* trapped beneath arctic permafrost. Significant amounts of methane would be released into the troposphere if much of the permafrost in tundra and boreal forest soils melts, as is occurring in parts of Canada, Alaska, China, and Mongolia. The resulting tropospheric warming could lead to more methane release and still more warming.

Another large contributor of methane is the world's 1.3 billion cattle that burp and pass gas. Methane is 25 times more potent per molecule as a greenhouse gas than CO_2, but it remains in the troposphere for only about 12 years, compared to about 60 years for CO_2.

EFFECTS OF GLOBAL WARMING

Effects of a Warmer Troposphere: An Overview (Science)

A warmer climate would have beneficial and harmful effects, but poor nations in the tropics would suffer the most.

A warmer global climate could have a number of harmful and beneficial effects (Figure 12-20, p. 272) for humans, other species, and ecosystems, depending

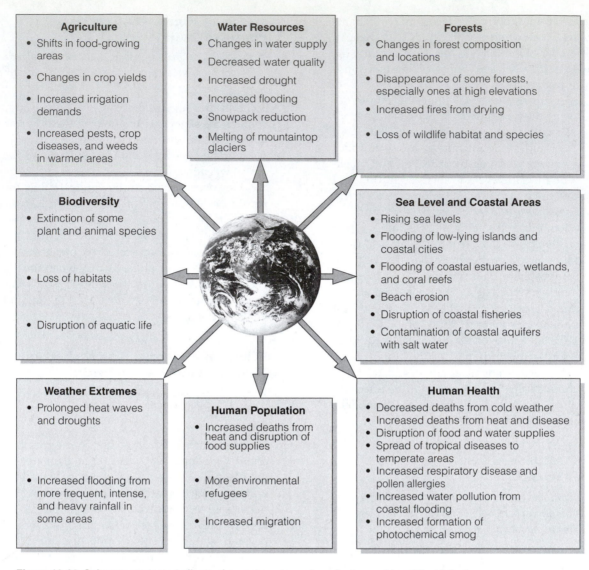

Agriculture
- Shifts in food-growing areas
- Changes in crop yields
- Increased irrigation demands
- Increased pests, crop diseases, and weeds in warmer areas

Water Resources
- Changes in water supply
- Decreased water quality
- Increased drought
- Increased flooding
- Snowpack reduction
- Melting of mountaintop glaciers

Forests
- Changes in forest composition and locations
- Disappearance of some forests, especially ones at high elevations
- Increased fires from drying
- Loss of wildlife habitat and species

Biodiversity
- Extinction of some plant and animal species
- Loss of habitats
- Disruption of aquatic life

Sea Level and Coastal Areas
- Rising sea levels
- Flooding of low-lying islands and coastal cities
- Flooding of coastal estuaries, wetlands, and coral reefs
- Beach erosion
- Disruption of coastal fisheries
- Contamination of coastal aquifers with salt water

Weather Extremes
- Prolonged heat waves and droughts
- Increased flooding from more frequent, intense, and heavy rainfall in some areas

Human Population
- Increased deaths from heat and disruption of food supplies
- More environmental refugees
- Increased migration

Human Health
- Decreased deaths from cold weather
- Increased deaths from heat and disease
- Disruption of food and water supplies
- Spread of tropical diseases to temperate areas
- Increased respiratory disease and pollen allergies
- Increased water pollution from coastal flooding
- Increased formation of photochemical smog

Figure 12-20 Science: projected effects of a warmer atmosphere for the world and the United States. Most of these effects could be harmful or beneficial depending on where one lives. Current models of the earth's climate cannot make reliable projections about where such effects might take place at a regional level and how long they might last. QUESTION: *Which five of these projected effects do you think are the most serious?* (Data from Intergovernmental Panel on Climate Change, U.S. Global Climate Change Research Program, and U.S. National Academy of Sciences)

mostly on their locations and on how rapidly the temperature changes.

Some areas will benefit because of less severe winters, more precipitation in dry areas, less precipitation in wet areas, and increased food production. Other areas will suffer harm from excessive heat, lack of water, and decreased food production. Here are a few of the effects projected by IPCC scientists based on current climate models.

The largest temperature increases will take place at the earth's poles and are likely to cause melting of glaciers and floating ice, which is now happening at an accelerating rate, alarming environmental scientists. The light-colored surfaces of ice and snow help cool the earth by reflecting 80–90% of incoming sunlight back into space. Melting of some of the earth's ice caps, float-

ing ice, and mountain glaciers exposes darker and less reflective surfaces of water and land, resulting in a warmer troposphere. As more ice melts, the troposphere becomes even warmer, which melts more ice and raises the temperature of the troposphere even more.

A related problem is a rise in global sea level, caused by runoff from the melting of land-based snow and ice and by the fact that water expands slightly when heated. During the last century the world's mean sea level rose about 19 centimeters (7.5 inches). But according to the 2001 IPCC report, the world's average sea level is *very likely* (90–99% certainty) to rise 9–88 centimeters (4–35 inches) during this century.

According to the IPCC, the highest projected rise in sea level of 88 centimeters (35 inches) would produce some dire effects. It could do the following:

- Threaten half the world's coastal estuaries, wetlands (one-third of those in the United States), and coral reefs.

- Disrupt many of the world's coastal fisheries.

- Flood low-lying barrier islands and cause gently sloping coastlines (especially along the U.S. East Coast) to erode and retreat inland by about 1.3 kilometers (0.8 mile).

- Flood agricultural lowlands and deltas in parts of Bangladesh, India, China, and Vietnam, where much of the world's rice is grown.

- Contaminate freshwater coastal aquifers with saltwater.

- Submerge some low-lying islands in the Pacific Ocean (the Marshall Islands), the Caribbean Sea, and the Indian Ocean (the Maldives, a chain of 1,200 small islands).

- Flood coastal areas with large human populations, including some of the world's largest cities. The U.S. Federal Emergency Management Agency (FEMA) estimates that one-fourth of the buildings within 153 meters (500 feet) of U.S. coastlines will be threatened by sea-level rise in the next 60 years. Within 50 years, 15–20% of low-lying Bangladesh will probably be underwater and perhaps one-third of this densely populated country will be submerged by 2100.

Because of the long lifetimes of greenhouse gases in the troposphere, if we magically stopped all greenhouse gas emissions from human activities today, the world's sea level would continue rising for at least 200 years, although at lower rates.

Global warming could also alter surface and deep ocean currents, which are connected and act like a gigantic conveyor belt to store CO_2 and heat in the deep sea and to transfer hot and cold water from the tropics to the poles (Figure 12-21). This helps keep much of the northern hemisphere fairly warm by pulling warm tropical water north, pushing cold water south, and releasing much of the heat stored in the water into the troposphere.

Scientists are concerned that in a warmer world an influx of freshwater from increased rain in the North Atlantic and thawing ice in the Arctic region might slow or disrupt this conveyor belt. If this happened, northern Europe and the northeast coast of North America could experience severe regional cooling. In other words, *global warming can lead to significant global cooling in some parts of the world,* with the climate of western Europe resembling that of Siberia.

A warmer climate could expand ranges and populations of some plant and animal species that can adapt to warmer climates. But this would include certain weeds, insect pests, and disease-carrying organisms. Other species will not fare so well. A 2004 report by the UN Environment Programme estimated that at

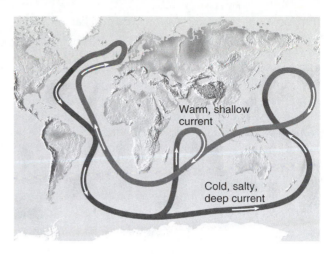

Figure 12-21 Natural capital: a connected loop of shallow and deep ocean currents stores CO_2 in the deep sea and transmits warm and cool water to various parts of the earth. It occurs when ocean water in the North Atlantic near Iceland is dense enough (because of its salt content and cold temperature) to sink to the ocean bottom, flow southward, and then move eastward to well up in the warmer Pacific. A shallower return current aided by winds then brings warmer and less salty—and thus less dense—water to the Atlantic. This water can cool and sink to begin the cycle again. A warmer planet would be a rainier one, which, coupled with melting glaciers, would increase the amount of freshwater flowing into the North Atlantic. This could slow or even jam this loop by diluting the saltwater and making it more buoyant (less dense) and less prone to sinking.

least 1 million species could face premature extinction by 2050 from global warming unless greenhouse gas emissions are drastically reduced.

The ecosystems *most likely* to suffer disruption and species loss are coral reefs, polar seas, coastal wetlands, arctic and alpine tundra, and high-elevation mountaintops. Shifts in regional climate would also threaten many parks, wildlife reserves, wilderness areas, and wetlands—wiping out the positive effects of current efforts to stem the loss of biodiversity.

In a warmer world, agricultural productivity may increase in some areas and decrease in others. For example, some agricultural production in the midwestern United States may shift northward to Canada. But overall food production could decrease because soils in midwestern Canada are generally less fertile than those in the midwestern United States. Crop production could also increase in Russia and Ukraine.

On the other hand, models predict a decline in agricultural productivity in tropical and subtropical regions, especially in Southeast Asia and Central America, where many of the world's poorest people live. And a decrease in high-elevation snowfall and glaciers could lead to a sharp decline in agricultural productivity in some heavily irrigated areas such as central and southern California.

What about the effects of global warming on people? According to the IPCC, the largest burden will fall on people and economies in poorer tropical and sub-tropical nations without the economic and technological resources needed to adapt to those harmful effects. A 2003 World Health Organization study estimated that climate change prematurely kills at least 160,000 people per year and that this number could double by 2030. By the end of this century, the annual death toll from global warming could be in the millions. And environmental scientist Norman Myers estimates that global warming during this century could produce at least 150 million and perhaps 250 million environmental refugees. See his Guest Essay on this topic on the website for this chapter.

DEALING WITH GLOBAL WARMING

Solutions: What Are Our Options? (Science, Economics, and Politics)

There is disagreement over what we should do about global warming.

There are two basic ways to deal with global warming. One is to reduce greenhouse gas emissions to slow down the rate of temperature increase and buy time to learn more about how the earth's climate system works and to shift to other noncarbon energy options. The other is to recognize that some warming is unavoidable and devise strategies to reduce its harmful effects. Most analysts believe we need a mix of both approaches.

There are two major schools of thought concerning what we should do now to reduce the effects of global warming. The first group calls for a *wait-and-see strategy*, with some scientists and economists calling for more research and a better understanding of the earth's climate system before making far-reaching and controversial economic and political decisions such as phasing out fossil fuels. The U.S. government currently advocates this approach.

A second and rapidly growing group of scientists, economists, business leaders, and political leaders (especially in the European Union) believe that we should *act now to reduce the risks from climate change brought about by global warming.* They argue that the potential for harmful economic, ecological, and social consequences is so great that action to slow the rate of change should not be delayed.

In 2005, national academies of sciences from the U.S., U.K., Germany, Italy, France, Russia, Japan, Canada, Brazil, China, and India issued an unprecedented joint statement saying that the scientific evidence on global climate change is clear enough for government leaders to commit to prompt action now. Any delay, they said, "will increase environmental damage and likely incur a greater cost."

In 1997, 2,700 economists led by eight Nobel laureates declared, "As economists, we believe that global climate change carries with it significant environmental, economic, social, and geopolitical risks and that preventive steps are justified."

X *HOW WOULD YOU VOTE?* Should we take serious action now to help slow global warming? Cast your vote online at www.thomsonedu.com/biology/miller.

Solutions: Reducing the Threat (Science, Economics, and Politics)

We can improve energy efficiency, rely more on carbon-free renewable energy resources, and find ways to keep much of the CO_2 we produce out of the troposphere.

Figure 12-22 presents a variety of prevention and cleanup solutions that climate analysts have suggested for slowing the rate and degree of global warming.

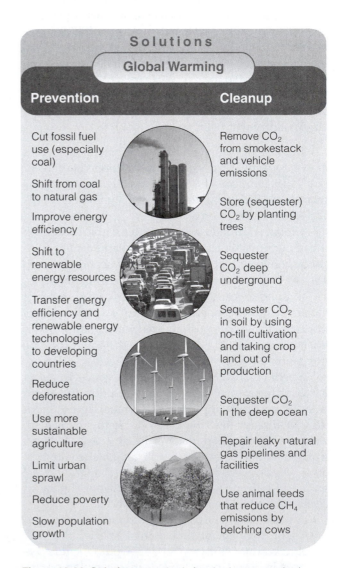

Solutions
Global Warming

Prevention	Cleanup
Cut fossil fuel use (especially coal)	Remove CO_2 from smokestack and vehicle emissions
Shift from coal to natural gas	
Improve energy efficiency	Store (sequester) CO_2 by planting trees
Shift to renewable energy resources	Sequester CO_2 deep underground
Transfer energy efficiency and renewable energy technologies to developing countries	Sequester CO_2 in soil by using no-till cultivation and taking crop land out of production
Reduce deforestation	
Use more sustainable agriculture	Sequester CO_2 in the deep ocean
Limit urban sprawl	Repair leaky natural gas pipelines and facilities
Reduce poverty	Use animal feeds that reduce CH_4 emissions by belching cows
Slow population growth	

Figure 12-22 Solutions: methods for slowing atmospheric warming during this century. QUESTION: *Which five of these mitigation solutions do you think are the most important?*

The efforts to reduce the impact of global warming come down to three major strategies: *improve energy efficiency to reduce fossil fuel use; shift from carbon-based fossil fuels to carbon-free renewable energy resources; and sequester as much CO_2 as possible in soil, vegetation, underground, and in the deep ocean*. The effectiveness of these three strategies would be enhanced by *reducing population*, which would decrease the number of fossil fuel consumers and CO_2 emitters, and by *reducing poverty*, which would decrease the need of the poor to clear more land for crops and fuelwood. However, the key solution to the climate change problem is similar to the solution to the air pollution problem: *stop using fossil fuels*.

X *HOW WOULD YOU VOTE?* Should we phase out the use of fossil fuels over the next fifty years? Cast your vote online at www.thomsonedu.com/biology/miller.

Removing and Storing CO_2 (Science and Economics)

We can prevent some of the CO_2 we produce from circulating in the troposphere, but the costs may be high and the effectiveness of various approaches is unknown.

Figure 12-23 shows several techniques for removing CO_2 from the troposphere or from smokestacks, and sequestering it in other parts of the environment. One way is to establish plantations of fast-growing trees such as pine and eucalyptus that remove CO_2 from the troposphere and store it in biomass. But this is a temporary approach because trees release their stored CO_2 back into the atmosphere when they die and decompose or if they are burned. And a 2005 study by a team of scientists found that plantations of pine and eucalyptus soak up lots of water and can dry up nearby streams and also change the salinity and acidity of the soil.

A second approach is to use plants such as switchgrass to remove CO_2 from the air and store it in the soil. But warmer temperatures can increase decomposition in soils and return some of the stored CO_2 to the atmosphere.

A third strategy is to *reduce the release of carbon dioxide and nitrous oxide from soil*. Ways to do this include *no-till cultivation* and setting aside depleted crop fields as conservation reserves.

A fourth approach is to remove CO_2 from smokestacks and *pump it deep underground* into unminable coal seams and abandoned oil fields, or *to inject it into*

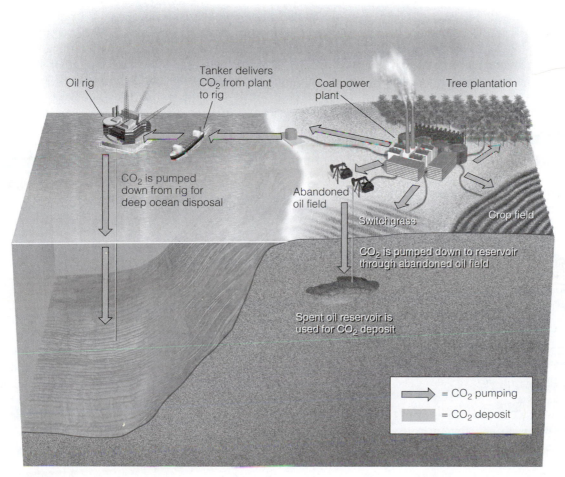

Figure 12-23
Solutions: methods for removing carbon dioxide from the atmosphere or from smokestacks and storing (sequestering) it in plants, soil, deep underground reservoirs, and the deep ocean. QUESTION: *Which two of these mitigation solutions do you think are the most effective?*

the deep ocean, as shown in Figure 12-23. There are several problems with this strategy. One is that current methods can remove only about 30% of the CO_2 from smokestack emissions and would increase the cost of producing coal-fired electricity by one-third to one-half, according to a 2005 IPCC study. In addition, the DOE estimates that the cost of sequestering carbon dioxide in various underground and deep-ocean repositories must be reduced by at least 90% to make this approach economically feasible. Finally, injecting large quantities of CO_2 into the ocean could upset the global carbon cycle, seawater acidity, and some forms of deep-sea life in unpredictable ways.

Government Roles in Reducing the Threat of Climate Change (Economics and Politics)

Governments can tax greenhouse gas emissions and energy use, increase subsidies and tax breaks for saving energy and using renewable energy, and decrease subsidies and tax breaks for fossil fuels.

Governments could use three major methods to promote the solutions listed in Figures 12-22 and 12-23. One is to phase in *carbon taxes* on each unit of CO_2 emitted by fossil fuels or *energy taxes* on each unit of fossil fuel that is burned. Decreasing taxes on income, labor, and profits to offset increases in consumption taxes on carbon emissions or fossil fuel use could help make such a strategy more politically acceptable.

A second strategy is to *level the economic playing field* by greatly increasing government subsidies for energy-efficiency, carbon-free renewable-energy technologies, carbon sequestration, and more sustainable agriculture, and by phasing out subsidies and tax breaks for using fossil fuels, nuclear power, and unsustainable agriculture.

The third strategy focuses on *technology transfer.* Governments of developed countries could help fund the transfer of energy-efficiency, carbon-free renewable-energy, carbon-sequestration, and more sustainable agriculture technologies to developing countries. Increasing the current tax on each international currency transaction by a quarter of a penny could finance this technology transfer, which would then generate wealth for developing countries.

International Climate Negotiations: The Kyoto Protocol (Politics and Economics)

Getting countries to agree on reducing their greenhouse gas emissions is difficult.

In December 1997, more than 2,200 delegates from 161 nations met in Kyoto, Japan, to negotiate a treaty to help slow global warming. The first phase of the resulting *Kyoto Protocol* went into effect on January 2005 with 157 countries (not including the United States and Australia) and the U.S states of California and Maine participating. It requires 38 developed countries to cut emissions of CO_2, CH_4, and N_2O to an average of at least 5.2% below their 1990 levels by 2012.

Developing countries were excluded from having to reduce their greenhouse gas emissions in the first phase because such reductions would curb much needed economic growth. The protocol also allows for trading greenhouse gas emissions among participating countries, which began in 2005. For example, a country or business that reduced its CO_2 emissions or planted trees would receive a certain number of credits. It can use these credits to avoid having to reduce its emissions in other areas, bank them for future use, or sell them to other countries or businesses.

Some analysts praise the Kyoto agreement as a small but important step in attempting to slow projected global warming and hope that rapidly growing nations such as China, Brazil, and India will agree to reduce their greenhouse gas in the second phase of the protocol, which is supposed to go into effect after 2012. Others see the agreement as a weak response to an urgent global problem.

In 2001, President George W. Bush withdrew U.S. participation from the Kyoto Protocol, arguing that it would harm the U.S. economy and did not require emissions reductions by developing countries such as China and India that have large and increasing emissions of greenhouse gases. This decision set off strong protests by many scientists, citizens, and leaders throughout most of the world. They pointed out that strong leadership from the United States is needed because it has the world's highest emissions of greenhouse gases.

X *HOW WOULD YOU VOTE?* Should the United States participate in the Kyoto Protocol? Cast your vote online at www.thomsonedu.com/biology/miller.

Moving beyond the Kyoto Protocol (Politics)

Countries could work together to develop a new international approach to slowing global warming.

In 2004, environmental law experts Richard B. Stewart and Jonathan B. Wiener proposed that countries work together to develop a new strategy for slowing global warming. They concluded that the Kyoto Protocol will have little effect on future global warming without support and action by the world's largest greenhouse gas emitters—the United States, China, and India.

They urge the development of a new climate treaty by the United States, China, India, Russia, Australia, Brazil, and other major greenhouse gas emitters. The treaty would also create an effective emissions trading program that includes developing countries omitted from such trading by the Kyoto Protocol. It would set achievable targets for reducing emissions for each 10 of the next 40 years, and evaluate global and national strategies for adapting to the harmful ecological and economic effects of global warming.

Actions by Some Countries, States, Cities, Businesses, Schools, and Individuals (Economics and Ethics)

Many countries, states, cities, companies, schools, and individuals are reducing their greenhouse gas emissions, improving energy efficiency, and increasing their use of carbon-free renewable energy.

Many countries are reducing their greenhouse gas emissions. For example, by 2000 Great Britain had reduced its CO_2 emissions to its 1990 level, well ahead of its Kyoto target goal. It did this mostly by relying more on natural gas than on coal, improving energy efficiency in industry and homes, and reducing gasoline use by raising the tax on gasoline. Between 2000 and 2050, Great Britain aims to cut its CO_2 emissions by 60%, mostly by improving energy efficiency and relying on renewable resources for 20% of its energy by 2030.

The Chinese are attempting to cut their greenhouse gas emissions by phasing out coal subsidies, shutting down inefficient coal-fired electric plants, stepping up its 20-year commitment to increase energy efficiency, and restructuring its economy to increase the use of renewable energy resources.

In 2005, New York, New Jersey, and five other northeastern U.S. states signed a mandatory agreement to reduce CO_2 emissions from their power plants by 10% by 2019. And since 1990, local governments in more than 500 cities around the world (including 165 in the United States) have established programs to reduce their greenhouse gas emissions. In addition, a growing number of major global companies, such as Alcoa, DuPont, IBM, Toyota, General Electric, British Petroleum (BP), and Shell have established targets to reduce, by 2010, their greenhouse-gas emissions by 10–65% from 1990 levels.

Students and faculty at Oberlin College in the U.S. state of Ohio have asked their board of trustees to reduce the college's CO_2 emissions to zero by 2020 by buying renewable energy or producing it. Twenty-five Pennsylvania colleges have joined to purchase wind power and other forms of carbon-free renewable energy. What is your school doing to help slow global warming?

Figure 12-24 lists some things you can do to cut CO_2 emissions.

What Can You Do?
Reducing CO₂ Emissions

- Drive a fuel-efficient car, walk, bike, carpool, and use mass transit
- Use energy-efficient windows
- Use energy-efficient appliances and lights
- Heavily insulate your house and seal all drafts
- Reduce garbage by recycling and reuse
- Insulate hot water heater
- Use compact fluorescent bulbs
- Plant trees to shade your house during summer
- Set water heater no higher than 49°C (120°F)
- Wash laundry in warm or cold water
- Use a low-flow shower head

Figure 12-24 Individuals matter: ways to reduce your annual emissions of CO_2. QUESTION: *Which three of these actions do you think are the most important?*

Solutions: Preparing for Global Warming (Science, Economics, and Ethics)

Many countries and cities are looking for ways to cope with the harmful effects of climate change.

According to the latest global climate models, the world needs to reduce 1990 emissions of greenhouse gases by 30–60% to stabilize their concentrations in the troposphere by 2050 (some say by 2020). Such a large reduction in emissions is unlikely because it would require widespread changes in industrial processes, energy sources, transportation options, and individual lifestyles.

As a result, many analysts suggest we should begin preparing for the possible harmful effects of long-term atmospheric warming and climate change, mostly for long-term economic and ethical reasons. Figure 12-25 (p. 278) shows some ways to implement this *adaptation strategy.*

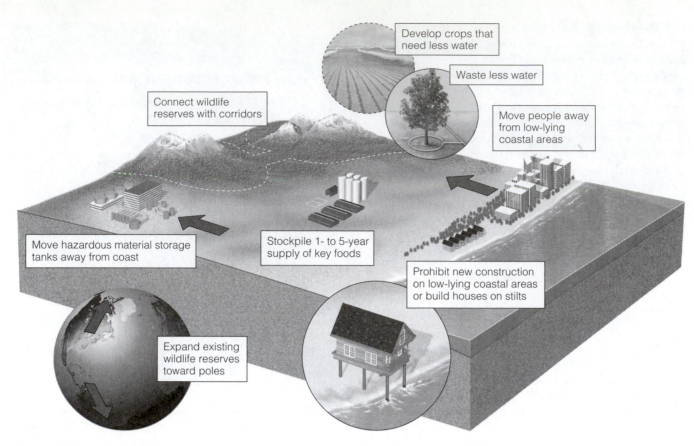

Figure 12-25 Solutions: ways to prepare for the possible long-term effects of climate change. QUESTION: *Which three of these adaptation solutions do you think are the most important?*

OZONE DEPLETION IN THE STRATOSPHERE

Threats to Ozone Levels in the Stratosphere (Science)

Less ozone in the stratosphere would allow more harmful UV radiation to reach the earth's surface.

A layer of ozone in the lower stratosphere (Figure 12-1) keeps about 95% of the sun's harmful UV radiation from reaching the earth's surface. Measuring instruments on balloons, aircraft, and satellites show considerable seasonal depletion (thinning) of ozone concentrations in the stratosphere above Antarctica and the Arctic. Similar measurements reveal a lower overall thinning everywhere except over the tropics.

Based on these measurements and mathematical and chemical models, the overwhelming consensus of researchers in this field is that ozone depletion (thinning) in the stratosphere is a serious threat to humans, other animals, and some of the plants and one-celled phytoplankton that use sunlight to support the earth's food webs.

What Causes Ozone Depletion? (Science)

Widespread use of several long-lived chemicals has reduced ozone levels in the stratosphere.

Thomas Midgley, Jr., a General Motors chemist, discovered the first chlorofluorocarbon (CFC) in 1930. Chemists soon developed similar compounds to create a family of highly useful CFCs, known by their trade name as Freons.

These chemically unreactive, odorless, nonflammable, nontoxic, and noncorrosive compounds seemed to be dream chemicals. Inexpensive to manufacture, they became popular as coolants in air conditioners and refrigerators, propellants in aerosol spray cans, cleaners for electronic parts such as computer chips, fumigants for granaries and ship cargo holds, and bubbles in plastic foam used for insulation and packaging.

But it turned out that CFCs were too good to be true. In 1974, calculations by chemists Sherwood Rowland and Mario Molina at the University of California–Irvine indicated that CFCs were lowering the average concentration of ozone in the stratosphere. They shocked both the scientific community and the $28-billion-per-year CFC industry by calling for an immediate ban of CFCs in spray cans (for which substitutes were available).

Rowland and Molina's research led them to four major conclusions. *First*, CFCs remain in the troposphere because they are insoluble in water and chemically unreactive. *Second*, over 11–20 years these heavier-than-air compounds rise into the stratosphere—mostly through convection, random drift, and the turbulent mixing of air in the troposphere. While in the troposphere they act as greenhouse gases.

Third, once they reach the stratosphere, the CFC molecules break down under the influence of high-energy UV radiation. This releases highly reactive chlorine atoms (Cl), as well as atoms of fluorine (F) and bromine (Br) from other related compounds. These atoms accelerate the breakdown of ozone (O_3) into O_2 and O in a cyclic chain of chemical reactions. As a consequence, ozone is destroyed faster than it forms in some parts of the stratosphere.

Finally, each CFC molecule can last in the stratosphere for 65–385 years, depending on its type. During that time, each chlorine atom released during the breakdown of CFC can convert hundreds of molecules of O_3 to O_2.

Overall, according to Rowland and Molina's calculations and later models and atmospheric measurements of CFCs in the stratosphere, these dream molecules had turned into global ozone destroyers.

The CFC industry (led by DuPont), a powerful, well-funded adversary with a lot of profits and jobs at stake, attacked Rowland and Molina's calculations and conclusions. The two researchers held their ground, expanded their research, and explained their results to other scientists, elected officials, and the media. After 14 years of delaying tactics, DuPont officials acknowledged in 1988 that CFCs were depleting the ozone layer and agreed to stop producing them once they found substitutes.

In 1995, Rowland and Molina received the Nobel Prize in chemistry for their work. In awarding the prize, the Royal Swedish Academy of Sciences said that they contributed to "our salvation from a global environmental problem that could have catastrophic consequences."

Measurements and models indicate that 75–85% of the observed ozone losses in the stratosphere since 1976 are the result of CFCs and other ozone-depleting chemicals (ODCs) released into the atmosphere by human activities beginning in the 1950s.

Annual Drops in Ozone Levels over the Earth's Poles (Science)

During four months of each year, up to half of the ozone in the stratosphere over Antarctica and a smaller amount over the Arctic is depleted.

In 1984, researchers analyzing satellite data discovered that 40–50% of the ozone in the upper stratosphere over Antarctica disappeared during October and November. This observed loss of ozone above Antarctica has been called an *ozone hole*. A more accurate term is *ozone thinning* because the ozone depletion varies with altitude and location. In some years, the total area involved has been greater than the area of North America.

When partial sunlight returns to Antarctica in October, huge masses of ozone-depleted air there flow northward and linger for a few weeks over parts of Australia, New Zealand, South America, and South Africa. This raises biologically damaging UV-B levels in these areas by 3–10% and in some years as much as 20%.

In 1988, scientists discovered that similar but usually less severe ozone thinning occurs over the Arctic from February to June, with a typical ozone loss of 11–38% (compared to a typical 50% loss above Antarctica). When this mass of air above the Arctic breaks up each spring, large masses of ozone-depleted air flow south to linger over parts of Europe, North America, and Asia.

Models indicate that the Arctic is unlikely to develop the large-scale ozone thinning found over the Antarctic. They also project ozone depletion over the Antarctic and the Arctic will be at its worst between 2010 and 2019.

Why Should We Be Worried about Ozone Depletion? (Science)

Increased UV radiation reaching the earth's surface from ozone depletion in the stratosphere is harmful to human health, crops, forests, animals, and materials such as paints and plastics.

Why should we care about ozone loss? Figure 12-26 (p. 280) lists some of the expected effects of decreased levels of ozone in the stratosphere. With less ozone in the stratosphere, more biologically damaging UV-A and UV-B radiation will reach the earth's surface. This will give people worse sunburns, more eye cataracts (a clouding of the eye's lens that reduces vision and can cause blindness if not corrected), and more skin cancers (Case Study, p. 280).

Natural Capital Degradation

Effects of Ozone Depletion

Human Health

- Worse sunburn
- More eye cataracts
- More skin cancers
- Immune system suppression

Food and Forests

- Reduced yields for some crops
- Reduced seafood supplies from reduced phytoplankton
- Decreased forest productivity for UV-sensitive tree species

Wildlife

- Increased eye cataracts in some species
- Decreased population of aquatic species sensitive to UV radiation
- Reduced population of surface phytoplankton
- Disrupted aquatic food webs from reduced phytoplankton

Air Pollution and Materials

- Increased acid deposition
- Increased photochemical smog
- Degradation of outdoor paints and plastics

Global Warming

- Accelerated warming because of decreased ocean uptake of CO_2 from atmosphere by phytoplankton and CFCs acting as greenhouse gases

Figure 12-26 Natural capital degradation: expected effects of decreased levels of ozone in the stratosphere. QUESTION: *Which five of these effects do you think are the most important?*

Case Study: Skin Cancer (Science)

Exposure to UV radiation is a major cause of skin cancers.

Research indicates that years of exposure to UV-B ionizing radiation in sunlight is the primary cause of *squamous cell* and *basal cell skin cancers* that together account for 95% of all skin cancers. Typically, a 15- to 40-year lag separates excessive exposure to UV-B and development of these cancers.

Caucasian children and adolescents who experience only a single severe sunburn double their chances of getting these two types of cancers. Some 90–95% of these types of skin cancer can be cured if detected early enough, although their removal may leave disfiguring scars. These cancers kill 1–2% of their victims, which amounts to about 2,300 deaths in the United States each year.

A third type of skin cancer, *malignant melanoma,* occurs in pigmented areas such as moles anywhere on the body. Within a few months, this type of cancer can spread to other organs. It kills about one-fourth of its victims (most under age 40) within 5 years, despite surgery, chemotherapy, and radiation treatments. Each year it kills about 100,000 people (including 7,700 Americans), mostly Caucasians.

A 2003 study found that women who used tanning parlors once a month or more increased their chances of developing malignant melanoma by 55%. The risk was highest for young adults. And a 2004 study at Dartmouth College found that people using tanning beds were more likely to develop basal cell and squamous cell skin cancers.

People (especially Caucasians) who experience three or more blistering sunburns before age 20 are five times more likely to develop malignant melanoma than those who have never had severe sunburns. About one of every ten people who get malignant melanoma has an inherited gene that makes them especially susceptible to the disease. Figure 12-27 lists ways for you to protect yourself from harmful UV radiation.

What Can You Do?

Reducing Exposure to UV-Radiation

- Stay out of the sun, especially between 10 A.M. and 3 P.M.
- Do not use tanning parlors or sunlamps.
- When in the sun, wear protective clothing and sunglasses that protect against UV-A and UV-B radiation.
- Be aware that overcast skies do not protect you.
- Do not expose yourself to the sun if you are taking antibiotics or birth control pills.
- Use a sunscreen with a protection factor of 15 or 30 anytime you are in the sun if you have light skin.
- Examine your skin and scalp at least once a month for moles or warts that change in size, shape, or color or sores that keep oozing, bleeding, and crusting over. If you observe any of these signs, consult a doctor immediately.

Figure 12-27 Individuals matter: ways to reduce your exposure to harmful UV radiation. QUESTION: *Which three of these actions do you think are the most important?*

PROTECTING THE OZONE LAYER

How Can We Protect the Ozone Layer? (Science and Politics)

To reduce ozone depletion, we must stop producing ozone-depleting chemicals.

According to researchers in this field, we should immediately stop producing all ozone-depleting chemicals (ODCs). However, even with immediate action, models indicate it will take about 60 years for the ozone layer to return to 1980 levels and at least 100 years for recovery to pre-1950 levels. *Good news:* Substitutes are available for most uses of CFCs, and others are being developed (Individuals Matter, at right).

In 1987, representatives of 36 nations meeting in Montreal, Canada, developed the *Montreal Protocol.* This treaty's goal was to cut emissions of CFCs (but not other ODCs) by about 35% between 1989 and 2000. After hearing more bad news about seasonal ozone thinning above Antarctica in 1989, representatives of 93 countries met in London in 1990 and in Copenhagen, Denmark, in 1992. They adopted the *Copenhagen Protocol,* an amendment that accelerated the phaseout of key ODCs.

These landmark international agreements, now signed by 177 countries, are important examples of global cooperation in response to a serious global environmental problem based on applying the precautionary principle. Without them, ozone depletion would be a much more serious threat. If nations continue to follow these agreements, ozone levels should return to 1980 levels by 2065 and to 1950 levels by 2100.

The ozone protocols set an important precedent by using *prevention* to solve a serious environmental problem. Nations and companies agreed to work together to solve this global problem for three reasons. *First,* there was convincing and dramatic scientific evidence of a serious problem. *Second,* CFCs were produced by a small number of international companies. *Third,* the certainty that CFC sales would decline over a period of years unleashed the economic and creative resources of the private sector to find even more profitable substitute chemicals.

According to a 1998 study by the World Meteorological Organization, ozone depletion in the stratosphere has been cooling the troposphere and has helped offset or disguise as much as 30% of the global warming from our emissions of greenhouse gases. Thus, restoring the ozone layer could lead to an increase in global warming. But the alternative is worse.

The four principles of sustainability can be used to help reduce the problems of global warming and stratospheric ozone depletion. We can reduce inputs of greenhouse gases and ODCs into the troposphere by relying more on direct and indirect forms of solar energy than on fossils fuels; reducing the waste of matter

INDIVIDUALS MATTER

Ray Turner and His Refrigerator

Ray Turner, an aerospace manager at Hughes Aircraft in California, made an important low-tech, ozone-saving discovery by using his head—and his refrigerator.

His concern for the environment led him to look for a cheap and simple substitute for the CFCs used as cleaning agents to remove films of oxidation from the electronic circuit boards manufactured at his plant.

Turner started by looking in his refrigerator. He decided to put drops of various substances on a corroded penny to see whether any of them removed the film of oxidation. Then he used his soldering gun to see whether solder would stick to the surface of the penny, indicating the film had been cleaned off.

First, he tried vinegar. No luck. Then he tried some ground-up lemon peel, also a failure. Next he tried a drop of lemon juice and watched as the solder took hold. The rest, as they say, is history.

Today, Hughes Aircraft uses inexpensive CFC-free citrus-based solvents to clean circuit boards. This new cleaning technique has reduced circuit board defects by about 75% at Hughes. And Turner got a hefty bonus. Now other companies, such as AT&T, clean computer boards and chips using acidic chemicals extracted from cantaloupes, peaches, and plums. Maybe you can find a solution to an environmental problem in your refrigerator, grocery store, drugstore, or backyard.

and energy resources and recycling and reusing matter resources; relying on diverse strategies, including finding substitutes for ODCs, increasing energy efficiency, using a variety of carbon-free renewable energy resources, and emphasizing pollution prevention; and reducing human population growth and wasteful resource consumption per person.

The atmosphere is the key symbol of global interdependence. If we can't solve some of our problems in the face of threats to this global commons, then I can't be very optimistic about the future of the world.

MARGARET MEAD

CRITICAL THINKING

1. Identify climate and topographic factors in your local community that **(a)** intensify air pollution and **(b)** help reduce air pollution.

2. Do you agree or disagree with the possible weaknesses of the U.S. Clean Air Acts listed on p. 263? Defend each of your positions. Can you identify other weaknesses?

3. Explain why you agree or disagree with each of the proposals listed in Figure 12-14 (p. 266) for shifting the emphasis to preventing air pollution over the next several decades. Which two of these proposals do you think are the most important?

4. In preparation for the 1992 UN Conference on the Human Environment in Rio de Janeiro, President George H. W. Bush's top economic adviser addressed representatives of governments from a number of countries in Williamsburg, Virginia. He told his audience not to worry about global warming because the average temperature increases scientists are predicting were much less than the temperature increase he experienced in coming from Washington, D.C., to Williamsburg. What is the fundamental flaw in this reasoning?

5. What changes might occur in **(a)** the global hydrologic cycle (Figure 2-25, p. 45) and **(b)** the global carbon cycle (Figure 2-26, p. 46) if the atmosphere experiences significant warming? Explain.

6. Of the two schools of thought on what should be done about possible global warming (p. 274), which do you favor? Explain.

7. Of the proposals in Figure 12-22 (p. 274) for reducing emissions of greenhouse gases into the troposphere, with which, if any, do you disagree? Why?

8. What are three consumption patterns or other features of your lifestyle that directly add greenhouse gases to the atmosphere? Explain how each of these features of your lifestyle violates one or more of the four principles of sustainability (Figure 1-11, p. 17). Which, if any, of these things would you be willing to give up to slow global warming and reduce other forms of air pollution?

9. Congratulations! You are in charge of the world. List your three most important actions for dealing with the problems of **(a)** air pollution, **(b)** global warming, and **(c)** depletion of ozone in the stratosphere.

10. List two questions that you would like to have answered as a result of reading this chapter.

LEARNING ONLINE

The website for this book contains helpful study aids and many ideas for further reading and research. They include a chapter summary, review questions for the entire chapter, flash cards for key terms and concepts, a multiple-choice practice quiz, interesting Internet sites, references, information about green careers, and a guide for accessing thousands of InfoTrac® College Edition articles. Log into

www.thomsonedu.com/biology/miller

Then choose Chapter 12, and select a learning resource. For access to animations, additional quizzes, chapter outlines and summaries, register and log into

at **www.thomsonedu.com** using the access code card in the front of your book.

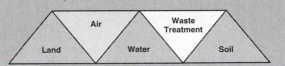

13 SOLID AND HAZARDOUS WASTE

Solid wastes are only raw materials we're too stupid to use.

ARTHUR C. CLARKE

WASTING RESOURCES

Types of Waste (Science)

We will always produce some waste, but we can produce much less.

In nature, there is essentially no waste because the wastes of one organism become nutrients for other organisms, according to a principle of sustainability. Humans, on the other hand, produce huge amounts of waste that go unused and pollute the environment.

One major category of waste is **solid waste**—any unwanted or discarded material that is not a liquid or a gas. Another type is **hazardous waste,** or **toxic waste,** which threatens human health or the environment because it is toxic, chemically active, corrosive, or flammable. Solid waste can be divided into two types: **municipal solid waste**—often called garbage or trash—produced directly by homes and workplaces; and **industrial solid waste** produced indirectly by mines, factories, refineries, food growers, and businesses that supply people with goods and services.

There are two reasons for being concerned about the amount of solid waste we produce directly and indirectly. One is that at least three-fourths of it represents an unnecessary waste of the earth's precious resources. The other is that in producing the solid products we use and often discard, we create huge amounts of air pollution (including greenhouse gases), water pollution, land degradation, and more solid and hazardous waste.

Solid Waste in the United States: Affluenza in Action

The United States produces about a third of the world's solid waste and buries more than half of it in landfills.

The United States leads the world in producing solid waste. With only 4.6% of the world's population, the United States produces about one-third of the world's solid waste—a glaring symptom of affluenza (p. 13). About 98.5% of U.S. solid waste is *industrial solid waste* from mining, oil and natural gas production, agriculture, and industrial activities that provide consumers with goods and services (Figure 13-1).

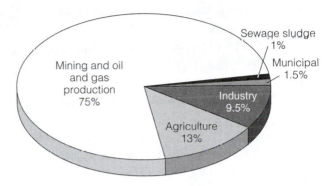

Figure 13-1 Natural capital degradation: sources of the estimated 11 billion metric tons (12 billion tons) of solid waste produced each year in the United States. Mining, oil and gas production, agricultural, and industrial activities produce 65 times as much solid waste as household activities. (Data from U.S. Environmental Protection Agency and U.S. Bureau of Mines)

The remaining 1.5% of solid waste is *municipal solid waste* or MSW. This small percentage of the overall solid waste problem is still huge. Each year, the United States generates enough MSW to fill a bumper-to-bumper convoy of garbage trucks encircling the globe almost eight times!

The United States also leads the world in trash production (by weight) per person. Americans produce about twice as much solid waste per person as do other industrial countries such as Japan, Germany, and France and about 5 to 10 times more per person than most developing countries do. Each day the average American produces over 2.0 kilograms (4.5 pounds) of MSW—70% more than in 1960. Paper makes up about 38% of the trash buried in U.S. landfills, followed by yard waste (12%), food waste (11%), and plastics (11%). Since 1990, the weight of MSW per American has leveled off, mostly because of increased recycling and the use of lighter products.

According to the U.S. Environmental Protection Agency (EPA), about 55% of the MSW produced in the United States is dumped in landfills, 30% is recycled or composted, and 15% is burned in incinerators.

Here are a few startling facts about the solid wastes consumers throw away in the high-waste economy of the United States:

- Enough aluminum to rebuild the country's entire commercial airline fleet every 3 months

- Enough tires each year to encircle the planet almost three times

- Enough disposable diapers per year that if linked end to end would reach to the moon and back seven times

- Discarded carpet each year that would cover the U.S. state of Delaware

- About 2.5 million nonreturnable plastic bottles every hour

- About 25 billion throwaway Styrofoam cups per year used mostly for drinking coffee

- About 25 million metric tons (27 million tons) of edible food per year

- Enough office paper each year to build a wall 3.5 meters (11 feet) high across the country from New York City to San Francisco, California

- Some 186 billion pieces of junk mail (an average of 660 pieces per American) each year, about 45% of which are thrown in the trash unopened

A special problem is *electronic waste* or *e-waste*, the fastest growing solid waste problem in the United States and the world. Each year Americans discard an estimated 130 million cell phones, 50 million computers, and 8 million television sets. E-waste is a source of toxic and hazardous wastes including compounds containing lead, mercury, and cadmium that can contaminate the air, surface water, groundwater, and soil.

According to a 2005 report by the Basel Action Network, about 50–80% of U.S. e-waste is shipped to China, India, Pakistan, Nigeria, and other developing countries where labor is cheap and environmental regulations are weak. Workers there, many of them children, dismantle the products to recover reusable parts and are thus exposed to toxic metals such as lead, mercury, and cadmium. The remaining scrap is dumped in waterways and fields, or burned in open fires, which exposes the workers to toxic dioxins.

The European Union (EU) leads the way in dealing with e-waste. It requires manufacturers to take back electronic products (as well as refrigerators and other appliances) at the end of their useful lives for repair, remanufacture, or recycling and bans e-waste in MSW. Japan is also adopting such cradle-to-grave standards for electronic devices and appliances.

INTEGRATED WASTE MANAGEMENT

Solutions: Waste Management—Prevention versus Waste Reduction (Science)

We can manage the solid wastes we produce or reduce or prevent their production.

We can deal with solid wastes in two ways. One is *waste management*—a *high-waste approach* (Figure 4-10, p. 82) that attempts to manage the wastes in ways that reduce environmental harm. In effect, it mixes the wastes we produce together and then transfers them from one part of the environment to another, usually by burying them, burning them, or shipping them off to another state or country.

The second approach is *waste reduction*, a *low-waste approach* that views most solid waste as potential resources that we should be reusing, recycling, or composting. With this approach, we should be taught to think of trash cans and garbage trucks as *resource containers* on their way to recycling or composting facilities.

There is no single solution to the solid waste problem. Most analysts call for using **integrated waste**

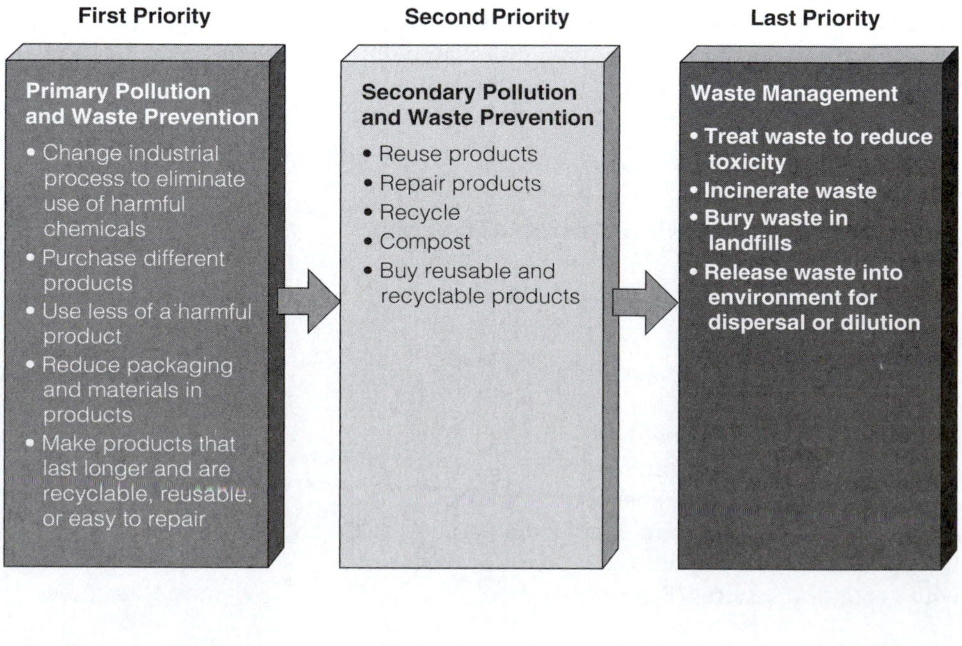

First Priority

Primary Pollution and Waste Prevention
- Change industrial process to eliminate use of harmful chemicals
- Purchase different products
- Use less of a harmful product
- Reduce packaging and materials in products
- Make products that last longer and are recyclable, reusable, or easy to repair

Second Priority

Secondary Pollution and Waste Prevention
- Reuse products
- Repair products
- Recycle
- Compost
- Buy reusable and recyclable products

Last Priority

Waste Management
- Treat waste to reduce toxicity
- Incinerate waste
- Bury waste in landfills
- Release waste into environment for dispersal or dilution

Figure 13-2 Integrated waste management: priorities suggested by prominent scientists for dealing with solid waste. To date, these waste-reduction priorities have not been followed in the United States and in most other countries. Instead, most efforts are devoted to waste management (bury it or burn it). (Data from U.S. Environmental Protection Agency and U.S. National Academy of Sciences)

management—a variety of strategies for both waste reduction and waste management to deal with the solid wastes we produce (Figure 13-2).

Currently, the order of priorities for dealing with solid waste in the United States and in most countries is the reverse of the order suggested in Figure 13-2 by prominent scientists. But some scientists and economists estimate that 75–90% of the solid waste we produce can be eliminated by a combination of *reducing waste production, reusing and recycling materials* (including composting), and *redesigning* manufacturing processes and buildings to produce less waste.

Solutions: Reducing Solid Waste (Science)

Reducing consumption and redesigning the products we produce are the best ways to cut waste production and promote sustainability.

Here are eight ways in which societies can reduce resource use, waste, and pollution, based on the 5 Rs—reduce, reuse, refuse, recycle, and redesign. First, *consume less.* Before buying anything, ask questions such as: Do I really *need* this or do I just *want* it? Can I buy it secondhand, or borrow or rent it?

Second, *redesign manufacturing processes and products to use less material and energy.* For example, the weight of cars has been reduced by about one-fourth by using such steel along with lightweight plastics and composite materials.

Third, *redesign manufacturing processes to produce less waste and pollution.* Most toxic organic solvents can be recycled within factories or replaced with water-based or citrus-based solvents, for example (Individuals Matter, p. 281).

Fourth, *develop products that are easy to repair, reuse, remanufacture, compost, or recycle.* A new Xerox photocopier with every part reusable or recyclable for easy remanufacturing should eventually save Xerox $1 billion in manufacturing costs.

Fifth, *shift from selling goods to selling the services they provide.* For example, instead of selling photocopiers, sell photocopying services, thereby saving materials and energy used to manufacture more copiers.

Sixth, *eliminate or reduce unnecessary packaging.* Use the following hierarchy for packaging: *no packaging, minimal packaging, reusable packaging,* and *recyclable packaging.*

Seventh, *use fee-per-bag* waste collection systems that charge consumers according to the amount of waste they throw away but provide free pickup of recyclable items.

Eighth, establish *cradle-to-grave responsibility* laws requiring companies to take back consumer products such as electronic equipment, appliances, and motor vehicles, as many European countries and Japan do.

What Can You Do?
Solid Waste

- Follow the four R's of resource use: Refuse, Reduce, Reuse, and Recycle.

- Ask yourself whether you relly need a particular item.

- Rent, borrow, or barter goods and services when you can.

- Buy things that are reusable, recyclable, or compostable, and be sure to reuse, recycle, and compost them.

- Do not use throwaway paper and plastic plates, cups, and eating utensils, and other disposable items when reusable or refillable versions are available.

- Use e-mail in place of conventional paper mail.

- Read newspapers and magazines online.

- Buy products in concentrated form whenever possible.

Figure 13-3 Individuals matter: ways to save resources and reduce your output of solid waste and pollution. QUESTION: *Which three of these actions do you think are the most important? Which things in this list do you now do or plan to do?*

Figure 13-3 lists some ways you can reduce your output of solid waste.

REUSE

How People Reuse Materials (Science)

Reusing products is an important way to reduce resource use, waste, and pollution in developed countries, but it can create hazards for the poor in developing countries.

In today's high-throughput societies, many goods are made for one-time use and are then discarded. We have increasingly substituted throwaway tissues for reusable handkerchiefs, disposable paper towels and napkins for reusable cloth ones, throwaway paper plates, cups, and plastic utensils for reusable plates, cups, and silverware, and throwaway beverage containers for refillable ones.

Reuse involves cleaning and using materials over and over and thus increasing the typical life span of a product. This form of waste reduction reduces the use of matter and energy resources, cuts pollution and waste, creates local jobs, and saves money. Traditional forms of reuse include salvaging automobile parts from older cars in junkyards and recovering bricks, doors, fine woodwork, and other items from old houses and buildings. Other strategies involve using yard sales, flea markets, secondhand stores, traditional and online auctions, and classified newspaper ads to sell and buy used items.

Reuse is alive and well in most developing countries, but it can be a health hazard for the poor who scavenge in open dumps for food scraps and items they can reuse or sell. They can be exposed to toxins and infectious diseases. And workers—many of them children—dismantling e-wastes for parts that can be reused or recycled can be exposed to toxic chemicals.

Case Study: Using Refillable Containers (Science and Economics)

Refilling and reusing containers uses less resources and energy, produces less waste, saves money, and creates local jobs.

Two examples of reuse are refillable glass beverage bottles and refillable soft drink bottles made of polyethylene terephthalate (PET) plastic. Typically, such bottles make 15 round-trips before they become too damaged for reuse and then are recycled. Reusing these containers stimulates local economies by creating local jobs related to their collection and refilling. Moreover, studies by Coca-Cola and PepsiCo of Canada show that their soft drinks in 0.5-liter (16-ounce) bottles cost one-third less in refillable bottles than in throwaway bottles.

But big companies make more money by producing and shipping throwaway beverage and food containers at centralized facilities. This shift has put many small local bottling companies, breweries, and canneries out of business.

Some call for banning all beverage containers that cannot be reused, as Denmark and Canada's Prince Edward Island have done. To encourage use of refillable glass bottles, Ecuador levies a refundable beverage container deposit fee that amounts to 50% of the cost of the drink. In Finland, 95% of the soft drink, beer, wine, and spirits containers are refillable, and in Germany, about three-fourths are refillable.

X *How Would You Vote?* Do you support banning all beverage containers that cannot be reused, as Denmark has done? Cast your vote online at www.thomsonedu.com/biology/miller.

Solutions: Other Ways to Reuse Things

We can use reusable shopping bags, food containers, and shipping pallets, and borrow tools from tool libraries.

Cloth bags can be used instead of paper or plastic bags to carry groceries and other items. Both paper and plastic bags are environmentally harmful, and the question of which is more damaging has no clear-cut answer. To encourage people to bring reusable bags, stores in the Netherlands and Ireland charge for plastic shopping bags. As a result, the use of such bags has dropped by 90–95% in both countries. Since 2004, supermarkets in Shanghai, China's largest city, have been charging shoppers for plastic bags in an attempt to reduce waste.

X *How Would You Vote?* Should consumers have to pay for plastic or paper bags at grocery and other stores? Cast your vote online at www.thomsonedu.com/biology/miller.

Another example is the use of reusable ceramic, metal, or plastic coffee mugs instead of throwaway paper or Styrofoam cups. An increasing number of coffeehouses and university food services are offering discounts to customers who bring their own mugs.

Manufacturers can use shipping pallets made of recycled plastic waste instead of throwaway wood pallets. In 1991, Toyota shifted entirely to reusable shipping containers. A similar move by the Xerox Corporation saves the company more than $3 million per year.

Another example of reuse involves *tool libraries* (such as those in the U.S. cities of Berkeley, California, and Takoma Park, Maryland) where people can check out a variety of power and hand tools. Figure 13-4

What Can You Do?

Reuse

- Buy beverages in refillable glass containers instead of cans or throwaway bottles.

- Use reusable plastic or metal lunchboxes.

- Carry sandwiches and store food in the refrigerator in reusable containers instead of wrapping them in aluminum foil or plastic wrap.

- Use rechargeable batteries and recycle them when their useful life is over.

- Carry groceries and other items in a reusable basket, a canvas or string bag, or a small cart.

- Use reusable sponges and washable cloth napkins, dishtowels, and handkerchiefs instead of throwaway paper ones.

- Buy used furniture, computers, cars, and other items.

- Give or sell items you no longer use to others.

Figure 13-4 Individuals matter: ways to reuse some of the items you buy. QUESTION: *Which three of these actions do you think are the most important?*

lists several ways for you to reuse some of the items you buy.

RECYCLING

Two Types of Recycling (Science)

Recycling is an important way to collect waste materials and turn them into useful products that can be sold in the marketplace.

Recycling involves reprocessing discarded solid materials into new, useful products. Households and workplaces produce five major types of materials that can be recycled: *paper products* (including newspaper, magazines, office paper, and cardboard), *glass, aluminum, steel,* and some *plastics.*

Materials collected for recycling can be reprocessed in two ways. In *primary* or *closed-loop* recycling, these materials are recycled into new products of the same type—turning used aluminum cans into new aluminum cans, for example. In *secondary recycling,* such materials are converted into different products. For example, used tires can be shredded and converted into rubberized road surfacing and newspapers can be transformed into cellulose insulation.

Scientists distinguish between two types of wastes that can be recycled—*preconsumer* or *internal waste* generated in a manufacturing process and *postconsumer* or *external waste* generated by consumer use of products. Preconsumer waste makes up more than three-fourths of the total.

In theory, just about anything is recyclable, but for recycling to work, two things must happen. *First,* the items separated for recycling must actually be recycled. Sometimes, mostly when prices for recycled raw materials fall sharply, they are mixed with other wastes and sent to landfills or incinerated.

Second, businesses and individuals must complete the recycling loop by buying products that are made from recycled materials. If we do not buy those products, recycling does not work.

Switzerland and Japan recycle about half of their MSW. The United States recycles about 30% of its MSW—up from 6.4% in 1960. This impressive increase was boosted by almost 9,000 curbside pickup recycling programs that serve about half the U.S. population.

Solutions: Two Ways to Recycle Municipal Solid Waste

There is disagreement over whether to send mixed urban wastes to centralized resource recovery plants or to sort recyclables before collection.

One way to recycle is to send mixed urban wastes to a centralized *materials-recovery facility* (MRF). There, machines or workers separate the mixed waste to recover valuable materials for sale to manufacturers as raw materials. The remaining paper, plastics, and other combustible wastes are recycled or burned to produce steam or electricity to run the recovery plant or to sell to nearby industries or homes. There are about 480 MRFs operating in the United States.

Such plants are expensive to build, operate, and maintain. If not operated properly, they can emit toxic air pollutants, and they produce a toxic ash that must be disposed of safely, usually in landfills.

MRFs are hungry beasts that require a steady diet of garbage to make them financially successful. Thus, their owners have a vested interest in increasing *throughput* of matter and energy resources to produce more trash—the reverse of what prominent scientists believe we should be doing (Figure 13-2).

To many experts, it makes more sense economically and environmentally for households and businesses to separate their trash into recyclable categories such as glass, paper, metals, certain types of plastics, and compostable materials. These segregated wastes can be collected and sold to scrap dealers, compost plants, and manufacturers.

The *source separation* approach has several advantages over the centralized approach. It produces much less air and water pollution and has low start-up costs and moderate operating costs. It also saves more energy, provides more jobs per unit of material, and yields cleaner and usually more valuable recyclables. In addition, it educates people about the need for waste reduction, reuse, and recycling.

To promote separation of wastes for recycling, over 4,000 communities in the United States use a *pay-as-you-throw* (PAUT) or *fee-per-bag* waste collection system. It charges households and businesses for the amount of mixed waste picked up but does not charge for pickup of materials separated for recycling.

X *How Would You Vote?* Should households and businesses be charged for the amount of mixed waste picked up but not charged for pickup of materials separated for recycling? Cast your vote online at www.thomsonedu.com /biology/miller.

Composting: Recycling by Copying Nature (Science)

Composting biodegradable organic waste mimics nature by recycling plant nutrients to the soil.

Composting is a simple process in which we copy nature to recycle some of the yard trimmings, food scraps, and other biodegradable organic wastes we produce. The organic material produced by composting can be added to soil to supply plant nutrients, slow soil erosion, retain water, and improve crop yields. For details on composting, see the website for this chapter.

Such wastes can be collected and composted in centralized community facilities. Some cities in Canada and many European Union countries compost more than 85% of their biodegradable wastes, compared to an average of 35% of such wastes in the United States. The resulting compost can be used as an organic soil fertilizer, topsoil, or landfill cover. It can also be used to help restore eroded soil on hillsides and along highways, and on strip-mined land, overgrazed areas, and eroded cropland.

To be successful, a large-scale composting program must be located carefully and odors must be controlled, because people do not want to live near a giant compost pile or plant. Composting programs must also exclude toxic materials that can contaminate the compost and make it unsafe for fertilizing crops and lawns.

Case Study: Recycling Plastics (Science and Economics)

Recycling many plastics is chemically and economically difficult.

Currently, only about 10% by weight of all plastic wastes in the United States are recycled, for three reasons. *First,* many plastics are difficult to isolate from other wastes because the many different resins used to make them are often difficult to identify, and some plastics are composites of different resins. Most plastics also contain stabilizers and other chemicals that must be removed before recycling.

Second, recovering individual plastic resins does not yield much material because only small amounts of any given resin are used per product. *Third,* the inflation-adjusted price of oil used to produce petrochemicals for making plastic resins is so low that the cost of virgin plastic resins is much lower than that of recycled resins. An exception is PET (polyethylene terephthalate), used mostly in plastic drink bottles.

Thus, mandating that plastic products contain a certain amount of recycled plastic resins is unlikely to work. It could also hinder the use of recycled plastics in reducing the resource content and weight of many widely used items such as plastic bags and bottles.

Cargill Dow is manufacturing biodegradable and recyclable plastic containers made from a polymer called polyactide (ACT), made from the sugar in corn syrup. Instead of being sent to landfills, containers made from such *bioplastic* could be composted to produce a soil conditioner.

Toyota, the world's No. 2 automaker, is investing $38 billion in a process that makes plastics from plants. By 2020, it expects to control two-thirds of the world's supply of such bioplastics.

Trade-Offs: Advantages and Disadvantages of Recycling (Science and Economics)

Recycling materials such as paper and metals has environmental and economic benefits.

Figure 13-5 lists the advantages and disadvantages of recycling. Whether recycling makes economic sense depends on how you look at the economic and environmental benefits and costs of recycling.

Critics say recycling does not make sense if it costs more to recycle materials than to send them to a landfill or incinerator. They also point out that recycling is often not needed to save landfill space because many areas are not running out of it. Critics concede that recycling may make economic sense for valuable and easy-to-recycle materials such as aluminum, paper, and steel, but probably not for cheap or plentiful resources such as glass from silica and most plastics that are expensive to recycle. They also argue that recycling should pay for itself.

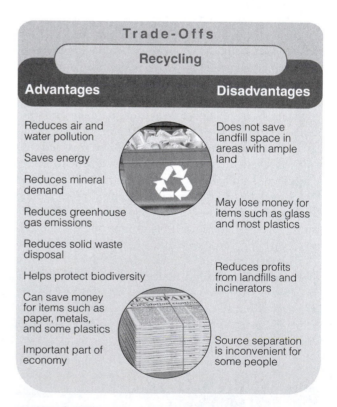

Figure 13-5 Trade-offs. advantages and disadvantages of recycling solid waste. QUESTION: *What are the single advantage and the single disadvantage that you think are the most important?*

But proponents of recycling point out that conventional garbage disposal systems are paid for by charges to households and businesses. So why should recycling be held to a different standard and forced to compete on an uneven playing field? Proponents also point out that reducing the use of landfills and incinerators is not as important as the other benefits of recycling (Figure 13-5). And they point to studies showing that the net economic, health, and environmental benefits of recycling far outweigh the costs.

X HOW WOULD YOU VOTE? Should we place much greater emphasis on recycling with the goal of recycling at least 60% of the municipal solid waste that we produce? Cast your vote online at www.thomsonedu.com/biology/miller.

Encouraging Reuse and Recycling (Economics, Politics, and Stewardship)

Reuse and recycling are hindered by prices of goods that do not reflect their harmful environmental impacts, too few government subsidies and tax breaks, and price fluctuations.

Three factors hinder reuse and recycling. *First,* we have a faulty accounting system in which the market price of a product does not include the harmful environmental and health costs associated with the product during its life cycle.

Second, there is an uneven economic playing field because in most countries, resource-extracting industries receive more government tax breaks and subsidies than recycling and reuse industries.

Third, the demand and thus the price paid for recycled materials fluctuate, mostly because buying goods made with recycled materials is not a priority for most governments, businesses, and individuals.

How can we encourage recycling and reuse? Proponents say that leveling the economic playing field is the best way to start. Governments can *increase* subsidies and tax breaks for reuse and recycling materials (the carrot) and *decrease* subsidies and tax breaks for making items from virgin resources (the stick).

Other strategies are to greatly increase use of the *fee-per-bag* system and to encourage or require government purchases of recycled products to help increase demand and lower prices. Governments can also pass laws requiring companies to take back and recycle or reuse packaging and electronic waste discarded by consumers, as is done in some European countries and in Japan.

X HOW WOULD YOU VOTE? Should governments pass laws requiring manufacturers to take back and reuse or recycle all packaging waste, appliances, electronic equipment, and motor vehicles at the end of their useful lives? Cast your vote online at www.thomsonedu.com/biology/miller.

One reason that recycling is popular is that it helps soothe the conscience of a throwaway society. Many people think that recycling their newspapers and aluminum cans is all they need to do to meet their environmental responsibility. They also do not realize that recycling is an output approach that deals with things we have produced. This takes away from resource reduction and reuse, which are much more important input approaches for reducing the flow and waste of resources.

BURNING AND BURYING SOLID WASTE

Trade-Offs: Burning Solid Waste (Science and Economics)

Japan and a few European countries incinerate most of their municipal solid waste.

Globally, municipal solid waste is burned in more than 1,000 large *waste-to-energy incinerators*, which boil water to make steam for heating water or space, or for producing electricity. Trace the flow of materials through this process as diagrammed in Figure 13-6 (p. 290).

Great Britain burns about 90% of its MSW in incinerators compared to 16% in the United States and about 8% in Canada. Figure 13-7 (p. 290) lists the advantages and disadvantages of using incinerators to burn solid and hazardous waste.

To be economically feasible incinerators must be fed huge volumes of trash daily. This encourages trash production and discourages waste reduction, reuse, and recycling. Since 1985, more than 280 new incinerator projects have been delayed or canceled in the United States because of high costs, concern over air pollution, and intense citizen opposition.

X HOW WOULD YOU VOTE? Do the advantages of incinerating solid waste outweigh the disadvantages? Cast your vote online at www.thomsonedu.com/biology/miller.

Burying Solid Waste (Science and Economics)

Most of the world's municipal solid waste is buried in landfills that eventually are expected to leak toxic liquids into the soil and underlying aquifers.

About 55% by weight of the MSW in the United States is buried in sanitary landfills, compared to 80% in Canada, 15% in Japan, and 12% in Switzerland.

There are two types of landfills. **Open dumps** are essentially fields or holes in the ground where garbage is deposited and sometimes covered with soil. They are rare in developed countries, but widely used in many developing countries.

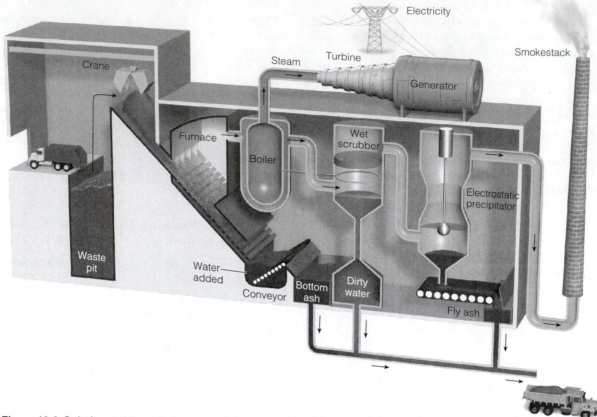

Figure 13-6 **Solutions:** this *waste-to-energy incinerator* with pollution controls burns mixed solid waste and recovers some of the energy to produce steam used for heating or producing electricity. (Adapted from EPA, *Let's Reduce and Recycle*)

Ash for treatment, disposal in landfill, or use as landfill cover

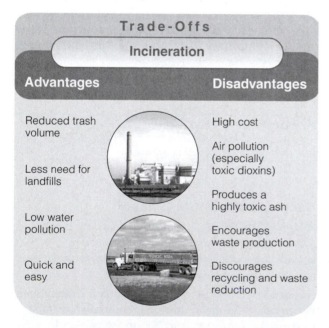

Figure 13-7 **Trade-offs:** advantages and disadvantages of incinerating solid waste. These trade-offs also apply to the incineration of hazardous waste. QUESTION: *What are the single advantage and the single disadvantage that you think are the most important?*

In newer landfills, called **sanitary landfills** (Figure 13-8), solid wastes are spread out in thin layers, compacted, and covered daily with a fresh layer of clay or plastic foam which helps keep the material dry and reduces leakage of contaminated water (leachate) from the landfill. This covering also lessens the risk of fire, decreases odor, and reduces accessibility of vermin to the landfill.

U.S. federal regulations require that all landfills be located on geologically suitable sites and away from lakes, rivers, floodplains, aquifer recharge zones, and earthquake-prone faults. They are lined with clay and plastic before being filled with garbage. The landfill bottom is covered with a second impermeable liner, usually made of several layers of clay, thick plastic, and sand. This liner collects *leachate* (rainwater contaminated as it percolates through the solid waste) and is intended to prevent its leakage into groundwater. Wells are drilled around the landfill to monitor any leakage.

Collected leachate is pumped from the bottom of the landfill, stored in tanks, and sent to a municipal or on-site sewage treatment plant. When full, the landfill

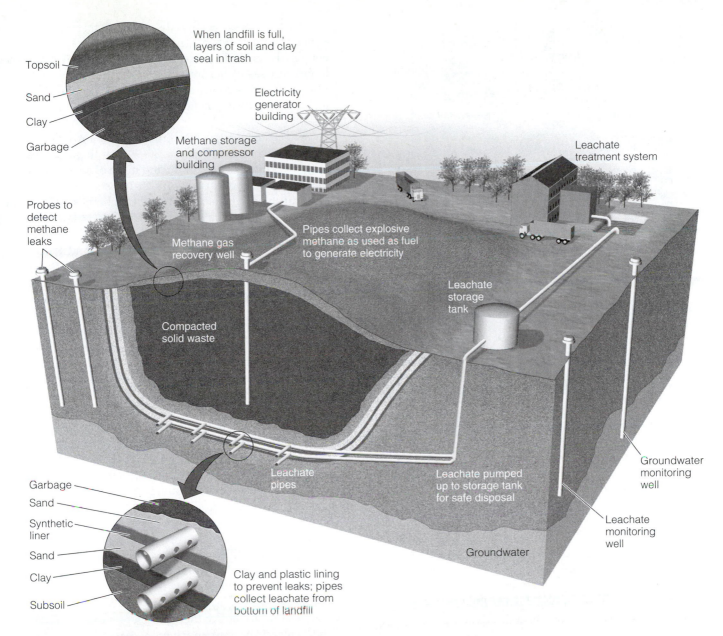

Figure 13-8 Solutions: state-of-the-art *sanitary landfill*, which is designed to eliminate or mini-mize environmental problems that plague older landfills. Even these landfills are expected to leak eventually, passing both the effects of contamination and cleanup costs on to future generations. Since 1997, only modern sanitary landfills are allowed in the United States. As a result, many older and small landfills have been closed and replaced with larger local and regional modern landfills.

Labels in figure:

Topsoil
Sand
Clay
Garbage

When landfill is full, layers of soil and clay seal in trash

Electricity generator building

Methane storage and compressor building

Leachate treatment system

Probes to detect methane leaks

Methane gas recovery well

Pipes collect explosive methane as used as fuel to generate electricity

Leachate storage tank

Compacted solid waste

Garbage
Sand
Synthetic liner
Sand
Clay
Subsoil

Leachate pipes

Clay and plastic lining to prevent leaks; pipes collect leachate from bottom of landfill

Leachate pumped up to storage tank for safe disposal

Groundwater monitoring well

Leachate monitoring well

Groundwater

---

is covered with clay, sand, gravel, and topsoil to pre-vent water from seeping in. Some filled landfills have been converted to city parks and golf courses. In Japan, golfers who smoke are in danger of blowing themselves up by igniting methane vented from a golf course built over a landfill.

Modern landfills are equipped with a network of vent pipes to collect landfill gas (consisting mostly of two potent greenhouse gases, methane and carbon dioxide) released by the underground decomposition of wastes anaerobically (in the absence of oxygen). The methane can be filtered out and burned in small gas turbines to produce steam or electricity for nearby facil-ities or sold to utilities for use as a fuel. But thousands of older and abandoned landfills in the United States (and elsewhere) do not have gas collection systems and

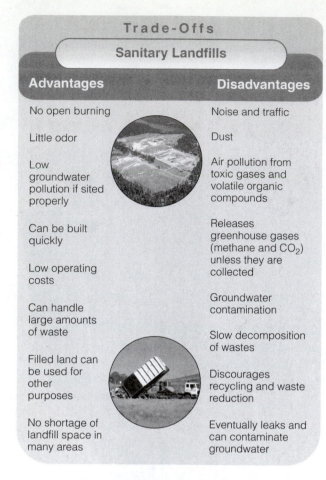

Trade-Offs

Sanitary Landfills

Advantages	Disadvantages
No open burning	Noise and traffic
Little odor	Dust
Low groundwater pollution if sited properly	Air pollution from toxic gases and volatile organic compounds
Can be built quickly	Releases greenhouse gases (methane and CO_2) unless they are collected
Low operating costs	Groundwater contamination
Can handle large amounts of waste	Slow decomposition of wastes
Filled land can be used for other purposes	Discourages recycling and waste reduction
No shortage of landfill space in many areas	Eventually leaks and can contaminate groundwater

Figure 13-9 Trade-offs: advantages and disadvantages of using sanitary landfills to dispose of solid waste. QUESTION: *What are the single advantage and the single disadvantage that you think are the most important?*

will emit methane and carbon dioxide for decades. Figure 13-9 lists the advantages and disadvantages of using sanitary landfills to dispose of solid waste. According to the EPA, all landfills eventually leak.

✗ *HOW WOULD YOU VOTE?* Do the advantages of burying solid waste in sanitary landfills outweigh the disadvantages? Cast your vote online at www.thomsonedu.com/biology/miller.

HAZARDOUS WASTE

What Is Hazardous Waste? (Science and Politics)

Developed countries produce about 80–90% of the world's harmful solid and liquid wastes; most of these hazardous and toxic wastes are not regulated.

Hazardous waste is any discarded solid or liquid material that is *toxic, ignitable, corrosive,* or *reactive* enough to explode or release toxic fumes. The two largest classes of hazardous wastes are organic compounds (such as solvents, pesticides, PCBs, and dioxins) and toxic heavy metals (such as lead, mercury, and arsenic).

According to the UN Environment Programme, developed countries produce 80–90% of these wastes, with the United States producing more of such wastes than any other country. Figure 13-10 lists some of the harmful chemicals found in many U.S. homes.

In the United States, two major federal laws regulate the management and disposal of hazardous waste. One is the Resource Conservation and Recovery Act (RCRA), which regulates only about 5% of the hazardous waste produced in the United States. In most other countries, especially developing countries, an even smaller percentage of hazardous waste is regulated.

A second U.S. federal law deals with hazardous wastes produced in the past. In 1980, the U.S. Congress passed the *Comprehensive Environmental Response, Compensation, and Liability Act,* commonly known as the *CERCLA* or *Superfund* program. The goals of this law are to identify abandoned hazardous waste sites and clean up such sites on a priority basis. The worst sites that represent an immediate and severe threat to human health are put on a *National Priorities List* (NPL) and scheduled for total cleanup using the most cost-effective method.

The Superfund law was designed to make polluters pay for cleaning up abandoned hazardous waste sites. The EPA points out that the strict polluter-pays principle in the Superfund Act has been effective in making illegal dumpsites virtually relics of the past. It has also forced waste producers, fearful of future liability claims, to reduce their production of such waste and to recycle or reuse much more of it.

So far, more than 70% of the cleanup costs have come from polluters identified as the responsible parties with the rest coming from a trust fund (the Superfund) financed until 1995 by taxes on chemical raw materials and oil. But under pressure from polluters and their insurance companies Congress refused to renew the tax on oil and chemical companies that financed the Superfund after it expired in 1995. The Superfund is now broke and taxpayers, not polluters, are footing the bill for future cleanups when the responsible parties cannot be found. As a result, the pace of cleanup has slowed.

About one of every four Americans lives near one of the country's 450,000 potentially harmful hazardous waste sites identified by the EPA. Epidemiological studies indicate that living within 2.9 kilometers (1.8

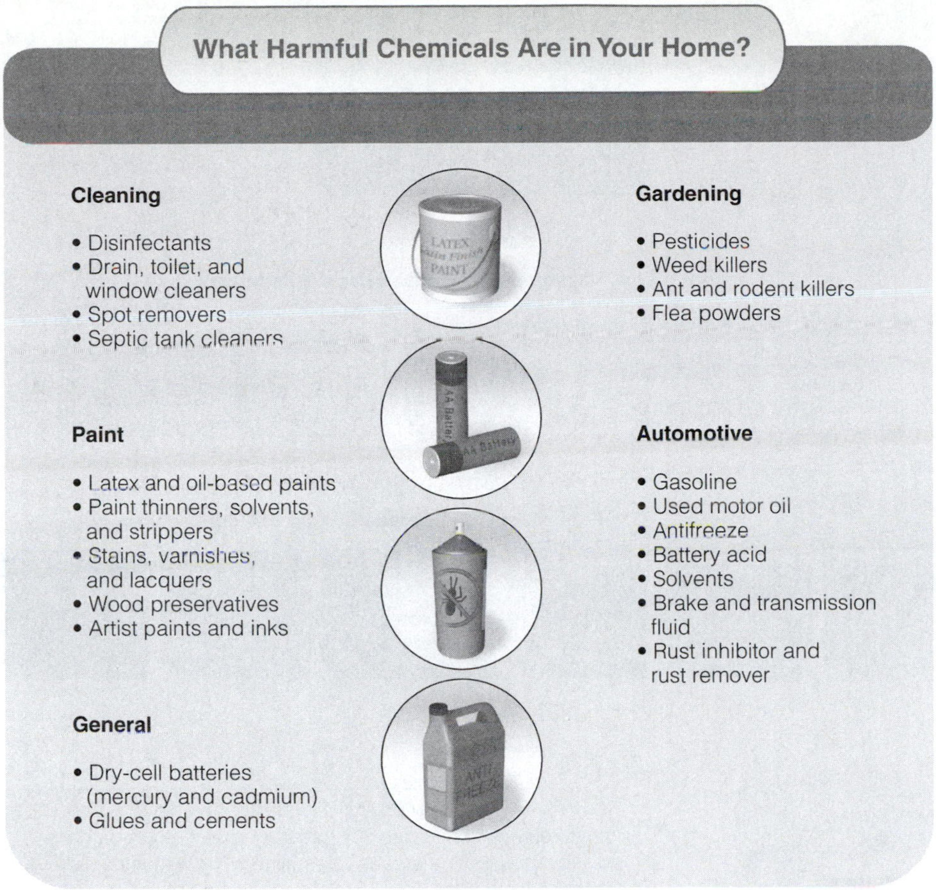

What Harmful Chemicals Are in Your Home?

Cleaning

- Disinfectants
- Drain, toilet, and window cleaners
- Spot removers
- Septic tank cleaners

Paint

- Latex and oil-based paints
- Paint thinners, solvents, and strippers
- Stains, varnishes, and lacquers
- Wood preservatives
- Artist paints and inks

General

- Dry-cell batteries (mercury and cadmium)
- Glues and cements

Gardening

- Pesticides
- Weed killers
- Ant and rodent killers
- Flea powders

Automotive

- Gasoline
- Used motor oil
- Antifreeze
- Battery acid
- Solvents
- Brake and transmission fluid
- Rust inhibitor and rust remover

Figure 13-10 Science: harmful chemicals found in many U.S. homes. The U.S. Congress has exempted disposal of these materials from government regulation. QUESTION: *Which of these chemicals are in your home?*

miles) of a hazardous waste site results in a significantly increased risk of several types of cancers and birth defects.

The U.S. Congress and various state legislatures have also passed laws that encourage the cleanup of *brownfields*—abandoned industrial and commercial sites that in most cases are contaminated with hazardous wastes. Examples include factories, junkyards, older landfills, and gas stations. By 2005, more than 40,000 former brownfield sites had been redeveloped in the United States and many other projects are under way.

Solutions: Integrated Management of Hazardous Waste (Science)

We can produce less hazardous waste and recycle, reuse, detoxify, burn, and bury it.

Figure 13-11 (p. 294) shows an integrated management approach suggested by the U.S. National Academy of Sciences that establishes three levels of priorities for dealing with hazardous wastes: produce less of them, convert them to less hazardous substances, and after these two priorities have been fulfilled, put what is left in long-term storage.

In Denmark, all hazardous and toxic waste from industries and households is delivered to 21 transfer stations throughout the country. The waste is then transferred to a large treatment facility. There, three-fourths of the waste is detoxified by physical, chemical, and biological methods and the rest is buried in a carefully designed and monitored landfill.

Some scientists and engineers consider biological treatment of hazardous waste as the wave of the future for cleaning up some types of toxic and hazardous waste. One approach is *bioremediation,* in which bacteria and enzymes help destroy toxic or hazardous substances or convert them to harmless compounds. See the Guest Essay by John Pichtel on this topic on the website for this chapter.

Another approach is *phytoremediation,* which involves using natural or genetically engineered plants to absorb, filter, and remove contaminants from polluted soil and water. Various plants have been identified as "pollution sponges" to help clean up soil and water contaminated with chemicals such as pesticides, organic solvents, radioactive metals, and toxic metals such as lead, mercury, and arsenic. Figure 13-12 (p. 294) lists advantages and disadvantages of phytoremediation.

Figure 13-11 Integrated hazardous waste management: priorities suggested by prominent scientists for dealing with hazardous waste. To date, these priorities have not been followed in the United States and in most other countries.

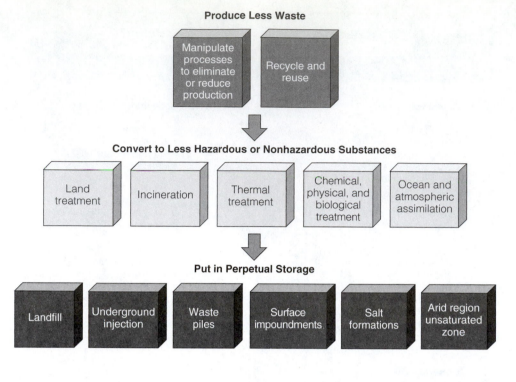

Produce Less Waste

Manipulate processes to eliminate or reduce production

Recycle and reuse

Convert to Less Hazardous or Nonhazardous Substances

Land treatment

Incineration

Thermal treatment

Chemical, physical, and biological treatment

Ocean and atmospheric assimilation

Put in Perpetual Storage

Landfill

Underground injection

Waste piles

Surface impoundments

Salt formations

Arid region unsaturated zone

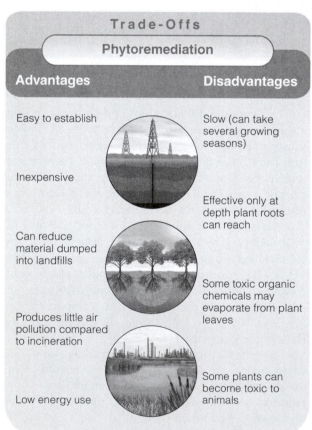

Trade-Offs

Phytoremediation

Advantages	Disadvantages
Easy to establish	Slow (can take several growing seasons)
Inexpensive	
	Effective only at depth plant roots can reach
Can reduce material dumped into landfills	
	Some toxic organic chemicals may evaporate from plant leaves
Produces little air pollution compared to incineration	
	Some plants can become toxic to animals
Low energy use	

Figure 13-12 Trade-offs: advantages and disadvantages of using *phytoremediation* to remove or detoxify hazardous waste. QUESTION: *What are the single advantage and the single disadvantage that you think are the most important?*

☒ HOW WOULD YOU VOTE? Do the advantages of using phytoremediation to detoxify hazardous waste outweigh the disadvantages? Cast your vote online at www.thomsonedu .com/biology/miller.

Hazardous waste can also be incinerated. This has the same advantages and disadvantages as burning solid wastes (Figure 13-7). Two major disadvantages of incinerating hazardous waste are that it releases air pollutants such as toxic dioxins and produces a highly toxic ash that must be safely and permanently stored in a landfill especially designed for hazardous waste.

Most hazardous waste in the United States is disposed of on land in deep underground wells; surface impoundments such as ponds, pits, or lagoons; and state-of-the-art landfills. In *deep-well disposal*, liquid hazardous wastes are pumped under pressure through a pipe into dry, porous geologic formations or zones of rock far beneath aquifers tapped for drinking and irrigation water. Theoretically, these liquids soak into the porous rock material and are isolated from overlying groundwater by essentially impermeable layers of rock.

Figure 13-13 lists the advantages and disadvantages of deep-well disposal of liquid hazardous wastes. Many scientists believe current regulations for deep-well disposal are inadequate and should be improved.

☒ HOW WOULD YOU VOTE? Do the advantages of deep-well disposal of hazardous waste from soil and water outweigh the disadvantages? Cast your vote online at www.thomsonedu .com/biology/miller.

Surface impoundments are excavated depressions such as ponds, pits, or lagoons into which liquid hazardous wastes are stored (Figure 9-22, p. 191). As water evaporates, the waste settles and becomes more concentrated. Figure 13-14 lists the advantages and disadvantages of this method.

EPA studies found that 70% of these storage basins in the United States have no liners, and as many as 90% may threaten groundwater. According to the EPA, all liners are likely to leak eventually and can contaminate groundwater.

X *How Would You Vote?* Do the advantages of storing hazardous wastes in surface impoundments outweigh the disadvantages? Cast your vote online at **www.thomsonedu.com/biology/miller**.

Hazardous wastes can also be stored in carefully designed *aboveground buildings*. This is especially useful in areas where the water table is close to the surface or areas that are above aquifers used for drinking water. These structures are built to withstand storms and to prevent the release of toxic gases. Leaks are monitored and any leakage is collected and treated.

Sometimes liquid and solid hazardous waste are put into drums or other containers and buried in carefully designed and monitored *secure hazardous waste landfills* (Figure 13-15, p. 296). This is the least used method because of the expense involved. In the United States, there are only 23 commercial hazardous waste landfills, although the EPA licenses some companies to store their hazardous waste in approved landfill sites.

To most environmental scientists, the only solution to the hazardous waste problem is to produce as little as possible in the first place (Figure 13-11)—a prevention approach. Figure 13-16 (p. 296) lists some ways in which each of us can help reduce our hazardous waste inputs into the environment.

Case Study: The Threat from Lead (Science)

Lead is especially harmful to children and is still used in leaded gasoline and household paints in about 100 countries.

Because it is a chemical element, lead (Pb) does not break down in the environment. This potent neurotoxin can harm the nervous system, especially in young children. Each year, 12,000–16,000 American children under age 9 are treated for acute lead poisoning, and about 200 die. About 30% of the survivors suffer from palsy, partial paralysis, blindness, and mental retardation.

Children under age 6 and unborn fetuses even with low blood levels of lead are especially vulnerable to nervous system impairment, lowered IQ (by an average of 7.4 points), shortened attention span, hyperactivity, hearing damage, and various behavior disorders.

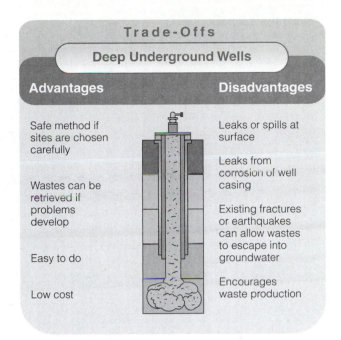

Figure 13-13 Trade-offs: advantages and disadvantages of injecting liquid hazardous wastes into deep underground wells. QUESTION: *What are the single advantage and the single disadvantage that you think are the most important?*

Figure 13-14 Trade-offs: advantages and disadvantages of storing liquid hazardous wastes in surface impoundments. QUESTION: *What are the single advantage and the single disadvantage that you think are the most important?*

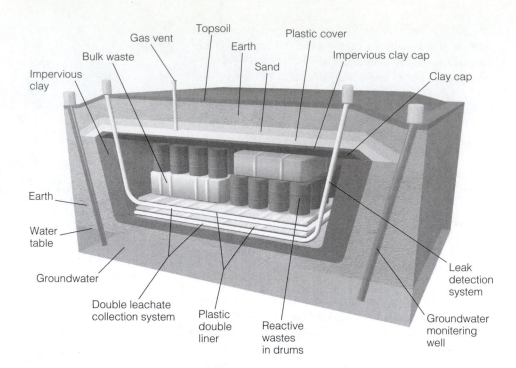

Figure 13-15 Solutions: secure hazardous waste landfill.

Good news: Between 1976 and 2000, the percentage of U.S. children ages 1 to 5 with blood lead levels above the safety standard dropped from 85% to 2.2%, preventing at least 9 million childhood lead poisonings. The primary reason for this drop was that government regulations banned leaded gasoline in 1976 (with complete phaseout by 1986) and lead-based paints in 1970 (but illegal use continued until about 1978). This is an excellent example of the power of pollution prevention.

Bad news: The U.S. Centers for Disease Control and Prevention estimate that at least 400,000 U.S. children still have unsafe blood levels of lead caused by exposure from a number of sources. A major source is inhalation or ingestion of lead particles from peeling lead-based paint found in about 38 million houses built before 1960. Lead can also leach from water lines and pipes and faucets containing lead. In addition, a 1993 study by the U.S. National Academy of Sciences and numerous other studies indicate *there is no safe level of lead in children's blood.*

Health scientists have proposed a number of ways to help protect children from lead poisoning, as listed in Figure 13-17.

Although the threat from lead has been reduced in the United States, this is not the case in many developing countries. About 80% of the gasoline sold in the world today is unleaded, but about 100 countries still use leaded gasoline. The WHO estimates that 130–200 million children around the world are at risk from lead poisoning, and 15–18 million children in developing countries have permanent brain damage because of lead poisoning—mostly from use of leaded gasoline. *Good news.* China recently phased out leaded gasoline in less than three years.

ACHIEVING A LOW-WASTE SOCIETY

Grassroots Action for Better Solid and Hazardous Waste Management (Politics)

In the United States, citizens have kept large numbers of incinerators, landfills, and hazardous waste treatment plants from being built in their local areas.

What Can You Do?

Hazardous Waste

- Use pesticides in the smallest amount possible.

- Use less harmful substances instead of commercial chemicals for most household cleaners. For example, use liquid ammonia to clean appliances and windows; vinegar to polish metals, clean surfaces, and remove stains and mildew; baking soda to clean household utensils, deodorize, and remove stains; borax to remove stains and mildew.

- Do not dispose of pesticides, paints, solvents, oil, antifreeze, or other products containing hazardous chemicals by flushing them down the toilet, pouring them down the drain, burying them, throwing them into the garbage, or dumping them down storm drains.

Figure 13-16 Individuals matter: ways to reduce your input of hazardous waste into the environment. QUESTION: *Which two of the things in this list do you plan to do?*

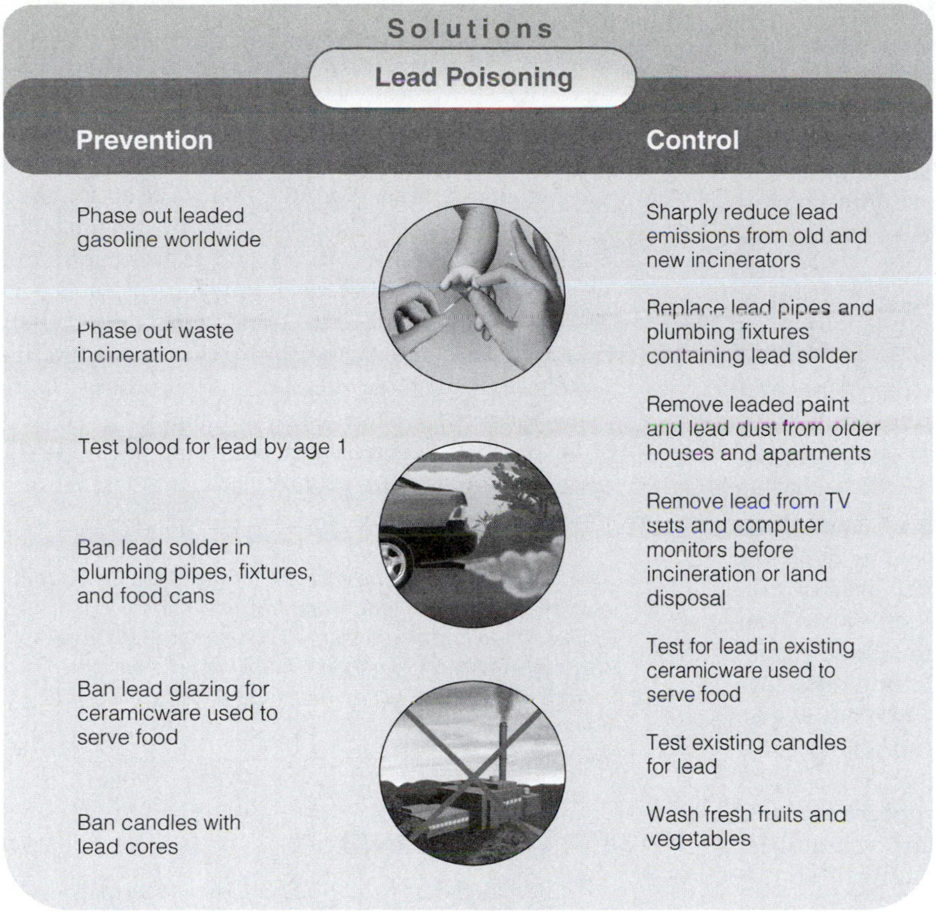

Figure 13-17 Solutions: ways to help protect children from lead poisoning. QUESTION: *Which two of these solutions do you think are the most important?*

Solutions

Lead Poisoning

Prevention	Control
Phase out leaded gasoline worldwide	Sharply reduce lead emissions from old and new incinerators
Phase out waste incineration	Replace lead pipes and plumbing fixtures containing lead solder
Test blood for lead by age 1	Remove leaded paint and lead dust from older houses and apartments
Ban lead solder in plumbing pipes, fixtures, and food cans	Remove lead from TV sets and computer monitors before incineration or land disposal
Ban lead glazing for ceramicware used to serve food	Test for lead in existing ceramicware used to serve food
	Test existing candles for lead
Ban candles with lead cores	Wash fresh fruits and vegetables

In the United States, individuals have organized to prevent hundreds of incinerators, landfills, and treatment plants for hazardous and radioactive wastes from being built in or near their communities.

Health risks from incinerators and landfills, when averaged over the entire country, are quite low, but the risks for people living near these facilities are much higher. They, not the rest of the population, are the ones whose health, lives, and property values are being threatened.

Manufacturers and waste industry officials point out that something must be done with the toxic and hazardous wastes produced as by-products of providing certain goods and services. They contend that if local citizens adopt a "not in my back yard" (NIMBY) approach, the waste still ends up in someone's back yard.

Many citizens do not accept this argument. To them, the best way to deal with most toxic or hazardous wastes is to produce much less of them, as suggested by the U.S. National Academy of Sciences (Figure 13-11). For such materials, they believe that the goal should be "not in anyone's back yard" (NIABY) or "not on planet Earth" (NOPE) with emphasis on pollution prevention and use of the precautionary principle.

Opposition has grown as numerous studies have shown that a disproportionate share of polluting factories, hazardous waste dumps, incinerators, and landfills are located in communities populated mostly by African Americans, Asian Americans, Latinos, Native Americans, and the working poor in general. This practice has been cited as an example of *environmental injustice*. See the Guest Essay on this subject by Robert Bullard on the website for this chapter.

Global Outlook: International Action to Reduce Hazardous Waste (Science and Politics)

An international treaty calls for phasing out the use of harmful persistent organic pollutants (POPs).

In 2000, delegates from 122 countries completed a global treaty to control 12 *persistent organic pollutants* (POPs). These widely used toxic chemicals are insoluble in water and soluble in fat. This means that in the fatty tissues of humans and other organisms feeding at high trophic levels in food webs, POPs can become concentrated to levels hundreds of thousand times higher than in the general environment (Figure 7-12, p. 140). POPs can also be transported long distances by wind and water.

The list of 12 chemicals, called the *dirty dozen*, includes DDT and 8 other chlorine-containing persistent pesticides, PCBs, dioxins, and furans. The goals of the treaty, which has gone into effect, are to ban or phase out use of these chemicals and detoxify or isolate stockpiles of them. The treaty allows 25 countries to continue using DDT to combat malaria until safer alternatives are available. The United States has not ratified this treaty.

Environmental scientists consider the POPs treaty an important milestone in international environmental law because it uses the *precautionary principle* to manage and reduce the risks from toxic chemicals. This list is expected to grow as scientific studies uncover more evidence of toxic and environmental damage from some of the chemicals we use.

In 2000, the Swedish Parliament enacted a law that by 2020 will ban all chemicals that are persistent and can bioaccumulate in living tissue. This law also requires an industry to perform risk assessments on all old and new chemicals and to show that these chemicals are safe to use, as opposed to requiring the government to show they are dangerous. In other words, chemicals are assumed guilty until proven innocent—the reverse of the current policy in the United States and most countries. There is strong opposition to this approach in the United States, especially by industries producing potentially dangerous chemicals.

Making the Transition to a Low-Waste Society: A New Vision

A number of principles and programs can be used to make the transition to a low-waste society during this century.

According to physicist Albert Einstein, "A clever person solves a problem; a wise person avoids it." To prevent pollution and reduce waste, many environmental scientists urge us to understand and live by four key principles:

- Everything is connected.

- There is no "away" for the wastes we produce.

- Dilution is not always the solution to pollution.

- The best and cheapest ways to deal with solid and hazardous wastes are waste reduction and pollution prevention.

Good news. The governments of Norway, Austria, and the Netherlands have committed themselves to reducing their resource waste by 75%. Other countries are following their lead. And at least 24 countries have *eco-labeling programs* that certify a product or service as having met specified environmental standards.

 Learn more about how shifting to a low-waste (low-throughput) economy would be the best long-term solution to environmental and resource problems at ThomsonNOW.

Making the transition to a low-waste society also requires applying the four principles of sustainability. Shifting to reliance on direct and indirect forms of *solar energy* will reduce our outputs of solid and hazardous waste, as will reusing and recycling materials by mimicking nature's *chemical cycling* processes. Integrated management of solid and hazardous waste using a *diversity* of approaches is another goal. And reducing the human *population* would mean fewer people using resources and producing wastes and pollution.

The key to addressing the challenge of toxics use and wastes rests on a fairly straightforward principle: harness the innovation and technical ingenuity that has characterized the chemicals industry from its beginning and channel these qualities in a new direction that seeks to detoxify our economy.

ANNE PLATT MCGINN

CRITICAL THINKING

1. Collect all the trash (excluding food waste) that you generate in a typical week. Measure its total weight and volume. Sort it into major categories such as paper, plastic, metal, and glass. Then weigh each category and calculate the percentage by weight in each category. What percentage by weight of this waste consists of materials that could be recycled or reused? What percentage by weight of the items could you have done without? Tally and compare the results for your entire class.

2. Would you oppose or support a *pay-as-you-throw* or *fee-per-bag* approach to financing disposal of solid waste? Explain.

3. Find three items you regularly use that are designed to be used once and thrown away. Are there other reusable products that you could use in place of these disposable items? Compare the cost of using the disposable option for a year versus the cost of buying the alternatives.

4. Would you oppose having a hazardous waste landfill, waste treatment plant, deep-injection well, or incinerator in your community? Explain. If you oppose these disposal facilities, how do you think the hazardous waste generated in your community and your state should be managed?

5. Give your reasons for agreeing or disagreeing with each of the following proposals for dealing with hazardous waste:

a. Reduce the production of hazardous waste and encourage recycling and reuse of hazardous materials by charging producers a tax or fee for each unit of waste generated.

b. Ban all land disposal and incineration of hazardous waste to encourage recycling, reuse, and treatment, and to protect air, water, and soil from contamination.

c. Provide low-interest loans, tax breaks, and other financial incentives to encourage industries that produce hazardous waste to reduce, recycle, reuse, treat, and decompose such waste.

6. Congratulations! You are in charge of the world. List the three most important components of your strategy for dealing with (a) solid waste and (b) hazardous waste.

7. List two questions that you would like to have answered as a result of reading this chapter.

LEARNING ONLINE

The website for this book contains helpful study aids and many ideas for further reading and research. They include a chapter summary, review questions for the entire chapter, flash cards for key terms and concepts, a multiple-choice practice quiz, interesting Internet sites, references, information about green careers, and a guide for accessing thousands of InfoTrac® College Edition articles. Log into

www.thomsonedu.com/biology/miller

Then choose Chapter 13, and select a learning resource. For access to animations, additional quizzes, chapter outlines and summaries, register and log into

at www.thomsonedu.com using the access code card in the front of your book.

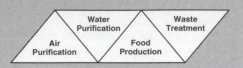

The main ingredients of an environmental ethic are caring about the planet and all of its inhabitants, allowing unselfishness to control the immediate self-interest that harms others, and living each day so as to leave the lightest possible footprints on the planet.

ROBERT CAHN

ECONOMIC SYSTEMS AND SUSTAINABILITY

Economic Resources

An economic system produces goods and services by using natural, human, and manufactured resources.

An **economic system** is a social institution through which goods and services are produced, distributed, and consumed to satisfy people's unlimited wants in the most efficient possible way.

Three types of resources or capital are used to produce goods and services (Figure 14-1). **Natural resources,** or **natural capital,** include goods and services produced by the earth's natural processes, which support all economies and all life (Figure 1-2, p. 6). See the Guest Essay on natural capital by Paul Hawken on the website for this chapter.

Human resources or **human capital,** include people's physical and mental talents that provide labor, innovation, culture, and organization. **Manufactured resources,** or **manufactured capital,** are items such as machinery, equipment, and factories made from natural resources with the help of human resources.

Environmentally Sustainable Economic Development

Ecological economists see economic systems as a component of nature's economy.

Neoclassical economists view the earth's natural capital as a subset or part of a human economic system. They assume that because of our ingenuity and technology, if we run out of any resources, we will find substitutes and that the potential for economic growth is essentially unlimited. They believe that the environmental and health costs of producing goods and services should be passed on to society and to future generations because including these costs in market prices would hinder economic growth.

Ecological economists disagree with this model and its assumptions. They view economic systems as subsystems of the environment that depend heavily on the earth's irreplaceable natural resources (Figure 14-2). They point out that there are no substitutes for many natural resources, such as air, water, fertile soil, and biodiversity.

They also believe that conventional economic growth eventually is unsustainable because it can deplete or degrade the natural resources on which economic systems depend. They argue that environmental and health effects of producing economic goods and services should be included in their market prices so that consumers will have accurate information about the environmental effects of the goods and services they buy. *Environmental economists* take a position between these two schools.

Ecological economists and many environmental economists distinguish between unsustainable economic growth and environmentally sustainable economic development (Figure 14-3). They call for making a shift from our current economy based on unlimited economic growth to a more *environmentally sustainable economy,* or *eco-economy* over the next several decades. See the Guest Essay on this topic by Herman Daly on the website for this chapter. Figure 14-4 (p. 302) shows

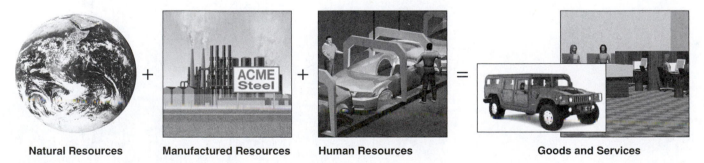

Natural Resources   **Manufactured Resources**   **Human Resources**   **Goods and Services**

Figure 14-1 Three types of resources are used to produce goods and services.

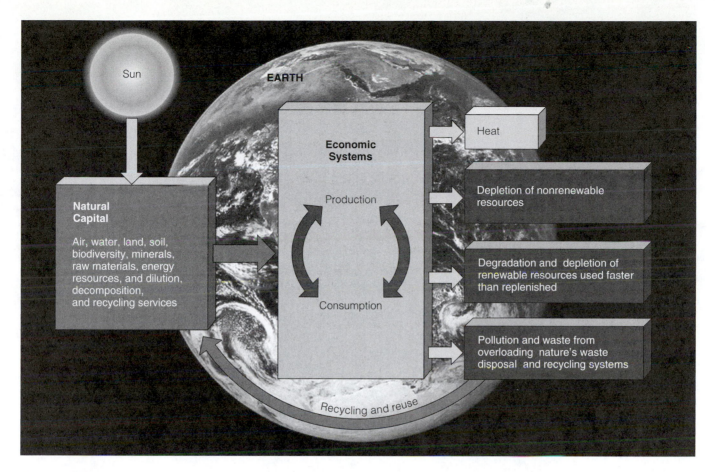

ThomsonNOW Active Figure 14-2 Solutions: *ecological economists* see all economies as human subsystems that depend on natural resources and services provided by the sun and earth. *See an animation based on this figure and take a short quiz on the concept.*

some of the goods and services these economists would encourage in an eco-economy.

 Learn more about how ecological economists view market-based systems and contrast their views with those of conventional economists at ThomsonNOW.

Ecological and environmental economists have suggested several strategies to help make the transition to a more sustainable eco-economy over the next several decades:

■ Use indicators that monitor economic and environmental health

■ Include in the market prices of goods and services their estimated harmful effects on the environment and human health (*full-cost pricing*)

Figure 14-3 Comparison of unsustainable economic growth and environmentally sustainable economic development according to ecological economists and many environmental economists.

Characteristic	Unsustainable Economic Growth	Environmentally Sustainable Economic Development
Production emphasis	Quantity	Quality
Natural resources	Not very important	Very important
Resource productivity	Inefficient (high waste)	Efficient (low waste)
Resource throughput	High	Low
Resource type emphasized	Nonrenewable	Renewable
Resource fate	Matter discarded	Matter recycled, reused, or composted
Pollution control	Cleanup (output reduction)	Prevention (input reduction)
Guiding principles	Risk–benefit analysis	Prevention and precaution

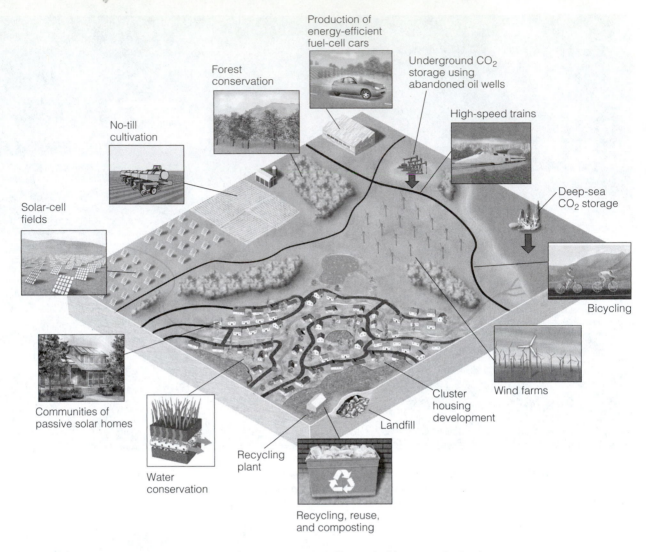

Figure 14-4 Solutions: some components of more environmentally sustainable economic development favored by ecological and environmental economists. The goal is to have economic systems put more emphasis on conserving and sustaining the air, water, soil, biodiversity, and other natural resources that sustain all life and all economies. Such a shift toward more efficient resource use, clean energy, clean production, and natural capital preservation can stimulate economies and create jobs.

■ Use *eco-labeling* to identify products produced by environmentally sound methods and thus help consumers make informed choices

■ Phase out environmentally harmful government subsidies and tax breaks while increasing such breaks for environmentally beneficial goods and services (*subsidy shifting*)

■ Decrease taxes on wages, income, and wealth while increasing taxes on pollution, resource waste, and environmentally harmful goods and services (*tax shifting*)

■ Use laws and regulations to prevent pollution and resource depletion

■ Use tradable permits to pollute or use resources to limit overall pollution and resource use

■ Sell services instead of things

■ Reduce poverty, one of the basic causes of environmental degradation, pollution, poor health, and premature death

Let us look at some of these proposed solutions in more detail.

USING ECONOMICS TO IMPROVE ENVIRONMENTAL QUALITY

Hidden Harmful Costs

The direct price you pay for something does not include indirect environmental, health, and other harmful costs associated with its production and use.

All economic goods and services have *internal* or *direct costs* associated with producing them. For example, if you buy a car the direct price you pay includes the costs of raw materials, labor, and shipping, as well as a markup to allow the car company and its dealers some profits. Once you buy the car you must pay additional direct costs for gasoline, maintenance, and repair.

Making, distributing, and using economic goods or services also involve *indirect* or *external costs* that are not included in their market prices and that affect people other than the buyer and seller. For example, to extract and process raw materials to make a car, we use nonrenewable energy and mineral resources, produce solid and hazardous wastes, disturb land, and pollute air and water. These external costs not included in the price of the car can have short- and long-term harmful effects on other people and on the earth's life-support systems.

Because these harmful costs are not included in the market price, most people do not connect them with car ownership. Still, the car buyer and other people in a society pay these hidden costs sooner or later, through poorer health, higher costs of health care and insurance, higher taxes for pollution control, traffic congestion, and land used for highways and parking.

Environmental and Economic Indicators

We need indicators that accurately reflect changing levels of environmental quality and human health.

Gross domestic product (GDP), and *per capita GDP* indicators provide a standardized and useful method for measuring and comparing the economic outputs of nations. The GDP is deliberately designed to measure the annual economic value of all goods and services produced within a country without attempting to distinguish between goods and services that are environmentally or socially beneficial and those that are harmful. Thus, economists who developed the GDP many decades ago never intended it to be used for measuring environmental quality or human well-being.

Environmental and ecological economists and environmental scientists call for the development and widespread use of new indicators to help monitor environmental quality and human well-being. One approach is to develop indicators that *add* to the GDP things not counted in the marketplace that enhance environmental quality and human well-being. They would also *subtract* from the GDP the costs of things that lead to a lower quality of life and deplete or degrade natural resources.

One such indicator is the *genuine progress indicator* (GPI), introduced in 1995 by Redefining Progress, a nonprofit organization that develops economics and

policy tools to help evaluate and promote sustainability. (This group also developed the concept of ecological footprints, Figure 1-6, p. 10.) Within the GPI, the estimated value of beneficial transactions that meet basic needs, but in which no money changes hands, are added to the GDP. Examples are unpaid volunteer work, healthcare for family members, childcare, and housework. Then the estimated harmful environmental costs (such as pollution and resource depletion and degradation) and social costs (such as crime) are subtracted from the GDP.

$$\begin{matrix} \text{Genuine} \\ \text{progress} \\ \text{indicator} \end{matrix} = \text{GDP} + \begin{matrix} \text{benefits not} \\ \text{included in} \\ \text{market transactions} \end{matrix} - \begin{matrix} \text{harmful} \\ \text{environmental} \\ \text{and social costs} \end{matrix}$$

Figure 14-5 compares the per capita GDP and GPI for the United States between 1950 and 2002. While the per capita GDP rose sharply, the per capita GPI stayed nearly flat and declined slightly between 1975 and 2002.

The GPI and other environmental and social indicators under development are far from perfect and include many crude estimates. But without such indicators, we do not know much about what is happening to people, the environment, and the planet's natural resource base. With them we have a way to help us determine what policies work. In effect, according to ecological and environmental economists, we are trying to guide national and global economics through treacherous economic and environmental waters at ever-increasing speeds without a good radar system.

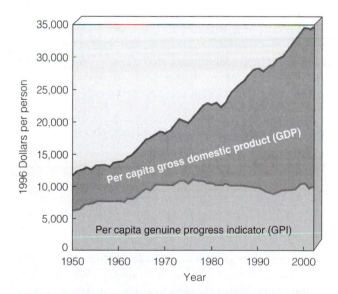

Figure 14-5 Monitoring environmental progress. Comparison of the per capita gross domestic product (GDP) and the per capita genuine progress indicator (GPI) in the United States between 1950 and 2002. (Data from Redefining Progress, 2004)

Full-Cost Pricing

Including external costs in market prices informs consumers about the harmful impacts their purchases have on the earth's life-support systems and on human health.

For most economists, creating an *environmentally transparent or honest market system* is a way to deal with the harmful costs of goods and services. It requires including the harmful indirect or external costs of goods and services in their market prices, so that the prices come as close as possible to their **full cost**—internal costs plus external costs. This system would reduce pollution and waste, improve human health, and allow consumers to make more informed choices because they would be aware of most or all of the costs involved when they buy something.

Full-cost pricing would encourage producers to invent more resource-efficient and less-polluting methods of production, thereby cutting their production costs. Jobs would be lost in environmentally harmful businesses as more consumers would choose green products, but jobs would be created in environmentally beneficial businesses.

If a shift to full-cost pricing took place over several decades, most environmentally harmful businesses would have time to transform themselves into environmentally beneficial businesses. And consumers would have time to adjust their purchases and buying habits to favor more environmentally beneficial products and services.

Full-cost pricing seems to make a lot of sense. So why is it not used more widely? There are two major reasons. First, many producers of harmful and wasteful goods would have to charge more and some would go out of business. Naturally, they oppose such pricing. Second, it is difficult to put a price tag on many environmental and health costs. But to ecological and environmental economists, making the best possible estimates is far better than not including such costs in what we pay for most goods and services.

X *HOW WOULD YOU VOTE?* Should full-cost pricing be used in setting the market prices of goods and services? Cast your vote online at www.thomsonedu.com/biology/miller.

Phasing in such a system would require government action. Few if any companies will volunteer to reduce short-term profits to become more environmentally responsible. Governments can use several strategies to encourage or force producers to work toward full-cost pricing. They include phasing out environmentally harmful subsidies, levying taxes on environmentally harmful goods and services, passing laws to regulate pollution and resource depletion, and using tradable permits for pollution or resource use. Let us look at these strategies in more detail.

Government Subsidies and Tax Breaks (Science, Economics, and Politics)

We can improve environmental quality and human health and help phase in full-cost pricing by removing environmentally harmful government subsidies and tax breaks.

One way to encourage a shift to full-cost pricing is to *phase out* environmentally harmful subsidies and tax breaks, which cost the world's governments almost $1 trillion a year and create a huge economic incentive for resource depletion and degradation. Examples include depletion subsidies and tax breaks for extracting minerals and oil from the ground, cutting timber on public lands, and providing farmers with low-cost water.

On paper, phasing out such subsidies may seem like a great idea. In reality, the economically and politically powerful interests receiving such subsidies want to keep—and if possible increase—these benefits and often lobby against subsidies and tax breaks for more environmentally beneficial competitors.

Such opposition will only be overcome when enough individuals work together to force elected officials to stop such practices. Consumers can also vote with their pocketbooks by not buying environmentally harmful goods and services and by buying more environmentally sustainable products. Consumer power works.

Some countries are beginning to reduce environmentally harmful subsidies. Japan, France, and Belgium have phased out all coal subsidies. Germany has cut coal subsidies in half and plans to phase them out completely by 2010. China has cut coal subsidies by about 73% and has imposed a tax on high-sulfur coals.

Shifting Taxes from Wages and Profits to Pollution and Waste: Green Taxes

Taxes on pollution and resource use can move us closer to full-cost pricing, and shifting taxes from wages and profits to pollution and waste would help make this feasible.

Another way to discourage pollution and resource waste is to *use green taxes* or *ecotaxes* to help include many of the harmful environmental costs of production and consumption in market prices. Taxes can be levied on a per-unit basis on the amount of pollution and hazardous waste produced, and on the use of fossil fuels, timber, and minerals.

Economists point out two requirements for successful implementation of *green taxes* and *tax shifting*. *First,* they should reduce or replace income, payroll, or other taxes. *Second,* the poor and lower middle class need a safety net to reduce the regressive nature of consumption taxes on essentials such as food, fuel, and housing.

To many analysts, the tax system in most countries is backwards. It can *discourage* what we want more of—jobs, income, and profit-driven innovation—and *encourage* what we want less of—pollution, resource waste, and environmental degradation. A more environmentally sustainable economic system would *lower* taxes on labor, income, and wealth and *raise* taxes on environmentally harmful activities.

With such a shift, for example, a tax on coal would include the increased health costs of breathing polluted air, damages from acid deposition, and estimated costs from climate change. Then taxes on wages and wealth could be reduced by the amount produced by the coal tax.

Some 2,500 economists, including eight Nobel Prize winners, have endorsed the concept of tax shifting. Nine western European countries have begun trial versions of such tax shifting. So far, only a small amount of revenue has been shifted by taxes on emissions of CO_2 and toxic metals, garbage production, and vehicles entering congested cities. But such experience shows that this idea works.

A 1998 public opinion poll found that 71% of American voters support a shift from taxing work (payroll taxes) and production (corporate taxes) to taxing pollution and consumption. But the public must be convinced that such shifts represent a real shift of the tax burden, not a ploy for an overall tax increase.

X *How Would You Vote?* Do you favor shifting taxes on wages and profits to pollution and waste? Cast your vote online at www.thomsonedu.com/biology/miller.

Using Environmental Laws and Regulations to Encourage Innovation

Environmental laws and regulations work best if they motivate companies to find innovative ways to control and prevent pollution and reduce resource waste.

Regulation is a form of government intervention in the marketplace that is widely used to help control or prevent pollution and reduce resource waste. It involves enacting and enforcing laws that set pollution standards, regulate harmful activities such as the release of toxic chemicals into the environment, and require that certain irreplaceable or slowly replenished resources be protected from unsustainable use.

According to many environmental and business leaders, *innovation-friendly regulations* can motivate companies to develop eco-friendly products and industrial processes that can create jobs and increase profits and competitiveness in national and international markets. But they also point out that some pollution control regulations are too costly and discourage innovation. Examples are regulations that concentrate on cleanup instead of prevention, are too prescriptive, mandating specific technologies, for example, or set compliance deadlines that are too short to allow companies to find innovative solutions. Experience shows that an innovation-friendly regulatory process sets goals, frees industries to meet them in any way that works, and allows enough time for innovation.

For many years, companies mostly resisted environmental regulation and developed an adversarial relationship with government regulators. In recent years, a growing number of companies have realized the economic and competitive advantages of making environmental improvements and recognized that their shareholder values depend in part on having good environmental records. As a result, a number of firms have begun looking for innovative and profitable ways to reduce resource use, pollution, and waste. They also realize that if they avoid producing a waste or pollutant, they don't have to spend money on controlling it and filling out government reports, and cannot be sued for any harm their outputs may cause. At the same time, many consumers have begun buying green products.

Using the Marketplace to Reduce Pollution and Resource Use

Governments can set a limit on pollution emissions or use of a resource, give pollution or resource-use permits to users, and allow them to trade their permits in the marketplace.

In a market approach to pollution control and resource use, the government grants *tradable pollution and resource-use permits* and sets a limit, or cap, on total emissions of a pollutant or use of a resource such as a fishery. Then it issues or auctions permits that distribute the total among manufacturers or users.

A permit holder not using its entire allocation can use the permit as a credit against future expansion, use it in another part of its operation, or sell it to other companies. In the United States, this approach has been used to reduce the emissions of sulfur dioxide and several other air pollutants, as discussed on p. 000. Tradable rights can also be established among countries to help preserve biodiversity and reduce emissions of greenhouse gases and other regional or global pollutants.

The effectiveness of such programs depends on how high or low the initial cap is set and on the rate at which the cap is reduced.

Green Economics: Selling Services instead of Things

Some businesses can greatly decrease their resource use, pollution, and waste by shifting from selling goods to selling the services the goods provide.

German chemist Michael Braungart and Swiss industry analyst Walter Stahel independently proposed in the

Ray Anderson is CEO of Interface, a company based in the U.S. city of Atlanta, Georgia. The company is the world's largest commercial manufacturer of carpet tiles, with 26 factories in six countries, customers in 110 countries, and more than $1 billion in annual sales.

Anderson changed the way he viewed the world and his business after reading Paul Hawken's book, *The Ecology of Commerce.* In 1994, he announced plans to develop the nation's first totally sustainable green corporation.

Since then, he has implemented hundreds of projects with the goals of zero waste, greatly reduced energy use, and eventually, zero use of fossil fuels, by relying on renewable solar energy. By 1999, his company had reduced resource waste by almost 30% and reduced energy waste enough to save $100 million. One of Interface's factories in California runs on solar cells to produce the world's first solar-made carpet.

To achieve the goal of zero waste, Interface plans to stop selling carpet and lease it as a way to encourage recycling. For a monthly fee, the company will install, clean, and inspect the carpet on a monthly basis, repair worn carpet tiles overnight, and recycle worn-out tiles into new carpeting. As Anderson puts it, "We want to harvest yesterday's carpets and recycle them with zero scrap going to the landfill and zero emissions into the ecosystem— and run the whole thing on sunlight."

Anderson is one of a growing number of business leaders committed to finding more economically and ecologically sustainable ways to do business while still making a profit for stockholders. Between 1993 and 1998, the company's revenues doubled and profits tripled, mostly because the company saved $130 million in material costs with an investment of less than $40 million. Andersen says he is having a blast.

---

mid-1980s a new economic model that would provide profits while greatly reducing resource use and waste. Their idea for more sustainable economies focuses on shifting from the current *material-flow economy* (Figure 4-10, p. 82) to a *service-flow economy.* Instead of buying most goods outright, customers *eco-lease* or rent the *services* that such goods provide.

In a service-flow economy, a manufacturer makes more money if its product uses the minimum amount of materials, lasts as long as possible, and is easy to maintain, repair, remanufacture, reuse, or recycle.

Such an economic shift based on eco-leasing is under way in some businesses. Since 1992, Xerox has been leasing most of its copy machines as part of its mission to provide *document services* instead of selling photocopiers. When the service contract expires, Xerox takes the machine back for reuse or remanufacture and has a goal of sending no material to landfills or incinerators. To save money, machines are designed to use recycled paper, have few parts, be energy efficient, and emit as little noise, heat, ozone, and copier chemical waste as possible. Canon in Japan and Fiat in Italy are taking similar measures.

Dow and several other chemical companies are doing a booming business in leasing organic solvents (used mostly to remove grease from surfaces), photographic developing chemicals, and dyes and pigments. In this *chemical service* business, the company delivers the chemicals, helps the client set up a recovery system, takes away the recovered chemicals, and delivers new chemicals as needed.

Ray Anderson, CEO of a large carpet and tile company, plans to lease rather than sell carpet (Individuals Matter, above). This reduces the company's resource use and harmful environmental impacts.

REDUCING POVERTY TO IMPROVE ENVIRONMENTAL QUALITY AND HUMAN WELL-BEING

Global Outlook: Distribution of the World's Wealth

Since 1960, most of the financial benefits of global economic growth have flowed up to the rich rather than down to the poor.

Poverty usually is defined as the inability to meet one's basic economic needs. According to the World Bank and the United Nations, 1.1 billion people—almost four times the entire U.S. population—struggle to survive on an income of less than $1 (U.S.) a day. About half of the world's people live on a daily income of $1–2 (U.S.).

Poverty has numerous harmful health and environmental effects (Figure 1-9, p. 13, and Figure 11-4, p. 239) and has been identified as one of the five major causes of the environmental problems we face.

Most neoclassical economists believe a growing economy can help the poor by creating more jobs, enabling more of the increased wealth to reach workers, and providing greater tax revenues that can be used to help the poor help themselves. Economists call this the *trickle-down* effect.

However, since 1960, most of the benefits of global economic growth as measured by income have flowed up to the rich rather than down to the poor (Figure 14-6). Since 1980, this *wealth gap* has grown. According to Ismail Serageldin, who has served as a poverty expert for the World Bank, the planet's richest three people have more wealth than the combined GDP of the world's 47 poorest countries and their 600 million people.

South African President Thabo Mbeki told delegates at the 2003 Johannesburg World Summit on Sustainable Development, "A global human society based on poverty for many and prosperity for a few, characterized by islands of wealth, surrounded by a sea of poverty, is unsustainable."

Solutions: Reducing Poverty (Economics, Politics, and Ethics)

We can sharply cut poverty by forgiving the international debts of the poorest countries and greatly increasing international aid and small individual loans to help the poor help themselves.

Analysts point out that reducing poverty requires the governments of most developing countries to make policy changes. One is to shift more of the national

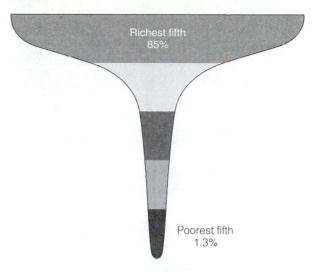

Richest fifth 85%

Poorest fifth 1.3%

Figure 14-6 Global outlook: the *global distribution of income* shows that most of the world's income flows up; the richest 20% of the world's population receive more of the world's income than all of the remaining 80%. Each horizontal band in this diagram represents one-fifth of the world's population. This upward flow of global income has accelerated since 1960 and especially since 1980. This trend can increase environmental degradation by increasing average per capita consumption by the richest 20% of the population and causing the poorest 20% of the world's people to survive by using renewable resources faster than they are replenished. (Data from UN Development Programme and Ismail Serageldin, "World Poverty and Hunger—A Challenge for Science," *Science* 296 (2002): 54–58)

Microloans to the Poor

SOLUTIONS

Most of the world's poor want to earn more, become more self-reliant, and have a better life. But they have no credit record. Also, they lack the assets that could be used for collateral to secure a loan to buy seeds and fertilizer for farming or tools and materials for a small business.

For almost three decades, an innovative tool called *microlending*, or *microfinance*, has helped deal with this problem. For example, since economist Muhammad Yunus started it in 1976, the Grameen (Village) Bank in Bangladesh has provided more than $4 billion in microloans (varying from $50 to $500) to several million mostly poor, rural, and landless women in 40,000 villages. About 94% of the loans are to women who start their own small businesses as seamstresses,

weavers, bookbinders, vendors, and phone service providers.

To stimulate repayment and provide support, the Grameen Bank organizes microborrowers into five-member "solidarity" groups. If one member of the group misses a weekly payment or defaults on the loan, the other members of the group must make the payments.

The Grameen Bank's experience has shown that microlending is both successful and profitable. Less than 3% of microloan repayments to the Grameen Bank are late, and the repayment rate on its loans is an astounding 90–95%. About half of Grameen's borrowers move above the poverty line within 5 years, and domestic violence, divorce, and birth rates are lower among most borrowers. These microloans have helped more than 4 million borrowers work their way out of poverty.

Microloans to the poor by the Grameen Bank are being used to develop day care centers, health clinics, reforestation projects, drinking water supply projects, literacy programs, and group insurance programs. They are also being used to bring small-scale solar and wind power systems to rural villages.

Grameen's model has inspired the development of microcredit projects in more than 58 countries that have reached 36 million people (including dependents), and the number is growing rapidly.

Critical Thinking

Why do you think international development and lending agencies such as the World Bank and the International Monetary Fund have largely ignored microloans? What two things would you do to change this situation?

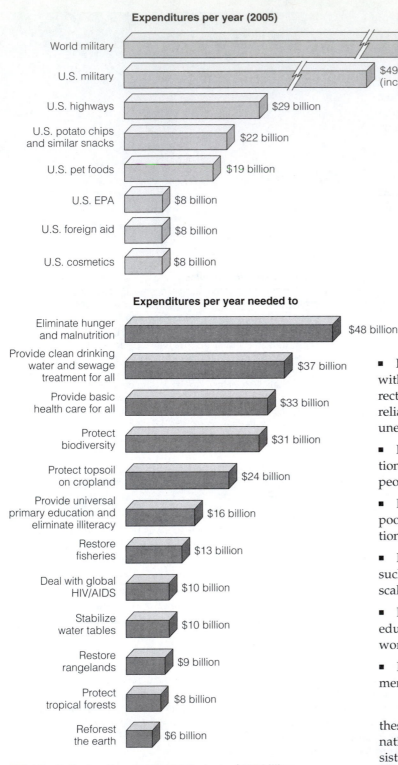

Expenditures per year (2005)

World military	$1 trillion
U.S. military	$492 billion (including Iraq)
U.S. highways	$29 billion
U.S. potato chips and similar snacks	$22 billion
U.S. pet foods	$19 billion
U.S. EPA	$8 billion
U.S. foreign aid	$8 billion
U.S. cosmetics	$8 billion

Expenditures per year needed to

Eliminate hunger and malnutrition	$48 billion
Provide clean drinking water and sewage treatment for all	$37 billion
Provide basic health care for all	$33 billion
Protect biodiversity	$31 billion
Protect topsoil on cropland	$24 billion
Provide universal primary education and eliminate illiteracy	$16 billion
Restore fisheries	$13 billion
Deal with global HIV/AIDS	$10 billion
Stabilize water tables	$10 billion
Restore rangelands	$9 billion
Protect tropical forests	$8 billion
Reforest the earth	$6 billion

Total Earth Restoration and Social Budget = $245 billion

Figure 14-7 Ethics: what should our priorities be? (Data from United Nations, World Health Organization, U.S. Department of Commerce, and U.S. Office of Management and Budget)

budget to help the rural and urban poor work their way out of poverty. Another is to give villages and villagers title to common lands and to crops and trees they plant on them.

Some analysts suggest that we could also help reduce global poverty by forgiving at least 60% of the $2.4 trillion debt that developing countries owe to developed countries and international lending agencies, and all of the $422 billion debt of the poorest and most heavily indebted countries. This would be done on the condition that money saved on debt interest be spent on meeting basic human needs. Currently, developing countries pay almost $300 billion per year in interest to developed countries to service this debt. According to environmental economist John Peet, this inability to "service their debt assures perpetual poverty for the poor nations, and, effectively, perpetual servitude to the rich nations."

Developed countries can also:

■ Increase nonmilitary government and private aid with mechanisms to assure that most of it goes directly to the poor to help them become more self-reliant and to help provide social safety nets such as unemployment payments and access to health care

■ Mount a massive global effort to combat malnutrition and the infectious diseases that kill millions of people prematurely and help perpetuate poverty

■ Encourage lending agencies to make small loans to poor people who want to increase their income (Solutions, p. 307)

■ Make investments in small-scale infrastructure such as solar-cell power facilities in villages, small scale irrigation projects, and farm-to-market roads

■ Mount a global effort to achieve universal primary education, promote gender equality, and empower women.

■ Help developing countries create more environmentally sustainable economies.

According to poverty expert Jeffrey D. Sachs, these goals could be met if the world's developed nations provided $160 billion per year in financial assistance, double the current $80 billion a year being provided by such countries. This amount is less than 0.1% of the world's annual income and is only a fraction of what the world devotes each year to military spending (Figure 14-7).

Making the Transition to Eco-Economies

Nature's four principles of sustainability and various environmental and economic strategies can be used to develop more environmentally sustainable economies.

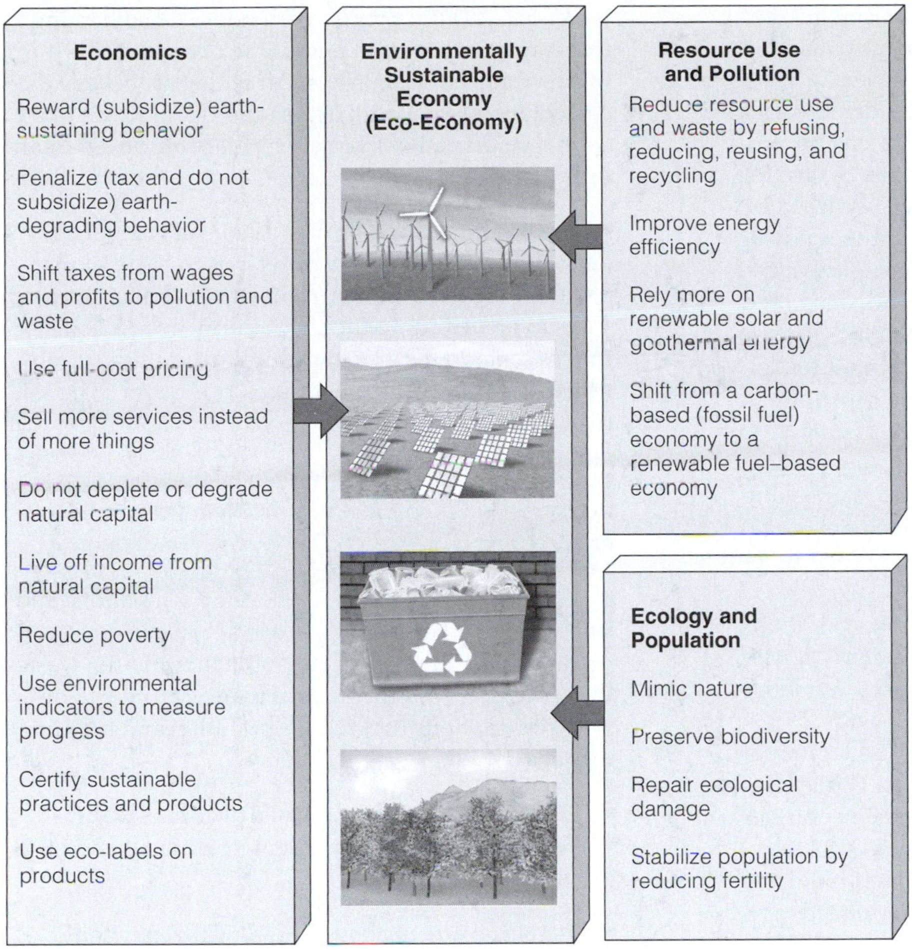

Economics	Environmentally Sustainable Economy (Eco-Economy)	Resource Use and Pollution
Reward (subsidize) earth-sustaining behavior		Reduce resource use and waste by refusing, reducing, reusing, and recycling
Penalize (tax and do not subsidize) earth-degrading behavior		Improve energy efficiency
Shift taxes from wages and profits to pollution and waste		Rely more on renewable solar and geothermal energy
Use full-cost pricing		Shift from a carbon-based (fossil fuel) economy to a renewable fuel–based economy
Sell more services instead of more things		
Do not deplete or degrade natural capital		
Live off income from natural capital		Ecology and Population
Reduce poverty		Mimic nature
Use environmental indicators to measure progress		Preserve biodiversity
Certify sustainable practices and products		Repair ecological damage
Use eco-labels on products		Stabilize population by reducing fertility

Figure 14-8 Solutions: principles for shifting to more environmentally sustainable economies or eco-economies during this century. QUESTION: *Which five of these solutions do you think are the most important?*

Figure 14-8 summarizes ways that Paul Hawken and several other environmental and business leaders have suggested for using nature's four principles of sustainability (Figure 1-11, p. 17) and the various economic tools discussed in this chapter to make the transition to more environmentally sustainable economies over the next several decades. Hawken's simple golden rule for such an economy is this: *"Leave the world better than you found it, take no more than you need, try not to harm life or the environment, and make amends if you do."*

Making this transition is already helping the economy. Environmental protection is a major growth industry that creates new jobs. According to Worldwatch Institute estimates, annual sales of global environmental industries are almost $700 billion—on a par with the global car industry—and employ more than 11 million people (2 million in the United States).

A growing number of economists and CEOs of major international companies point out that shifting to more environmentally sustainable economies over the next several decades will create immense profits and huge numbers of jobs in eco-friendly or more environmentally sustainable businesses. Making such a shift by applying the four principles of sustainability will require bold leadership by business leaders and elected officials and bottom-up political pressure from concerned citizens. Forward-looking investors, corporate executives, and political leaders are recognizing that sustainability is good for the environment and the economy.

However, much of this depends squarely on politics. It involves the difficult task of convincing more business leaders, elected officials, and voters to begin changing current government systems of economic rewards and penalties, as discussed in the following section.

POLITICS AND ENVIRONMENTAL POLICY

Dealing with Environmental Problems in Democracies

Democracies are designed to deal mostly with short-term, isolated problems, and they find it difficult to deal with long-term, interrelated environmental problems.

Democracy is government by the people through elected officials and representatives. In a *constitutional* *democracy,* a constitution provides the basis of government authority, limits government power by mandating free elections, and guarantees free speech.

Political institutions in constitutional democracies are designed to allow gradual change to ensure economic and political stability. In the United States, for example, rapid and destabilizing change is curbed by a system of checks and balances that distributes power among the three branches of government—*legislative, executive,* and *judicial*—and among federal, state, and local governments.

In passing laws, developing budgets, and formulating regulations, elected and appointed government officials must deal with pressure from many competing *special-interest groups.* Each group advocates passing laws, providing subsidies or tax breaks, or establishing regulations favorable to its cause, and weakening or repealing laws, subsidies, taxes, and regulations unfavorable to its position.

Some special-interest groups such as corporations are *profit-making organizations.* Others are *nonprofit nongovernmental organizations* (NGOs) such as labor unions and environmental organizations.

The deliberate design for stability and gradual change in democracies is highly desirable. But several related features of democratic governments hinder their ability to deal with environmental problems. Many problems such as climate change and biodiversity loss have long-range effects, are interrelated, and require integrated long-term solutions emphasizing prevention.

But because elections are held every few years, most politicians seeking re-election focus on short-term, isolated issues rather than on complex, interrelated, time-consuming, and long-term problems. Also, too many political leaders have too little understanding of how the earth's natural systems work and how they support all life, economies, and societies.

Principles for Making Environmental Policy Decisions

Several principles can guide us in developing and implementing more environmentally and economically sustainable and just environmental policies.

An **environmental policy** consists of laws, rules, and regulations related to an environmental problem that are developed, implemented, and enforced by one or more government agencies. Analysts have suggested that legislators and individuals evaluating existing or proposed environmental policy should be guided by several principles:

- *The humility principle:* Our understanding of nature and of the consequences of our actions is quite limited.

- *The reversibility principle:* Try not to do something that cannot be reversed later if the decision turns out to be wrong. For example, most biologists believe the current large-scale destruction and degradation of the earth's biodiversity is largely irreversible on a human time scale.

- *The precautionary principle:* When substantial evidence indicates an activity threatens to harm human health or the environment, take precautionary measures to prevent or reduce such harm, even if some of the cause-and-effect relationships are not fully established scientifically. In such cases, it is better to be safe than sorry.

- *The prevention principle:* Whenever possible, make decisions that help prevent a problem from occurring or becoming worse.

- *The polluter pays principle:* Develop regulations and use economic tools such as green taxes to insure that polluters bear the cost of the pollutants and the wastes they produce. This encourages manufacturers to develop innovations that reduce pollution and resource waste.

- *The integrative principle:* Make decisions that involve integrated solutions to environmental and other problems.

- *The public participation principle:* Citizens should have open access to environmental data and information and the right to participate in developing, criticizing, and modifying environmental policies.

- *The human rights principle:* All people have a right to an environment that does not harm their health and well-being.

- *The environmental justice principle:* Establish environmental policy so that no group of people bears an unfair share of the harms from pollution, degradation, or the execution of environmental laws. See the Guest Essay on this subject by Robert D. Bullard on the website for this chapter.

Influencing Environmental Policy: Individuals Matter

Most improvements in environmental quality result from citizens putting pressure on elected officials and from individuals developing innovative solutions to environmental problems.

A major theme of this book is that *individuals matter.* History shows that significant change usually comes from the *bottom up* when individuals join with others to bring about change. Without grassroots political ac-

What Can You Do?

Influencing Environmental Policy

- Become informed on issues
- Run for office (especially at local level)
- Make your views known at public hearings
- Make your views known to elected representatives
- Contribute money and time to candidates for office
- Vote
- Form or join nongovernment organizations (NGOs) seeking change
- Support reform of election campaign financing

Figure 14-9 Individuals matter: ways you can influence environmental policy. QUESTION: *Which three of these actions do you think are the most important?*

tion by millions of individual citizens and organized groups, the air you breathe and the water you drink today would be much more polluted, and much more of the earth's biodiversity would have disappeared. Figure 14-9 lists ways you can influence and change government policies in constitutional democracies.

Many people recycle, buy environmentally friendly products, and take part in other important and responsible activities that help the environment. But in order to influence environmental policy people need to actively work together to improve communities and neighborhoods, as the citizens of Curitiba, Brazil, (Case Study, p. 106) have done. This and other cases have demonstrated the validity of the insight of Aldo Leopold: "All ethics rest upon a single premise: that the individual is a member of a community of interdependent parts."

Not only can we participate, but each of us can provide *environmental leadership* in four ways. First, we can *lead by example,* using our own lifestyles and values to show others that change is possible and beneficial.

Second, we can *work within existing economic and political systems to bring about improvement* by campaigning and voting for candidates and by communicating with elected officials. We can also send a message to companies producing environmentally harmful products or services by *voting with our wallets* and letting them know the choices we have made and why. And we can work within the system by choosing environmental careers (See the information about environmental careers found on the website for this book).

Third, we can *run for some sort of local office.* Look in the mirror. Maybe you are one who can make a difference as an office holder.

Fourth, we can *propose and work for better solutions to environmental and other problems.* Leadership is more than being against something. It also involves coming up with better ways to accomplish goals and persuading people to work together to achieve such goals. If we care enough to get involved, each of us can make a difference.

Environmental Policy in the United States

Formulating, legislating, and executing environmental policy in the United States is a complex, difficult, and controversial process.

The major function of the federal government in the United States is to develop and implement *policy* for dealing with various issues. Policy is typically composed of *laws* passed by the legislative branch, *regulations* instituted by the executive branch to put laws into effect, and *funding* to implement and enforce the laws and regulations. Figure 14-10 (p. 312) is a greatly simplified overview of how individuals and lobbyists for and against a particular environmental law interact with the three branches of government in the United States. Figure 14-11 (p. 313) lists some of the major environmental laws passed in the United States since 1969.

Passing a law is not enough to make policy. The next step involves trying to get enough funds appropriated to implement and enforce the law. Indeed, developing and adopting a budget is the most important and controversial activity of the executive and legislative branches.

Once a law has been passed and funded, the appropriate government department or agency must draw up regulations for implementing it. An affected group may take the agency to court for failing to implement and enforce the regulations or for enforcing them too rigidly.

Politics plays an important role in the policies and staffing of environmental regulatory agencies—depending on what political party is in power and the prevailing environmental attitudes. Industries facing environmental regulations often put political pressure on regulatory agencies and lobby to have the president appoint people to high positions in such agencies who come from the industries being regulated. In other words, the regulated try to take over the agencies and become the regulators—described by some as "putting foxes in charge of the henhouse."

In addition, people in regulatory agencies work closely with and often develop friendships with officials in the industries they are regulating. Some industries and other groups offer regulatory agency

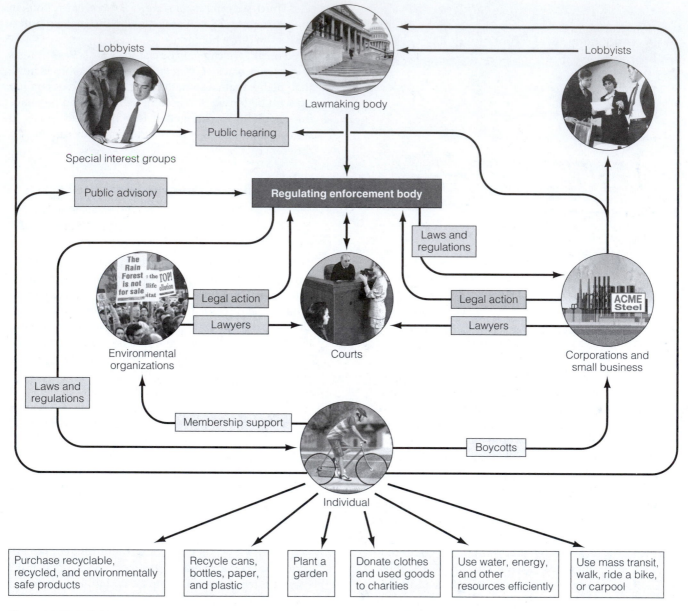

Figure 14-10 Individuals matter: greatly simplified overview of how individuals and lobbyists for and against a particular environmental law interact with the legislative, executive, and judicial branches of the U.S. government. The bottom of this diagram also shows some ways in which individuals can bring about environmental change through their own lifestyles. See the website for this book for details on contacting elected representatives.

employees high-paying jobs in an attempt to influence their regulatory decisions. This can lead to a *revolving door,* as employees move back and forth between industry and government.

Case Study: Managing Public Lands in the United States—Politics in Action

Since the 1800s, controversy has swirled over how publicly owned lands in the United States should be used and managed because of the valuable resources they contain.

No nation has set aside as much of its land for public use, resource extraction, enjoyment, and wildlife as has the United States. The federal government manages roughly 35% of the country's land that belongs to every American. About three-fourths of this federal public land is in Alaska, and another fifth is in the western states (Figure 14-12, p. 314).

Some federal public lands are used for many purposes. For example, the *National Forest System* comprises 155 national forests and 22 national grasslands. These forests, managed by the U.S. Forest Service (USFS), are used for logging, mining, livestock grazing, farming, oil and gas extraction, recreation, and conservation of watershed, soil, and wildlife resources.

The Bureau of Land Management (BLM) manages *National Resource Lands.* These lands are used primarily for mining, oil and gas extraction, and livestock grazing.

The U.S. Fish and Wildlife Service (USFWS) manages 542 *National Wildlife Refuges.* Most refuges protect habitats and breeding areas for waterfowl and big game to provide a harvestable supply for hunters; a few protect endangered species from extinction. Permitted activities in most refuges include hunting, trapping, fishing, oil and gas development, mining, logging, grazing, some military activities, and farming.

Uses of other public lands are more restricted. The *National Park System* is managed by the National Park Service (NPS). It includes 58 major parks (mostly in the West) and 331 national recreation areas, monuments, memorials, battlefields, historic sites, parkways, trails, rivers, seashores, and lakeshores. Only camping, hiking, sport fishing, and boating can take place in the national parks, but sport hunting, mining, and oil and gas drilling are allowed in National Recreation Areas.

The most restricted public lands are 660 roadless areas that make up the *National Wilderness Preservation System.* These areas lie within other public lands and are managed by agencies in charge of those lands. Most of these areas are open only for recreational activities such as hiking, sport fishing, camping, and nonmotorized boating.

Many federal public lands contain valuable oil, natural gas, coal, timber, and mineral resources. Since the 1800s, debates have focused on how the resources on these lands should be used and managed.

Most conservation biologists, environmental economists, and many free-market economists believe that four principles should govern use of public land:

- Protecting biodiversity, wildlife habitats, and the ecological functioning of public land ecosystems should be the primary goal.

- No one should receive government subsidies or tax breaks for using or extracting resources on public lands.

- The American people deserve fair compensation for the use of their property.

- All users or extractors of resources on public lands should be fully responsible for any environmental damage they cause.

There is strong and effective opposition to these ideas. Economists, developers, and resource extractors tend to view public lands in terms of their usefulness in providing mineral, timber, and other resources and their ability to increase short-term economic growth.

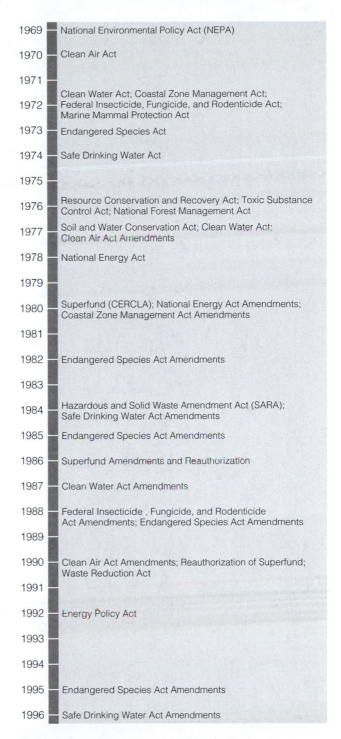

Year	Law
1969	National Environmental Policy Act (NEPA)
1970	Clean Air Act
1971	
1972	Clean Water Act; Coastal Zone Management Act; Federal Insecticide, Fungicide, and Rodenticide Act; Marine Mammal Protection Act
1973	Endangered Species Act
1974	Safe Drinking Water Act
1975	
1976	Resource Conservation and Recovery Act; Toxic Substance Control Act; National Forest Management Act
1977	Soil and Water Conservation Act; Clean Water Act; Clean Air Act Amendments
1978	National Energy Act
1979	
1980	Superfund (CERCLA); National Energy Act Amendments; Coastal Zone Management Act Amendments
1981	
1982	Endangered Species Act Amendments
1983	
1984	Hazardous and Solid Waste Amendment Act (SARA); Safe Drinking Water Act Amendments
1985	Endangered Species Act Amendments
1986	Superfund Amendments and Reauthorization
1987	Clean Water Act Amendments
1988	Federal Insecticide, Fungicide, and Rodenticide Act Amendments; Endangered Species Act Amendments
1989	
1990	Clean Air Act Amendments; Reauthorization of Superfund; Waste Reduction Act
1991	
1992	Energy Policy Act
1993	
1994	
1995	Endangered Species Act Amendments
1996	Safe Drinking Water Act Amendments

Figure 14-11 Solutions: some major environmental laws and their amended versions enacted in the United States since 1969. A more detailed list is found on the website for this chapter.

They have succeeded in blocking implementation of the four principles just listed. For example, in recent years, the government has given more than $1 billion a year in subsidies to privately owned mining, fossil fuel extraction, logging, and grazing interests using U.S. public lands.

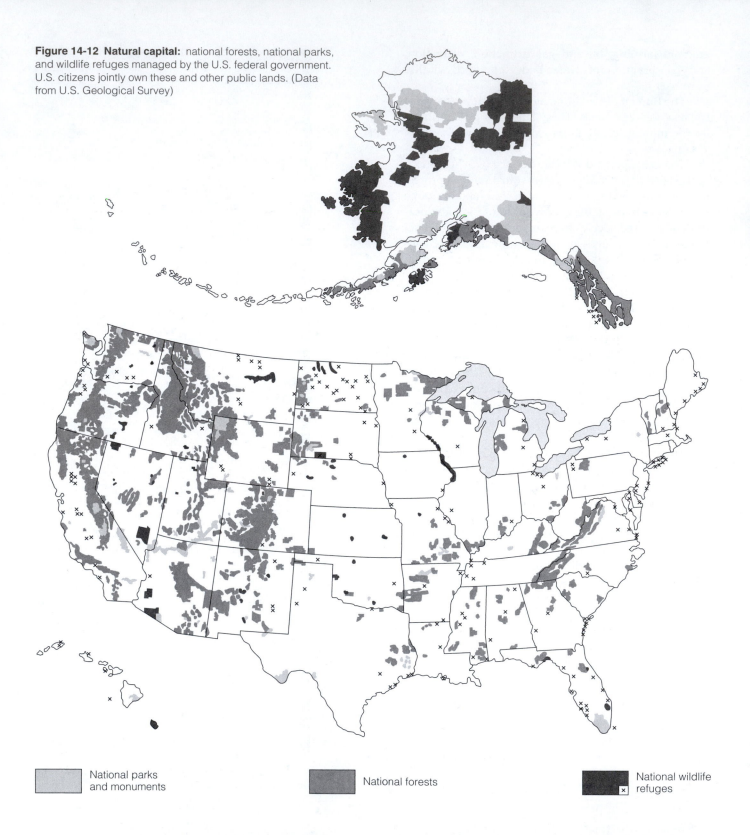

Figure 14-12 Natural capital: national forests, national parks, and wildlife refuges managed by the U.S. federal government. U.S. citizens jointly own these and other public lands. (Data from U.S. Geological Survey)

National parks and monuments

National forests

National wildlife refuges

Some developers and resource extractors have sought to go further, mounting a campaign to get the U.S. Congress to pass laws that would

- Sell public lands or their resources to corporations or individuals, usually at less than market value, or turn over their management to state and local governments

- Slash federal funding for regulatory administration of public lands

- Cut all old-growth forests in the national forests and replace them with tree plantations

- Open all national parks, national wildlife refuges, and wilderness areas to oil drilling, mining, off-road vehicles, and commercial development

- Do away with the National Park Service and launch a 20-year construction program of new concessions and theme parks run by private firms in the national parks

- Continue mining on public lands under the provisions of the 1872 Mining Law, which allows mining interests to pay no royalties to taxpayers for hard-rock minerals they remove and to avoid taking full responsibility for any damage they cause

- Repeal the Endangered Species Act or modify it to allow economic factors to override protection of endangered and threatened species

- Redefine government-protected wetlands so about half of them would no longer be protected

- Prevent individuals or groups from legally challenging these uses of public land for private financial gain

Since 2002, the U.S. Congress has expanded the extraction of mineral, timber, and fossil fuel resources on U.S. public lands and weakened environmental laws and regulations protecting such lands from environmental abuse and exploitation. Conservation biologists and environmental scientists argue that this depletes the country's irreplaceable natural capital by increasing environmental degradation and decreasing biodiversity.

X *HOW WOULD YOU VOTE?* Should much more of U.S. public lands (or government-owned lands in the country where you live) be opened up to the extraction of timber, mineral, and energy resources? Cast your vote online at www.thomsonedu .com/biology/miller.

Major Environmental Groups and Networks

Major environmental groups and networks monitor environmental activities, work to pass and strengthen environmental laws, and work with corporations to find solutions to environmental problems.

The spearheads of the global conservation, environmental, and environmental justice movements are the more than 100,000 nonprofit NGOs working at the international, national, state, and local levels—up from about 2,000 in 1970. The growing influence of these organizations is one of the most important changes influencing environmental decisions and policies.

NGOs range from grassroots groups with just a few members to global organizations like the 5-million-member World Wide Fund for Nature, with offices in 48 countries. Other international groups with large memberships include Greenpeace, the Nature Conservancy, Grameen Bank (Solutions, p. 307), and Conservation International. Using e-mail and the Internet, some environmental NGOs have organized themselves into an array of powerful international networks.

In the United States, more than 8 million citizens belong to more than 30,000 NGOs dealing with environmental and environmental justice issues. They range from small *grassroots* groups to large heavily *funded* groups staffed by expert lawyers, scientists, and economists.

The large groups have become powerful and important forces within the political system. They have helped persuade the U.S. Congress to pass and strengthen environmental laws (Figure 14-11) and to fight off attempts to weaken or repeal such laws. Some industries and environmental groups are working together to find solutions to environmental problems. For example, Environmental Defense worked with McDonald's to redesign its packaging system to eliminate its plastic hamburger containers, and with General Motors to remove high-pollution cars from the road.

The base of the environmental movement in the United States and throughout the world consists of thousands of grassroots citizens' groups organized to improve environmental quality, often at the local level. According to political analyst Konrad von Moltke, "There isn't a government in the world that would have done anything for the environment if it weren't for the citizen groups." Taken together, a loosely connected worldwide network of grassroots NGOs working for bottom-up political, social, economic, and environmental change can be viewed as an emerging citizen-based *global sustainability movement.*

These groups have worked with individuals and communities to oppose harmful projects such as landfills, waste incinerators, nuclear waste dumps, clear-cutting of forests, and various potentially harmful development projects. They have also taken action against environmental injustice. See the Guest Essay on this topic by Robert D. Bullard on the website for this chapter. Grassroots groups have also formed land trusts and other local organizations to save wetlands, forests, farmland, and ranchland from development, and they have helped restore forests (Individuals Matter, p. 118), degraded rivers, and wetlands.

Some grassroots environmental groups use the nonviolent and nondestructive tactics of protest marches, tree sitting (Individuals Matter, p. 114), and other devices for generating publicity to help educate and sway members of the public to oppose various environmentally harmful activities. Much more controversial are militant environmental groups that break into labs to free animals used to test drugs or that destroy property such as bulldozers and SUVs. Most environmentalists strongly oppose such tactics.

X *HOW WOULD YOU VOTE?* Do you support the use of nonviolent and nondestructive civil disobedience tactics by some environmental groups? Cast your vote online at www .thomsonedu.com/biology/miller.

Case Study: Environmental Action by Students in the United States

Many student environmental groups work to bring about environmental improvements in their schools and local communities.

Since 1988, there has been a boom in environmental awareness at a number of college campuses and public schools across the United States. Most student environmental groups work with members of the faculty and administration to bring about environmental improvements in their schools and local communities.

Many of these groups make environmental audits of their campuses or schools. Then they use the data gathered to propose changes that will make their campus or school more ecologically sustainable, usually saving money in the process. As a result, students have helped convince almost 80% of universities and colleges in the United States to develop recycling programs.

Students at Oberlin College in Ohio helped design a more sustainable environmental studies building. At Northland College in Wisconsin, students helped design a "green" dorm that features a large wind generator, panels of solar cells, recycled furniture, and waterless (composting) toilets.

At Minnesota's St. Olaf College, students have carried out sustainable agriculture and ecological restoration projects. Students at Brown University studied the impacts of lead and other toxic pollutants in low-income neighborhoods in their city of Providence, Rhode Island.

Such student-spurred environmental activities and research studies are spreading to universities in at least 42 other countries. See Noel Perrin's Guest Essay on environmental activities at U.S. colleges on this chapter's website.

Case Study: Threats to the U.S. Environmental Legal and Regulatory Structure

Environmental groups have helped pass environmental laws and educate the public and business and political leaders about environmental issues, but an organized movement has undermined many of these efforts.

Since 1970, a variety of environmental groups in the United States and other countries have helped increase understanding of environmental issues and have gained public support for an array of environmental and resource-use laws in the United States and other countries. In addition, they have helped individuals deal with a number of local environmental problems.

Since 1980, a well-organized and well-funded movement has mounted a strong campaign to weaken or repeal existing U.S. environmental laws and regula-

tions, change the way in which public lands are used (p. 315), and destroy the reputation and effectiveness of the U.S. environmental movement.

Three major groups are strongly opposed to many environmental proposals, laws, and regulations: some corporate leaders and other powerful people who see environmental laws and regulations as threats to their wealth and power; citizens who see environmental laws and regulations as threats to their private property rights and jobs; and state and local government officials who resent having to implement federal environmental laws and regulations with little if any federal funding (unfunded mandates) or who disagree with certain regulations.

One problem is that the focus of environmental issues has shifted from easy-to-see dirty smokestacks and burning rivers to more complex, controversial, and often invisible environmental problems such as climate change and biodiversity loss. Explaining such complex issues to the public and mobilizing support for often controversial, long-range solutions to such problems is difficult. See the Guest Essay on environmental reporting by Andrew C. Revkin on the website for this chapter.

Another problem is that some environmentalists have mostly brought bad news about the state of the environment. History shows that bearers of bad news are not received well, and opponents of environmentalists have used this to undermine environmental concerns.

History also shows that people are moved to bring about change mostly by an inspiring, positive vision of what the world could be like, one that provides hope for the future. So far, environmentalists with a variety of beliefs and goals have not worked together to develop broad, compelling, and positive visions that can be used as road maps of hope for a more sustainable future for humans and other species.

Since 1980, the well-funded anti-environmental movement in the United States with specific goals, values, and an overarching vision of accelerating exploitation of the earth's natural capital for short-term economic benefit has succeeded in weakening the U.S. environmental movement and the laws and regulations that provide environmental protection for U.S. citizens.

And since 2000, these efforts have escalated. Most major U.S. federal environmental laws and regulatory agencies are now being weakened by a combination of executive rulings, congressional actions, staffing of federal agencies with officials who favor weakening those agencies, decreasing funding, ignoring sound scientific consensus, and stifling dissent. Environmental leaders warn that the entire U.S. legal and regulatory environmental structure, built up by decades of bipartisan consensus, is being systematically undermined, much like the foundation and walls of a house being eaten away by termites.

Polls show that about 80% of the U.S. public strongly supports environmental laws and regulations and do not want them weakened. But polls also show that less than 10% of the U.S. public views the environment as one of the nation's most pressing problems. As a result, environmental concerns often do not get transferred to the ballot box. As one political scientist put it, "Environmental concerns are like the Florida Everglades, a mile wide but only a few inches deep."

Preventing further weakening of the U.S. environmental laws and regulations and repairing the damage that has been done will require a combination of education about what is happening and organized bottom-up political pressure from concerned citizens, which led to passage of important environmental laws in the first place. People on all sides of important environmental issues must listen to one another's concerns, try to find areas of agreement, work together to find solutions, and improve environmental laws and regulations.

GLOBAL ENVIRONMENTAL POLICY

National and Global Security

Many analysts believe that environmental security is as important as military and economic security.

Countries are legitimately concerned with *military security* and *economic security*. However, ecologists and environmental and ecological economists point out that all economies are supported by the earth's natural capital.

According to environmental expert Norman Myers,

> If a nation's environmental foundations are degraded or depleted, its economy may well decline, its social fabric deteriorate, and its political structure become destabilized as growing numbers of people seek to sustain themselves from declining resource stocks. Thus, national security is no longer about fighting forces and weaponry alone. It relates increasingly to watersheds, croplands, forests, genetic resources, climate, and other factors that, taken together, are as crucial to a nation's security as are military factors.

Proponents of this view call for all countries to make environmental security a major focus of diplomacy and government policy at all levels. This could be implemented by a council of advisers made up of highly qualified experts in environmental, economic, and military security, who integrate all three security concerns in making major decisions.

Global Outlook: International Environmental Organizations

International environmental organizations gather and evaluate environmental data, help develop environmental treaties, and provide funds and loans for sustainable economic development.

A variety of international environmental organizations helps shape and set environmental policy. Perhaps the most influential is the United Nations, which houses a large family of organizations such as the UN Environment Programme (UNEP), the World Health Organization (WHO), United Nations Development Programme (UNDP), and the Food and Agriculture Organization (FAO).

Other organizations that make or influence environmental decisions are the World Bank, the Global Environment Facility (GEF), and the World Conservation Union (IUCN). The website for this chapter has a list of major international environmental organizations. Despite their often limited funding, these and other organizations have played important roles in

- Expanding understanding of environmental issues
- Gathering and evaluating environmental data
- Developing and monitoring international environmental treaties
- Providing grants and loans for sustainable economic development and reducing poverty
- Helping more than 100 nations develop environmental laws and institutions

Progress in Developing International Environmental Cooperation and Policy

Earth summits and international environmental treaties have played important roles in dealing with global environmental problems, but most treaties are not effectively monitored and enforced.

Since the 1972 UN Conference on the Human Environment in Stockholm, Sweden, progress has been made in addressing environmental issues at the global level. Figure 14-13 (p. 318) lists some of the good and bad news about international efforts to deal with global environmental problems such as poverty, climate change, biodiversity loss, and ocean pollution.

In 2004, environmental leader Gus Speth argued that global environmental problems are getting worse and that international efforts to solve them are inadequate. He and several other environmental leaders call

Trade-Offs

Global Efforts on Environmental Problems

Good News	Bad News

Good News

Environmental protection agencies in 115 nations

Over 500 international environmental treaties and agreements

UN Environment Programme (UNEP) created in 1972 to negotiate and monitor international environmental treaties

1992 Rio Earth Summit adopted key principles for dealing with global environmental problems

2002 Johannesburg Earth Summit attempted to implement policies and goals of 1992 Rio summit and find ways to reduce poverty

Bad News

Most international environmental treaties lack criteria for monitoring and evaluating their effectiveness

1992 Rio Earth Summit led to nonbinding agreements without enough funding to implement them

By 2003 there was little improvement in the major environmental problems discussed at the 1992 Rio summit

2002 Johannesburg Earth Summit failed to provide adequate goals, deadlines, and funding for dealing with global environmental problems such as climate change, biodiversity loss, and poverty

Figure 14-13 Trade-offs: good and bad news about international efforts to deal with global environmental problems. QUESTION: *What are the single piece of good news and the single piece of bad news that you think are the most important?*

for the creation of a World Environmental Organization (WEO), on the order of the World Health Organization and the World Trade Organization, to deal with global environmental challenges.

ENVIRONMENTAL WORLDVIEWS AND ETHICS: CLASHING VALUES AND CULTURES

What Is an Environmental Worldview?

Your environmental worldview encompasses how you think the world works, what you believe your environmental role in the world should be, and what you believe is right and wrong environmental behavior.

People disagree about how serious our environmental problems are and what we should do about them. These conflicts arise mostly out of differing **environmental worldviews:** how people think the world works; what they believe their role in the world should be; and what they believe is right and wrong environmental behavior (**environmental ethics**). People with widely differing environmental worldviews can take the same data, be logically consistent, and arrive at quite different conclusions because they start with different assumptions and values.

There are many different environmental worldviews. Some are *human-centered*. Others are *life-centered* with the primary focus on individual species or on sustaining the earth's natural life forms (biodiversity) and life-support systems (biosphere) for the benefit of humans and other forms of life.

Human-Centered Environmental Worldviews

This type of worldview holds that humans are the planet's most important species and should become managers or stewards of the earth.

Some people have a **planetary management worldview.** According to this human-centered worldview, humans are the planet's most important and dominant species, and we can and should manage the earth mostly for our own benefit. Other species and parts of nature are seen as having only *instrumental value* based on how useful they are to us. Figure 14-14 (left) summarizes the four major beliefs or assumptions of one version of this worldview.

Another human-centered environmental worldview is the **stewardship worldview.** It assumes that we have an ethical responsibility to be caring and responsible managers, or *stewards*, of the earth. Figure 14-14 (center) summarizes the major beliefs of this worldview.

According to the stewardship view, as we use the earth's natural capital, we are borrowing from the earth and from future generations. As a result, we have an ethical responsibility to pay the debt by leaving the earth in at least as good a condition as we enjoy. In thinking about our responsibility toward future generations, some analysts believe we should consider the wisdom of the 18th century Iroquois Confederation of Native Americans: *In our every deliberation, we must consider the impact of our decisions on the next seven generations.*

Some people believe any human-centered worldview will eventually fail because it wrongly assumes we now have or can gain enough knowledge to become effective managers or stewards of the earth. These critics point out that we do not even know how many species live on the earth, much less what their roles are and how they interact with one another and

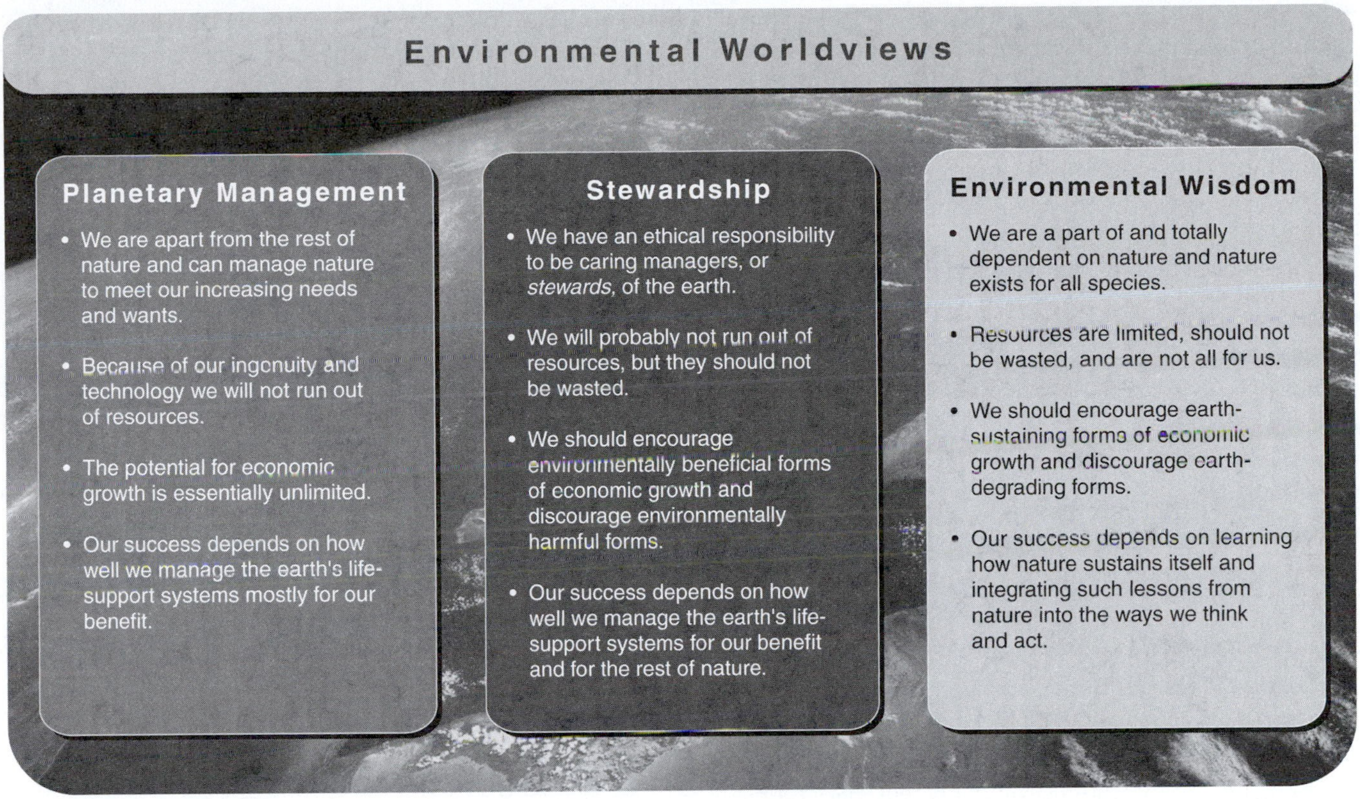

Environmental Worldviews

Planetary Management

- We are apart from the rest of nature and can manage nature to meet our increasing needs and wants.

- Because of our ingenuity and technology we will not run out of resources.

- The potential for economic growth is essentially unlimited.

- Our success depends on how well we manage the earth's life-support systems mostly for our benefit.

Stewardship

- We have an ethical responsibility to be caring managers, or *stewards*, of the earth.

- We will probably not run out of resources, but they should not be wasted.

- We should encourage environmentally beneficial forms of economic growth and discourage environmentally harmful forms.

- Our success depends on how well we manage the earth's life-support systems for our benefit and for the rest of nature.

Environmental Wisdom

- We are a part of and totally dependent on nature and nature exists for all species.

- Resources are limited, should not be wasted, and are not all for us.

- We should encourage earth-sustaining forms of economic growth and discourage earth-degrading forms.

- Our success depends on learning how nature sustains itself and integrating such lessons from nature into the ways we think and act.

Figure 14-14 Comparison of three major environmental worldviews.

their nonliving environment. We have only an inkling of what goes on in a handful of soil, a meadow, an ocean, or any other part of the earth.

Life-Centered and Earth-Centered Environmental Worldviews

Some believe that we have an ethical responsibility not to degrade the earth's ecosystems, biodiversity, and biosphere for all forms of life.

Critics of human-centered environmental worldviews believe that these views should be expanded to recognize the *inherent* or *intrinsic value* of all forms of life, regardless of their potential or actual use to humans. Most people with a life-centered worldview believe we have an ethical responsibility to avoid causing the premature extinction of species through our activities for two reasons. *First*, each species is a unique storehouse of genetic information that should be respected and protected because it exists (*intrinsic value*). *Second*, each species is a potential economic good for human use (*instrumental value*).

Some people think we should go beyond focusing mostly on species. They believe we have an ethical responsibility not to degrade the earth's ecosystems, biodiversity, and biosphere for this and future generations of humans and other species. This *earth-centered* environmental worldview is devoted to preserving the earth's biodiversity and the functioning of its life-support systems for all forms of life now and in the future.

One earth-centered worldview is called the **environmental wisdom worldview.** Figure 14-14 (right) summarizes its major beliefs. In many respects, it is the opposite of the planetary management worldview (Figure 14-14, left). According to this worldview, we are part of—not apart from—the community of life and the ecological processes that sustain all life.

This worldview is based on the ethic that we should strive to care for all species and all of humanity. It also suggests that the earth does not need us managing it in order to go on, whereas we need the earth in order to survive. *We cannot save the earth because it does not need saving.* What we need to save is the existence of our own species and other species that may become extinct because of our activities. See Lester Milbrath's Guest Essay on this topic on the website for this chapter.

Global and national polls indicate a cultural shift toward the stewardship and environmental wisdom worldviews. For example, a 2002 global survey found that 52% of the respondents and 74% of those in the United States and other leading industrial countries agreed that "protecting the environment should be given priority over economic growth and creating jobs." And a 2002 national survey of the United States found that Americans strongly agreed that humans (1) should adapt to nature rather than modify it to suit them (76%), **(2)** have moral duties and obligations to plant and other animal species (87%), and **(3)** have moral duties and obligations to nonliving nature such as air, water, and rocks (80%). And 90% of Americans responding to this poll believed that nature has value within itself regardless of any value humans place on it.

The key is to convert these value shifts into individual and collective actions that implement these environmental beliefs at the individual, local, national, and global levels.

X *HOW WOULD YOU VOTE?* Which one of the following comes closest to your environmental worldview: planetary management, stewardship, or environmental wisdom? Cast your vote online at www.thomsonedu.com/biology/miller.

LIVING MORE SUSTAINABLY

Environmental Literacy

Environmentally literate citizens and leaders are needed to build more environmentally sustainable and socially just societies.

Learning how to live more sustainably requires a foundation of environmental education. Here are some key goals of environmental education or ecological literacy:

- Develop respect or reverence for all life

- Understand as much as we can about how the earth works and sustains itself (Figure 1-11, p. 17), and use such knowledge to guide our lives, communities, and societies

- Look for connections within the biosphere and between our actions and the biosphere

- Use critical thinking skills (pp. 3–4) to become seekers of environmental wisdom instead of overfilled vessels of environmental information

- Understand and evaluate our environmental worldviews and continue this as a lifelong process

- Learn how to evaluate the beneficial and harmful consequences to the earth of our choices of lifestyle and profession, today and in the future

- Foster a desire to make the world a better place and act on this desire

Specifically, an ecologically literate person should have a basic comprehension of:

- Concepts such as environmental sustainability, natural capital, exponential growth, carrying capacity, risk, and risk analysis

- Environmental history to help keep us from repeating past mistakes

- The two laws of thermodynamics and the law of conservation of matter

- Basic principles of ecology, such as food webs, nutrient cycling, biodiversity, ecological succession, and population dynamics

- Human population dynamics

- Ways to sustain biodiversity

- Sustainable agriculture and forestry

- Soil conservation

- Sustainable water use

- Nonrenewable mineral resources

- Nonrenewable and renewable energy resources

- Climate change and ozone depletion

- Pollution prevention and waste reduction

- Sustainable cities

- Environmentally sustainable economic and political systems

- Environmental worldviews and ethics

According to environmental educator Mitchell Thomashow, four basic questions should be at the heart of environmental literacy. *First,* where do the things I consume come from? *Second,* what do I know about the place where I live? *Third,* how am I connected to the earth and other living things? *Fourth,* what is my purpose and responsibility as a human being? How we answer these questions determines our *ecological identity*.

Figure 14-15 summarizes guidelines and strategies that have been discussed throughout this book for achieving more sustainable societies.

Learning from the Earth

In addition to formal learning, we need to learn by experiencing nature directly.

Formal environmental education is important, but is it enough? Many analysts say no and urge us to take the time to escape the cultural and technological body armor we use to insulate ourselves from nature and to experience nature directly.

They suggest we kindle a sense of awe, wonder, mystery, excitement, and humility by standing under the stars, sitting in a forest, or taking in the majesty and power of an ocean.

We might pick up a handful of soil and try to sense the teeming microscopic life in it that keeps us alive. We might look at a tree, mountain, rock, or bee and try to sense how they are a part of us and we a part of them as interdependent participants in the earth's life-sustaining recycling processes.

Many psychologists believe that consciously or unconsciously, we spend much of our lives in a search for roots: something to anchor us in a bewildering and frightening sea of change. As philosopher Simone Weil observed, "To be rooted is perhaps the most important and least recognized need of the human soul."

Solutions

Developing Environmentally Sustainable Societies

Guidelines	Strategies

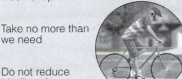

Guidelines

Learn from and copy nature

Do not degrade or deplete the earth's natural capital, and live off the natural income it provides

Take no more than we need

Do not reduce biodiversity

Try not to harm life, air, water, soil

Do not change the world's climate

Help maintain the earth's capacity for self-repair

Do not overshoot the earth's carrying capacity

Repair past ecological damage

Leave world in as good a shape as—or better than—we found it

Strategies

Sustain biodiversity

Eliminate poverty

Develop eco-economies

Build sustainable communities

Do not use renewable resources faster than nature can replace them

Use sustainable agriculture

Depend more on locally available renewable energy from the sun, wind, flowing water, and sustainable biomass

Emphasize pollution prevention and waste reduction

Do not waste matter and energy resources

Recycle, reuse, and compost 60–80% of matter resources

Maintain a human population size such that needs are met without threatening life-support systems

Emphasize ecological restoration

Figure 14-15 Solutions: guidelines and strategies for achieving more sustainable societies. QUESTION: *Which three guidelines and which three strategies do you think are the most important?*

Earth-focused philosophers say that to be rooted, each of us needs to find a *sense of place:* a stream, a mountain, a yard, a neighborhood lot, or any piece of the earth we feel at one with as a place we know, experience emotionally, and love. When we become part of a place, it becomes a part of us. Then we are driven to defend it from harm and to help heal its wounds.

This might lead us to recognize that the healing of the earth and the healing of the human spirit are one and the same. We might discover and tap into what Aldo Leopold calls "the green fire that burns in our hearts" and use this as a force for respecting and working with the earth and with one another.

Living More Simply

Some affluent people are voluntarily adopting lifestyles in which they enjoy life more by consuming less.

Many analysts urge us to *learn how to live more simply.* Seeking happiness through the pursuit of material things is considered folly by almost every major religion and philosophy. Yet this message is preached incessantly by modern advertising, which encourages us to buy more and more things as a way to achieve happiness. As a result, polls reveal that too many people are working too many hours to buy too much stuff that gives them too little true happiness.

Some affluent people in developed countries are adopting a lifestyle of *voluntary simplicity*, doing and enjoying more with less by learning to live more simply. Voluntary simplicity is a form of *environmentally ethical consumption* that is based on Mahatma Gandhi's *principle of enoughness:* "The earth provides enough to satisfy every person's need but not every person's greed. . . . When we take more than we need, we are simply taking from each other, borrowing from the future, or destroying the environment and other species."

Most of the world's major religions have similar teachings: "Why do you spend your money for that which is not bread, and your labor for that which does not satisfy?" (Christianity: Old Testament, Isaiah 55:2). "Eat and drink, but waste not by excess" (Islam: Koran 7.31). "One should abstain from acquisitiveness" (Hinduism: Acarangastura 2.119). "He who knows he has enough is rich" (Taoism: Tao Te Ching, Chapter 33).

Implementing this principle means asking oneself, "How much is enough?" This is not easy because people in affluent societies are conditioned to want more and more and to think of wants as vital needs.

Becoming Better Environmental Citizens

We can help make the world a better place by not falling into mental traps that lead to denial and inaction, and by keeping our empowering feelings of hope slightly ahead of any immobilizing feelings of despair.

We all make some direct or indirect contributions to the environmental problems we face. However, because we do not want to feel guilty or bad about the environmental harm we cause, we try not to think about it too much—a path that can lead to denial and inaction.

Some suggest that we move beyond blame, guilt, fear, and apathy by recognizing and avoiding common mental traps that lead to denial, indifference, and inaction. These traps include:

- *gloom-and-doom pessimism* (it is hopeless),

- *blind technological optimism* (science and technofixes will save us),

- *fatalism* (we have no control over our actions and the future),

- *extrapolation to infinity* (if I cannot change the entire world quickly, I will not try to change any of it),

- *paralysis by analysis* (searching for the perfect worldview, philosophy, solutions, and scientific information before doing anything), and

- *faith in simple, easy answers*.

Some of these traps lead to unrealistic hopes and some lead to unrealistic fears. To avoid them, it helps us to keep our empowering feelings of hope slightly ahead of any immobilizing feelings of despair.

Recognizing that there is no single correct or best solution to the environmental problems we face is also important. Indeed, one of nature's principles of sustainability holds that preserving diversity—in this case being flexible and adaptable in trying a variety of solutions to our environmental problems—is the best way to adapt to the earth's and life's largely unpredictable, ever-changing conditions.

Finally, we should have fun and take time to enjoy life. Laugh every day and enjoy nature, beauty, friendship, and love. This helps empower us to become good earth citizens who practice *good earthkeeping*.

Components of the Sustainability Revolution

We need hope, a positive vision of the future, and commitment to making the world a better place to live.

The industrial revolution took place during the past 275 years. Now in the 21st Century, environmental leaders say it is time for an *environmental* or *sustainability revolution*. It would have several components:

- A *biodiversity protection revolution* devoted to protecting and sustaining the genes, species, natural systems, and chemical and biological processes that make up the earth's biodiversity

- An *efficiency revolution* that sharply reduces our waste of matter and energy resources

- An *energy revolution* based on decreasing our dependence on carbon-based nonrenewable fossil fuels and increasing our dependence on forms of renewable energy such as the sun, wind, flowing water, biomass, and geothermal energy

- A *pollution prevention revolution* that reduces pollution and environmental degradation by applying the precautionary principle and preventing it from happening

- A *sufficiency revolution* dedicated to meeting the basic needs of all people on the planet while affluent societies learn to live more sustainably by living with less

- A *demographic revolution* based on bringing the size and growth rate of the human population into balance with the earth's ability to support humans and other species sustainably

- An *economic and political revolution* in which we use economic systems to reward environmentally beneficial behavior and discourage environmentally harmful behavior

We possess an incredible array of scientific, technological, and economic solutions to the environmental problems we face. Our challenge is to implement such solutions by converting environmental knowledge, wisdom, and beliefs into political action.

As we explore different paths to sustainability we must first understand that our lives and societies depend on *natural capital* and that one of the biggest threats to our ways of life is our own active role in *natural capital degradation*. With that understanding, we begin the search for *solutions* to difficult environmental problems. Competing interests working together to find the solutions must make *trade-offs* because this is the essence of the political process.

This requires understanding that *individuals matter*. Virtually all of the environmental progress we have made during the last few decades occurred because individuals banded together to insist that we can do better.

This journey begins in your own community because in the final analysis *all sustainability is local*. It really begins with your own lifestyle. This is the meaning of the motto: "Think globally, act locally."

It is an incredibly exciting time to be alive as we struggle to implement such ideals by entering into a new relationship with the earth that keeps us all alive and supports our economies. The cultural transition to a more sustainable world will not be easy, but it can be done if enough of us care.

Envision the earth's life-sustaining processes as a beautiful and diverse web of interrelationships—a kaleidoscope of patterns, rhythms, and connections whose very complexity and multitude of possibilities remind us that cooperation, sharing, honesty, humility, and love should be the guidelines for our behavior toward one another and the earth.

When there is no dream, the people perish.

PROVERBS 29:18

CRITICAL THINKING

1. Should we attempt to maximize economic growth by producing and consuming more and more economic goods? Explain. What are the alternatives?

2. Suppose that over the next 20 years the current harmful environmental and health costs of goods and services are internalized so their market prices reflect their total costs. List three harmful and three beneficial effects that such full-cost pricing might have on your lifestyle. Do you favor making this shift?

3. List all the goods you use, and then identify those that meet your basic needs and those that satisfy your wants. Identify any economic wants that you **(a)** would be willing to give up, **(b)** believe you should give up but are unwilling to give up, and **(c)** hope to give up in the future. Relate the results of this analysis to your personal impact on the environment. Compare your results with those of your classmates.

4. Are you for or against shifting to a service-flow economy in the country and community where you live? Explain. Do you think it will be possible to shift to such an economy over the next two to three decades? Explain. What are the three most important strategies for doing this?

5. What are the greatest strengths and weaknesses of the system of government in your country with respect to **(a)** protecting the environment and **(b)** ensuring environmental justice for all? What three major changes, if any, would you make in this system?

6. This chapter has summarized three major environmental worldviews (Figure 14-14, p. 319). Go through

these worldviews and find the beliefs you agree with to describe your own environmental worldview. Which of your beliefs, if any, were added or modified as a result of taking this course? Compare your answer with those of your classmates.

7. Would you, or do you, use the principle of voluntary simplicity in your life? How?

8. Explain why you agree or disagree with the following ideas: **(a)** everyone has the right to have as many children as they want, **(b)** each member of the human species has a right to use as many resources as they want, **(c)** individuals should have the right to do anything they want with land they own, **(d)** other species exist to be used by humans, and **(e)** all forms of life have an intrinsic value and therefore have a right to exist. Are your answers consistent with the beliefs of your environmental worldview that you described in question 6?

9. List two questions that you would like to have answered as a result of reading this chapter.

LEARNING ONLINE

The website for this book contains helpful study aids and many ideas for further reading and research. They include a chapter summary, review questions for the entire chapter, flash cards for key terms and concepts, a multiple-choice practice quiz, interesting Internet sites, references, information about green careers, and a guide for accessing thousands of InfoTrac® College Edition articles. Log into

www.thomsonedu.com/biology/miller

Then choose Chapter 14, and select a learning resource. For access to animations, additional quizzes, chapter outlines and summaries, register and log into

at **www.thomsonedu.com** using the access code card in the front of your book

Supplement 1

MEASUREMENT UNITS

LENGTH

Metric

1 kilometer (km) = 1,000 meters (m)
1 meter (m) = 100 centimeters (cm)
1 meter (m) = 1,000 millimeters (mm)
1 centimeter (cm) = 0.01 meter (m)
1 millimeter (mm) = 0.001 meter (m)

English

1 foot (ft) = 12 inches (in)
1 yard (yd) = 3 feet (ft)
1 mile (mi) = 5,280 feet (ft)
1 nautical mile = 1.15 miles

Metric–English

1 kilometer (km) = 0.621 mile (mi)
1 meter (m) = 39.4 inches (in)
1 inch (in) = 2.54 centimeters (cm)
1 foot (ft) = 0.305 meter (m)
1 yard (yd) = 0.914 meter (m)
1 nautical mile = 1.85 kilometers (km)

AREA

Metric

1 square kilometer (km^2) = 1,000,000 square meters (m^2)
1 square meter (m^2) = 1,000,000 square millimeters (mm^2)
1 hectare (ha) = 10,000 square meters (m^2)
1 hectare (ha) = 0.01 square kilometer (km^2)

English

1 square foot (ft^2) = 144 square inches (in^2)
1 square yard (yd^2) = 9 square feet (ft^2)
1 square mile (mi^2) = 27,880,000 square feet (ft^2)
1 acre (ac) = 43,560 square feet (ft^2)

Metric–English

1 hectare (ha) = 2.471 acres (ac)
1 square kilometer (km^2) = 0.386 square mile (mi^2)
1 square meter (m^2) = 1.196 square yards (yd^2)
1 square meter (m^2) = 10.76 square feet (ft^2)
1 square centimeter (cm^2) = 0.155 square inch (in^2)

VOLUME

Metric

1 cubic kilometer (km^3) = 1,000,000,000 cubic meters (m^3)
1 cubic meter (m^3) = 1,000,000 cubic centimeters (cm^3)
1 liter (L) = 1,000 milliliters (mL) = 1,000 cubic centimeters (cm^3)
1 milliliter (mL) = 0.001 liter (L)
1 milliliter (mL) = 1 cubic centimeter (cm^3)

English

1 gallon (gal) = 4 quarts (qt)
1 quart (qt) = 2 pints (pt)

Metric–English

1 liter (L) = 0.265 gallon (gal)
1 liter (L) = 1.06 quarts (qt)
1 liter (L) = 0.0353 cubic foot (ft^3)
1 cubic meter (m^3) = 35.3 cubic feet (ft^3)
1 cubic meter (m^3) = 1.30 cubic yards (yd^3)
1 cubic kilometer (km^3) = 0.24 cubic mile (mi^3)
1 barrel (bbl) = 159 liters (L)
1 barrel (bbl) = 42 U.S. gallons (gal)

MASS

Metric

1 kilogram (kg) = 1,000 grams (g)
1 gram (g) = 1,000 milligrams (mg)
1 gram (g) = 1,000,000 micrograms (μg)
1 milligram (mg) = 0.001 gram (g)
1 microgram (μg) = 0.000001 gram (g)
1 metric ton (mt) = 1,000 kilograms (kg)

English

1 ton (t) = 2,000 pounds (lb)
1 pound (lb) = 16 ounces (oz)

Metric–English

1 metric ton (mt) = 2,200 pounds (lb) = 1.1 tons (t)
1 kilogram (kg) = 2.20 pounds (lb)
1 pound (lb) = 454 grams (g)
1 gram (g) = 0.035 ounce (oz)

ENERGY AND POWER

Metric

1 kilojoule (kJ) = 1,000 joules (J)
1 kilocalorie (kcal) = 1,000 calories (cal)
1 calorie (cal) = 4,184 joules (J)

Metric–English

1 kilojoule (kJ) = 0.949 British thermal unit (Btu)
1 kilojoule (kJ) = 0.000278 kilowatt-hour (kW-h)
1 kilocalorie (kcal) = 3.97 British thermal units (Btu)
1 kilocalorie (kcal) = 0.00116 kilowatt-hour (kW-h)
1 kilowatt-hour (kW-h) = 860 kilocalories (kcal)
1 kilowatt-hour (kW-h) = 3,400 British thermal units (Btu)
1 quad (Q) = 1,050,000,000,000,000 kilojoules (kJ)
1 quad (Q) = 2,930,000,000,000 kilowatt-hours (kW-h)

TEMPERATURE CONVERSIONS

Fahrenheit(°F) to Celsius(°C):
$$°C = (°F − 32.0) ÷ 1.80$$
Celsius(°C) to Fahrenheit(°F):
$$°F = (°C × 1.80) + 32.0$$

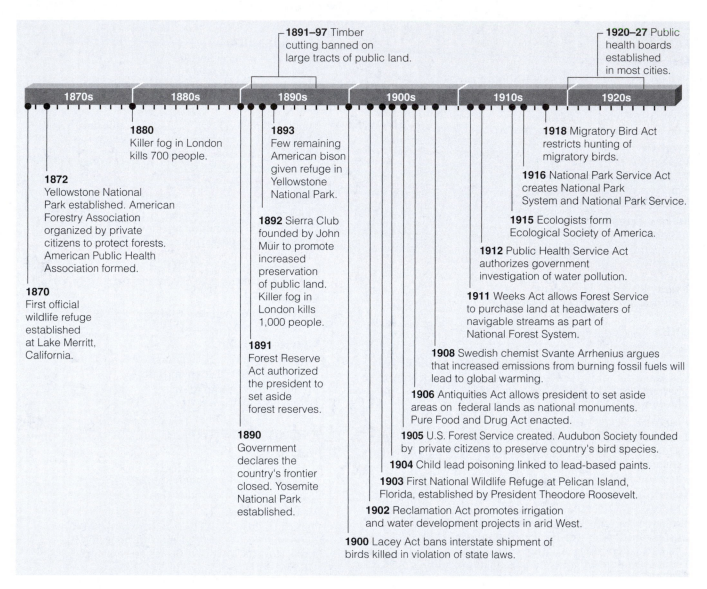

1891–97 Timber cutting banned on large tracts of public land.

1920–27 Public health boards established in most cities.

1870s 1880s 1890s 1900s 1910s 1920s

1880 Killer fog in London kills 700 people.

1893 Few remaining American bison given refuge in Yellowstone National Park.

1872 Yellowstone National Park established. American Forestry Association organized by private citizens to protect forests. American Public Health Association formed.

1892 Sierra Club founded by John Muir to promote increased preservation of public land. Killer fog in London kills 1,000 people.

1870 First official wildlife refuge established at Lake Merritt, California.

1891 Forest Reserve Act authorized the president to set aside forest reserves.

1890 Government declares the country's frontier closed. Yosemite National Park established.

1918 Migratory Bird Act restricts hunting of migratory birds.

1916 National Park Service Act creates National Park System and National Park Service.

1915 Ecologists form Ecological Society of America.

1912 Public Health Service Act authorizes government investigation of water pollution.

1911 Weeks Act allows Forest Service to purchase land at headwaters of navigable streams as part of National Forest System.

1908 Swedish chemist Svante Arrhenius argues that increased emissions from burning fossil fuels will lead to global warming.

1906 Antiquities Act allows president to set aside areas on federal lands as national monuments. Pure Food and Drug Act enacted.

1905 U.S. Forest Service created. Audubon Society founded by private citizens to preserve country's bird species.

1904 Child lead poisoning linked to lead-based paints.

1903 First National Wildlife Refuge at Pelican Island, Florida, established by President Theodore Roosevelt.

1902 Reclamation Act promotes irrigation and water development projects in arid West.

1900 Lacey Act bans interstate shipment of birds killed in violation of state laws.

1870–1930

Figure 1 Examples of the increased role of the federal government in resource conservation and public health and establishment of key private environmental groups, 1870–1930.

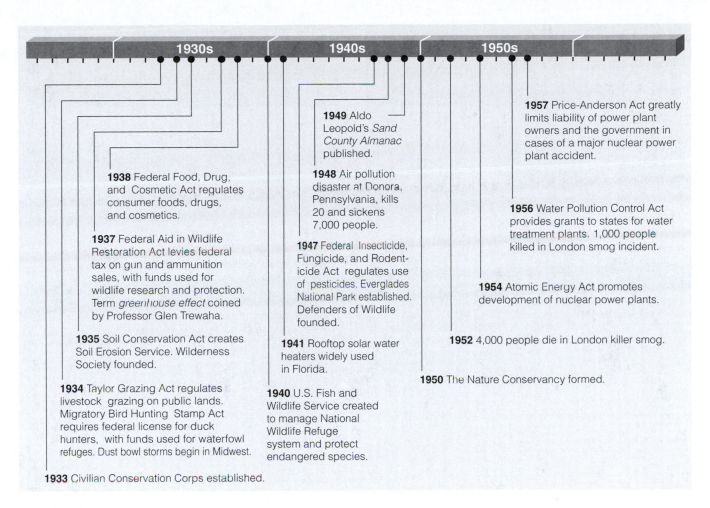

1957 Price-Anderson Act greatly limits liability of power plant owners and the government in cases of a major nuclear power plant accident.

1949 Aldo Leopold's *Sand County Almanac* published.

1948 Air pollution disaster at Donora, Pennsylvania, kills 20 and sickens 7,000 people.

1956 Water Pollution Control Act provides grants to states for water treatment plants. 1,000 people killed in London smog incident.

1938 Federal Food, Drug, and Cosmetic Act regulates consumer foods, drugs, and cosmetics.

1937 Federal Aid in Wildlife Restoration Act levies federal tax on gun and ammunition sales, with funds used for wildlife research and protection. Term *greenhouse effect* coined by Professor Glen Trewaha.

1947 Federal Insecticide, Fungicide, and Rodent-icide Act regulates use of pesticides. Everglades National Park established. Defenders of Wildlife founded.

1954 Atomic Energy Act promotes development of nuclear power plants.

1935 Soil Conservation Act creates Soil Erosion Service. Wilderness Society founded.

1941 Rooftop solar water heaters widely used in Florida.

1952 4,000 people die in London killer smog.

1934 Taylor Grazing Act regulates livestock grazing on public lands. Migratory Bird Hunting Stamp Act requires federal license for duck hunters, with funds used for waterfowl refuges. Dust bowl storms begin in Midwest.

1940 U.S. Fish and Wildlife Service created to manage National Wildlife Refuge system and protect endangered species.

1950 The Nature Conservancy formed.

1933 Civilian Conservation Corps established.

1930–1960

Figure 2 Some important conservation and environmental events, 1930–1960.

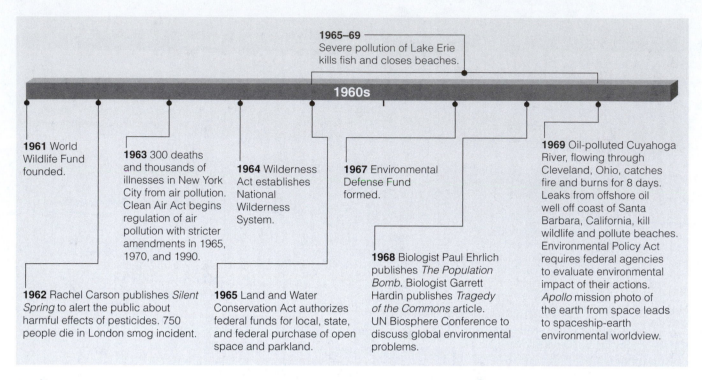

1960s

Figure 3 Some important environmental events during the 1960s.

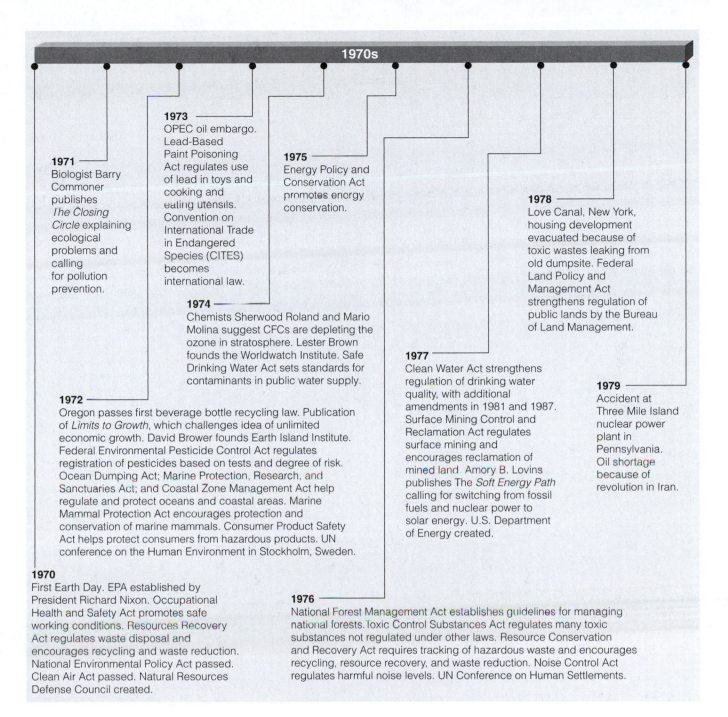

1970s

1971
Biologist Barry Commoner publishes *The Closing Circle* explaining ecological problems and calling for pollution prevention.

1973
OPEC oil embargo. Lead-Based Paint Poisoning Act regulates use of lead in toys and cooking and eating utensils. Convention on International Trade in Endangered Species (CITES) becomes international law.

1975
Energy Policy and Conservation Act promotes energy conservation.

1978
Love Canal, New York, housing development evacuated because of toxic wastes leaking from old dumpsite. Federal Land Policy and Management Act strengthens regulation of public lands by the Bureau of Land Management.

1974
Chemists Sherwood Roland and Mario Molina suggest CFCs are depleting the ozone in stratosphere. Lester Brown founds the Worldwatch Institute. Safe Drinking Water Act sets standards for contaminants in public water supply.

1977
Clean Water Act strengthens regulation of drinking water quality, with additional amendments in 1981 and 1987. Surface Mining Control and Reclamation Act regulates surface mining and encourages reclamation of mined land. Amory B. Lovins publishes The *Soft Energy Path* calling for switching from fossil fuels and nuclear power to solar energy. U.S. Department of Energy created.

1979
Accident at Three Mile Island nuclear power plant in Pennsylvania. Oil shortage because of revolution in Iran.

1972
Oregon passes first beverage bottle recycling law. Publication of *Limits to Growth*, which challenges idea of unlimited economic growth. David Brower founds Earth Island Institute. Federal Environmental Pesticide Control Act regulates registration of pesticides based on tests and degree of risk. Ocean Dumping Act; Marine Protection, Research, and Sanctuaries Act; and Coastal Zone Management Act help regulate and protect oceans and coastal areas. Marine Mammal Protection Act encourages protection and conservation of marine mammals. Consumer Product Safety Act helps protect consumers from hazardous products. UN conference on the Human Environment in Stockholm, Sweden.

1970
First Earth Day. EPA established by President Richard Nixon. Occupational Health and Safety Act promotes safe working conditions. Resources Recovery Act regulates waste disposal and encourages recycling and waste reduction. National Environmental Policy Act passed. Clean Air Act passed. Natural Resources Defense Council created.

1976
National Forest Management Act establishes guidelines for managing national forests. Toxic Control Substances Act regulates many toxic substances not regulated under other laws. Resource Conservation and Recovery Act requires tracking of hazardous waste and encourages recycling, resource recovery, and waste reduction. Noise Control Act regulates harmful noise levels. UN Conference on Human Settlements.

1970s

Figure 4 Some important environmental events during the 1970s, sometimes called the *environmental decade*.

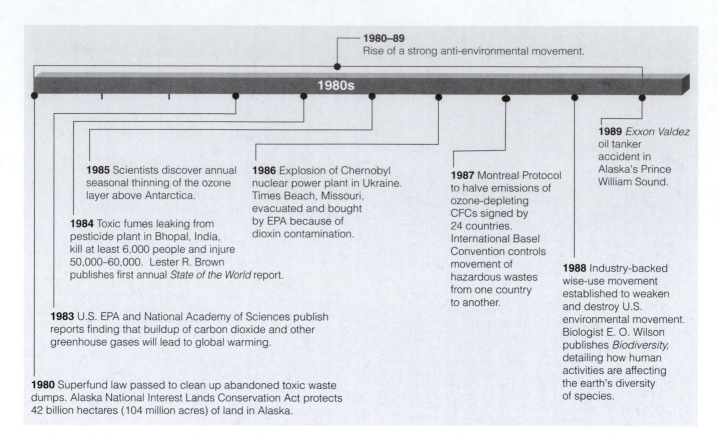

1980–89
Rise of a strong anti-environmental movement.

1980s

1985 Scientists discover annual seasonal thinning of the ozone layer above Antarctica.

1984 Toxic fumes leaking from pesticide plant in Bhopal, India, kill at least 6,000 people and injure 50,000–60,000. Lester R. Brown publishes first annual *State of the World* report.

1986 Explosion of Chernobyl nuclear power plant in Ukraine. Times Beach, Missouri, evacuated and bought by EPA because of dioxin contamination.

1987 Montreal Protocol to halve emissions of ozone-depleting CFCs signed by 24 countries. International Basel Convention controls movement of hazardous wastes from one country to another.

1989 *Exxon Valdez* oil tanker accident in Alaska's Prince William Sound.

1988 Industry-backed wise-use movement established to weaken and destroy U.S. environmental movement. Biologist E. O. Wilson publishes *Biodiversity,* detailing how human activities are affecting the earth's diversity of species.

1983 U.S. EPA and National Academy of Sciences publish reports finding that buildup of carbon dioxide and other greenhouse gases will lead to global warming.

1980 Superfund law passed to clean up abandoned toxic waste dumps. Alaska National Interest Lands Conservation Act protects 42 billion hectares (104 million acres) of land in Alaska.

1980s

Figure 5 Some important environmental events during the 1980s.

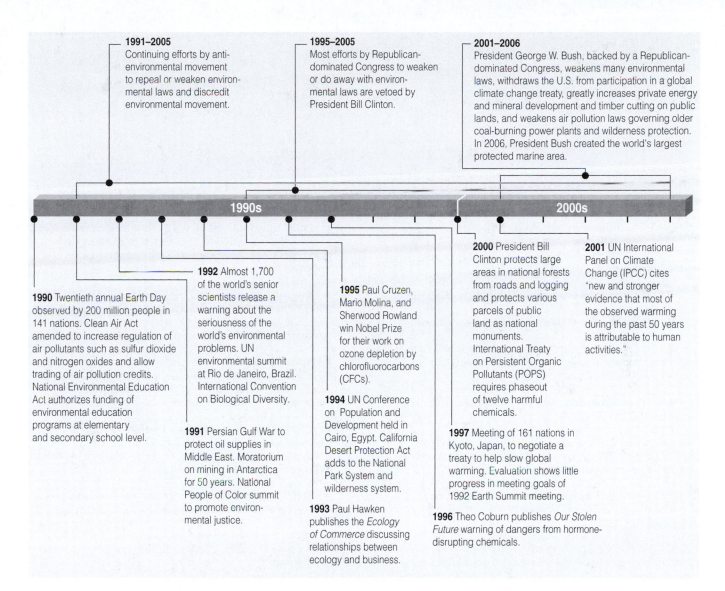

1991–2005
Continuing efforts by anti-environmental movement to repeal or weaken environmental laws and discredit environmental movement.

1995–2005
Most efforts by Republican-dominated Congress to weaken or do away with environmental laws are vetoed by President Bill Clinton.

2001–2006
President George W. Bush, backed by a Republican-dominated Congress, weakens many environmental laws, withdraws the U.S. from participation in a global climate change treaty, greatly increases private energy and mineral development and timber cutting on public lands, and weakens air pollution laws governing older coal-burning power plants and wilderness protection. In 2006, President Bush created the world's largest protected marine area.

1990s

2000s

1990 Twentieth annual Earth Day observed by 200 million people in 141 nations. Clean Air Act amended to increase regulation of air pollutants such as sulfur dioxide and nitrogen oxides and allow trading of air pollution credits. National Environmental Education Act authorizes funding of environmental education programs at elementary and secondary school level.

1991 Persian Gulf War to protect oil supplies in Middle East. Moratorium on mining in Antarctica for 50 years. National People of Color summit to promote environmental justice.

1992 Almost 1,700 of the world's senior scientists release a warning about the seriousness of the world's environmental problems. UN environmental summit at Rio de Janeiro, Brazil. International Convention on Biological Diversity.

1993 Paul Hawken publishes the *Ecology of Commerce* discussing relationships between ecology and business.

1994 UN Conference on Population and Development held in Cairo, Egypt. California Desert Protection Act adds to the National Park System and wilderness system.

1995 Paul Cruzen, Mario Molina, and Sherwood Rowland win Nobel Prize for their work on ozone depletion by chlorofluorocarbons (CFCs).

1996 Theo Coburn publishes *Our Stolen Future* warning of dangers from hormone-disrupting chemicals.

1997 Meeting of 161 nations in Kyoto, Japan, to negotiate a treaty to help slow global warming. Evaluation shows little progress in meeting goals of 1992 Earth Summit meeting.

2000 President Bill Clinton protects large areas in national forests from roads and logging and protects various parcels of public land as national monuments. International Treaty on Persistent Organic Pollutants (POPS) requires phaseout of twelve harmful chemicals.

2001 UN International Panel on Climate Change (IPCC) cites "new and stronger evidence that most of the observed warming during the past 50 years is attributable to human activities."

1990–2006

Figure 6 Some important environmental events, 1990–2006.

BRIEF HISTORY OF THE AGE OF OIL

Some milestones in the Age of Oil:

- **1857:** First commercial oil well is drilled near Titusville, Pennsylvania.

- **1905:** Oil supplies 10% of U.S. energy.

- **1925:** United States produces 71% of the world's oil.

- **1930:** Because of an oil glut, oil sells for 10¢ a barrel.

- **1953:** U.S. oil companies account for about half of the world's oil production, and the United States is the world's leading oil exporter.

- **1955:** United States has 20% of the world's estimated oil reserves.

- **1960:** OPEC formed so developing countries, with most of the world's known oil and projected oil reserves, can get a higher price for their oil.

- **1973:** United States uses 30% of the world's oil, imports 36% of this oil, and has only 5% of the world's proven oil reserves.

- **1973–1974:** OPEC reduces oil imports to the West and bans oil exports to the U.S. because of its support for Israel in the 18-day Yom Kippur War with Egypt and Syria. World oil prices rise sharply and lead to double-digit inflation in the United States and many other countries and a global economic recession.

- **1975:** Production of estimated U.S. oil reserves peaks.

- **1979:** Iran's Islamic Revolution shuts down most of Iran's oil production and reduces world oil production.

- **1981:** the Iran–Iraq war pushes global oil prices to a historic high.

- **1983:** Facing an oil glut, OPEC cuts its oil prices.

- **1985:** U.S. domestic oil production begins to decline and is not expected to increase enough to affect the global price of oil or to reduce U.S. dependence on oil imports.

- **August 1990–June 1991:** The United States and its allies fight the Persian Gulf War to oust Iraqi invaders from Kuwait and to protect Western access to Saudi Arabian and Kuwaiti oil supplies.

- **2004:** The United States and a small number of allies fight a second Persian Gulf War to oust Sadam Hussein from power in Iraq and to protect Western access to Saudi Arabia, Kuwait, and Iraq.

- **2006:** OPEC has 67% of world oil reserves and produces 40% of the world's oil. The United States has only 2.9% of oil reserves, uses 26% of the world's oil production, and imports 60% of its oil.

- **2010:** U.S. could be importing at least 61% of the oil it uses as consumption continues to exceed production.

- **2020:** The United States could be importing at least 70% of the oil it uses, as consumption continues to exceed production.

- **2010–2030:** Production of oil from the world's estimated oil reserves is expected to peak. Oil prices are expected to increase gradually as the demand for oil increasingly exceeds the supply—unless the world decreases its demand by wasting less energy and shifting to other sources of energy.

- **2010–2048:** Domestic U.S. oil reserves are projected to be 80% depleted.

- **2042–2083:** A gradual decline in dependence on oil.

SUPPLEMENT 4

HOW TO ANALYZE A SCIENTIFIC ARTICLE

This journal article reports on the movements of a female wolf during the summer of 2002 in northwestern Canada. It also reports on a scientific process of inquiry, observation and interpretation to learn where, how and why the wolf traveled as she did. In some ways, this article reflects the story of "how to do science" told on pp. 20–22 of this textbook. These notes are intended to help you read and understand how scientists work and how they report on their work.

(1) A R C T I C

(2) VOL. 57, NO. 2 (JUNE 2004) P. 196–203

(3) ## Long Foraging Movement of a Denning Tundra Wolf

(4) Paul F. Frame,[1,2] David S. Hik,[1] H. Dean Cluff,[3] and Paul C. Paquet[4]

(5) (Received 3 September 2003; accepted in revised form 16 January 2004)

(6) ABSTRACT. Wolves (*Canis lupus*) on the Canadian barrens are intimately linked to migrating herds of barren-ground caribou (*Rangifer tarandus*). We deployed a Global Positioning System (GPS) radio collar on an adult female wolf to record her movements in response to changing caribou densities near her den during summer. This wolf and two other females were observed nursing a group of 11 pups. She traveled a minimum of 341 km during a 14-day excursion. The straight-line distance from the den to the farthest location was 103 km, and the overall minimum rate of travel was 3.1 km/h. The distance between the wolf and the radio-collared caribou decreased from 242 km one week before the excursion to 8 km four days into the excursion. We discuss several possible explanations for the long foraging bout.

(7) Key words: wolf, GPS tracking, movements, *Canis lupus*, foraging, caribou, Northwest Territories

(8) RÉSUMÉ. Les loups (*Canis lupus*) dans la toundra canadienne sont étroitement liés aux hardes de caribous des toundras (*Rangifer tarandus*). On a équipé une louve adulte d'un collier émetteur muni d'un système de positionnement mondial (GPS) afin d'enregistrer ses déplacements en réponse au changement de densité du caribou près de sa tanière durant l'été. On a observé cette louve ainsi que deux autres en train d'allaiter un groupe de 11 louveteaux. Elle a parcouru un minimum de 341 km durant une sortie de 14 jours. La distance en ligne droite de la tanière à l'endroit le plus éloigné était de 103 km, et la vitesse minimum durant tout le voyage était de 3,1 km/h. La distance entre la louve et le caribou muni du collier émetteur a diminué de 242 km une semaine avant la sortie à 8 km quatre jours après la sortie. On commente diverses explications possibles pour ce long épisode de recherche de nourriture.

Mots clés: loup, repérage GPS, déplacements, *Canis lupus*, recherche de nourriture, caribou, Territoires du Nord-Ouest

Traduit pour la revue *Arctic* par Nésida Loyer.

(9) ### Introduction

Wolves (*Canis lupus*) that den on the central barrens of mainland Canada follow the seasonal movements of their main prey, migratory barren-ground caribou (*Rangifer tarandus*) (Kuyt, 1962; Kelsall, 1968; Walton et al., 2001). However, most wolves do not den near caribou calving grounds, but select sites farther south, closer to the tree line (Heard and Williams, 1992). Most caribou migrate beyond primary wolf denning areas by mid-June and do not return until mid-to-late July (Heard et al., 1996; Gunn et al., 2001). Conse-

quently, caribou density near dens is low for part of the summer.

(10) During this period of spatial separation from the main caribou herds, wolves must either search near the homesite for scarce caribou or alternative prey (or both), travel to where prey are abundant, or use a combination of these strategies.

(11) Walton et al. (2001) postulated that the travel of tundra wolves outside their normal summer ranges is a response to low caribou availability rather than a pre-dispersal exploration like that observed in territorial wolves (Fritts and Mech, 1981; Messier, 1985). The authors postulated this because most such travel was directed toward caribou calving grounds. We report details of such a long-distance excursion by a breeding female tundra wolf wearing a GPS radio collar. We discuss the relationship of the excursion to movements of satellite-collared caribou (Gunn et al., 2001), supporting the hypothesis that tundra wolves make directional, rapid, long-distance movements in response to seasonal prey availability.

[1] Department of Biological Sciences, University of Alberta, Edmonton, Alberta T6G 2E9, Canada
[2] Corresponding author: pframe@ualberta.ca
[3] Department of Resources, Wildlife, and Economic Development, North Slave Region, Government of the Northwest Territories, P.O. Box 2668, 3803 Bretzlaff Dr., Yellowknife, Northwest Territories X1A 2P9, Canada; Dean_Cluff@gov.nt.ca
[4] Faculty of Environmental Design, University of Calgary, Calgary, Alberta T2N 1N4, Canada; current address: P.O. Box 150, Meacham, Saskatchewan S0K 2V0, Canada

196

---

1 Title of the journal, which reports on science taking place in Arctic regions.

2 Volume number, issue number and date of the journal, and page numbers of the article.

3 Title of the article: a concise but specific description of the subject of study—one episode of long-range travel by a wolf hunting for food on the Arctic tundra.

4 Authors of the article: scientists working at the institutions listed in the footnotes below. Note #2 indicates that P. F. Frame is the *corresponding author*—the person to contact with questions or comments. His email address is provided.

5 Date on which a draft of the article was received by the journal editor, followed by date on which a revised draft was accepted for publication. Between these dates, the article was reviewed and critiqued by other scientists, a process called peer review. The authors revised the article to make it clearer, according to those reviews.

6 ABSTRACT: A brief description of the study containing all basic elements of this report. First sentence summarizes the *background* material. Second sentence encapsulates the *methods* used. The rest of the paragraph sums up the *results*. Authors introduce the main *subject* of the study—a female wolf (#388) with pups in a den—and refer to later *discussion* of possible explanations for her behavior.

7 Key words are listed to help researchers using computer databases. Searching the databases using these key words will yield a list of studies related to this one.

8 RÉSUMÉ: The French translation of the abstract and key words. Many researchers in this field are French Canadian. Some journals provide such translations in French or in other languages.

9 INTRODUCTION: Gives the background for this wolf study. This paragraph tells of known or suspected wolf behavior that is important for this study. Note that (a) major species mentioned are always accompanied by scientific names, and (b) statements of fact or *postulations* (claims or assumptions about what is likely to be true) are followed by references to studies that established those facts or supported the postulations.

10 This paragraph focuses directly on the wolf behaviors that were studied here.

11 This paragraph starts with a statement of the *hypothesis* being tested, one that originated in other studies and is supported by this one. The hypothesis is restated more succinctly in the last sentence of this paragraph. This is the *inquiry* part of the scientific process—asking questions and suggesting possible answers.

12 This map shows the study area and depicts wolf and caribou locations and movements during one summer. Some of this information is explained below.

13 STUDY AREA: This section sets the stage for the study, locating it precisely with latitude and longitude coordinates and describing the area (illustrated by the map in Figure 1).

14 Here begins the story of how prey (caribou) and predators (wolves) interact on the tundra. Authors describe movements of these nomadic animals throughout the year.

15 We focus on the denning season (summer) and learn how wolves locate their dens and travel according to the movements of caribou herds.

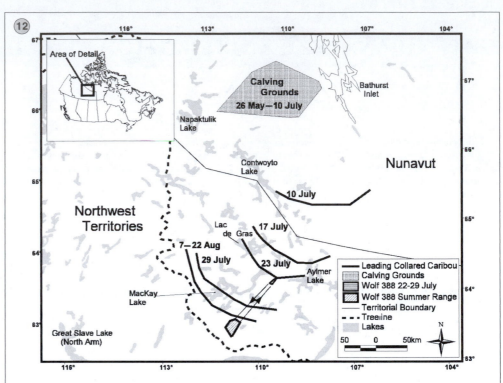

Figure 1. Map showing the movements of satellite radio-collared caribou with respect to female wolf 388's summer range and long foraging movement, in summer 2002.

13 Study Area

Our study took place in the northern boreal forest–low Arctic tundra transition zone (63° 30′ N, 110° 00′ W; Figure 1; Timoney et al., 1992). Permafrost in the area changes from discontinuous to continuous (Harris, 1986). Patches of spruce (*Picea mariana, P. glauca*) occur in the southern portion and give way to open tundra to the northeast. Eskers, kames, and other glacial deposits are scattered throughout the study area. Standing water and exposed bedrock are characteristic of the area.

14 *Details of the Caribou-Wolf System*

The Bathurst caribou herd uses this study area. Most caribou cows have begun migrating by late April, reaching calving grounds by June (Gunn et al., 2001;

Figure 1). Calving peaks by 15 June (Gunn et al., 2001), and calves begin to travel with the herd by one week of age (Kelsall, 1968). The movement patterns of bulls are less known, but bulls frequent areas near calving grounds by mid-June (Heard et al., 1996; Gunn et al., 2001). In summer, Bathurst caribou cows generally travel south from their calving grounds and then, parallel to the tree line, to the northwest. The rut usually takes place at the tree line in October (Gunn et al., 2001). The winter range of the Bathurst herd varies among years, ranging through the taiga and along the tree line from south of Great Bear Lake to southeast of Great Slave Lake. Some caribou spend the winter on the tundra (Gunn et al., 2001; Thorpe et al., 2001).

15 In winter, wolves that prey on Bathurst caribou do not behave territorially. Instead, they follow the herd throughout its winter range (Walton et al., 2001; Musiani, 2003). However, during denning (May–

Table 1. Daily distances from wolf 388 and the den to the nearest radio-collared caribou during a long excursion in summer 2002.

Date (2002)	Mean distance from caribou to wolf (km)	Daily distance from closest caribou to den
12 July	242	241
13 July	210	209
14 July	200	199
15 July	186	180
16 July	163	162
17 July	151	148
18 July	144	137
19 July[1]	126	124
20 July	103	130
21 July	73	130
22 July[2]	40	110
23 July[2]	9	104
29 July[3]	16	43
30 July	32	43
31 July	28	44
1 August	29	46
2 August[4]	54	52
3 August	53	53
4 August	74	74
5 August	75	75
6 August	74	75
7 August	72	75
8 August	76	75
9 August	79	79

[1] Excursion starts.
[2] Wolf closest to collared caribou.
[3] Previous five days' caribou locations not available.
[4] Excursion ends.

August, parturition late May to mid-June), wolf movements are limited by the need to return food to the den. To maximize access to migrating caribou, many wolves select den sites closer to the tree line than to caribou calving grounds (Heard and Williams, 1992). Because of caribou movement patterns, tundra denning wolves are separated from the main caribou herds by several hundred kilometers at some time during summer (Williams, 1990:19; Figure 1; Table 1).

 Muskoxen do not occur in the study area (Fournier and Gunn, 1998), and there are few moose there (H.D. Cluff, pers. obs.). Therefore, alternative prey for wolves includes waterfowl, other ground-nesting birds, their eggs, rodents, and hares (Kuyt, 1972; Williams, 1990:16; H.D. Cluff and P.F. Frame, unpubl. data). During 56 hours of den observations, we saw no ground squirrels or hares, only birds. It appears that the abundance of alternative prey was relatively low in 2002.

Methods

Wolf Monitoring

 We captured female wolf 388 near her den on 22 June 2002, using a helicopter net-gun (Walton et al., 2001). She was fitted with a releasable GPS radio collar (Merrill et al., 1998) programmed to acquire locations at 30-minute intervals. The collar was electronically released (e.g., Mech and Gese, 1992) on 20 August 2002. From 27 June to 3 July 2002, we observed 388's den with a 78 mm spotting scope at a distance of 390 m.

Caribou Monitoring

In spring of 2002, ten female caribou were captured by helicopter net-gun and fitted with satellite radio collars, bringing the total number of collared Bathurst cows to 19. Eight of these spent the summer of 2002 south of Queen Maud Gulf, well east of normal Bathurst caribou range. Therefore, we used 11 caribou for this analysis. The collars provided one location per day during our study, except for five days from 24 to 28 July. Locations of satellite collars were obtained from Service Argos, Inc. (Landover, Maryland).

Data Analysis

Location data were analyzed by ArcView GIS software (Environmental Systems Research Institute Inc., Redlands, California). We calculated the average distance from the nearest collared caribou to the wolf and the den for each day of the study.

Wolf foraging bouts were calculated from the time 388 exited a buffer zone (500 m radius around the den) until she re-entered it. We considered her to be traveling when two consecutive locations were spatially separated by more than 100 m. Minimum distance traveled was the sum of distances between each location and the next during the excursion.

We compared pre- and post-excursion data using Analysis of Variance (ANOVA; Zar, 1999). We first tested for homogeneity of variances with Levene's test (Brown and Forsythe, 1974). No transformations of these data were required.

Results

Wolf Monitoring

Pre-Excursion Period: Wolf 388 was lactating when captured on 22 June. We observed her and two other females nursing a group of 11 pups between 27 June and 3 July. During our observations, the pack consisted of at least four adults (3 females and 1 male) and 11 pups. On 30 June, three pups were moved to a location 310 m from the other eight and cared for by an uncollared female. The male was not seen at the den after the evening of 30 June.

Before the excursion, telemetry indicated 18 foraging bouts. The mean distance traveled during these bouts was 25.29 km (± 4.5 SE, range 3.1–82.5 km). Mean greatest distance from the den on foraging

16 Other variables are considered—prey other than caribou and their relative abundance in 2002.

17 METHODS: There is no one scientific method. Procedures for each and every study must be explained carefully.

18 Authors explain when and how they tracked caribou and wolves, including tools used and the exact procedures followed.

19 This important subsection explains what data were calculated (average distance ...) and how, including the software used and where it came from. (The calculations are listed in Table 1.) Note that the behavior measured (traveling) is carefully defined.

20 RESULTS: The heart of the report and the *observation* part of the scientific process. This section is organized parallel to the Methods section.

21 This subsection is broken down by periods of observation. Pre-excursion period covers the time between 388's capture and the start of her long-distance travel. The investigators used visual observations as well as telemetry (measurements taken using the global positioning system (GPS)) to gather data. They looked at how 388 cared for her pups, interacted with other adults, and moved about the den area.

22 The key in the lower right-hand corner of the map shows areas (shaded) within which the wolves and caribou moved, and the dotted trail of 388 during her excursion. From the results depicted on this map, the investigators tried to determine when and where 388 might have encountered caribou and how their locations affected her traveling behavior.

23 The wolf's excursion (her long trip away from the den area) is the focus of this study. These paragraphs present detailed measurements of daily movements during her two-week trip—how far she traveled, how far she was from collared caribou, her time spent traveling and resting, and her rate of speed. Authors use the phrase "minimum distance traveled" to acknowledge they couldn't track every step but were measuring samples of her movements. They knew that she went at least as far as they measured. This shows how scientists try to be exact when reporting results. Results of this study are depicted graphically in the map in Figure 2.

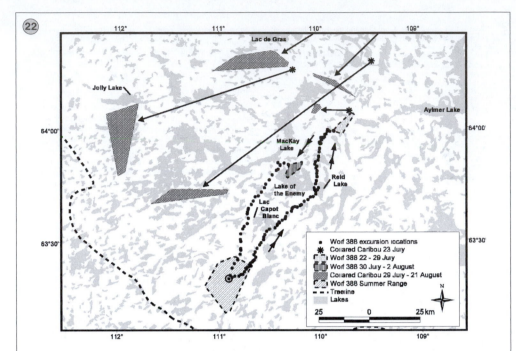

Figure 2. Details of a long foraging movement by female wolf 388 between 19 July and 2 August 2002. Also shown are locations and movements of three satellite radio-collared caribou from 23 July to 21 August 2002. On 23 July, the wolf was 8 km from a collared caribou. The farthest point from the den (103 km distant) was recorded on 27 July. Arrows indicate direction of travel.

bouts was 7.1 km (± 0.9 SE, range 1.7–17.0 km). The average duration of foraging bouts for the period was 20.9 h (± 4.5 SE, range 1–71 h).

The average daily distance between the wolf and the nearest collared caribou decreased from 242 km on 12 July, one week before the excursion period, to 126 km on 19 July, the day the excursion began (Table 1).

23 **Excursion Period:** On 19 July at 2203, after spending 14 h at the den, 388 began moving to the northeast and did not return for 336 h (14 d; Figure 2). Whether she traveled alone or with other wolves is unknown. During the excursion, 476 (71%) of 672 possible locations were recorded. The wolf crossed the southeast end of Lac Capot Blanc on a small land bridge, where she paused for 4.5 h after traveling for 19.5 h (37.5

km). Following this rest, she traveled for 9 h (26.3 km) onto a peninsula in Reid Lake, where she spent 2 h before backtracking and stopping for 8 h just off the peninsula. Her next period of travel lasted 16.5 h (32.7 km), terminating in a pause of 9.5 h just 3.8 km from a concentration of locations at the far end of her excursion, where we presume she encountered caribou. The mean duration of these three movement periods was 15.7 h (± 2.5 SE), and that of the pauses, 7.3 h (± 1.5). The wolf required 72.5 h (3.0 d) to travel a minimum of 95 km from her den to this area near caribou (Figure 2). She remained there (35.5 km2) for 151.5 h (6.3 d) and then moved south to Lake of the Enemy, where she stayed (31.9 km^2) for 74 h (3.1 d) before returning to her den. Her greatest distance from the den, 103 km, was recorded 174.5 h (7.3 d) after the excursion began, at 0433 on 27 July. She was 8 km from a collared caribou on 23 July, four days after the excursion began (Table 1).

The return trip began at 0403 on 2 August, 318 h (13.2 d) after leaving the den. She followed a relatively direct path for 18 h back to the den, a distance of 75 km.

The minimum distance traveled during the excursion was 339 km. The estimated overall minimum travel rate was 3.1 km/h, 2.6 km/h away from the den and 4.2 km/h on the return trip.

Post-Excursion Period: We saw three pups when recovering the collar on 20 August, but others may have been hiding in vegetation.

Telemetry recorded 13 foraging bouts in the post-excursion period. The mean distance traveled during these bouts was 18.3 km (+ 2.7 SE, range 1.2–47.7 km), and mean greatest distance from the den was 7.1 km (+ 0.7 SE, range 1.1–11.0 km). The mean duration of these post-excursion foraging bouts was 10.9 h (+ 2.4 SE, range 1 33 h).

When 388 reached her den on 2 August, the distance to the nearest collared caribou was 54 km. On 9 August, one week after she returned, the distance was 79 km (Table 1).

Pre- and Post-Excursion Comparison

We found no differences in the mean distance of foraging bouts before and after the excursion period (F = 1.5, df = 1, 29, p = 0.24). Likewise, the mean greatest distance from the den was similar pre- and post-excursion (F = 0.004, df = 1, 29, p = 0.95). However, the mean duration of 388's foraging bouts decreased by 10.0 h after her long excursion (F = 3.1, df = 1, 29, p = 0.09).

Caribou Monitoring

Summer Movements: On 10 July, 5 of 11 collared caribou were dispersed over a distance of 10 km, 140 km south of their calving grounds (Figure 1). On the same day, three caribou were still on the calving grounds, two were between the calving grounds and the leaders, and one was missing. One week later (17 July), the leading radio-collared cows were 100 km farther south (Figure 1). Two were within 5 km of each other in front of the rest, who were more dispersed. All radio-collared cows had left the calving grounds by this time. On 23 July, the leading radio-collared caribou had moved 35 km farther south, and all of them were more widely dispersed. The two cows closest to the leader were 26 km and 33 km away, with 37 km between them. On the next location (29 July), the most southerly caribou were 60 km

farther south. All of the caribou were now in the areas where they remained for the duration of the study (Figure 2).

A Minimum Convex Polygon (Mohr and Stumpf, 1966) around all caribou locations acquired during the study encompassed 85 119 km^2.

Relative to the Wolf Den: The distance from the nearest collared caribou to the den decreased from 241 km one week before the excursion to 124 km the day it began. The nearest a collared caribou came to the den was 43 km away, on 29 and 30 July. During the study, four collared caribou were located within 100 km of the den. Each of these four was closest to the wolf on at least one day during the period reported.

28 Discussion

Prey Abundance

Caribou are the single most important prey of tundra wolves (Clark, 1971; Kuyt, 1972; Stephenson and James, 1982; Williams, 1990). Caribou range over vast areas, and for part of the summer, they are scarce or absent in wolf home ranges (Heard et al., 1996). Both the long distance between radio-collared caribou and the den the week before the excursion and the increased time spent foraging by wolf 388 indicate that caribou availability near the den was low. Observations of the pups' being left alone for up to 18 h, presumably while adults were searching for food, provide additional support for low caribou availability locally. Mean foraging bout duration decreased by 10.0 h after the excursion, when collared caribou were closer to the den, suggesting an increase in caribou availability nearby.

Foraging Excursion

One aspect of central place foraging theory (CPFT) deals with the optimality of returning different-sized food loads from varying distances to dependents at a central place (i.e., the den) (Orians and Pearson, 1979). Carlson (1985) tested CPFT and found that the predator usually consumed prey captured far from the central place, while feeding prey captured nearby to dependants. Wolf 388 spent 7.2 days in one area near caribou before moving to a location 23 km back towards the den, where she spent an additional 3.1 days, likely hunting caribou. She began her return trip from this closer location, traveling directly to the den. While away, she may have made one or more successful kills and spent time meeting her own energetic needs before returning to the den. Alternatively, it may have taken several attempts to make a kill,

24 Post-excursion measurements of 388's movements were made to compare with those of the pre-excursion period. In order to compare, scientists often use *means*, or averages, of a series of measurements—mean distances, mean duration, etc.

25 In the comparison, authors used statistical calculations (F and df) to determine that the differences between pre- and post-excursion measurements were *statistically insignificant*, or close enough to be considered essentially the same or similar.

26 As with wolf 388, the investigators measured the movements of caribou during the study period. The areas within which the caribou moved are shown in Figure 2 by shaded polygons mentioned in the second paragraph of this subsection.

27 This subsection summarizes how distances separating predators and prey varied during the study period.

28 DISCUSSION: This section is the *interpretation* part of the scientific process.

29 This subsection reviews observations from other studies and suggests that this study fits with patterns of those observations.

30 Authors discuss a prevailing *theory* (CBFT) which might explain why a wolf would travel far to meet her own energy needs while taking food caught closer to the den back to her pups. The results of this study seem to fit that pattern.

31 Here our authors note other possible explanations for wolves' excursions presented by other investigators, but this study does not seem to support those ideas.

32 Authors discuss possible reasons for why 388 traveled directly to where caribou were located. They take what they learned from earlier studies and apply it to this case, suggesting that the lay of the land played a role. Note that their description paints a clear picture of the landscape.

33 Authors suggest that 388 may have learned in traveling during previous summers where the caribou were. The last two sentences suggest ideas for future studies.

34 Or maybe 388 followed the scent of the caribou. Authors acknowledge difficulties of proving this, but they suggest another area where future studies might be done.

35 Authors suggest that results of this study support previous studies about how fast wolves travel to and from the den. In the last sentence, they speculate on how these observed patterns would fit into the theory of evolution.

36 Authors also speculate on the fate of 388's pups while she was traveling. This leads to . . .

---

which she then fed on before beginning her return trip. We do not know if she returned food to the pups, but such behavior would be supported by CPFT.

31 Other workers have reported wolves' making long round trips and referred to them as "extraterritorial" or "pre-dispersal" forays (Fritts and Mech, 1981; Messier, 1985; Ballard et al., 1997; Merrill and Mech, 2000). These movements are most often made by young wolves (1–3 years old), in areas where annual territories are maintained and prey are relatively sedentary (Fritts and Mech, 1981; Messier, 1985). The long excursion of 388 differs in that tundra wolves do not maintain annual territories (Walton et al., 2001), and the main prey migrate over vast areas (Gunn et al., 2001).

Another difference between 388's excursion and those reported earlier is that she is a mature, breeding female. No study of territorial wolves has reported reproductive adults making extraterritorial movements in summer (Fritts and Mech, 1981; Messier, 1985; Ballard et al., 1997; Merrill and Mech, 2001). However, Walton et al. (2001) also report that breeding female tundra wolves made excursions.

Direction of Movement

32 Possible explanations for the relatively direct route 388 took to the caribou include landscape influence and experience. Considering the timing of 388's trip and the locations of caribou, had the wolf moved northwest, she might have missed the caribou entirely, or the encounter might have been delayed.

A reasonable possibility is that the land directed 388's route. The barrens are crisscrossed with trails worn into the tundra over centuries by hundreds of thousands of caribou and other animals (Kelsall, 1968; Thorpe et al., 2001). At river crossings, lakes, or narrow peninsulas, trails converge and funnel towards and away from caribou calving grounds and summer range. Wolves use trails for travel (Paquet et al., 1996; Mech and Boitani, 2003; P. Frame, pers. observation). Thus, the landscape may direct an animal's movements and lead it to where cues, such as the odor of caribou on the wind or scent marks of other wolves, may lead it to caribou.

33 Another possibility is that 388 knew where to find caribou in summer. Sexually immature tundra wolves sometimes follow caribou to calving grounds (D. Heard, unpubl. data). Possibly, 388 had made such journeys in previous years and killed caribou. If this were the case, then in times of local prey scarcity she might travel to areas where she had hunted successfully before. Continued monitoring of tundra wolves may answer questions about how their food needs are met in times of low caribou abundance near dens.

34 Caribou often form large groups while moving south to the tree line (Kelsall, 1968). After a large aggregation of caribou moves through an area, its scent can linger for weeks (Thorpe et al., 2001:104). It is conceivable that 388 detected caribou scent on the wind, which was blowing from the northeast on 19–21 July (Environment Canada, 2003), at the same time her excursion began. Many factors, such as odor strength and wind direction and strength, make systematic study of scent detection in wolves difficult under field conditions (Harrington and Asa, 2003). However, humans are able to smell odors such as forest fires or oil refineries more than 100 km away. The olfactory capabilities of dogs, which are similar to wolves, are thought to be 100 to 1 million times that of humans (Harrington and Asa, 2003). Therefore, it is reasonable to think that under the right wind conditions, the scent of many caribou traveling together could be detected by wolves from great distances, thus triggering a long foraging bout.

Rate of Travel

35 Mech (1994) reported the rate of travel of Arctic wolves on barren ground was 8.7 km/h during regular travel and 10.0 km/h when returning to the den, a difference of 1.3 km/h. These rates are based on direct observation and exclude periods when wolves moved slowly or not at all. Our calculated travel rates are assumed to include periods of slow movement or no movement. However, the pattern we report is similar to that reported by Mech (1994), in that homeward travel was faster than regular travel by 1.6 km/h. The faster rate on return may be explained by the need to return food to the den. Pup survival can increase with the number of adults in a pack available to deliver food to pups (Harrington et al., 1983). Therefore, an increased rate of travel on homeward trips could improve a wolf's reproductive fitness by getting food to pups more quickly.

Fate of 388's Pups

36 Wolf 388 was caring for pups during den observations. The pups were estimated to be six weeks old, and were seen ranging as far as 800 m from the den. They received some regurgitated food from two of the females, but were unattended for long periods. The excursion started 16 days after our observations, and it is improbable that the pups could have traveled the distance that 388 moved. If the pups died, this would have removed parental responsibility, allowing the long movement.

Our observations and the locations of radio-collared caribou indicate that prey became scarce in the area of the den as summer progressed. Wolf 388 may have abandoned her pups to seek food for herself. However, she returned to the den after the excursion, where she was seen near pups. In fact, she foraged in a similar pattern before and after the excursion, suggesting that she again was providing for pups after her return to the den.

(37) A more likely possibility is that one or both of the other lactating females cared for the pups during 388's absence. The three females at this den were not seen with the pups at the same time. However, two weeks earlier, at a different den, we observed three females cooperatively caring for a group of six pups. At that den, the three lactating females were observed providing food for each other and trading places while nursing pups. Such a situation at the den of 388 could have created conditions that allowed one or more of the lactating females to range far from the den for a period, returning to her parental duties afterwards. However, the pups would have been weaned by eight weeks of age (Packard et al., 1992), so nonlactating adults could also have cared for them, as often happens in wolf packs (Packard et al., 1992; Mech et al., 1999).

Cooperative rearing of multiple litters by a pack could create opportunities for long-distance foraging movements by some reproductive wolves during summer periods of local food scarcity. We have recorded multiple lactating females at one or more tundra wolf dens per year since 1997. This reproductive strategy may be an adaptation to temporally and (38) spatially unpredictable food resources. All of these possibilities require further study, but emphasize both the adaptability of wolves living on the barrens and their dependence on caribou.

Long-range wolf movement in response to caribou (39) availability has been suggested by other researchers (Kuyt, 1972; Walton et al., 2001) and traditional ecological knowledge (Thorpe et al., 2001). Our report demonstrates the rapid and extreme response of wolves to caribou distribution and movements in summer. Increased human activity on the tundra (mining, road building, pipelines, ecotourism) may influence caribou movement patterns and change the interactions between wolves and caribou in the region. Continued monitoring of both species will help us to assess whether the association is being affected adversely by anthropogenic change.

(40) Acknowledgements

This research was supported by the Department of Resources, Wildlife, and Economic Development, Government of the Northwest Territories; the Department of Biological Sciences at the University of Alberta; the Natural Sciences and Engineering Research Council of Canada; the Department of Indian and Northern Affairs Canada; the Canadian Circumpolar Institute; and DeBeers Canada, Ltd. Lorna Ruechel assisted with den observations. A. Gunn provided caribou location data. We thank Dave Mech for the use of GPS collars. M. Nelson, A. Gunn, and three anonymous reviewers made helpful comments on earlier drafts of the manuscript. This work was done under Wildlife Research Permit – WL002948 issued by the Government of the Northwest Territories, Department of Resources, Wildlife, and Economic Development.

(41) References

BALLARD, W.B., AYRES, L.A., KRAUSMAN, P.R., REED, D.J., and FANCY, S.G. 1997. Ecology of wolves in relation to a migratory caribou herd in northwest Alaska. Wildlife Monographs 135. 47 p.

BROWN, M.B., and FORSYTHE, A.B. 1974. Robust tests for the equality of variances. Journal of the American Statistical Association 69:364–367.

CARLSON, A. 1985. Central place foraging in the red-backed shrike (*Lanius collurio* L.): Allocation of prey between forager and sedentary consumer. Animal Behaviour 33:664–666.

CLARK, K.R.F. 1971. Food habits and behavior of the tundra wolf on central Baffin Island. Ph.D. Thesis, University of Toronto, Ontario, Canada.

ENVIRONMENT CANADA. 2003. National climate data information archive. Available online: http://www.climate.weatheroffice.ec.gc.ca/Welcome_e.html

FOURNIER, B., and GUNN, A. 1998. Musk ox numbers and distribution in the NWT, 1997. File Report No. 121. Yellowknife: Department of Resources, Wildlife, and Economic Development, Government of the Northwest Territories. 55 p.

FRITTS, S.H., and MECH, L.D. 1981. Dynamics, movements, and feeding ecology of a newly protected wolf population in northwestern Minnesota. Wildlife Monographs 80. 79 p.

GUNN, A., DRAGON, J., and BOULANGER, J. 2001. Seasonal movements of satellite-collared caribou from the Bathurst herd. Final Report to the West Kitikmeot Slave Study Society, Yellowknife, NWT. 80 p. Available online: http://www.wkss.nt.ca/HTML/08_ProjectsReports/PDF/Seasonal MovementsFinal.pdf

HARRINGTON, F.H., and ASA, C.S. 2003. Wolf communication. In: Mech, L.D., and Boitani, L., eds. Wolves: Behavior, ecology, and conservation. Chicago: University of Chicago Press. 66–103.

HARRINGTON, F.H., MECH, L.D., and FRITTS, S.H. 1983. Pack size and wolf pup survival: Their relationship under varying ecological conditions. Behavioral Ecology and Sociobiology 13:19–26.

HARRIS, S.A. 1986. Permafrost distribution, zonation and stability along the eastern ranges of the cordillera of North America. Arctic 39(1):29–38.

HEARD, D.C., and WILLIAMS, T.M. 1992. Distribution of wolf dens on migratory caribou ranges in the Northwest

37 Discussion of cooperative rearing of pups and, in turn, to speculation on how this study and what is known about cooperative rearing might fit into the animal's strategies for survival of the species. Again, the authors approach the broader theory of evolution and how it might explain some of their results.

38 And again, they suggest that this study points to several areas where further study will shed some light.

39 In conclusion, the authors suggest that their study supports the hypothesis being tested here. And they touch on the implications of increased human activity on the tundra predicted by their results.

40 ACKNOWLEDGEMENTS: Authors note the support of institutions, companies, and individuals. They thank their reviewers and list permits under which their research was carried on.

41 REFERENCES: List of all studies cited in the report. This may seem tedious, but is a vitally important part of scientific reporting. It is a record of the sources of information on which this study is based. It provides readers with a wealth of resources for further reading on this topic. Much of it will form the foundation of future scientific studies like this one.

Territories, Canada. Canadian Journal of Zoology 70:1504–1510.

HEARD, D.C., WILLIAMS, T.M., and MELTON, D.A. 1996. The relationship between food intake and predation risk in migratory caribou and implication to caribou and wolf population dynamics. Rangifer Special Issue No. 2:37–44.

KELSALL, J.P. 1968. The migratory barren-ground caribou of Canada. Canadian Wildlife Service Monograph Series 3. Ottawa: Queen's Printer. 340 p.

KUYT, E. 1962. Movements of young wolves in the Northwest Territories of Canada. Journal of Mammalogy 43:270–271.

———. 1972. Food habits and ecology of wolves on barren-ground caribou range in the Northwest Territories. Canadian Wildlife Service Report Series 21. Ottawa: Information Canada. 36 p.

MECH, L.D. 1994. Regular and homeward travel speeds of Arctic wolves. Journal of Mammalogy 75:741–742.

MECH, L.D., and BOITANI, L. 2003. Wolf social ecology. In: Mech, L.D., and Boitani, L., eds. Wolves: Behavior, ecology, and conservation. Chicago: University of Chicago Press. 1–34.

MECH, L.D., and GESE, E.M. 1992. Field testing the Wildlink capture collar on wolves. Wildlife Society Bulletin 20:249–256.

MECH, L.D., WOLFE, P., and PACKARD, J.M. 1999. Regurgitative food transfer among wild wolves. Canadian Journal of Zoology 77:1192–1195.

MERRILL, S.B., and MECH, L.D. 2000. Details of extensive movements by Minnesota wolves (*Canis lupus*). American Midland Naturalist 144:428–433.

MERRILL, S.B., ADAMS, L.G., NELSON, M.E., and MECH, L.D. 1998. Testing releasable GPS radiocollars on wolves and white-tailed deer. Wildlife Society Bulletin 26:830–835.

MESSIER, F. 1985. Solitary living and extraterritorial movements of wolves in relation to social status and prey abundance. Canadian Journal of Zoology 63:239–245.

MOHR, C.O., and STUMPF, W.A. 1966. Comparison of methods for calculating areas of animal activity. Journal of Wildlife Management 30:293–304.

MUSIANI, M. 2003. Conservation biology and management of wolves and wolf-human conflicts in western North America. Ph.D. Thesis, University of Calgary, Calgary, Alberta, Canada.

ORIANS, G.H., and PEARSON, N.E. 1979. On the theory of central place foraging. In: Mitchell, R.D., and Stairs, G.F., eds. Analysis of ecological systems. Columbus: Ohio State University Press. 154–177.

PACKARD, J.M., MECH, L.D., and REAM, R.R. 1992. Weaning in an arctic wolf pack: Behavioral mechanisms. Canadian Journal of Zoology 70:1269–1275.

PAQUET, P.C., WIERZCHOWSKI, J., and CALLAGHAN, C. 1996. Summary report on the effects of human activity on gray wolves in the Bow River Valley, Banff National Park, Alberta. In: Green, J., Pacas, C., Bayley, S., and Cornwell, L., eds. A cumulative effects assessment and futures outlook for the Banff Bow Valley. Prepared for the Banff Bow Valley Study. Ottawa: Department of Canadian Heritage.

STEPHENSON, R.O., and JAMES, D. 1982. Wolf movements and food habits in northwest Alaska. In: Harrington, F.H., and Paquet, P.C., eds. Wolves of the world. New Jersey: Noyes Publications. 223–237.

THORPE, N., EYEGETOK, S., HAKONGAK, N., and QITIRMIUT ELDERS. 2001. The Tuktu and Nogak Project: A caribou chronicle. Final Report to the West Kitikmeot/Slave Study Society, Ikaluktuuttiak, NWT. 160 p.

TIMONEY, K.P., LA ROI, G.H., ZOLTAI, S.C., and ROBINSON, A.L. 1992. The high subarctic forest-tundra of northwestern Canada: Position, width, and vegetation gradients in relation to climate. Arctic 45(1):1–9.

WALTON, L.R., CLUFF, H.D., PAQUET, P.C., and RAMSAY, M.A. 2001. Movement patterns of barren-ground wolves in the central Canadian Arctic. Journal of Mammalogy 82:867–876.

WILLIAMS, T.M. 1990. Summer diet and behavior of wolves denning on barren-ground caribou range in the Northwest Territories, Canada. M.Sc. Thesis, University of Alberta, Edmonton, Alberta, Canada.

ZAR, J.H. 1999. Biostatistical analysis. 4th ed. New Jersey: Prentice Hall. 663 p.

GLOSSARY

abiotic Nonliving. Compare *biotic*.

acid See *acid solution*.

acidic solution Any water solution that has more hydrogen ions (H^+) than hydroxide ions (OH^-); any water solution with a pH less than 7. Compare *basic solution, neutral solution*.

adaptation Any genetically controlled structural, physiological, or behavioral characteristic that helps an organism survive and reproduce under a given set of environmental conditions. It usually results from a beneficial mutation. See *biological evolution, differential reproduction, mutation, natural selection*.

adaptive radiation Process in which numerous new species evolve to fill vacant and new ecological niches in changed environments, usually after a mass extinction. Typically, this process takes millions of years.

adaptive trait See *adaptation*.

aerobic respiration Complex process that occurs in the cells of most living organisms, in which nutrient organic molecules such as glucose ($C_6H_{12}O_6$) combine with oxygen (O_2) to produce carbon dioxide (CO_2), water (H_2O), and energy. Compare *photosynthesis*.

affluenza Unsustainable addiction to overconsumption and materialism exhibited in the lifestyles of affluent consumers.

age structure Percentage of the population (or number of people of each gender) at each age level in a population.

alien species See *nonnative species*.

alpha particle Positively charged matter, consisting of two neutrons and two protons, that is emitted as radioactivity from the nuclei of some radioisotopes. See also *beta particle, gamma ray*.

altitude See *elevation*.

anaerobic respiration Form of cellular respiration in which some decomposers get the energy they need through the breakdown of glucose (or other nutrients) in the absence of oxygen. Compare *aerobic respiration*.

ancient forest See *old-growth forest*.

annual Plant that grows, sets seed, and dies in one growing season. Compare *perennial*.

anthropocentric Human-centered.

aquatic Pertaining to water. Compare *terrestrial*.

aquatic life zone Marine and freshwater portions of the biosphere. Examples include freshwater life zones (such as lakes and streams) and ocean or marine life zones (such as estuaries, coastlines, coral reefs, and the deep ocean).

aquifer Porous, water-saturated layers of sand, gravel, or bedrock that can yield an economically significant amount of water.

arid Dry. A desert or other area with an arid climate has little precipitation.

artificial selection Process by which humans select one or more desirable genetic traits in the population of a plant or animal species and then use *selective breeding* to produce populations containing many individuals with the desired traits. Compare *genetic engineering, natural selection*.

asexual reproduction Reproduction in which a mother cell divides to produce two identical daughter cells that are clones of the mother cell. This type of reproduction is common in single-celled organisms. Compare *sexual reproduction*.

atmosphere Mass of air surrounding the earth. See *stratosphere, troposphere*.

atom The basic building block of all chemical elements and thus all matter; the smallest unit of an element that can exist and still have the unique characteristics of that element; made of *subatomic particles*. Compare *ion, molecule*.

atomic number Number of protons in the nucleus of an atom. Compare *mass number*.

autotroph See *producer*.

background extinction Normal extinction of various species as a result of changes in local environmental conditions. Compare *mass depletion, mass extinction*.

bacteria Prokaryotic, one-celled organisms. Some transmit diseases. Most act as decomposers and get the nutrients they need by breaking down complex organic compounds in the tissues of living or dead organisms into simpler inorganic nutrient compounds. See *prokaryotic*.

barrier islands Long, thin, low offshore islands of sediment that generally run parallel to the shore along some coasts.

basic solution Water solution with more hydroxide ions (OH^-) than hydrogen ions (H^+); water solution with a pH greater than 7. Compare *acidic solution, neutral solution*.

benthos Bottom-dwelling organisms. Compare *decomposer, nekton, plankton*.

beta particle Swiftly moving electron emitted by the nucleus of a radioactive isotope. See also *alpha particle, gamma ray*.

biocentric Life-centered. Compare *anthropocentric*.

biodegradable Capable of being broken down by decomposers.

biodegradable pollutant Material that can be broken down into simpler substances (elements and compounds) by bacteria or other decomposers. Paper and most organic wastes such as animal manure are biodegradable but can take decades to biodegrade in modern landfills. Compare *degradable pollutant, nondegradable pollutant, slowly degradable pollutant.*

biodiversity Variety of different species (*species diversity*); genetic variability among individuals within each species (*genetic diversity*); variety of ecosystems (*ecological diversity*); and functions such as energy flow and matter cycling needed for the survival of species and biological communities (*functional diversity*).

biogeochemical cycle Natural processes that recycle nutrients in various chemical forms from the nonliving environment to living organisms and then back to the nonliving environment. Examples include the carbon, oxygen, nitrogen, phosphorus, sulfur, and hydrologic cycles.

biological community See *community.*

biological diversity See *biodiversity.*

biological evolution Change in the genetic makeup of a population of a species in successive generations. If continued long enough, it can lead to the formation of a new species. Note that populations—not individuals—evolve. See also *adaptation, differential reproduction, natural selection, theory of evolution.*

biomass Organic matter produced by plants and other photosynthetic producers; total dry weight of all living organisms that can be supported at each trophic level in a food chain or web; dry weight of all organic matter in plants and animals in an ecosystem; plant materials and animal wastes used as fuel.

biome Terrestrial regions inhabited by certain types of life, especially vegetation. Examples include various types of deserts, grasslands, and forests.

biosphere Zone of the earth where life is found. It consists of parts of the atmosphere (the troposphere), hydrosphere (mostly surface water and groundwater), and lithosphere (mostly soil and surface rocks and sediments on the bottoms of oceans and other bodies of water) where life is found.

biotic Living organisms. Compare *abiotic.*

biotic pollution Harmful ecological and economic effects from the presence of accidentally or deliberately introduced species into ecosystems.

biotic potential Maximum rate at which the population of a given species can increase when there are no limits on its rate of growth. See *environmental resistance.*

birth rate See *crude birth rate.*

broadleaf deciduous plants Plants such as oak and maple trees that survive drought and cold by shedding their leaves and becoming dormant. Compare *broadleaf evergreen plants, coniferous evergreen plants.*

broadleaf evergreen plants Plants that keep most of their broad leaves year-round. An example is the trees found in the canopies of tropical rain forests. Compare *broadleaf deciduous plants, coniferous evergreen plants.*

calorie Unit of energy; amount of energy needed to raise the temperature of 1 gram of water by $1C°$ (unit on Celsius temperature scale). See also *kilocalorie.*

carbon cycle Cyclic movement of carbon in different chemical forms from the environment to organisms and then back to the environment.

carnivore Animal that feeds on other animals. Compare *herbivore, omnivore.*

carrying capacity (K) Maximum population of a particular species that a given habitat can support over a given period.

cell Smallest living unit of an organism. Each cell is encased in an outer membrane or wall and contains genetic material (DNA) and other parts to perform its life function. Organisms such as bacteria consist of only one cell, but most organisms contain many cells.

chain reaction Multiple nuclear fissions, taking place within a certain mass of a fissionable isotope, which release an enormous amount of energy in a short time.

chemical One of the millions of different elements and compounds found naturally and synthesized by humans. See *compound, element.*

chemical change Interaction between chemicals in which the chemical composition of the elements or compounds involved changes. Compare *nuclear change, physical change.*

chemical evolution Formation of the earth and its early crust and atmosphere, evolution of the biological molecules necessary for life, and evolution of systems of chemical reactions needed to produce the first living cells. These processes are believed to have occurred about 1 billion years before biological evolution. Compare *biological evolution.*

chemical formula Shorthand way to show the number of atoms (or ions) in the basic structural unit of a compound. Examples include H_2O, $NaCl$, and $C_6H_{12}O_6$.

chemical reaction See *chemical change.*

chemosynthesis Process in which certain organisms (mostly specialized bacteria) extract inorganic compounds from their environment and convert them into organic nutrient compounds without the presence of sunlight. Compare *photosynthesis.*

chlorinated hydrocarbon Organic compound made up of atoms of carbon, hydrogen, and chlorine. Examples include DDT and PCBs.

chromosome Grouping of genes and associated proteins in plant and animal cells that carry certain types of genetic information. See *genes.*

clear-cutting Method of timber harvesting in which all trees in a forested area are removed in a single cutting. Compare *selective cutting, strip-cutting.*

climate Physical properties of the troposphere of an area based on analysis of its weather records over a long period (at least 30 years). The two main factors determining an area's climate are the *temperature,* with its seasonal variations, and the amount and distribution of *precipitation.* Compare *weather.*

climax community See *mature community.*

coastal wetland Land along a coastline, extending inland from an estuary that is covered with saltwater all or part of the year. Examples include marshes, bays, lagoons, tidal flats, and mangrove swamps. Compare *inland wetland.*

coastal zone Warm, nutrient-rich, shallow part of the ocean that extends from the high-tide mark on land to the edge of a shelf-like extension of continental land masses known as the continental shelf. Compare *open sea.*

coevolution Evolution in which two or more species interact and exert selective pressures on each other that can lead each species to undergo adaptations. See *evolution, natural selection.*

cold front Leading edge of an advancing mass of cold air. Compare *warm front*.

commensalism Interaction between organisms of different species in which one type of organism benefits and the other type is neither helped nor harmed to any great degree. Compare *mutualism*.

commercial extinction Depletion of the population of a wild species used as a resource to a level at which it is no longer profitable to harvest the species.

commercial inorganic fertilizer Commercially prepared mixture of plant nutrients such as nitrates, phosphates, and potassium applied to the soil to restore fertility and increase crop yields.

common-property resource Resource that people normally are free to use; each user can deplete or degrade the available supply. Most such resources are renewable and owned by no one. Examples include clean air, fish in parts of the ocean not under the control of a coastal country, migratory birds, gases of the lower atmosphere, and the ozone content of the upper atmosphere (stratosphere). See *tragedy of the commons*.

community Populations of all species living and interacting in an area at a particular time.

competition Two or more individual organisms of a single species (*intraspecific competition*) or two or more individuals of different species (*interspecific competition*) attempting to use the same scarce resources in the same ecosystem.

complex carbohydrates Two or more monomers of *simple sugars* (such as glucose) linked together.

compound Combination of atoms, or oppositely charged ions, of two or more elements held together by attractive forces called chemical bonds. Compare *element*.

concentration Amount of a chemical in a particular volume or weight of air, water, soil, or other medium.

coniferous evergreen plants Cone-bearing plants (such as spruces, pines, and firs) that keep some of their narrow, pointed leaves (needles) all year. Compare *broadleaf deciduous plants, broadleaf evergreen plants*.

coniferous trees Cone-bearing trees, mostly evergreens, that have needle-shaped or scale-like leaves. They produce wood known commercially as softwood. Compare *deciduous plants*.

consensus science See *sound science*.

conservation Sensible and careful use of natural resources by humans. People with this view are called *conservationists*.

conservation biology Multidisciplinary science created to deal with the crisis of maintaining the genes, species, communities, and ecosystems that make up earth's biological diversity. Its goals are to investigate human impacts on biodiversity and to develop practical approaches to preserving biodiversity.

conservationist Person concerned with using natural areas and wildlife in ways that sustain them for current and future generations of humans and other forms of life.

constancy Ability of a living system, such as a population, to maintain a certain size. Compare *inertia, resilience*.

consumer Organism that cannot synthesize the organic nutrients it needs and gets its organic nutrients by feeding on the tissues of producers or of other consumers; generally divided into *primary consumers* (herbivores), *secondary consumers* (carnivores), *tertiary (higher-level) consumers, omnivores,* and *detritivores* (decomposers and detritus feeders). In economics, one who uses economic goods. Compare *producer*.

controlled burning Deliberately set, carefully controlled surface fires that reduce flammable litter and decrease the chances of damaging crown fires. See *ground fire, surface fire*.

convergent plate boundary Area where the earth's lithospheric plates are pushed together. See *subduction zone*. Compare *divergent plate boundary, transform fault*.

coral reef Formation produced by massive colonies containing billions of tiny coral animals, called polyps, that secrete a stony substance (calcium carbonate) around themselves for protection. When the corals die, their empty outer skeletons form layers and cause the reef to grow. Coral reefs are found in the coastal zones of warm tropical and subtropical oceans.

core Inner zone of the earth. It consists of a solid inner core and a liquid outer core. Compare *crust, mantle*.

corrective feedback loop See *negative feedback loop*.

critical mass Amount of fissionable nuclei needed to sustain a nuclear fission chain reaction.

crown fire Extremely hot forest fire that burns ground vegetation and treetops. Compare *controlled burning, ground fire, surface fire*.

crude birth rate Annual number of live births per 1,000 people in the population of a geographic area at the midpoint of a given year. Compare *crude death rate*.

crude death rate Annual number of deaths per 1,000 people in the population of a geographic area at the midpoint of a given year. Compare *crude birth rate*.

crust Solid outer zone of the earth. It consists of oceanic crust and continental crust. Compare *core, mantle*.

data See *scientific data*.

DDT Dichlorodiphenyltrichloroethane, a chlorinated hydrocarbon that has been widely used as an insecticide but is now banned in some countries.

death rate See *crude death rate*.

debt-for-nature swap Agreement in which a certain amount of foreign debt is canceled in exchange for local currency investments that will improve natural resource management or protect certain areas in the debtor country from harmful development.

deciduous plants Trees, such as oaks and maples, and other plants that survive during dry seasons or cold seasons by shedding their leaves. Compare *coniferous trees, succulent plants*.

decomposer Organism that digests parts of dead organisms and fragments and wastes cast-off by living organisms breaking down the complex organic molecules in those materials into simpler inorganic compounds and then absorbing the soluble nutrients. Decomposers consist of various bacteria and fungi. Compare *consumer, detritivore, producer*.

deductive reasoning Using logic to arrive at a specific conclusion based on a generalization or premise.

deforestation Removal of trees from a forested area without adequate replanting.

degradable pollutant Potentially polluting chemical that is broken down completely or reduced to acceptable levels by natural physical, chemical, and biological processes. Compare *biodegradable pollutant, nondegradable pollutant, slowly degradable pollutant*.

demographic transition Hypothesis that countries, as they become industrialized, have declines in death rates followed by declines in birth rates.

desert Biome in which evaporation exceeds precipitation and the average amount of precipitation is less than 25 centimeters (10 inches) per year. Such areas have little vegetation or have widely spaced, mostly low vegetation. Compare *forest, grassland*.

detritivore Consumer organism that feeds on detritus, parts of dead organisms, and cast-off fragments and wastes of living organisms. The two principal types are *detritus feeders* and *decomposers*.

detritus Parts of dead organisms and cast-off fragments and wastes of living organisms.

detritus feeder Organism that extracts nutrients from fragments of dead organisms and their cast-off parts and organic wastes. Examples include earthworms, termites, and crabs. Compare *decomposer*.

deuterium (D; hydrogen-2) Isotope of the element hydrogen, with a nucleus containing one proton and one neutron and a mass number of 2.

developed country Country that is highly industrialized and has a high per capita GDP. Compare *developing country*.

developing country Country that has low to moderate industrialization and low to moderate per capita GDP. Most are located in Africa, Asia, and Latin America. Compare *developed country*.

dieback Sharp reduction in the population of a species when its numbers exceed the carrying capacity of its habitat. See *carrying capacity*.

differential reproduction Phenomenon in which individuals with adaptive genetic traits produce more living offspring than do individuals without such traits. See *natural selection*.

dissolved oxygen (DO) content Amount of oxygen gas (O_2) dissolved in a given volume of water at a particular temperature and pressure, often expressed as a concentration in parts of oxygen per million parts of water.

distribution Area over which a species can be found.

disturbance Event that disrupts an ecosystem or community. Examples of *natural disturbances* include fires, hurricanes, tornadoes, droughts, and floods. Examples of *human-caused disturbances* include deforestation, overgrazing, and plowing.

divergent plate boundary Area where the earth's lithospheric plates move apart in opposite directions. Compare *convergent plate boundary, transform fault*.

DNA (deoxyribonucleic acid) Large molecules in the cells of organisms that carry genetic information in living organisms.

domesticated species Wild species tamed or genetically altered by crossbreeding for use by humans for food (cattle, sheep, and food crops), pets (dogs and cats), or enjoyment (animals in zoos and plants in gardens). Compare *wild species*.

doubling time Time it takes (usually in years) for the quantity of something growing exponentially to double. It can be calculated by dividing the annual percentage growth rate into 70.

drought Condition in which an area does not get enough water because of lower-than-normal precipitation or higher-than-normal temperatures that increase evaporation.

durability See *sustainability*.

earthquake Shaking of the ground resulting from the fracturing and displacement of rock, which produces a fault, or from subsequent movement along the fault.

ecological diversity Variety of forests, deserts, grasslands, oceans, streams, lakes, and other biological communities interacting with one another and with their nonliving environment. See *biodiversity*. Compare *functional diversity, genetic diversity, species diversity*.

ecological efficiency Percentage of energy transferred from one trophic level to another in a food chain or web.

ecological footprint Amount of biologically productive land and water needed to supply a population with the renewable resources it uses and to absorb or dispose of the wastes from such resource use. It measures the average environmental impact of populations in different countries and areas. See *per capita ecological footprint*

ecological niche Total way of life or role of a species in an ecosystem. It includes all physical, chemical, and biological conditions that a species needs to live and reproduce in an ecosystem. See *fundamental niche, realized niche*.

ecological restoration Deliberate alteration of a degraded habitat or ecosystem to restore as much of its ecological structure and function as possible.

ecological succession Process in which communities of plant and animal species in a particular area are replaced over time by a series of different and often more complex communities. See *primary succession, secondary succession*.

ecologist Biological scientist who studies relationships between living organisms and their environment.

ecology Biological science that studies the relationships between living organisms and their environment; study of the structure and functions of nature.

economic development Improvement of human living standards by economic growth. Compare *economic growth, environmentally sustainable economic development*.

economic growth Increase in the capacity to provide people with goods and services; an increase in gross domestic product (GDP). Compare *economic development, environmentally sustainable economic development*. See *gross domestic product*.

ecosphere See *biosphere*.

ecosystem Community of different species interacting with one another and with the chemical and physical factors making up its nonliving environment.

ecosystem services Natural services or natural capital that support life on the earth and are essential to the quality of human life and the functioning of the world's economies. Compare *natural capital*.

electromagnetic radiation Forms of kinetic energy traveling as electromagnetic waves. Examples include radio waves, TV waves, microwaves, infrared radiation, visible light, ultraviolet radiation, X rays, and gamma rays. Compare *ionizing radiation, nonionizing radiation*.

electron (e) Tiny particle moving around outside the nucleus of an atom. Each electron has one unit of negative charge and almost no mass. Compare *neutron, proton*.

element Chemicals, such as hydrogen (H), iron (Fe), sodium (Na), carbon (C), nitrogen (N), or oxygen (O), whose distinctly different atoms serve as the basic building blocks of all matter. Two or more elements combine to form the compounds that make up most of the world's matter. Compare *compound*.

elevation Distance above sea level. Compare *latitude*.

emigration Movement of people out of a specific geographic area. See *migration*. Compare *immigration*.

endangered species Wild species with so few individual survivors that the species could soon become extinct in all or most of its natural range. Compare *threatened species*.

endemic species Species that is found in only one area. Such species are especially vulnerable to extinction.

energy Capacity to do work by performing mechanical, physical, chemical, or electrical tasks or to cause a heat transfer between two objects at different temperatures.

energy quality Ability of a form of energy to do useful work. High-temperature heat and the chemical energy in fossil fuels and nuclear fuels are concentrated high-quality energy. Low-quality energy such as low-temperature heat is dispersed or diluted and cannot do much useful work. See *high-quality energy, low-quality energy*.

environment All external conditions and factors, living and nonliving, that affect any living organism or other specified system.

environmental degradation Depletion or destruction of a potentially renewable resource such as soil, grassland, forest, or wildlife. If used faster than it is naturally replenished, the resource becomes nonrenewable (on a human time scale) or nonexistent (extinct). See also *sustainable yield*.

environmental ethics Human beliefs about what is right or wrong with how we treat the environment.

environmentalism Social movement dedicated to protecting the earth's life-support systems for us and other species.

environmentalist Person who is concerned about the impact of people on environmental quality and believes that some human actions are degrading parts of the earth's life-support systems for humans and many other forms of life.

environmentally sustainable economic development Development that *encourages* forms of economic growth that meet the basic needs of the current generations of humans and other species without preventing future generations of humans and other species from meeting their basic needs and *discourages* environmentally harmful and unsustainable forms of economic growth. It is the economic component of an *environmentally sustainable society*. Compare *economic development, economic growth*.

environmentally sustainable society Society that meets the current and future basic needs of its people for basic resources in a just and equitable manner without compromising the ability of future generations of humans and other species to meet their basic needs.

environmental movement Social movement that flourished in the United States in the 1960s when a growing number of citizens organized to demand that political leaders enact laws and develop policies to curtail pollution, clean up polluted environments, and protect unspoiled areas from environmental degradation.

environmental resistance All of the limiting factors that act together to limit the growth of a population. See *biotic potential, limiting factor*.

environmental revolution Cultural change involving the halting of population growth and altering of lifestyles, political and economic systems, and the way we treat the environment so that we can help sustain the earth for ourselves and other species. It requires working with the rest of nature by learning more about how nature sustains itself. See *environmental wisdom worldview*.

environmental science Interdisciplinary study that uses information from the physical and social sciences to learn how the earth works, how we interact with the earth, and how to deal with environmental problems.

environmental scientist Scientist who uses information from the physical and social sciences to understand how the earth works, to learn how humans interact with the earth, and to develop solutions to environmental problems.

environmental wisdom worldview The belief that humans are part of and totally dependent on nature; that nature exists for all species; that people should encourage earth-sustaining forms of economic growth and development and discourage earth-degrading forms; and that our success depends on learning how the earth sustains itself and integrating such environmental wisdom into the ways we think and act. Compare *planetary management worldview, stewardship worldview*.

environmental worldview Set of assumptions and beliefs about how the world works, what a person's role in the world should be, and what is right and wrong environmental behavior. Compare *environmental ethics*.

EPA U.S. Environmental Protection Agency; responsible for managing federal efforts to control air and water pollution, radiation and pesticide hazards, environmental research, hazardous waste, and solid waste disposal.

epiphyte Plant that uses its roots to attach itself to branches high in trees, especially in tropical forests.

estuary Partially enclosed coastal area at the mouth of a river where its freshwater, carrying silt and runoff from the land, mixes with salty seawater.

eukaryotic organism Classification of cell structure in which the cell is surrounded by a membrane and has a distinct nucleus and several other internal parts. Most organisms consist of eukaryotic cells. Compare *prokaryotic organism*.

euphotic zone Upper layer of a body of water through which sunlight can penetrate and support photosynthesis.

eutrophic lake Lake with a large or excessive supply of plant nutrients, mostly nitrates and phosphates. Compare *mesotrophic lake, oligotrophic lake*.

evaporation Conversion of a liquid to a gas.

even-aged management Method of forest management in which trees, sometimes of a single species in a given stand, are maintained at roughly the same age and size and are harvested all at once. Compare *uneven-aged management*.

evergreen plants Plants that keep some of their leaves or needles throughout the year. Examples include ferns and cone-bearing trees (conifers) such as firs, spruces, pines, redwoods, and sequoias. Compare *deciduous plants, succulent plants*.

evolution See *biological evolution*.

exhaustible resource See *nonrenewable resource*.

exotic species See *nonnative species*.

experiment Procedure a scientist uses to study a phenomenon under known conditions. Scientists conduct some experiments in the laboratory and others in nature.

exponential growth Growth in which some quantity, such as population size or economic output, increases at a constant rate per unit of time. An example is the growth sequence 2, 4, 8, 16, 32, 64, and so on. When the increase in quantity over time is plotted, this type of growth yields a curve shaped like the letter J. Compare *linear growth*.

extinction Complete disappearance of a species from the earth. It happens when a species cannot adapt and successfully reproduce under new environmental conditions or when a species evolves into one or more new species. Compare *speciation*. See also *endangered species, mass depletion, mass extinction, threatened species*.

family planning Providing information, clinical services, and contraceptives to help people choose the number and spacing of children they want to have.

feedback loop Occurs when an output of matter, energy, or information is fed back into the system as an input and leads to changes in that system.

fermentation See *anaerobic respiration*.

fertility Number of births that occur to an individual woman.

first law of thermodynamics In any physical or chemical change, no detectable amount of energy is created or destroyed, but energy can be changed from one form to another; you cannot get more energy out of something than you put in; in terms of energy quantity, you cannot get something for nothing. This law does not apply to nuclear changes, in which energy can be produced from small amounts of matter. See *second law of thermodynamics*.

fissionable isotope Isotope that can split apart when hit by a neutron at the right speed and thus undergo nuclear fission. Examples include uranium-235 and plutonium-239.

flows See *throughputs*.

food chain Series of organisms in which each eats or decomposes the preceding one. Compare *food web*.

food web Complex network of many interconnected food chains and feeding relationships. Compare *food chain*.

forest Biome with enough average annual precipitation (at least 76 centimeters, or 30 inches) to support the growth of tree species and smaller forms of vegetation. Compare *desert, grassland*.

fossils Skeletons, bones, shells, body parts, leaves, seeds, or impressions of such items that provide recognizable evidence of organisms that lived long ago.

foundation species Species that plays a major role in shaping communities by creating and enhancing a habitat that benefits other species. Compare *indicator species, keystone species, native species, nonnative species*.

free-access resource See *common-property resource*.

freshwater life zones Aquatic systems where water with a dissolved salt concentration of less than 1% by volume accumulates on or flows through the surfaces of terrestrial biomes. Examples include *standing* (lentic) bodies of fresh-

water such as lakes, ponds, and inland wetlands and *flowing* (lotic) systems such as streams and rivers. Compare *biome*.

front Boundary between two air masses with different temperatures and densities. See *cold front, warm front*.

frontier environmental worldview View, such as that held by European colonists settling North America in the 1600s, that an area has vast resources to be conquered by settlers dominating the land.

frontier science Preliminary scientific data, hypotheses, and models that have not been widely tested and accepted. Compare *junk science, sound science*.

functional diversity Biological and chemical processes or functions such as energy flow and matter cycling needed for the survival of species and biological communities. See *biodiversity, ecological diversity, genetic diversity, species diversity*.

fundamental niche Full potential range of the physical, chemical, and biological factors a species can use if it does not face any competition from other species. See *ecological niche*. Compare *realized niche*.

gamma ray Form of ionizing electromagnetic radiation with a high energy content emitted by some radioisotopes. It readily penetrates body tissues. See also *alpha particle, beta particle*.

GDP See *gross domestic product*.

gene mutation See *mutation*.

gene pool Sum total of all genes found in the individuals of the population of a particular species.

generalist species Species with a broad ecological niche that can live in many different places, eat a variety of foods, and tolerate a wide range of environmental conditions. Examples include flies, cockroaches, mice, rats, and humans. Compare *specialist species*.

genes Coded units of information about specific traits that are passed from parents to offspring during reproduction. They consist of segments of DNA molecules found in chromosomes.

gene splicing See *genetic engineering*.

genetic adaptation Changes in the genetic makeup of organisms of a species that allow the species to reproduce and gain a competitive advantage under changed environmental conditions. See *differential reproduction, evolution, mutation, natural selection*.

genetically modified organism (GMO) Organism whose genetic makeup has been altered by genetic engineering.

genetic diversity Variability in the genetic makeup among individuals within a single species. See *biodiversity*. Compare *ecological diversity, functional diversity, species diversity*.

genetic engineering Insertion of an alien gene into an organism to give it a beneficial genetic trait. Compare *artificial selection, natural selection*.

geographic isolation Separation of populations of a species for long times into different areas.

geology Study of the earth's dynamic history. Geologists study and analyze rocks and the features and processes of the earth's interior and surface.

global climate change Changes in any aspects of the earth's climate, including temperature, precipitation, and storm intensity and patterns.

global warming Warming of the earth's atmosphere because of increases in the concentrations of one or more greenhouse gases; a political issue in which people have argued about the role of human activities in global warming. See *greenhouse effect, greenhouse gases.* Compare *global climate change.*

grassland Biome found in regions where moderate annual average precipitation (25–76 centimeters, or 10–30 inches) is enough to support the growth of grass and small plants but not enough to support large stands of trees. Compare *desert, forest.*

greenhouse effect Natural effect that releases heat in the atmosphere (troposphere) near the earth's surface. Water vapor, carbon dioxide, ozone, and other gases in the lower atmosphere (troposphere) absorb some of the infrared radiation (heat) radiated by the earth's surface. Their molecules vibrate and transform the absorbed energy into longer-wavelength infrared radiation (heat) in the troposphere. If the atmospheric concentrations of these greenhouse gases increase and other natural processes do not remove them, the average temperature of the lower atmosphere will increase gradually. Compare *global warming.* See also *natural greenhouse effect.*

greenhouse gases Gases in the earth's lower atmosphere (troposphere) that cause the greenhouse effect. Examples include carbon dioxide, chlorofluorocarbons, ozone, methane, water vapor, and nitrous oxide.

gross domestic product (GDP) Annual market value of all goods and services produced by all firms and organizations, foreign and domestic, operating within a country. See *per capita GDP.*

gross primary productivity (GPP) Rate at which an ecosystem's producers capture and store a given amount of chemical energy as biomass in a given length of time. Compare *net primary productivity.*

ground fire Fire that burns decayed leaves or peat deep below the ground surface. Compare *crown fire, surface fire.*

groundwater Water that sinks into the soil and is stored in slowly flowing and slowly renewed underground reservoirs called aquifers; underground water in the zone of saturation, below the water table. Compare *runoff, surface water.*

habitat Place or type of place where an organism or population of organisms lives. Compare *ecological niche.*

habitat fragmentation Breakup of a habitat into smaller pieces, usually as a result of human activities.

half-life Time needed for one-half of the nuclei in a radioisotope to emit their radiation. Each radioisotope has a characteristic half-life, which may range from a few millionths of a second to several billion years. See *radioisotope.*

heat Total kinetic energy of all randomly moving atoms, ions, or molecules within a given substance, excluding the overall motion of the whole object. Heat always flows spontaneously from a hot sample of matter to a colder sample of matter. This is one way to state the second law of thermodynamics. Compare *temperature.*

herbivore Plant-eating organism. Examples include deer, sheep, grasshoppers, and zooplankton. Compare *carnivore, omnivore.*

heterotroph See *consumer.*

high Air mass with a high pressure. Compare *low.*

high-quality energy Energy that is concentrated and has great ability to perform useful work. Examples include high-temperature heat and the energy in electricity, coal, oil, gasoline, sunlight, and nuclei of uranium-235. Compare *low-quality energy.*

high-quality matter Matter that is concentrated and contains a high concentration of a useful resource. Compare *low-quality matter.*

high-throughput economy Situation in most advanced industrialized countries, in which ever-increasing economic growth is sustained by maximizing the rate at which matter and energy resources are used, with little emphasis on pollution prevention, recycling, reuse, reduction of unnecessary waste, and other forms of resource conservation. Compare *low-throughput economy, matter-recycling-and-reuse economy.*

high-waste economy See *high-throughput economy.*

host Plant or animal on which a parasite feeds.

human capital See *human resources.*

human resources People's physical and mental talents that provide labor, innovation, culture, and organization. Human capital also includes the cultural skills people pick up from their families and friends, the ability to be trustworthy, and the drive to achieve.

hunter–gatherers People who get their food by gathering edible wild plants and other materials and by hunting wild animals and fish.

hydrocarbon Organic compound of hydrogen and carbon atoms. The simplest hydrocarbon is methane (CH_4), the major component of natural gas.

hydrologic cycle Biogeochemical cycle that collects, purifies, and distributes the earth's fixed supply of water within the biosphere and from the environment to living organisms and then back to the environment.

hydrosphere Earth's *liquid water* (oceans, lakes, other bodies of surface water, and underground water), *frozen water* (polar ice caps, floating ice caps, and ice in soil, known as permafrost), and *water vapor* in the atmosphere. See also *hydrologic cycle.*

hypothesis See *scientific hypothesis.*

immature community Community at an early stage of ecological succession. It usually has a low number of species and ecological niches and cannot capture and use energy and cycle critical nutrients as efficiently as more complex, mature communities. Compare *mature community.*

immigrant species See *nonnative species.*

immigration Migration of people into a country or area to take up permanent residence. See *migration.* Compare *emigration.*

indicator species Species that serve as early warnings that a community or ecosystem is being degraded. Compare *foundation species, keystone species, native species, nonnative species.*

inductive reasoning Using specific observations and measurements to arrive at a general conclusion or hypothesis.

inertia Ability of a living system to resist being disturbed or altered.

infant mortality rate Number of babies out of every 1,000 born each year who die before their first birthday.

inland wetland Land away from the coast, such as a swamp, marsh, or bog, that is covered all or part of the time with freshwater. Compare *coastal wetland*.

inorganic compounds All compounds not classified as organic compounds. See *organic compounds*.

inorganic fertilizer See *commercial inorganic fertilizer*.

input Matter, energy, or information entering a system. Compare *output, throughput*.

input pollution control See *pollution prevention*.

instrumental value Value of an organism, species, ecosystem, or the earth's biodiversity based on its usefulness to humans. Compare *intrinsic value*.

interspecific competition Attempts by members of two or more species to use the same limited resources in an ecosystem. See *competition, intraspecific competition*.

intertidal zone The area of shoreline between low and high tides.

intraspecific competition Attempts by two or more organisms of a single species to use the same limited resources in an ecosystem. See *competition, interspecific competition*.

intrinsic rate of increase (r) Rate at which a population could grow if it had unlimited resources. Compare *environmental resistance*.

intrinsic value Value of an organism, species, ecosystem, or the earth's biodiversity based on its existence, regardless of whether it has any usefulness to humans. Compare *instrumental value*.

invasive species See *nonnative species*.

invertebrates Animals that have no backbones. Compare *vertebrates*.

ion Atom or group of atoms with one or more positive (+) or negative (−) electrical charges. Compare *atom, molecule*.

ionizing radiation Fast-moving alpha or beta particles or high-energy radiation (gamma rays) emitted by radioisotopes. They have enough energy to dislodge one or more electrons from atoms they hit, thereby forming charged ions in living tissue that can react with and damage the tissue. Compare *nonionizing radiation*.

isotopes Two or more forms of a chemical element that have the same number of protons but different mass numbers because they have different numbers of neutrons in their nuclei.

J-shaped curve Curve with a shape similar to that of the letter J; can represent prolonged exponential growth. See *exponential growth*.

junk science Scientific results or hypotheses presented as sound science without having undergone the rigors of the peer review process. Compare *frontier science, sound science*.

keystone species Species that play roles affecting many other organisms in an ecosystem. Compare *foundation species, indicator species, native species, nonnative species*.

kilocalorie (kcal) Unit of energy equal to 1,000 calories. See *calorie*.

kinetic energy Energy that matter has because of its mass and speed or velocity. Compare *potential energy*.

K-selected species Species that produce a few, often fairly large offspring but invest a great deal of time and energy to ensure that most of those offspring reach reproductive age. Compare *r-selected species*.

K-strategists See *K-selected species*.

lake Large natural body of standing freshwater formed when water from precipitation, land runoff, or groundwater flow fills a depression in the earth created by glaciation, earth movement, volcanic activity, or a giant meteorite. See *eutrophic lake, mesotrophic lake, oligotrophic lake*.

land degradation Decrease in the ability of land to support crops, livestock, or wild species in the future as a result of natural or human-induced processes.

latitude Distance from the equator. Compare *altitude*.

law of conservation of energy See *first law of thermodynamics*.

law of conservation of matter In any physical or chemical change, matter is neither created nor destroyed but merely changed from one form to another; in physical and chemical changes, existing atoms are rearranged into different spatial patterns (physical changes) or different combinations (chemical changes).

law of tolerance The existence, abundance, and distribution of a species in an ecosystem are determined by whether the levels of one or more physical or chemical factors fall within the range tolerated by the species. See *threshold effect*.

LDC See *developing country*.

less developed country (LDC) See *developing country*.

life-centered environmental worldview The belief that we have an ethical responsibility to prevent degradation of the earth's ecosystems, biodiversity, and biosphere, and that there is *inherent* or *intrinsic value* of all forms of life, regardless of their potential or actual use to humans.

life expectancy Average number of years a newborn infant can be expected to live.

limiting factor Single factor that limits the growth, abundance, or distribution of the population of a species in an ecosystem. See *limiting factor principle*.

limiting factor principle Too much or too little of any abiotic factor can limit or prevent growth of a population of a species in an ecosystem, even if all other factors are at or near the optimal range of tolerance for the species.

linear growth Growth in which a quantity increases by some fixed amount during each unit of time. An example is growth that increases in the sequence 2, 4, 6, 8, 10, and so on. Compare *exponential growth*.

lipids Chemically diverse group of large organic compounds that do not dissolve in water. Examples are *fats and oils* for storing energy, *waxes* for structure, and *steroids* for producing hormones.

lithosphere Outer shell of the earth, composed of the crust and the rigid, outermost part of the mantle outside the asthenosphere; material found in the earth's plates. See *crust, mantle*.

logistic growth Pattern in which exponential population growth occurs when the population is small, and population growth decreases steadily with time as the population approaches the carrying capacity. See *S-shaped curve*.

low An air mass with a low pressure. Compare *high*.

low-quality energy Energy that is dispersed and has little ability to do useful work. An example is low-temperature heat. Compare *high-quality energy*.

low-quality matter Matter that is dilute or dispersed or contains a low concentration of a useful resource. Compare *high-quality matter*.

low-throughput economy Economy based on working with nature by recycling and reusing discarded matter; preventing pollution; conserving matter and energy resources by reducing unnecessary waste and use; not degrading renewable resources; building things that are easy to recycle, reuse, and repair; not allowing population size to exceed the carrying capacity of the environment; and preserving biodiversity and ecological integrity. Compare *high-throughput economy*, *matter-recycling-and-reuse economy*.

low-waste economy See *low-throughput economy*.

mangrove swamps Swamps found on the coastlines in warm tropical climates. They are dominated by mangrove trees, any of about 55 species of trees and shrubs that can live partly submerged in the salty environment of coastal swamps.

mantle Zone of the earth's interior between its core and its crust. Compare *core*, *crust*. See *lithosphere*.

mass The amount of material in an object.

mass depletion A period during which widespread, often global extinction rates are higher than normal but not high enough to classify as a mass extinction. Compare *background extinction*, *mass extinction*.

mass extinction A catastrophic, widespread, often global event in which major groups of species are wiped out over a short time compared with normal (background) extinctions. Compare *background extinction*, *mass depletion*.

mass number Sum of the number of neutrons (n) and the number of protons (p) in the nucleus of an atom. It gives the approximate mass of that atom. Compare *atomic number*.

matter Anything that has mass (the amount of material in an object) and takes up space.

matter quality Measure of how useful a matter resource is, based on its availability and concentration. See *high-quality matter*, *low-quality matter*.

matter-recycling-and-reuse economy Economy that emphasizes maximizing the reuse and recycling of materials. The goal is to allow economic growth to continue without depleting matter resources and without producing excessive pollution and environmental degradation. Compare *high-throughput economy*, *low-throughput economy*.

mature community Fairly stable, self-sustaining community in an advanced stage of ecological succession; usually has a diverse array of species and ecological niches; captures and uses energy and cycles critical chemicals more efficiently than simpler, immature communities. Compare *immature community*.

maximum sustainable yield See *sustainable yield*.

MDC See *developed country*.

mesotrophic lake Lake with a moderate supply of plant nutrients. Compare *eutrophic lake*, *oligotrophic lake*.

metabolism Ability of a living cell or organism to capture and transform matter and energy from its environment to supply its needs for survival, growth, and reproduction.

microorganisms Organisms such as bacteria that are so small that it takes a microscope to see them.

migration Movement of people into and out of a specific geographic area. See *immigration*, *emigration*.

mineral Any naturally occurring inorganic substance found in the earth's crust as a crystalline solid. See *mineral resource*.

mineral resource Concentration of naturally occurring solid, liquid, or gaseous material in or on the earth's crust in a form and amount such that extracting and converting it into useful materials or items is currently or potentially profitable. Mineral resources are classified as *metallic* (such as iron and tin ores) or *nonmetallic* (such as fossil fuels, sand, and salt).

minimum viable population (MVP) Estimate of the smallest number of individuals necessary to ensure the survival of a population in a region for a specified time period, typically ranging from decades to 100 years.

mixture Combination of one or more elements and compounds.

model Approximate representation or simulation of a system being studied.

molecule Combination of two or more atoms of the same chemical element (such as O_2) or different chemical elements (such as H_2O) held together by chemical bonds. Compare *atom*, *ion*.

more developed country (MDC) See *developed country*.

mutation Random change in DNA molecules making up genes that can alter anatomy, physiology, or behavior in offspring.

mutualism Type of species interaction in which both participating species generally benefit. Compare *commensalism*.

nanotechnology Using atoms and molecules to build materials from the bottom up using the elements in the periodic table as its raw materials.

native species Species that normally live and thrive in a particular ecosystem. Compare *foundation species*, *indicator species*, *keystone species*, *nonnative species*.

natural capital Natural resources and natural services that keep us and other species alive and support our economies

natural greenhouse effect Heat buildup in the troposphere because of the presence of certain gases, called greenhouse gases. Without this effect, the earth would be nearly as cold as Mars, and life as we know it could not exist. Compare *global warming*.

natural law See *scientific law*.

natural radioactive decay Nuclear change in which unstable nuclei of atoms spontaneously shoot out particles (usually alpha or beta particles) or energy (gamma rays) at a fixed rate.

natural rate of extinction See *background extinction*.

natural resources See *natural capital*.

natural selection Process by which a particular beneficial gene (or set of genes) is reproduced in succeeding generations more than other genes. The result of natural selection is a population that contains a greater proportion of organisms better adapted to certain environmental conditions. See *adaptation*, *biological evolution*, *differential reproduction*, *mutation*.

negative feedback loop A process that causes a system to change in the opposite direction. See *feedback loop*. Compare *positive feedback loop*.

nekton Strongly swimming organisms found in aquatic systems. Compare *benthos, plankton*.

net primary productivity (NPP) Rate at which all the plants in an ecosystem produce net useful chemical energy; equal to the difference between the rate at which the plants in an ecosystem produce useful chemical energy (gross primary productivity) and the rate at which they use some of that energy through cellular respiration. Compare *gross primary productivity*.

neutral solution Water solution containing an equal number of hydrogen ions (H^+) and hydroxide ions (OH^-); water solution with a pH of 7. Compare *acidic solution, basic solution*.

neutron (n) Elementary particle in the nuclei of all atoms (except hydrogen-1). It has a relative mass of 1 and no electric charge. Compare *electron, proton*.

niche See *ecological niche*.

nitrogen cycle Cyclic movement of nitrogen in different chemical forms from the environment to organisms and then back to the environment.

nitrogen fixation Conversion of atmospheric nitrogen gas by lightning, bacteria, and cyanobacteria into forms useful to plants; part of the nitrogen cycle.

nondegradable pollutant Material that is not broken down by natural processes. Examples include the toxic elements lead and mercury. Compare *biodegradable pollutant, degradable pollutant, slowly degradable pollutant*.

nonionizing radiation Forms of radiant energy such as radio waves, microwaves, infrared light, and ordinary light that do not have enough energy to cause ionization of atoms in living tissue. Compare *ionizing radiation*.

nonnative species Species that migrate into an ecosystem or are deliberately or accidentally introduced into an ecosystem by humans. Compare *native species*.

nonpersistent pollutant See *degradable pollutant*.

nonpoint source Large or dispersed land areas such as crop fields, streets, and lawns that discharge pollutants into the environment over a large area. Compare *point source*.

nonrenewable resource Resource that exists in a fixed amount in the earth's crust and has the potential for renewal by geological, physical, and chemical processes taking place over hundreds of millions to billions of years. Examples include copper, aluminum, coal, and oil. We classify these resources as exhaustible because we are extracting and using them at a much faster rate than they are formed. Compare *renewable resource*.

nuclear change Process in which nuclei of certain isotopes spontaneously change, or are forced to change, into one or more different isotopes. The three principal types of nuclear change are natural radioactivity, nuclear fission, and nuclear fusion. Compare *chemical change, physical change*.

nuclear energy Energy released when atomic nuclei undergo a nuclear reaction such as the spontaneous emission of radioactivity, nuclear fission, or nuclear fusion.

nuclear fission Nuclear change in which the nuclei of certain isotopes with large mass numbers (such as uranium-235 and plutonium-239) are split apart into lighter nuclei when struck by a neutron. This process releases more neutrons and a large amount of energy. Compare *nuclear fusion*.

nuclear fusion Nuclear change in which two nuclei of isotopes of elements with a low mass number (such as hydrogen-2 and hydrogen-3) are forced together at extremely high temperatures until they fuse to form a heavier nucleus (such as helium-4). This process releases a large amount of energy. Compare *nuclear fission*.

nucleic acids Large polymer molecules made by linking hundreds to thousands of four types of monomers called *nucleotides*.

nucleus Extremely tiny center of an atom, making up most of the atom's mass. It contains one or more positively charged protons and one or more neutrons with no electrical charge (except for a hydrogen-1 atom, which has one proton and no neutrons in its nucleus).

nutrient Any food or element an organism must take in to live, grow, or reproduce.

nutrient cycle See *biogeochemical cycle*.

old-growth forest Virgin and old, second-growth forests containing trees that are often hundreds—sometimes thousands—of years old. Examples include forests of Douglas fir, western hemlock, giant sequoia, and coastal redwoods in the western United States. Compare *second-growth forest, tree plantation*.

oligotrophic lake Lake with a low supply of plant nutrients. Compare *eutrophic lake, mesotrophic lake*.

omnivore Animal that can use both plants and other animals as food sources. Examples include pigs, rats, cockroaches, and humans. Compare *carnivore, herbivore*.

open sea Part of an ocean that lies beyond the continental shelf. Compare *coastal zone*.

organic compounds Compounds containing carbon atoms combined with each other and with atoms of one or more other elements such as hydrogen, oxygen, nitrogen, sulfur, phosphorus, chlorine, and fluorine. All other compounds are called *inorganic compounds*.

organism Any form of life.

output Matter, energy, or information leaving a system. Compare *input, throughput*.

output pollution control See *pollution cleanup*.

ozone layer Layer of gaseous ozone (O_3) in the stratosphere that protects life on earth by filtering out most harmful ultraviolet radiation from the sun.

paradigm shift Shift in scientific thinking that occurs when the majority of scientists in a field or related fields agree that a new explanation or theory is better than the old one.

parasite Consumer organism that lives on or in, and feeds on, a living plant or animal, known as the host, over an extended period. The parasite draws nourishment from and gradually weakens its host; it may or may not kill the host. See *parasitism*.

parasitism Interaction between species in which one organism, called the parasite, preys on another organism, called the host, by living on or in the host. See *host, parasite*.

per capita ecological footprint Amount of biologically productive land and water needed to supply each person or population with the renewable resources they use and to absorb or dispose of the wastes from such resource use. It measures the average environmental impact of individuals or populations in different countries and areas. Compare *ecological footprint*.

per capita GDP Annual gross domestic product (GDP) of a country divided by its total population at midyear. It gives the average slice of the economic pie per person. Used to be called per capita gross national product (GNP). See *gross domestic product*.

perennial Plant that can live for more than 2 years. Compare *annual*.

permafrost Perennially frozen layer of the soil that forms when the water there freezes. It is found in arctic tundra.

perpetual resource A resource that is essentially inexhaustible on a human time scale because it is renewed continuously. Solar energy is an example. Compare *nonrenewable resource, renewable resource*.

persistence See *inertia*.

persistent pollutant See *slowly degradable pollutant*.

pH Numeric value that indicates the relative acidity or alkalinity of a substance on a scale of 0 to 14, with the neutral point at 7. Acid solutions have pH values lower than 7; basic or alkaline solutions have pH values greater than 7.

phosphorus cycle Cyclic movement of phosphorus in different chemical forms from the environment to organisms and then back to the environment.

photosynthesis Complex process that takes place in cells of green plants. Radiant energy from the sun is used to combine carbon dioxide (CO_2) and water (H_2O) to produce oxygen (O_2), carbohydrates (such as glucose, $C_6H_{12}O_6$), and other nutrient molecules. Compare *aerobic respiration, chemosynthesis*.

physical change Process that alters one or more physical properties of an element or a compound without changing its chemical composition. Examples include changing the size and shape of a sample of matter (crushing ice and cutting aluminum foil) and changing a sample of matter from one physical state to another (boiling and freezing water). Compare *chemical change, nuclear change*.

phytoplankton Small, drifting plants, mostly algae and bacteria, found in aquatic ecosystems. Compare *plankton, zooplankton*.

pioneer community First integrated set of plants, animals, and decomposers found in an area undergoing primary ecological succession. See *immature community, mature community*.

pioneer species First hardy species—often microbes, mosses, and lichens—that begin colonizing a site as the first stage of ecological succession. See *ecological succession, pioneer community*.

planetary management worldview The belief that humans are separate from nature, that nature exists mainly to meet our needs and increasing wants, and that we can use our ingenuity and technology to manage the earth's life-support systems, mostly for our benefit. It assumes that economic growth is unlimited. Compare *environmental wisdom worldview, stewardship worldview*.

plankton Small plant organisms (phytoplankton) and animal organisms (zooplankton) that float in aquatic ecosystems.

point source Single identifiable source that discharges pollutants into the environment. Examples include the smokestack of a power plant, drainpipe of a meatpacking plant,

chimney of a house, and exhaust pipe of an automobile. Compare *nonpoint source*.

pollutant Particular chemical or form of energy that can adversely affect the health, survival, or activities of humans or other living organisms. See *pollution*.

pollution An undesirable change in the physical, chemical, or biological characteristics of air, water, soil, or food that can adversely affect the health, survival, or activities of humans or other living organisms.

pollution cleanup Device or process that removes or reduces the level of a pollutant after it has been produced or has entered the environment. Examples include efforts to contain an oil spill or to decontaminate groundwater. Compare *pollution prevention*.

pollution prevention Device or process that prevents a potential pollutant from forming or entering the environment or sharply reduces the amount entering the environment. Examples include automobile emission control devices and sewage treatment plants. Compare *pollution cleanup*.

population Group of individual organisms of the same species living in a particular area.

population change Increase or decrease in the size of a population. It is equal to (Births + Immigration) − (Deaths + Emigration).

population density Number of organisms in a particular population found in a specified area or volume.

population dispersion General pattern in which the members of a population are arranged throughout its habitat.

population distribution Variation of population density over a particular geographic area. For example, a country has a high population density in its urban areas and a much lower population density in rural areas.

population dynamics Major abiotic and biotic factors that tend to increase or decrease the population size and affect the age and sex composition of a species.

population size Number of individuals making up a population's gene pool.

positive feedback loop A process that causes a system to change further in the same direction. See *feedback loop*. Compare *negative feedback loop*.

potential energy Energy stored in an object because of its position or the position of its parts. Compare *kinetic energy*.

poverty Inability to meet basic needs for food, clothing, and shelter.

prairies See *grasslands*.

precautionary principle When there is scientific uncertainty about potentially serious harm from chemicals or technologies, decision makers should act to prevent harm to humans and the environment. See *pollution prevention*.

precipitation Water in the form of rain, sleet, hail, and snow that falls from the atmosphere onto land and bodies of water.

predation Situation in which an organism of one species (the predator) captures and feeds on parts or all of an organism of another species (the prey).

predator Organism that captures and feeds on parts or all of an organism of another species (the prey).

predator–prey relationship Interaction between two organisms of different species in which one organism, called the *predator*, captures and feeds on parts or all of the other organism, called the *prey*.

prey Organism that is captured and serves as a source of food for an organism of another species (the predator).

primary consumer Organism that feeds on plants (herbivore) or on other producers. Compare *detritivore, omnivore, secondary consumer*.

primary productivity See *gross primary productivity, net primary productivity*.

primary succession Ecological succession in an area that has never been occupied by a community of organisms. See *ecological succession*. Compare *secondary succession*.

producer Organism that uses solar energy (green plants) or chemical energy (some bacteria) to manufacture the organic compounds it needs as nutrients from simple inorganic compounds obtained from its environment. Compare *consumer, decomposer*.

prokaryotic organism Classification of cell structure in which the cell contains no distinct nucleus or organelles enclosed by membranes. A prokaryotic cell is much simpler and usually much smaller than a eukaryotic cell. All bacteria are single-celled prokaryotic organisms. Compare *eukaryotic organism*.

proteins Large polymer molecules formed by linking together long chains of monomers called *amino acids*.

proton (p) Positively charged particle in the nuclei of all atoms. Each proton has a relative mass of 1 and a single positive charge. Compare *electron, neutron*.

pyramid of energy flow Diagram representing the flow of energy through each trophic level in a food chain or food web. With each energy transfer, only a small part (typically 10%) of the usable energy entering one trophic level is transferred to the organisms at the next trophic level.

radiation Fast-moving particles (particulate radiation) or waves of energy (electromagnetic radiation). See *alpha particle, beta particle, gamma ray*.

radioactive decay Change of a radioisotope to a different isotope by the emission of radioactivity.

radioactive isotope See *radioisotope*.

radioactive waste Waste products of nuclear power plants, research, medicine, weapon production, or other processes involving nuclear reactions. See *radioactivity*.

radioactivity Nuclear change in which unstable nuclei of atoms spontaneously shoot out "chunks" of mass, energy, or both at a fixed rate. The three principal types of radioactivity are gamma rays and fast-moving alpha particles and beta particles.

radioisotope Isotope of an atom that spontaneously emits one or more types of radioactivity (alpha particles, beta particles, or gamma rays).

rain shadow effect Low precipitation on the far side (leeward side) of a mountain when prevailing winds flow up and over a high mountain or range of high mountains. This creates semiarid and arid conditions on the leeward side of a high mountain range.

range See *distribution*.

range of tolerance Range of chemical and physical conditions that must be maintained for populations of a particular species to stay alive and grow, develop, and function normally. See *law of tolerance*.

rare species Species that has naturally small numbers of individuals (often because of limited geographic ranges or low population densities) or that has been locally depleted by human activities.

realized niche Parts of the fundamental niche of a species that are actually used by that species. See *ecological niche*. Compare *fundamental niche*.

recombinant DNA DNA that has been altered to contain genes or portions of genes from organisms of different species.

reconciliation ecology Science of inventing, establishing, and maintaining new habitats to conserve species diversity in places where people live, work, or play.

recycling Collecting and reprocessing a resource so that it can be made into new products. An example is collecting aluminum cans, melting them down, and using the aluminum to make new cans or other aluminum products. Compare *reuse*.

reforestation Renewal of trees and other types of vegetation on land where trees have been removed; can be done naturally by seeds falling from nearby trees or artificially by people planting seeds or seedlings.

reliable runoff Surface runoff of water that generally can be counted on as a stable source of water from year to year. See *runoff*.

renewable resource Resource that can be replenished rapidly (in hours to several decades) through natural processes as long as it is not used up faster than it is replaced. Examples include trees in forests, grasses in grasslands, wild animals, fresh surface water in lakes and streams, most groundwater, fresh air, and fertile soil. If such a resource is used faster than it is replenished, it can be depleted and converted into a nonrenewable resource. Compare *nonrenewable resource* and *perpetual resource*. See also *environmental degradation*.

replacement-level fertility Number of children a couple must have to replace themselves. The average for a country or the world usually is slightly higher than 2 children per couple (2.1 in the United States and 2.5 in some developing countries) because some children die before reaching their reproductive years. See also *total fertility rate*.

reproduction Production of offspring by one or more parents.

reproductive isolation Long-term geographic separation of members of a particular sexually reproducing species.

reproductive potential See *biotic potential*.

resilience Ability of a living system to bounce back and repair damage after a disturbance that is not too drastic.

resource Anything obtained from the environment to meet human needs and wants. It can also be applied to other species.

resource partitioning Process of dividing up resources in an ecosystem so that species with similar needs (overlapping ecological niches) use the same scarce resources at different times, in different ways, or in different places. See *ecological niche, fundamental niche, realized niche*.

respiration See *aerobic respiration*.

restoration ecology Research and scientific study devoted to restoring, repairing, and reconstructing damaged ecosystems.

reuse Using a product over and over again in the same form. An example is collecting, washing, and refilling glass beverage bottles. Compare *recycling*.

r-selected species Species that reproduce early in their life span and produce large numbers of usually small and short-lived offspring in a short period. Compare *K-selected species*.

r-strategists See *r-selected species*.

rule of 70 Doubling time (in years) = 70/(percentage growth rate). See *doubling time, exponential growth*.

runoff Freshwater from precipitation and melting ice that flows on the earth's surface into nearby streams, lakes, wetlands, and reservoirs. See *reliable runoff, surface runoff, surface water*. Compare *groundwater*.

salinity Amount of various salts dissolved in a given volume of water.

scavenger Organism that feeds on dead organisms that were killed by other organisms or died naturally. Examples include vultures, flies, and crows. Compare *detritivore*.

science Attempts to discover order in nature and to use that knowledge to make predictions about what should happen in nature. See *frontier science, scientific data, scientific hypothesis, scientific law, scientific methods, scientific model, scientific theory, sound science*.

scientific data Facts obtained by making observations and measurements. Compare *scientific hypothesis, scientific law, scientific methods, scientific model, scientific theory*.

scientific hypothesis Educated guess that attempts to explain a scientific law or certain scientific observations. Compare *scientific data, scientific law, scientific methods, scientific model, scientific theory*.

scientific law Description of what scientists find happening in nature repeatedly in the same way, without known exception. See *first law of thermodynamics, law of conservation of matter, second law of thermodynamics*. Compare *scientific data, scientific hypothesis, scientific methods, scientific model, scientific theory*.

scientific methods Ways in which scientists gather data and formulate and test scientific hypotheses, models, theories, and laws. See *scientific data, scientific hypothesis, scientific law, scientific model, scientific theory*.

scientific model Simulation of a complex process or system. Many are mathematical models that are run and tested using computers.

scientific theory Well-tested and widely accepted scientific hypothesis. Compare *scientific data, scientific hypothesis, scientific law, scientific methods, scientific model*.

secondary consumer Organism that feeds only on primary consumers. Compare *detritivore, omnivore, primary consumer*.

secondary succession Ecological succession in an area in which natural vegetation has been removed or destroyed but the soil is not destroyed. See *ecological succession*. Compare *primary succession*.

second-growth forest Stands of trees resulting from secondary ecological succession. Compare *old-growth forest, tree farm*.

second law of energy See *second law of thermodynamics*.

second law of thermodynamics In any conversion of heat energy to useful work, some of the initial energy input is always degraded to lower-quality, more dispersed, less useful energy—usually low-temperature heat that flows into the environment. See *first law of thermodynamics*.

selective cutting Cutting of intermediate-aged, mature, or diseased trees in an uneven-aged forest stand, either singly or in small groups. This encourages the growth of younger trees and maintains an uneven-aged stand. Compare *clear-cutting, strip-cutting*.

sexual reproduction Reproduction in organisms that produce offspring by combining sex cells or *gametes* (such as ovum and sperm) from both parents. It produces offspring that have combinations of traits from their parents. Compare *asexual reproduction*.

slowly degradable pollutant Material that is slowly broken down into simpler chemicals or reduced to acceptable levels by natural physical, chemical, and biological processes. Compare *biodegradable pollutant, degradable pollutant, nondegradable pollutant*.

social capital Positive force created when people with different views and values find common ground and work together to build understanding, trust, and informed shared visions of what their communities, states, nations, and the world could and should be. Compare *natural capital*.

soil Complex mixture of inorganic minerals (clay, silt, pebbles, and sand), decaying organic matter, water, air, and living organisms.

soil conservation Methods used to reduce soil erosion, prevent depletion of soil nutrients, and restore nutrients previously lost by erosion, leaching, and excessive crop harvesting.

soil erosion Movement of soil components, especially topsoil, from one place to another, usually by wind, flowing water, or both. This natural process can be greatly accelerated by human activities that remove vegetation from soil.

soil horizons Horizontal zones that make up a particular mature soil. Each horizon has a distinct texture and composition that varies with different types of soils. See *soil profile*.

soil profile Cross-sectional view of the horizons in a soil. See *soil horizons*.

solar capital Solar energy that warms the planet and supports photosynthesis, the process that plants use to provide food for themselves and for us and other animals. This direct input of solar energy also produces indirect forms of renewable solar energy such as wind and flowing water. Compare *natural capital*.

solar energy Direct radiant energy from the sun and a number of indirect forms of energy produced by the direct input of such radiant energy. Principal indirect forms of solar energy include wind, falling and flowing water (hydropower), and biomass (solar energy converted into chemical energy stored in the chemical bonds of organic compounds in trees and other plants).

sound science Concepts and ideas that are widely accepted by experts in a particular field of the natural or social sciences. These results of science are very reliable. Compare *frontier science, junk science*.

specialist species Species with a narrow ecological niche that may be able to live in only one type of habitat, tolerate only a narrow range of climatic and other environmental conditions, or use only one type or a few types of food. Compare *generalist species*.

speciation Formation of two species from one species because of divergent natural selection in response to changes in environmental conditions; usually takes thousands of years. Compare *extinction*.

species Group of organisms that resemble one another in appearance, behavior, chemical makeup and processes, and genetic structure. Organisms that reproduce sexually are classified as members of the same species only if they can actually or potentially interbreed with one another and produce fertile offspring.

species diversity Number of different species and their relative abundances in a given area. See *biodiversity*. Compare *ecological diversity, genetic diversity*.

species equilibrium model Widely accepted model which says that the number of different species found on an island is determined by a balance between two factors: the rate at which new species immigrate to the island and the rate at which existing species become extinct on the island.

species evenness Abundance of individuals within each species contained in a community.

species richness Number of different species contained in a community.

S-shaped curve Leveling off of an exponential, J-shaped curve when a rapidly growing population exceeds the carrying capacity of its environment and ceases to grow.

stewardship worldview The belief that humans can manage the earth for our benefit but that we have an ethical responsibility to be caring and responsible managers, or *stewards*, of the earth. It calls for encouraging environmentally beneficial forms of economic growth and discouraging environmentally harmful forms. Compare *environmental wisdom worldview, planetary management worldview*.

stratosphere Second layer of the atmosphere, extending about 17–48 kilometers (11–30 miles) above the earth's surface. It contains small amounts of gaseous ozone (O_3), which filters out about 95% of the incoming harmful ultraviolet radiation emitted by the sun. Compare *troposphere*.

stream Flowing body of surface water. Examples are creeks and rivers.

strip-cutting Variation of clear-cutting in which a strip of trees is clear-cut along the contour of the land, with the corridor being narrow enough to allow natural regeneration within a few years. After regeneration, another strip is cut above the first, and so on. Compare *clear-cutting, selective cutting*.

subatomic particles Extremely small particles—electrons, protons, and neutrons—that make up the internal structure of atoms.

subduction zone Area in which oceanic lithosphere is carried downward (subducted) under an island arc or continent at a convergent plate boundary. A trench ordinarily forms at the boundary between the two converging plates. See *convergent plate boundary*.

subpopulation The individuals of a species that live in a habitat patch.

succession See *ecological succession, primary succession, secondary succession*.

succulent plants Plants, such as desert cacti, that survive in dry climates by having no leaves, thus reducing the loss of scarce water. They store water and use sunlight to produce the food they need in the thick, fleshy tissue of their green stems and branches. Compare *deciduous plants, evergreen plants*.

sulfur cycle Cyclic movement of sulfur in various chemical forms from the environment to organisms and then back to the environment.

surface fire Forest fire that burns only undergrowth and leaf litter on the forest floor. Compare *crown fire, ground fire*. See *controlled burning*.

surface runoff Water flowing off the land into bodies of surface water. See *reliable runoff*.

surface water Precipitation that does not infiltrate the ground or return to the atmosphere by evaporation or transpiration. See *runoff*. Compare *groundwater*.

survivorship curve Graph showing the number of survivors in different age groups for a particular species.

sustainability Ability of earth's various systems, including human cultural systems and economies, to survive and adapt to changing environmental conditions indefinitely.

sustainable development See *environmentally sustainable economic development*.

sustainable living Taking no more potentially renewable resources from the natural world than can be replenished naturally and not overloading the capacity of the environment to cleanse and renew itself by natural processes.

sustainable society Society that manages its economy and population size without doing irreparable environmental harm by overloading the planet's ability to absorb environmental insults, replenish its resources, and sustain human and other forms of life over a specified period, usually hundreds to thousands of years. During this period, the society satisfies the needs of its people without depleting natural resources and thereby jeopardizing the prospects of current and future generations of humans and other species.

sustainable yield (sustained yield) Highest rate at which a potentially renewable resource can be used indefinitely without reducing its available supply. See also *environmental degradation*.

synergistic interaction Interaction of two or more factors or processes causing the combined effect to be greater than the sum of their separate effects.

synergy See *synergistic interaction*.

system Set of components that function and interact in some regular and theoretically predictable manner.

temperature Measure of the average speed of motion of the atoms, ions, or molecules in a substance or combination of substances at a given moment. Compare *heat*.

terrestrial Pertaining to land. Compare *aquatic*.

tertiary (higher-level) consumers Animals that feed on animal-eating animals. They feed at high trophic levels in food chains and webs. Examples include hawks, lions, bass, and sharks. Compare *detritivore, primary consumer, secondary consumer*.

theory of evolution Widely accepted scientific idea that all life forms developed from earlier life forms. Although this theory conflicts with the creation stories of many religions, it is the way biologists explain how life has changed over the past 3.6–3.8 billion years and why it is so diverse today.

theory of island biogeography See *species equilibrium model*.

third and higher level consumers Carnivores that feed on other carnivores.

threatened species Wild species that is still abundant in its natural range but is likely to become endangered because of a decline in numbers. Compare *endangered species*.

threshold effect Harmful or fatal effect of a small change in environmental conditions that exceeds the limit of tolerance of an organism or population of a species. See *law of tolerance*.

throughput Rate of flow of matter, energy, or information through a system. Compare *input, output*.

throwaway society See *high-throughput economy*.

time delays Amount of time in a feedback loop between the input of a stimulus and the response to it. See *feedback loop, negative feedback loop, positive feedback loop*.

tolerance limits Minimum and maximum limits for physical conditions (such as temperature) and concentrations of chemical substances beyond which no members of a particular species can survive. See *law of tolerance*.

total fertility rate (TFR) Estimate of the average number of children born alive to women in a given population during their lifetimes.

tragedy of the commons Depletion or degradation of a potentially renewable resource to which people have free and unmanaged access. An example is the depletion of commercially desirable fish species in the open ocean beyond areas controlled by coastal countries. See *common-property resource*.

transform fault Area where the earth's lithospheric plates move in opposite but parallel directions along a fracture (fault) in the lithosphere. Compare *convergent plate boundary, divergent plate boundary*.

transgenic organisms See *genetically modified organism* (GMOs).

transpiration Process in which water is absorbed by the root systems of plants, moves up through the plants, passes through pores (stomata) in their leaves or other parts, and evaporates into the atmosphere as water vapor.

tree farm See *tree plantation*.

tree plantation Site planted with one or only a few tree species in an even-aged stand. When the stand matures it is usually harvested by clear-cutting and then replanted. These farms normally raise rapidly growing tree species for fuelwood, timber, or pulpwood. See *even-aged management*. Compare *old-growth forest, second-growth forest, uneven-aged management*.

trophic level All organisms that are the same number of energy transfers away from the original source of energy (for example, sunlight) that enters an ecosystem. For example, all producers belong to the first trophic level, and all herbivores belong to the second trophic level in a food chain or a food web.

troposphere Innermost layer of the atmosphere. It contains about 75% of the mass of earth's air and extends about 17 kilometers (11 miles) above sea level. Compare *stratosphere*.

tsunami Series of large waves generated when part of the ocean floor suddenly rises or drops, usually because of an earthquake.

ultraplankton Huge populations of extremely small photosynthetic bacteria that may be responsible for 70% of the primary productivity near the ocean surface.

uneven-aged management Method of forest management in which trees of different species in a given stand are maintained at many ages and sizes to permit continuous natural regeneration. Compare *even-aged management*.

upwelling Movement of nutrient-rich bottom water to the ocean's surface. It can occur far from shore but usually takes place along certain steep coastal areas where the surface layer of ocean water is pushed away from shore and replaced by cold, nutrient-rich bottom water.

utilitarian value See *instrumental value*.

vertebrates Animals that have backbones. Compare *invertebrates*.

volcano Vent or fissure in the earth's surface through which magma, liquid lava, and gases are released into the environment.

warm front Boundary between an advancing warm air mass and the cooler one it is replacing. Because warm air is less dense than cool air, an advancing warm front rises over a mass of cool air. Compare *cold front*.

water cycle See *hydrologic cycle*.

watershed Land area that delivers water, sediment, and dissolved substances via small streams to a major stream (river).

weather Short-term changes in the temperature, barometric pressure, humidity, precipitation, sunshine, cloud cover, wind direction and speed, and other conditions in the troposphere at a given place and time. Compare *climate*.

wetland Land that is covered all or part of the time with saltwater or freshwater, excluding streams, lakes, and the open ocean. See *coastal wetland, inland wetland*.

wilderness Area where the earth and its community of life have not been seriously disturbed by humans and where humans are only temporary visitors.

wildlife All free, undomesticated species. Sometimes the term is used to describe animals only.

wildlife resources Wildlife species that have actual or potential economic value to people.

wild species Species found in the natural environment. Compare *domesticated species*.

worldview How people think the world works and what they think their role in the world should be. See *environmental wisdom worldview, planetary management worldview, stewardship worldview*.

zooplankton Animal plankton; small floating herbivores that feed on plant plankton (phytoplankton). Compare *phytoplankton*.

INDEX

Note: Page numbers in boldface indicate terms in boldface in the text. Page numbers followed by italicized *f* or *t* indicate figures or tables, and *b* indicates boxed material.

Abiotic components in ecosystems, **33**
Abortion, birth rate and availability of, 87
Aboveground buildings, hazardous waste storage in, 295
Abyssal zone of ocean, 65, 66*f*
Accuracy in making measurements, 21
Acid deposition (acid rain), 48, **257**–59
　formation of, 258*f*
　global regions affected by, 258*f*
　harmful effects of, 259
　solutions for preventing and cleaning up, 260*f*
Acid rain. *See* Acid deposition
Active solar heating system, **224***f*, 225*f*
Acute effect of toxic chemicals, 244
Adaptation, **54**
　limits on, 54
　strategy of, in response to global warming, 277, 278*f*
Adaptive radiations, **58**
Adaptive trait, **54**
Aerobic respiration, **35**, 46
Affluence, environmental quality and benefits of, 14
Affluenza, **13**–14
　simple living as response to, 321
　solid-waste production linked to, 283–84
Age structure, population, 88*t*, **89**–91
　AIDS epidemic and, 91*f*, 241*f*
　diagrams of, 89*f*
　effect of, on population growth, 89, 90*f*
　making population and economic projections based on, 89–91
　U.S. baby-boom generation, 89, 90*f*
Agribusiness, 150
Agricultural revolution, 15
Agriculture. *See also* Crop(s); Food production
　aquaculture, 161, 162–63
　effects of global warming on, 272*f*, 273
　effects of ozone depletion on, 280*f*
　gene revolution in, 158–60
　government policies on, 163
　green revolutions in, 156–58
　industrialized (high-input), 150–52
　irrigation and, 153–54, 173
　meat production, 160–61
　monoculture, 110
　pest management in, 164–69
　plantation, 150
　polyculture, 152
　soils and, 152–54, 154–56
　sustainable, 154–56, 169–70
　traditional low input, 152
　tree farms, 110
　in United States, 150–52
　in urban areas, 163
　water needs for, 173
　water pollution caused by, 187, 193*f*
Agroforestry, **152**, 155
AIDS (acquired immune deficiency syndrome), 241
　death rate, 241*f*
　HIV and, 240
　population declines attributed to, 91
Air circulation, climate and global, 253, 254*f*
Air pollutants
　formaldehyde as, 260
　indoor, 255, 260, 261*f*
　major types of, 255*t*
　primary, and secondary, 255
　radon gas as, 260–62
　soot and black carbon aerosols as, 271
Air pollution, **255**–62
　acid deposition and, 257–59
　atmospheric ozone depletion and increased, 280*f*
　climate change and effects of, 271
　factors influencing levels of outdoor, 257
　fossil-fuel burning as source of, 205, 206*f*, 209, 211*f*
　harmful effects of, on human health, 260, 261, 262
　indoor, 255, 260–62, 265*f*
　industrial smog as, 256
　motor vehicles as cause of, 256*f*, 257
　photochemical smog as, 256–57
　preventing and reducing, 262–66
　radioactive radon gas as, 260–62
　solutions for reducing, 264*f*, 265*f*, 266*f*
　types and sources of, 255–56
Alien (nonnative) species, **70**. *See also* Nonnative species
Alley cropping, **152**, 155
Altitude, climate, biomes, and effects of, 62*f*, 63
American Wind Energy Association, 228
Ammonification, nitrogen fixation and, 47
Amphibians, case study on disappearance of, 70–71*f*
Anderson, Ray, development of sustainable corporation by, 306*b*
Animal(s)
　as food source, 150
　giant pandas, 56
　gray wolf, 119*b*
　keystone species, 71–72
　speciation of Arctic and gray fox, 57*f*
　toxic chemical laboratory studies on, 245–46
　use of, in chemical toxicity lab experiments, 245–46
　zoos and aquariums for protection of endangered, 145
Animal factories, 150
Animal feedlots, 160, 161*f*
Animal manure, **156**
　as fuel, 229
Antarctic, food web in, 38*f*
Antibiotics, pathogen resistance to, 239–40
Aquaculture, 150, 161, **162**–63
　advantages and disadvantages of, 162*f*
　sustainable, 163*f*
Aquariums, protecting endangered species in, 145
Aquatic life zones, **64**–69
　biodiversity in, 125–27
　effects of acid deposition on, 259
　food web in, 38*f*
　freshwater, 64, 66–69
　layers of, 64
　limiting factors in, 34
　marine (saltwater), 64, 65, 66*f*
　net primary productivity in, 40*f*
　protecting biodiversity in, 108, 125–27
　range of tolerance for temperature in, 34*f*
Aquifers, 45, **172**. *See also* Groundwater
　withdrawing groundwater from, 179–80
Aral Sea water-diversion project, ecological disaster of, 181–82
Arboreta, 144
Arctic National Wildlife Refuge (ANWR), 205
Arctic region, oil supplies in, 205
Argentine fire ant, 137, 138*f*, 139*f*
Arsenic as water pollutant, 191–92
Artificial selection, **58**–59, 158
El-Ashry, Mohamed, 182, 236
Atmosphere, **30**, 31*f*, **252**–54
　air pollution in (*see* Air pollution)
　carbon dioxide levels in, over time, 254, 266, 267*f*, 268*f*
　chemical makeup of, 252
　climate and, 253–54 (*see also* Climate)
　layers and average temperature of, 252*f*, 267*f*, 270*f*
　natural greenhouse effect in, 254
　ozone in, 252 (*see also* Ozone (O_3); Ozone depletion)
　removal of carbon dioxide from, and storage, 275–76
　stratosphere of, 30, 252 (*see also* Stratosphere)
　temperature of, 266, 267*f*, 270*f* (*see also* Climate change; Global warming)
　troposphere of, 30, 252 (*see also* Troposphere)
　weather and, 253–54 (*see also* Weather)
Atom(s), **23**
Atomic number, **24**
Autotrophs (producers), 33, **34**, 39–41

Baby boom generation, U.S., 89, 90*f*
Background extinction, **57**
Bacon, Francis, 83
Bacteria, 29
　antibiotic resistant, 239–40
　as pathogens, 238–39

Balance of nature, 76
Bangladesh
    flooding in, 176–77
    microlending by Grameen Bank in, 307b
Basal cell skin cancer, 280
Basel Action Network, 284
Bathyal zone, ocean, 65, 66f
Bats, ecological role of, 133–34
Bicycles as transportation, 103, 104f
Bioaccumulation of toxic chemicals, 140f
Biodegradable waste, composting of, 287–88
Biodiversity, 36–37
    in aquatic life zones, 125–27
    changes in, over geological time, 58f
    components of, 36f
    ecosystems and preservation of (see
        Ecosystem approach to sustaining biodi-
        versity)
    effects of global warming on, 272f
    human impacts on, 108–9, 125–26
    instrumental value of, 109, 133
    intrinsic value of, 109, 134
    loss of (see Extinction of species)
    priorities for protecting, 126–27
    reasons for protecting, 109
    species, 36f, 37 (see also Species)
    species preservation (see Species approach
        to sustaining biodiversity)
    sustainability principles and protection of,
        16, 17f
Biodiversity hot spots, 121–22
    global, 122f
    in United States, 142f
Biofuels, 229
Biogas production, 230
Biogeochemical (nutrient) cycles, 31f,
    44–50
Biological community, 30. See also Commu-
    nity(ies)
Biological control of pests, 168
Biological diversity, 36. See also Biodiversity
Biological evolution, 52. See also Evolution
Biological extinction of species, 129. See also
    Extinction of species
Biological hazards, 237, 238–43. See also Dis-
    ease; Infectious disease
    deaths per year from, 239f
    HIV and AIDS epidemic, 240, 241
    malaria, 241–42
    pathogen resistance to antibiotics and,
        239–40
    reducing incidence of infectious disease,
        242–43
    transmissible and nontransmissible dis-
        ease as, 238–39
    tuberculosis as, 240
    viruses and viral disease as, 240–41
Biomass, 39
    energy production from incinerating, 202,
        229–30
    in food chains and food webs, 39
Biomass plantations, 229
Biome(s), 60–64
    climate and, 60–63
    desert, 61, 63f, 64f
    earth's major, 61f
    forests, 62f, 64f
    global air circulation and, 254f
    grasslands, 61, 64f
    human impact on, 63–64
    natural capital degradation in, 64f
    precipitation, temperature, and, 62f
Biopharming, 59
Bioplastic, 288
Bioremediation of hazardous waste, 293

Biosphere, 30
    components of, 30–31
    ecosystems of (see Ecosystem(s))
Biosphere reserve, 121f
Bioterrorism, 238
Biotic components in ecosystems, 33
Biotic potential of populations, 77
Birds
    effect of pesticides on, 165b
    extinction of passenger pigeon, 129–30
    as indicator species, 71, 136
    in North and South America, 136f
    protecting bluebirds, 146b
    red-cockaded woodpeckers, 56
    resource partitioning among warblers, 73f
Birth control methods, effectiveness of, 87
Birth rate (crude birth rate), 85
    fertility and, 86–87, 88t
    reducing, with family planning programs,
        93–94
    in United States, 86
Bitumen, 206
Black carbon aerosols, 271
Bluebirds, reconciliation ecology and protec-
    tion of, 146b
Blue revolution, 186
Boomerang effect, pesticides and, 167f
Botanical gardens, 144
Bottled water, 199
Brazil
    Curitiba, as sustainable city in, 106–7
    demographic factors in Nigeria, U.S., and,
        90f
    tropical forests in, 115
Broad-spectrum agents, 164
Bronchitis, chronic, 262
Brown, Lester R., 170
Brown-air smog, 257
Brownfields, 293
Buffer zone concept, 120
Buildings
    hazardous waste storage in aboveground,
        295
    net energy ratios for heating, 203f
    reducing water waste in, 184–85
    saving energy in, 222–23
    sick-building syndrome, 260
    solar-heated, 224f, 225f
Bureau of Land Management (BLM), U.S.,
    313
Burning
    of biomass, to produce energy, 229–30
    of hazardous waste, 294
    of solid wastes, 289, 290f
Buses as transportation, 103, 104f
Bush, George W., presidential administration
    of
    climate change policies, 276
    environmental policies of, 316–17
Business
    developing environmentally sustainable,
        306b
    reducing water use by, 184, 185f
Butterfly farms, 145

CAFE (Corporate Average Fuel Economy)
    standards, 220
California Water Project, 180, 181f
Canada, oil reserves in, 205
Cancer
    caused by toxic chemicals, 243
    lung, 262
    UV radiation and skin, 279, 280
Cap-and-trade emissions program, 263
Capital, natural. See Natural capital

Captive breeding, 145
Carbohydrates, 23
Carbon cycle, 45, 46–47f
    effects of human activities on, 46–47
Carbon dioxide (CO$_2$) as greenhouse gas, 254
    as air pollutant, 255t
    effects of higher levels of, on photosynthe-
        sis, 271
    fossil fuels and emissions of, 46, 47f, 205,
        206f, 209
    global temperature and levels of, 266, 267f,
        268f, 269f
    reducing emissions of, 276, 277f
    removal and storage of, 275f, 276
    storage of, in oceans, 270
Carbon monoxide as air pollutant, 255t
Carbon taxes, 276
Carcinogens, 243
Carnivores, 35
Carrying capacity (K), 78
    results of exceeding, 78–79
Car-sharing, 103
Carson, Rachel, work of, 165b
Case reports on chemical toxicity, 245
Case studies
    AIDS disease and HIV virus, 240, 241
    amphibians, 70b, 71f
    Aral Sea disaster, 181–82
    bats, 133–34
    California Water Project, 180f
    Chernobyl nuclear power plant accident,
        214–15
    deaths caused by smoking, 237–38
    flooding in Bangladesh, 176
    Great Lakes pollution, 190
    high-level radioactive waste in, 217b
    insects, 30
    integrated pest management, 168–69
    lead as threat to human health, 295–96,
        297f
    malaria, 241–42
    Mexico City as urban area, 101–2
    microorganisms, 28–29
    motor vehicle use in U.S., 102
    nature reserves in Costa Rica, 121
    reconciliation ecology for protection of
        bluebirds, 146b
    recycling plastics, 288
    reuse of refillable containers, 286
    sharks, 72b
    slowing population growth in India and
        China, 94–96
    student environmental action in U.S., 316
    tuberculosis, 240
    U.S. environmental laws and regulations,
        threats to, 316–17
    U.S. freshwater resources, 173f, 174f
    U.S. immigration and population size, 88
    U.S. industrial food production, 150–52
    U.S. national parks, 118–20
    U.S. oil supplies, 205
    U.S. soil erosion, 153
    U.S. storage of radioactive wastes, 217b
    water conflicts in Middle East, 174, 175f
Cash crops, 150
Cell(s), 24, 28
    relationship of chromosomes, genes, DNA
        and, 24f
Center-pivot low-pressure sprinkler, 183, 184f
Centers for Disease Control and Prevention
    (CDCs), 237
Central Arizona Project, 181f
Central receiver system, 225
Chain reaction, nuclear, 211, 212f
Channelization of streams, 177

Chateaubriand, François-Augusté-René de, 108
Chemical bonds, **23**
Chemical change in matter, **26**
Chemical energy, 26
Chemical formula, **23**
Chemical hazards, 243–47. *See also* Hazardous wastes; Toxic chemicals
  arsenic as, 191–92
  assessing toxicity of, 244–47
  dirty dozen, 247, 298
  effects of, on human body, 243–44
  in homes, 293f
  indoor air pollutants as, 260, 261f
  lack of sufficient knowledge about, 246–47
  lead, 295–96, 297f
  persistent organic pollutants (POPs), 247
  pollution prevention and precautionary principle applied to, 247
Chemical reactions, 26
Chernobyl nuclear power accident, radioactive pollutants from, 213, 214–15
Chesapeake Bay, pollution in, 193, 194f
Children
  cost of rearing and educating, 87
  in labor force, 86
  protecting from lead poisoning, 295–96, 297f
  reducing malnutrition and hunger-related deaths in, 149
China
  acid deposition in, 259
  basic demographic data for, 95f
  biogas production in, 230
  coal reserves in, and use of, 208, 209
  greenhouse gas-emission reductions in, 277
  population regulation in, 95–96
Chlorinated hydrocarbons, 23
Chlorofluorocarbons (CFCs), 278
  cutting emissions of, 281
  replacements for, 281b
  stratospheric ozone depletion and role of, 278–79
Chromosomes, 23
  relationship of cells, genes, DNA and, 24f
Chronic effects of toxic chemicals, 244
Chronic undernutrition, **148**–49
Cilia, respiratory tract, 262
Circle of poison, pesticides and, 167b
Cities. *See* Urban areas
Clarke, Arthur C., 283
Clean Air Acts, U.S., 255, 262–63
Cleanup approach
  to acid deposition, 260f
  to air pollution, 264f, 265f
  to coastal water pollution, 195f
  to global warming, 274f
  to groundwater pollution, 192f
  to pollution, 11
  to soil salinization, 154f
Clean Water Acts, U.S., 195, 198
Clear-cutting of forests, 112f
Climate, **60, 253**
  atmospheric gases, natural greenhouse effect and, 254
  biomes and, 254f (*see also* Biome(s))
  change in (*see* Climate change)
  earth's rotation on axis and, 254
  global air circulation and, 253, 254f
  models of, 268, 269f
  ocean currents and, 253f, 273f
  in urban areas, 100
Climate change, 266–77. *See also* Global warming

concentrations of greenhouse gases and, 266, 267f, 268f
  evolution and, 55
  factors affecting earth's average temperature and potential of, 270–71
  historic changes in earth's temperature and past, 266, 267f
  human activities leading to, 266–68
  possible effects of, 271–74
  reasons for concern about, 270
  scientific consensus on, 268–69
  solutions to deal with global warming and, 274–77
  as species extinction threat, 139–40
  tropospheric warming and, 266–68
  water pollution and, 187
Climax community, 76
Clone, 59
Cloud cover, climate and changes in, 270–71
Coal, **208**–10
  advantages and disadvantages of, as fuel, 209, 211f
  converting, into gaseous and liquid fuel, 210, 211f
  environmental impact of burning, 206f, 209f
  formation of, 209f
  nuclear energy versus, 216f
Coal-burning power plants, 210f
  energy waste in, 220
  trading emissions from, 263
Coal gasification, **210**
Coal liquefaction, **210**
Coastal wetlands, 65
Coastal zones, **65**
  Chesapeake Bay estuary, 193–94
  effects of global warming on, 272f
  integrated coastal management, 126, 194
  pollution in, 192, 193f
  protecting, 195f
Cockroaches, 55b
Cogeneration, **220**
Colinvaux, Paul A., 252
Combined heat and power (CHP) systems, **220**
Commensalism, **74**–75
Commercial energy, 201
  flow of, in U.S. economy, 219f
  fossil fuels as main source of, 203–10
  global sources of, 202f
  types of, 202
  U.S. sources of, 202f, 233f
  used by industrial agriculture in U.S., 152f
Commercial inorganic fertilizer, **156**
Commission on Immigration Reform, U.S., 88
Common-property resources, **9**
Community(ies), **30**, 70–76. *See also* Ecosystem(s)
  climax, 76
  diversity in, 37
  ecological succession in, 75–76, 125
  foundation species in, 72–73
  human impact on, 80–83
  indicator species in, 71
  keystone species in, 71–72
  populations dynamics in, 77–80
  species interactions in, 73–75
  types of species in, 70–71
Comparative risk analysis, 247, 248f
Competition, interspecific, 73
Competitor species, 79–80
Complex carbohydrates, 23
Compost and composting, **156**
  of biodegradable organic waste, 287–88
  toilet systems, 197

Compounds, **23**
  organic, and inorganic, 23f
Comprehensive Environmental Response, Compensation, and Liability Act (CERCLA Superfund program), U.S., 292
Conference on Population and Development (1994), 94
Coniferous forests, soils of, 43f
Consensus (sound) science, **6, 21**–22
Conservation concessions, 116–17
Conservation of energy, law of, **27**
Conservation of matter, law of, **27**
Conservation-tillage farming, **154**–55
Constitutional democracy, 309
Consumer education, 235
Consumers (heterotrophs), 33, **35**
Consumption
  excessive, 13–14
  excessive, and species extinctions, 139–40
  simple living, and reduced, 321
Containers, reuse of refillable, 286
Containment vessel, nuclear reactor, 212
Contour farming, **155**f
Control group, 21, 245, 246
Controlled experiment, 21
  on chemical toxicity, 246
Control rods, nuclear reactors, 212
Conventional natural gas, 207, 208, 209f
Conventional (light) oil, 203. *See also* Oil
Convention on Biological Diversity (CBD), 141
Convention on International Trade in Endangered Species (CITES), 141
Coolant, nuclear reactors, 212
Corporate Average Fuel Economy (CAFE) standards, 220
Costanza, Robert, 110
Costa Rica
  ecological restoration in, 124b
  nature reserves in, 121f
Costs
  hidden harmful, 302–3
  pricing based on full, 304
Cover crops, 155
Crash (dieback), population, 79
Critical habitat, 142
Critical thinking skills, developing, 3–4
Crop(s)
  alternative and natural pest-control methods for, 164, 167–69
  cash, 150
  cover, 155
  crossbreeding, 158
  genetic engineering of, 158–60
  grain, 150, 157f
  green revolution and increased yields in, 156–58
  monoculture of, 110
  pesticide use on, 164–67
  polyculture of, 152
  residues from, as biofuel, 229
Croplands, 149
  ecological and economic services provided by, 151f
Crop rotation, **156**
Crossbreeding of crops, 158
Crude birth rate, **85**. *See also* Birth rate
Crude death rate, **85**. *See also* Death rate
Crude oil (petroleum), 194, 203. *See also* Oil
Cultural eutrophication of lakes, 67, **189**–90
Cultural hazards, **237**
  smoking as, 237–38
Curitiba, Brazil, as sustainable ecocity, 106–7
Cuyahoga River, Ohio, cleanup of, 188–89

Dams and reservoirs, 178f, 179–80. *See also* Lakes
  advantages and disadvantages of, 179–80
  hydropower at, 227
DDT, 164–65
  bioaccumulation and biomagnification of, 140f
Death rate (crude death rate), **85**
  AIDS as cause of higher, 91
  air pollution as cause of human, 262f
  factors affecting, 87, 88t
  from select causes, 238f, 248f
  from select infectious diseases, 239f
Debt-for-nature swaps, 116
Deciduous forests, soils of, 43f
Decommissioned nuclear power plants, 216
Decomposers, 33, **35**, 36
Deep underground storage
  of hazardous waste, 294, 295f
  of high-level radioactive wastes, 217b
Deforestation
  causes of, 116f
  flooding caused by, 176f
  global extent of, 110–11
  tropical, 115–17, 135
Degradable wastes in water, 190
Democracy(ies), **309**
  dealing with environmental problems in, 309–10
  environmental action by student groups in, 315
  environmental leadership in, 311
  environmental policy in U.S. as, 311–12
  individual's role in influencing environmental policy in, 310, 311f, 312f
  principles for making environmental policy in, 310
  public lands management in U.S. as, 312–15
  role of environmental groups and networks in, 315
  threats to environmental laws and regulations in U.S. as, 316–17
Demographic transition, **92**–93
Demographic trap, 93
Demography, **85**
Denitrification, nitrogen fixation and, 47
Department of Energy (DOE), U.S., 205
Depletion time, nonrenewable resources, **11**
Desalination of water, **182**
Deserts, 61, 64f
  ecosystem in temperate, 63f
  human impacts on, 63–64
  natural capital degradation in, 64f
  soil of, 43f
Designer babies, 59
Detritivores, **35**f
Detritus, **35**f
Detritus feeders, **35**f
Deuterium-tritium nuclear fusion reactions, 218f
Developed countries, **8**. *See also names of individual countries, e.g.,* Canada
  actions of, to reduce global poverty, 308
  increasing freshwater supplies in, 178
  indoor air pollution in, 260
  population, consumption, and technology in, 14f, 15
  population, wealth/income, resource use, and pollution/waste in, 8f
  river and stream pollution in, 188–89
Developing countries, **8**. *See also names of individual countries, e.g.,* India
  increasing freshwater supplies in, 178
  indoor air pollution in, 260

Kyoto Protocol and obligations of, 276
  population, consumption, and technology in, 14f, 15
  population, wealth/income, resource use, and pollution/waste in, 8f
  river and stream pollution in, 189
  urban poverty in, 101
Dieback (crash), population, 79
Differential reproduction, **52**
Direct costs, 303
Dirty dozen (chemical hazards), 247, 298
Disease
  death rates from, 239f
  HIV and AIDS, 91, 240, 241
  malaria, 241–42
  nutritional deficiencies and, 149
  parasites, 74
  pathogen resistance to antibiotics and, 239–40
  reducing incidence of, 242–43
  relationship of, to poverty and malnutrition, 13, 148–49
  respiratory, 262
  tobacco, smoking, and, 237–38
  transmissible and nontransmissible, 238–39
  tuberculosis, 240
  viral, 240–41
Dissolved oxygen (DO) content, **34**
Distillation of water, 182
DNA (deoxyribonucleic acid), 23
  relationship of cells, chromosomes, genes, and, 24f
Dobzhansky, Theodosius, 69
Document services, 306
Dose, toxic chemical, **244**
  average lethal, for humans, 245t
Dose-response curve, **246**f
Drainage basin, **67**, 68f, **173**
Drinking water, 198–200. *See also* Freshwater
  arsenic in, 191–92
  bottled, 199
  purification of, 198–99
  reducing disease by disinfecting, 196
  water pollution and access to clean, 187
Drip irrigation systems, 183, 184f
Dry acid deposition, 257
Dry steam geothermal energy, 231
Dubos, René, 5

Earth
  atmosphere (*see* Atmosphere)
  biomes of, 61f
  climate and rotation of, 254 (*see also* Climate)
  climate change and global warming (*see* Climate change; Global warming)
  conditions on, as favorable for life, 32–33
  ecosystems of (*see* Ecosystem(s))
  energy flow between sun and, 32f
  human overpopulation and, 92
  life-support systems of (*see* Life-support systems, earth's)
  natural capital of (*see* Natural capital)
  as ocean planet, 65f
  poles of, ozone depletion at, 279
  structure of, 31f
  temperature of, 266, 267f, 270f
Earth-centered environmental worldviews, 319–20
Earthquakes, 55
Ecocities, 105–6
  Curitiba, Brazil, as, 106–7
Eco-labeling programs, 298, 302

Ecological diversity, 36f, **37**
Ecological economists, 300, 301f
Ecological efficiency, **39**
Ecological extinction of species, 129
Ecological footprint, human, 10
  on natural systems, 80, 81f
  per capita, in three countries, 10f
  total, in three countries, 10f
Ecological niche, **55**–56
  generalist, 55b, 56
  habitat versus, 56
  specialist, 56
  in tropical rain forests, 56f
Ecological restoration, 123–25
  in Costa Rica, 124b
  defined, **124**
  methods and principles of, 124–25
Ecological services
  provided by croplands, 151f
  provided by forests, 109f
  provided by freshwater systems, 67f
  provided by marine ecosystems, 65f
Ecological succession, **75**–76
  predictability of, 76
  primary, 75, 76f
  secondary, 75, 125
Ecology, **28**
  in communities, 70–76
  communities, ecosystems, and, 30
  defined, **5, 28**
  goal of, 31
  human impacts on natural systems, 80–83
  organisms and, 28 (*see also* Organisms)
  organizational levels of matter in nature and, 29f
  populations and, 30, 77–80
  principles of, and sustainable living, 320
  reconciliation, 145, 146b
  of species (*see* Species)
Economically-depleted resource, 10
Economic development, **8**
  demographic transition and, 92–93
  environmentally-sustainable, 300–302
  trade-offs of, 8f
Economic growth, **8**
  sustainable and unsustainable, 301f
Economics, 300–309
  improving environmental quality using, 302–6
  reducing poverty to improve environmental quality, 306–9
  selling services as green, 305–6
  sustainability and, 300–302
  sustainable energy and, 233–35
Economic security, 317
Economic services
  provided by croplands, 151f
  provided by forests, 109f
  provided by freshwater systems, 67f
  provided by marine ecosystems, 65f
Economic systems, **300**
  market-based, 302–5
  sustainability and 300–302
Economists, three schools of, 300–302
Economy(ies). *See also* Economics
  developing environmentally-sustainable, 300–302
  high-throughput (high-waste), 81, 82f
  as human subsystem, 301f
  low-throughput, 82f
  matter-recycling-and reuse, 81
  shift to service-flow, 305–6
  transition to environmentally-sustainable, 81–83, 301–2

Ecosystem(s), **30.** *See also* Community
  abiotic and biotic components of, 33
  aquatic (*see* Aquatic life zones)
  artificial, 124
  biodiversity and (*see* Biodiversity; Ecosystem approach to sustaining biodiversity)
  biological components of, 34–36
  biomes and (*see* Biome(s))
  energy flow in, 31f, 36f, 37–41
  feeding (trophic) levels in, 37f, 39
  of fields, 33f
  food chains and food webs in, 39
  gross and net primary productivity of, 39–41
  limiting factors in, 34
  major components of, 33–34
  matter cycling in, 31f, 36f
  population range of tolerance in, 33, 34f
  structural components of, 36f
  terrestrial (*see* Terrestrial life zones)
Ecosystem approach to sustaining biodiversity, 108–28
  in aquatic systems, 108, 125–27
  ecological restoration as, 123–25
  forest resource management, 109–15
  goals, strategies, and tactics of, 109f
  human impacts on biodiversity and, 108–9, 125–26
  national parks as, 118–20
  nature reserves as, 120–23
  priorities for, 127–28
  species approach versus, 109f (*see also* Species approach to sustaining biodiversity)
  tropical deforestation and, 115–17, 118b
Ecotourism, 133
Education
  of consumers, 234–35
  environmental literacy, 320
  sustainable energy and, 233–35
Egg pulling, 145
Egypt, water resources, 174, 175f
Electrical energy, 26
Electricity, generation of
  costs of, from different sources, 229t
  from flowing water, 227
  for nuclear energy, cost of, 212
  solar cells for, 226f
  solar generation of high-temperature, 225f
  from wind, 227–29
Electromagnetic radiation, **26**
Electromagnetic spectrum, 26f
Electronic waste (e-waste), 284
Electrons (e), **24**
Elements, **23**
  cycling of crucial, in biosphere, 31f
  isotopes of, 24
Elephants, reduction in ranges/habitat for, 135f
Emissions trading policy, 263, 305
Emphysema, 262
Endangered species, **130.** *See also* Extinction of species
  examples of, 131f
  gray wolf, 119b
  laws and treaties to protect, 119, 141–43
  sanctuaries for protection of, 144–46
Endangered Species Act of 1973, U.S., 119b, 141–42
  accomplishments of, 144b
  future of, 143
Endemic species, **57,** 136
Endocrine system, chemical hazards to human, 243–44

Energy, 23–28. *See also* Energy resources
  defined, **25**
  efficiency in use of (*see* Energy efficiency)
  electromagnetic radiation as, 26f
  flow of (*see* Energy flow)
  kinetic, and potential, 25–26
  laws of thermodynamics governing changes in, 27–28
  matter and, 23–25, 26
  net, 202–3
  one-way flow of, and life on earth, 31f
  productivity of, 28
  quality of, 26, 27
Energy conservation, **219**–20
Energy content, 26
Energy efficiency, **28,** 219–23
  in buildings, 222–23
  of common devices, 220
  energy waste and, 219–20
  in industry, 220
  in transportation, 220–22
Energy flow
  between earth and sun, 32f
  in ecosystems, 31f, 36f, 37–41
  pyramid of, 39f
Energy productivity, **28.** *See also* Energy efficiency
Energy quality, **26,** 27
Energy resources, 201–36
  biomass conversion into liquid and gaseous fuels, 230–31
  biomass incineration as renewable, 229–30
  commercial, 201, 202
  energy efficiency as, and reducing waste, 219–23
  evaluating, 201–3
  fossil fuels as nonrenewable, 203–10
  geothermal, as renewable, 231
  hydrogen as, 231–32
  hydropower as renewable, 227
  net energy and, 202–3
  nuclear energy as nonrenewable, 210–19
  solar, as renewable, 201, 223–27
  solid waste incineration to produce, 289, 290f
  wind power, as renewable, 227–29
  strategies for developing sustainable, 233–35
Energy taxes, 276
Energy waste, 219–20
Environment, defined, **5**
Environmental citizenship, 321–22
Environmental degradation, **9.** *See also* Environmental problems; Natural capital degradation
  food production linked to, 159f
  in urban areas, 98f, 99–102
Environmental economists, 300
Environmental ethics, **318**
  environmental worldviews and, 318–20
  expenditure priorities and, 308f
Environmental groups, 315
  grassroots, 296–97, 315
  international, 317
  student, 316
Environmental hazards. *See* Hazard(s)
Environmental history, events in U.S., S2–S7
Environmental indicators, 136. *See also* Indicator species
Environmentalism, **5.** *See also* Environmental groups
Environmental justice, 297
  environmental policy and principle of, 310

Environmental law. *See also* Environmental regulations
  air pollution, 262–63
  endangered species, 119b, 141–43, 144b
  forest management, 114–15
  hazardous wastes, 292
  improving environmental quality with, 305
  lobbying for, 312f
  major U.S., 313f
  passage of, 312f
  pesticides, 166–67
  public lands, 312–15
  water quality, 195–96, 198
  wilderness preservation, 123
  threats to U.S., 316–17
Environmental leadership, 311
Environmental literacy, 320
Environmentally-sustainable economic development, **9,** 300–302
Environmentally-sustainable economies, developing, 81–83
  making transition to, 308–9
  methods for evaluating and improving environmental quality in, 302–6
  reducing poverty in, 306–8
Environmentally-sustainable society, **7**
  components of, 322–23
  environmental citizenship and, 321–22
  environmental literacy in, 320
  guidelines and strategies for developing, 321f
  learning from the earth to create, 320–21
  simple living and, 321
Environmental pessimists, 16
Environmental policy, **310**–17
  democratic political systems, politics, and, 309–10
  environmental groups and development of, 315, 316
  global, 317–18
  individual's influence on, 310, 311f, 312f
  principles for making, 310
  on public lands management in U.S., 312–15
  threats to U.S. environmental laws and regulations developed from, 316–17
  in United States, 311–12
Environmental problems, 5–19
  affluence and solutions to, 14
  biodiversity loss (*see* Extinction of species)
  comparative risk analysis of, 247, 248f
  connections among, 14–15
  cultural changes and, 15
  economic growth and development, and, 8–9 (*see also* Economics)
  environmental worldviews and, 15–18
  five basic causes of, 12, 13f
  food production and, 159f
  human population growth (*see* Human population growth)
  international cooperation on, 317, 318f
  key, 12
  managing, in democratic political systems, 309–10
  pollution as, 11 (*see also* Air pollution; Pollution; Water pollution)
  population growth linked, 7
  poverty and, 12–13 (*see also* Poverty)
  resources, resource consumption, and, 9–11, 13–14
  sustainable living and, 5–7 (*see also* Sustainability; Sustainable living)
  of urban areas, 98f, 99–102
  waste and (*see* Waste)

Environmental Protection Agency (EPA), U.S., 164, 191, 198, 260, 283
Environmental quality
affluence and, 14
economic measurements and indicators of, 303
Environmental regulations. *See also* Environmental law
improving environmental quality with innovation-friendly, 305–6
lobbying for environmental laws and, 312*f*
Environmental resistance, **78**
Environmental revolution
components of, 322–23
sustainability principles and, 17
Environmental science, **1**
author's invitation about this book on, 4
defined, **5**
improving critical thinking skills about, 3–4
improving study and learning skills on, 1–2
limitations of, 22
reasons for studying, 1
Environmental wisdom worldview, **16,** 319*f*, 320
Environmental worldview(s), **16, 318**
environmental wisdom, 16, 319*f*, 320
planetary management, 16, 318, 319*f*
stewardship, 16, 318, 319*f*
Epidemic, 238
Epidemiological studies on toxic chemicals, 245
Epiphytes, 75
Erosion. *See* Soil erosion
Estuary, 65
Chesapeake Bay, 193, 194*f*
Ethanol as fuel, 230*f*
Ethics. *See* Environmental ethics
Ethiopia, water resources in, 174, 175*f*
Euphotic zone, ocean, 65, 66*f*
Euphrates River, 175*f*
Eutrophication, **189**
Eutrophic lake, **67**
Evergreen coniferous forests
resource partitioning by warblers in, 73*f*
soils of, 43*f*
Evolution, **52**–69
adaptation and, 52–54
defining biological, 52
future of, 58–59
geologic processes, climate change, catastrophes, and, 55
microevolution and, 52
through mutation, 52
myths about, 54
natural selection and, 52–54
overview of biological, 53*f*
of species, 56–58
species extinction and (*see* Extinction of species)
theory of, 52
Executive branch of democratic governments, 310
Experimental group, 21, 245
Experiments, scientific, 21
on toxic chemicals, 245–46
Exponential growth of populations, **78**
External costs, 303
Extinction of species, **57**–58, 129–32. *See also* Endangered species; Threatened (vulnerable) species
background, 57
case study of passenger pigeons, 129–30
characteristics of species prone to, 132*f*

endangered and threatened species and, 130
estimated rates of, 130
habitat destruction, degradation, and fragmentation as cause of, 134–36
human activities and premature, 108–9, 130–32, 139–40
illegal killing or sale of species as cause of, 140–41
legal and economic approaches to preventing, 141–43
mass, 57
nonnative species as cause of, 136–39
role of climate change and pollution in, 139–40
role of market in exotic pets and plants in, 141
role of predators and pest control in, 141
sanctuary approach to preventing, 144–46
three types of, 129
Extinction spasm, 130
Exxon Valdez oil spill, 194, 195

Family planning, **93**–94
empowering women for better, 94
in India and China, 94–96
Farms, protecting endangered plants on, 144–45
Federal Emergency Management Agency (FEMA), 273
Federal Water Pollution Control Act of 1972, 195
Feebate programs, 221, 263
Feedlots, 150, 160, 161*f*
Fee-per-bag waste collection, 287, 289
Fertility, human, **85**
changes rate of, in U.S., 86
declining rates of, 85
factors affecting, 86–87
reduced, and population declines, 91
replacement-level, 85
total rate of (TFR), 85, 86*f*
Fertilizers, inorganic and organic, 156
First law of thermodynamics, **27**
Fish
as food resource, 161–63
sharks, 72*b*
world harvest of, 162*f*
Fisheries, 126–27, **161**
aquaculture, 150, 161, 162–63
management of, 127*f*
oceanic, 150, 161, 162*f*
Fish farming, **162**
Fish ranching, **162**
Fitness, biological, 54
Flexcar, 103
Flooding, 175–76
in Bangladesh, 175–76
deforestation as cause of, 176*f*
reducing risk of, 177
in urban areas, 99–100
Flood irrigation, 183
Floodplains, 68, **176**
Food and Agricultural Organization (FAO), U.N., 111, 148, 317
Food and Drug Administration (FDA), U.S., 166
Food and food resources. *See also* Food production
genetically modified, 159, 160*f*
food security, 148
human nutrition, malnutrition, and chronic hunger, 148–49
human overnutrition, 149

most important animal and plant sources of, 150
new, 163–64
protecting, with alternative methods, 167–69
protecting, with chemical pesticides, 164–67
reducing childhood deaths caused by hunger and malnutrition, 149
solutions for global food security, 163–64
Food chains, **37**–39
bioaccumulation and biomagnification of toxins in, 140*f*
energy flows in, 37*f*, 39
Food production, 149–52. *See also* Agriculture; Food and food resources
effects of ozone depletion on, 280*f*
gene revolution in, 158–60
global types of, 151*f*
green revolution in, 156–58
increasing fish and shellfish harvest, 161–63
increasing meat production, 160–61
industrial, 150–52
most important animal and plant food sources, 150
pest management and, 164–69
soil degradation linked to, 152–54
success of modern, 149–50
sustainable, 154–56, 169–70
traditional methods of, 152
Food security, **148**
Food web, **39**
in Antarctic, 38*f*
energy flows in, 39
Forest(s), 63, 109–13
building roads into, 111
certifying sustainably-grown timber from, 113
deforestation of tropical, 115–17
ecological and economic services provided by, 109–10
effects of acid deposition on, 259
effects of global warming on, 272*f*
effects of ozone depletion on, 280*f*
human impact on, 63–64
latitude and, 62*f*
natural capital degradation in, 64*f*
nonviolent civil disobedience to protect, 114*b*
sustainable management of, 112–13
tree harvesting methods in, 111–12
types of, 110
in United States, 113–15
Forestry, sustainable, 112–13
Formaldehyde, 260, 261*f*
Fossil fuels, 201*f*
air emissions from burning, 206*f*
coal, 208–10
natural gas, 207–8
nuclear power as alternative to, 216–17
oil, 203–7
U.S. and global dependence upon, as energy source, 202*f*
Foundation species, 72–73
Fox, speciation of Arctic and gray, 57*f*
Free-access resources, **9**
Freshwater
ecological disasters involving supplies of, 181–82
conflict over supplies of, 174–75
global population and supplies of, 171*f*
groundwater, 171–72 (*see also* Groundwater)
increasing supplies of, 178–81

Freshwater (cont'd)
  reducing waste of, 182–86
  resources, in U.S., 173–74
  seawater desalination to create, 182
  shortages of, 174, 175f
  surface, 67, 68, 172–74
  uses of, 173
  water pollution in, 188–92
Freshwater life zones, 64, 66–69
  ecological and economic services of, 67f
  inland wetlands, 68
  lakes, 66, 67f
  streams and rivers, 67, 68f, 174, 175f
Frogs, life cycle of, 71f
Frontier science, **21**
Fuel(s)
  biofuels, 229
  carbon dioxide emissions, by select, 206f
  ethanol and methanol as, 230–31
  fossil fuels (see Fossil fuels)
  gasoline, 220f
  global energy sources by, 202f
  nuclear, 203, 214f, 215–16
Fuel cells, 231–32
Fuel cell vehicles, 222f, 264
Fuel economy, motor vehicle, 220, 263
Fuelwood shortage, 202, 229
Full-cost pricing, **304**
Full-cost pricing, 103
Functional diversity, 36f, **37**
Fungicides, 164

Gandhi, Mahatma, 321
Gasohol, 230
Gasoline
  inflation-adjusted cost of, 220f
Gender imbalance, 96
Gene(s), **23**
  relationship of cells, chromosomes, DNA and, 24f
Gene banks, 144
Gene pool, **52**
Generalist species, 55b, **56**
  in tropical rain forest, 56f
Gene splicing, **59.** See also Genetic engineering
Gene therapy, 59
Genetically-modified food (GMF), 159, 160f
Genetically-modified organisms (GMOs), **59,** 60f
Genetic diversity, 30, 36f, **37,** 52
Genetic engineering, 59, 60f
  concerns about, 59
  increasing crop yields with, 158–60
Genetic information, preserving species for, 133
Genetic resistance to pesticides, 166
Genetic variability (genetic diversity), 30, 36f, 37, 52
Genuine progress indicator (GPI), 303
Geographic isolation, speciation and, 57f
Geological time, changes in biodiversity over, 58f
Geologic processes, evolution and, 55
Geothermal energy, **231**
  carbon dioxide emissions from, 206f
Geothermal heat pump (GHP), 231
Germs (pathogens), 238–40. See also Infectious disease
Giant pandas, 56
Glacial and interglacial periods in earth's history, 266
Global climate change, **268.** See also Climate change
Global cooling, 266

Global Environment Facility (GEF), 317
Global sustainability movement, 315
Global warming, 47, 266, **268**–77. See also Climate change
  effects of, 271–74
  factors affecting earth's temperature and, 270–71
  government role in reducing threat of, 276
  individual and local responses to, 277
  international agreements regarding, 276–77
  model of processes involved in, 269f
  nuclear energy as alternative to reduce, 216–17
  options for dealing with, 274
  ozone depletion and, 280f
  preparing for, 276, 277f
  removal and storage of carbon dioxide to reduce, 275–76
  solutions for reducing threat of, 274–75
  in troposphere, 266–68, 269f
Global Water Policy Project, 189
Government(s)
  agricultural policies of, 163
  democratic, 309–10
  environmental policy and (see Environmental policy)
  role of, in reducing threats of climate change, 276
  subsidies and tax policies of, and environmental quality, 304–5
Grain crops, 150, 157f
  conversion of, in to meat, 161f
Grameen Bank, Bangladesh, 307b
Grasslands, 61
  human impact on, 63–64
  natural capital degradation in, 64f
  soils of, 43f
Grassroots environmental action, 296–97, 315
Gravity, 31
Gravity-flow irrigation, 184f
Gray-air smog, 256
Great Britain, reduction in greenhouse gas emissions in, 277
Great Lakes
  case study on pollution and cleanup in, 190
  primary ecological succession on Isle Royale in, 76f
Green Belt Movement, Kenya, 118b
Greenhouse effect, **254.** See also Climate change; Greenhouse gases
  natural, 254
Greenhouse gases, 32, **254**
  carbon dioxide as (see Carbon dioxide(CO_2) as greenhouse gas)
  global warming and, 266–68, 269f, 270–71
  increases of, in earth's troposphere, 266, 267f, 268f, 269f
  methane, 207, 254, 266, 268f, 271
  nitrous oxide, 47, 254, 266, 268f
Green manure, **156**
Green revolution in agriculture, **156**–58
Green roofs (living roofs), 223
Green taxes, 304
Gross domestic product (GDP), **8**
  genuine progress indicator (GPI) versus, 303
Gross primary productivity (GPP), **39,** 40f
Groundwater, **171**–73
  arsenic in, 191–92
  confined and unconfined aquifers, 172f
  pollution of, 179f, 190–92

  preventing and slowing withdrawal of, 179, 180f
  withdrawal of, 179, 179f

Habitat, **30, 56**
  designation of critical, in U.S., 142
  ecological niche versus, 56
  fragmentation, degradation, and destruction of, as extinction threat, 134–36
Habitat conservation plans (HCPs), 142
Habitat islands, 136
Hardin, Garrett, 9
Harmful algal blooms (HABs), 193
Hawken, Paul G., 84
Hazard(s), 237–51
  biological, 238–43 (see also Disease; Infectious disease)
  chemical, 243–47
  cultural, 237–38
  physical (earthquakes, volcanoes), 237
  risk, risk assessment, and, 237
  risk analysis and, 247–50
  types of, 237
Hazardous chemicals, **243.** See also Chemical hazards; Hazardous wastes
Hazardous wastes, **283, 292**–96
  defined, 292–93
  in homes, 283f
  individual's role in reducing, 296f
  integrated management of, 293–95
  lead as, 295–96, 297f
  phytoremediation of, 293, 294f
  U.S. laws regulating, 292
Health. See Human health
Healthy Forests Restoration Act (2003), U.S., 114–15
Heat
  as energy, 26
  high temperature, 225f
  ocean storage of carbon dioxide and, 270, 273f
Heating
  net energy ratios for, 203f
  saving energy in, 223
  solar, 224f, 225f, 226f
  space, 203f
Heavy crude oil, 203. See also Oil
Heliostats, 225
Hepatitis B (HBV), 240
Herbivores, **35**
Heterotrophs, **35**
Higher-level consumers, **35**
High-input (industrialized) agriculture, **150**–52
High-level radioactive wastes, 216
  storage of, in U.S., 217b
High-quality energy, **26**
  one-way flow of, and life on earth, 31f
High-quality matter, **25**
High-temperature heat and electricity solar generation, 225f
High-throughput (high-waste) economy, **81,** 82f
  waste management strategies for, 284
Hill, Julia Butterfly, 114b
HIPPO (Habitat destruction and fragmentation, Invasive (alien) species, Population growth, Pollution, Overharvesting), 134–36
Hirschorn, Joel, 250
HIV (human immunodeficiency virus), 240, 241
Homes
  indoor air pollution in, 255, 260, 261f
  net energy ratios for heating, 203f

pesticides use in, 167
reducing energy use in, 222–23
reducing water use in, 184–85
solar-heated, 224f, 225f
toxic and hazardous chemicals in, 293f
Hormone(s)
chemical hazards to human, 244
controlling insects by disrupting, 168
Human capital/human resources, **300**
Human-centered environmental worldview, 318, 319f
Human culture
birth rates and values of, 87
environment and changes in, 15
Human health. *See also* Human nutrition
biological hazards (disease) to, 237, 238–43
chemical hazards to, 237, 243–47
comparative risk analysis of, 247, 248f, 249f
cultural hazards to, 237–38
effects of acid deposition on, 259
effects of air pollution on, 260, 261, 262
effects of global warming on, 272f
effects of ozone depletion and ultraviolet radiation on, 279, 280f
pesticides and, 166
physical hazards to, 237
poisons and (*see* Poison)
poverty as risk to, 247, 248f, 249f (*see also* Poverty)
risk, probability, and categories of hazards to, 237
risk analysis of hazards to, 247–50
smoking, tobacco, and damage to, 237–38
toxic chemicals and, 243–44
toxicology and assessment of threats to, 244–27
urban living and, 99, 100
Human impact on environment, 22f, 80–83. *See also* Environmental problems; Natural capital degradation
air pollution (*see* Air pollution)
on aquatic biodiversity, 125–26
on biodiversity, 58, 108–9, 125–26
on carbon cycle, 46–47
characteristics of natural systems versus human-dominated systems and, 80f
climate change, greenhouse gases, and, 266–68
in deserts, grasslands, and forests, 63–64
ecological footprint and, 10, 80, 81f
extracting, processing of mineral resources, 207f
on marine systems, 66f
on natural systems, 80
on nitrogen cycle, 48–49
pesticide use and, 165–66
on phosphorus cycle, 49
precautionary principle and, 80–81
solutions for developing sustainable economies, 81–83
species extinction and, 130–32, 139–40
unintended results of, 22f
on water (hydrologic) cycle, 45
Human nutrition
food security and, 148
hunger, malnutrition and, 13, 148–49
overnutrition, 149
saving children from malnutrition and disease, 149
vitamins, minerals, and, 149
Human population
age structure of, 89–91
case studies on, in India and China, 94–96
effects of global warming on, 272f, 273–74
factors affecting size of, 85–88

freshwater supplies and, 171f
growth of (*see* Human population growth)
influencing size of, 92–94
making urban areas more sustainable and livable for, 104–7
rapid decline in, 91
sustainability principles and control of, 16, 17f
transportation, urban development, and, 102–3
urbanization, and distribution of, 96–99
urban resource and environmental problems linked to, 99–102
wealth/income, resource use, pollution/waste, and, 8f
Human population growth, 7, 84
age structure and, 89–91
benefits of slowing, 91
birth/death rates and, 85, 86–87
fertility rates and, 85, 86–87
immigration and, 88
poverty and, 13 (*see also* Poverty)
slowing, in China and India, 94–96
solutions for influencing, 92–94
species extinction linked to, 139–40
sustainability principles and control of, 16, 17f
Human rights principle, environmental policy and, 310
Humility principle, environmental policy and, 310
Humus, **42**
Hunger, chronic, **148**–49
reducing childhood death from, 149
Hybrid gas-electric vehicles, 221f, 264
Hydrocarbons, 23
Hydrogen as fuel, 231–32
advantages/disadvantages of, 232f
motor vehicles powered by, 222f
role of fuel cells, 222f, 232
Hydrological poverty, 174
Hydrologic (water) cycle, **44, 45f**
effects of human activities on, 45
Hydropower, 227
Hydrosphere, **30,** 31f
Hydrothermal reservoirs, 231

Igneous rock, **50**
Immigration into United States, 88
Immigration Reform and Control Act of 1986, 88f
Immune system, chemical hazards to human, 243–44
Incineration
of biomass, 229–30
of hazardous waste, 294
of solid wastes, 289, 290f
Income
global distribution of, 306, 307f
population, resource use/waste, and, 8f
India
basic demographic data for, 95f
ecological footprint of, 10f
population regulation in, 94–95
Indicator species, **71**
amphibians as, 70b, 71f
birds as, 71, 136
Indirect costs, 303
Individuals matter (influence and role of individuals), 6, 17
alternatives to ozone-depleting chemicals, 281b
creating green sustainable business, 306b
environmental policy and influence of, 310, 311f, 312f

protecting forests, 114b
protecting ozone layer, 281b
protecting wild species, 146
reducing air pollution, 265f
reducing CO_2 emissions, 277f
reducing energy use and waste, 235f
reducing exposure to UV radiation, 280f
reducing hazardous waste input, 296f
reducing pesticide exposure and effects, 167f
reducing solid waste, 285f
reducing threat of invasive species, 139f
reducing water pollution, 200f
reducing water use and waste, 186f
sustainable agriculture and, 170f
sustaining terrestrial biodiversity, 123f
water use and waste reduction and, 186f
Indoor air pollution, 255, 260–62
formaldehyde and, 260, 261f
harmful effects of, 260
individual's role in reducing, 265f
radon gas, 255t, 260–62
solutions for reducing, 265f, 266f
types and sources of, 260, 261f
Industrial heat, net energy ratios for high-temperature, 203f
Industrialized (high-input) agriculture, **150**
meat production, 160–61
in United States, 150–52
Industrial-medical revolution, 15
Industrial smog, **256**
Industrial solid wastes, **283**
Industry
brownfields and abandoned, 293
improving energy efficiency in, **220**
reducing water use by, 184, 185f
water pollution caused by, 187, 193f
Infant mortality rate, **87,** 88t
Infectious agents. *See* Pathogens
Infectious disease
antibiotic resistance in bacteria and, 239–40
deadliest, 239f
death rates from, 239f
hepatitis B, 240
HIV and AIDS, 91, 240, 241
influenza, 240
malaria, 241–42
malnutrition and, 13, 148–49
pathways for, in humans, 239f
reducing incidence of, 242–43
SARS, 240
tuberculosis, 240
in urban areas, 100
viral, 240–41
West Nile virus, 240
Infiltration of water in soils, **42**
Influenza (flu), 240
Information-globalization revolution, 15
Inland wetlands, freshwater, **68**
Innovation-friendly environmental regulations, 305
Inorganic compounds, **23**
Inorganic fertilizers, **156**
Input pollution control, **11.** *See also* Prevention approach
Insect(s)
Argentine fire ants, 137, 138f, 139f
cockroaches, 55b
control of, as pests, 164–69
ecological role of, 30
hormone disrupters of, 168
introduced nonnative, 138f
sex attractants (pheromones) for, 168
Insecticides, 164. *See also* Pesticides

Instrumental value, **109**, 133
Integrated coastal management, 126, 194
Integrated pest management (IPM), **168**–69
Integrated waste management
    of hazardous waste, 293–95
    of solid waste, **284**f, 285
Integrative principle, environmental policy and, 310
Intercropping, **152**
Interface Corporation, 306b
Intergovernmental Panel on Climate Change (IPCC), 266, 267, 268, 272, 273
Internal combustion engine, energy waste in, 220
Internal costs, 303
International agreements
    on protecting ozone layer, 281
    reducing greenhouse emissions, and responding to threat of global warning, 276
    reducing hazardous waste, 297–98
Internet, critical thinking applied to, 4
Interplanting, **152**
Interspecific competition, **73**
Intrinsic rate of increase (r), population growth, **77**, 78f
Intrinsic value, **109**
    of wild species, 134
Introduced species. *See also* Nonnative species
    accidentally introduced, 137, 138f
    deliberately introduced, 136–37, 138f
Invasive species, **70**. *See also* Nonnative species
Iodine, as micronutrient, 149
Ions, **23**
Iron, as micronutrient, 149
Irrigation
    major systems of, 183, 184f
    reducing water waste in, 183–84
    soil salinization and waterlogging caused by, 153–54
    water use for, 173
Isotopes, **24**
Israel, water resources in, 183

Janzen, Daniel, 124
Jordan River basin, 174, 175f
J-shaped curve of population growth, 7f, 78, 84
Judicial branch of democratic governments, 310
Junk science, **21**–22

Kenya
    ecotourism in, 133
    Green Belt Movement in, 118b
Kerogen, 207
Keystone species, **71**–72
Kinetic energy, **25**
Kyoto Protocol, 276

Laboratory experiments to test toxic chemicals, 245–46
Labor force, 86
Lakes, **66**. *See also* Dams and reservoirs
    Aral Sea water disaster, 181–82
    eutrophication and cultural eutrophication in, 67, 189–90
    Great Lakes, 190
    mesotrophic, 67
    oligotrophic, 67
    water pollution in, 189–90
    zones of, 67f

Land
    availability of, transportation, and urbanization, 102
    climate affected by mass of, 254
    global uses of, 108
    increasing amount of cultivated, 158
    marginal, 158
    protecting, in nature reserves, 120–21
    protecting endangered species on private, 142–43
Land degradation
    salinization and waterlogging as, 153–54
    soil erosion as, 152–53
Landfills
    open dumps, 289
    sanitary, 290, 291f, 292f
    secure hazardous waste, 295, 296f
Land trust groups, 120
Lateral recharge, 172
Latitude, climate, biomes, and effects of, 62f, **63**
Law. *See* Environmental law
Law of conservation of energy, **27**
Law of conservation of matter, **27**
Law of thermodynamics, 27–28
Law of tolerance, **34**
LD50 (median lethal dose) of chemicals, 244, 245t
Leaching of soil, **42**
Lead
    health threat of, to children, 295–96
    solutions for preventing/controlling poisoning by, 297f
Leadership, environmental, 311
Learning skills, 1–4
    importance of studying environmental science, 1
    improving critical thinking skills and, 3–4
    improving study and, 1–2
    on trade-offs in environmentalism, 4
Legislative branch of democratic governments, 310, 312f
Leopold, Aldo, 16, 128, 129, 148
Lethal dose (LD) of chemicals, 244, 245t
Life-centered environmental worldviews, 319–20
Life cycle costs, energy efficiency and, **220**
Life expectancy, **87**, 88t
Life-support systems, earth's, 28–37
    biodiversity as component of, 36–37
    Earth as favorable for life and, 32–33
    ecology and, 28
    four spheres of, 30–31
    gravity and, 31
    insects as, 30
    major components of ecosystems and, 33–36
    matter cycling and, 31
    microorganisms as, 28–29
    populations, communities, and ecosystems as, 30
    solar energy and, 31, 32
Light, speed of, 26
Light bulb
    energy efficiency, 223
    energy waste in incandescent, 220
Light pollution, 100
Light-water reactors (LWR), nuclear, 212, 213f
    nuclear fuel cycle in, 214f, 215–16
Limiting factor principle, **34**
Limiting factors in ecosystems, **34**
Lipids, 23
Liquefied natural gas (LNG), **207**
Liquefied petroleum gas (LPG), **208**
Lithosphere, **30**

Livestock
    methane emissions and, 271
    production of, 116, 160–61
Living machines, sewage treatment with, 197
Living sustainably. *See* Sustainable living
Living systems
    human impact on, 80, 81f (*see also* Human impact on environment)
    precautionary principle for protecting, 80–81
    second law of thermodynamics in, 28f
Local extinction of species, 129
Logistic growth of populations, 78f
Low-energy precision application (LEPA) sprinklers, 183, 184f
Low-quality energy, **26**
Low-quality matter, **25**
Low-throughput (low-waste) economy, **82**f
    achieving, 296–98
    integrated waste management for, 284–85
Lungs, emphysema and cancer in human, 262

Maathai, Wangari, Green Belt Movement founded by, 118b
McGinn, Anne Platt, 298
Mad cow disease, 161
Malaria, 241–42
    global distribution of, 241f
    life cycle of, 242f
Malignant melanoma, 280
Malnutrition, 13, **148**–49
    reducing childhood deaths from, 149
Mangrove forest swamps, 65
Manufactured resources/manufactured capital, **300**
Marginal land, 158
Marine life zones, 64–66. *See also* Ocean(s)
    coastal zone, 65 (*see also* Coastal zones)
    ecological and economic services provided by, 65f
    food web in Antarctic, 38f
    natural capital degradation in, 66f
    open sea, 65
    protecting biodiversity of, 126–27
Marine protected areas (MPAs), 126
Marine reserves, 126
Market-based economic systems
    environmental and economic indicators in, 303
    full-cost pricing in, 304
    hidden harmful costs in, 302–3
    subsidies, taxes, and, 304–5
Market forces. *See also* Economics
    food production and, 163
    improving environmental quality using, 305
    reducing air pollution using, 263–64
    reducing pollution and resource use relying on, 305
Marriage, birth rate and average age at, 87
Marsh, freshwater, 68
Marsh, George Perkins, 70
Mass depletion, **57**, 58
Mass extinctions, **57**, 58
Mass number, **24**
Mass transit rail, 103, 104f
Material efficiency, **25**
Materials-flow economy, shift from, 305–6
Materials-recovery facilities (MRFs), 287
Matter, **23**–28
    atoms and ions of, 23, 24–25
    elements and compounds of, 23
    energy and, 25–26
    law of conservation of, 27

levels of organization in, 28f
molecules, cells, and, 23–24
quality of, 25
physical and chemical changes in, 26
physical states of, 25
Matter cycling in ecosystems, 31f, 36f, 44–50
biogeochemical (nutrient) cycles and, 44
carbon cycle, 45–47
nitrogen cycle, 47–49
phosphorus cycle, 49
rock cycle, 50
water (hydrologic) cycle, 44–45
Matter efficiency (resource productivity), 25
Matter quality, 25
examples of, 25f
Matter-recycling-and-reuse economy, 81
Mead, Margaret, 17, 281
Measurement
uncertainty, accuracy, and precision in, 21
units of, S1
Meat production, 160–61
conversion of grain into meat, 161f
increased livestock production for, 160–61
rangelands and pasture for, 149
Mechanical energy, 26
Megacities/megalopolises, 96, 97f, 98f
Megareserve, 121
Menander, 51
Mesotrophic lake, 67
Metallic mineral resources, 10
Metamorphic rock, 50
Methane (CH_4), 207
global warming and emissions of, 266, 268f, 271
as greenhouse gas, 254
Methane hydrates, 208, 271
Methanol as fuel, 230f, 231
Metropolitan areas. See Urban areas
Mexico City, environmental problems of, 101–2
Microbes (microorganisms), 28–29. See also Bacteria
Microevolution, 52
Microirrigation, 183, 184f
Microlending (microfinance), 307b
Micronutrients, human nutrition and, 149
Microorganisms (microbes), ecological role of, 28–29. See also Bacteria
Micropower systems, 233
decentralized power system using, 234f
Middle East, water conflicts in, 174, 175f
Midgley, Thomas, Jr., 278
Military security, 317
Millennium Ecosystem Assessment, 15, 125, 126
Mineral resources, metallic and nonmetallic, 10
Minerals, human nutrition and, 149
Mining
water pollution caused by mineral, 187
of water resources, 172
Mississippi River, 177
Molecule, 23
Molina, Mario, 279
Monoculture, 110
Monomer, 23
Montague, Peter, 196
Montreal Protocol, 281
Mosquito, as disease vector for malaria and, 242f
Motor vehicles
advantages and disadvantages of, 102–3
air pollution caused by, 256f, 257
alternatives to cars, 103
ethanol and methanol as fuel for, 230f

fuel cell, 222f, 264
fuel-efficiency standards in, 220, 263
hybrid gas-electric, 221f, 264
plug-in, wind-powered, 221
reducing air pollution from, 264f
reducing use of, 103
saving energy in use of, 220–22
in United States, 102
MSW. See Municipal solid waste
MTBE (methyl tertiary butyl ether), 191
Muir, John, 20
Multivariable analysis, 21
Municipal solid waste (MSW), 283
recycling, 287
Mutagens, 52, 243
Mutations, 52
biological evolution and, 52
caused by toxic chemicals, 243
Mutualism, 74, 75f
Myers, Norman, 132, 317

Narrow-spectrum agents, 164
National Forest System, U.S., 114, 312–13, 314f
logging in, 114–15
nonviolent civil disobedience to protect, 114b
National Marine Fisheries Service (NMFS), 142
National parks, 118–20
threats to global, 118
U.S., 118–20, 313, 314f
National Park Service, U.S. (NPS), 313
National Park System, U.S., 313, 314f
National Priorities List (NPL), U.S., 292
National Resource Lands, U.S., 313
National security
environmental policy and, 317
nuclear energy and, 213, 215
National Wilderness Preservation System, U.S., 313
National Wildlife Refuges, U.S., 144
Native species, 70
Natural capital, 5, 6f, 300
atmosphere as, 252f, 269f
biodiversity as, 36f, 58f, 136f
biological evolution as, 53f
biomes and climate as, 61f, 62f, 63f, 254f
carbon cycle as, 46–47f
cells, nuclei, chromosomes, DNA, and genes as, 24f
climate as, 253f, 254f, 269f, 273f
commercial energy as, 202f
croplands as, 151f
detritus feeders and decomposers as, 35f
earth as ocean planet as, 65f
ecological niches as, 56f
ecosystem components as, 33f, 36f
food chains and trophic levels as, 37f
food production methods as, 151f
food web in Antarctic as, 38f
forests as, 109f
freshwater systems as, 67f, 68f
hydrologic cycle as, 45f
levels of organization in matter as, 29f
limits on population growth as, 78f
marine ecosystems as, 65f, 66f
natural greenhouse effect as, 32f
natural resources, natural services, and, 6f
net and gross primary productivity as, 40f
nitrogen cycle as, 48f
nonrenewable energy sources as, 201f, 209f
one-way flow of energy and crucial elements cycling as, 31f
pharmaceutical plants as, 133f

phosphorus cycle as, 49f
public lands as, 314f
pyramid of energy flow as, 39f
range of tolerance for population of organisms as, 34f
rock cycle as, 50f
soil, soil formation, and soil profiles as, 41f, 43f
solar energy as, 32f
water resources, 171f, 172f, 173f, 180f, 188f
Natural capital, endangered and threatened, 131f, 138f, 139f, 142f
in biodiversity hot spots, 122f, 142f
Natural capital, lost, 129f, 132f
Natural capital degradation, 6
acid deposition as, 258f
air pollution as, 256f
atmosphere, and global warming as, 270f
in biomes, 64f, 66f
of earth's biodiversity, 108f, 134f, 135f, 139f, 140f
from energy resource use, 206f, 207f
environmental problems, causes and connections, and, 12f
of fish catch, 162f
food production linked to, 159f
of forest resources, 110f, 111f, 116f
human activities and, 22f
human ecological footprint and, 10, 80, 81f
ozone depletion as, 280f
of soil resources, 153f, 154f
in urban areas, 100f
urban sprawl as, 98f, 99f
of water resources, 175f, 176f, 179f, 191f, 193f, 194f
Natural gas, 207–8
advantages and disadvantages of conventional, 208, 209f
carbon dioxide emissions produced by burning, 206f
Natural greenhouse effect, 32f, 47, 254
Natural recharge, 172
Natural (scientific) law, 21
Natural resources, 6f, 300. See also Natural capital; Resource(s)
Natural selection, 52–54
myths about evolution through, 54
Natural services, 6f
Nature Conservancy, 120
Nature reserves, 120–23
in Costa Rica, 121
management of U.S., 312–15
protecting global biodiversity in, 121–22, 123f
protecting land against human exploitation with, 120–21
wilderness as, 122–23, 313
wilderness protection in U.S., 123
Neem tree, 117
Neoclassical economists, 300
Nervous system, chemical hazards to human, 243–44
Net energy, 203–4
Net energy ratios, 203
Netherlands, ecological footprint of, 10f
Net primary productivity (NPP), 39–41
estimated average, of select ecosystems, 40f
Neurotoxins, 244
lead as, 295–96, 297f
Neutrons (n), 24
New urbanism, 104
Niche, 55. See also Ecological niche
Nigeria, demographic factors in Brazil, U.S., and, 90f

Nile River water resources, 174, 175f
Nitrate ions, 47
Nitric acid (HNO₃), 48
Nitrification, 47
Nitrite ions, 47
Nitrogen cycle, **47,** 48f
  human impact on, 48–49
Nitrogen dioxide (NO₂), 48
Nitrogen fixation, 47–48
Nitrogen oxide (NO), 48
  as air pollutant, 255t
Nitrous oxide (N₂O), 47, 254, 266
  global warming and, 266, 268f
Nondegradable pollutants in water, 191
Nonmetallic mineral resources, 10
Nonnative species, **70**
  human introduction of, 136–37, 138f
  reducing threats from, 137–39
Nonpoint sources of pollution, **11**
  reducing, 195
  water pollution, 186–87, 195
Nonprofit nongovernment organizations
  (NGOs), environmental policy
  and, 310
Nonrenewable resources, **10**–11
  as energy source, 201, 203–10
  environmental impact of using, 207f
Nonthreshold dose-response model, 246f
Nontransmissible disease, **238**–39
Nonviolent civil disobedience, 114b
North America, bird species in, 136f
No-till cultivation, 275
Nuclear energy, 26, 210–19
  advantages/disadvantages of conven-
    tional, 215–16
  Chernobyl nuclear plant accident, 213,
    214–15
  coal energy versus, 216f
  emissions produced by burning, 206f
  fuel cycle, 203, 214f, 215–16
  future of, 218–19
  history of, 212–13
  net energy ratio, 203
  nuclear fission as, **210**–11
  nuclear fusion and, 217–18
  potential of, for solving environmental
    problems, 216–17
  problem of aging and worn-out reactors,
    216
  reactors and electricity production from,
    211–12
  security issues, terrorism, and dirty
    bombs, 213, 215
  waste produced by, 216, 217b
Nuclear fission, **210**–11
  chain reaction, 212f
Nuclear fission reactors, 211–12
  accidents at, 213, 214–15b
  decommissioning old, 216
  energy waste in, 220
  light-water reactors, 212
  waste produced by, 216, 217f
Nuclear fuel cycle, 203, 214f, 215–16
Nuclear fusion, 217, **218**
Nuclear Regulatory Commission (NRC),
  U.S., 215
Nuclear weapons, 215
Nucleic acids, 23
Nucleus (atom), **24**
Nutrient(s), **30, 44**
Nutrient cycles (biogeochemical cycles), 31f,
  44–50
Nutrient recycling, sustainability principles
  and, 16, 17f
Nutrition. See Human nutrition

Ocean(s). See also Marine life zones
  as carbon dioxide-storage and heat reser-
    voir, 270, 273f
  climate linked to currents and drifts of,
    253f, 273f
  coastal zones, 65
  open sea of, 65, 66f
  pollution (see Ocean pollution)
  protecting biodiversity of, 126–27
  sea levels, and global warming, 272, 273
  sharks in ecology of, 72b
Oceanic fisheries, 150, 161, 162f
Ocean pollution, 192–95
  Chesapeake Bay estuary as case study,
    193–94
  coastal areas affected by, 192, 193f
  oil as pollutant in, 194–95
  protecting coastal waters, 195
  tolerance levels, 192
Ogallala Aquifer, 178, 179f
Oil, 203–7
  advantages/disadvantages of conven-
    tional, 205–6
  crude (petroleum), 203–4
  environmental impact of, 206f, 207f
  global supplies of, 204–5
  history of, S8
  nuclear energy as alternative to imported,
    216–17
  oil shale and oil sands as sources of heavy,
    206–7, 208f
  refining of crude, 204f
  U.S. supplies of, 205
  as water pollutant, 194–95
Oil sand, **206**–7, 208f
Oil shale, 206, **207,** 208f
Oil spills, 194–95
Old-growth forests, **110**
Oligotrophic lake, **67**
Omnivores, **35**
Open dumps, **289**
Open sea, **65**
  zones of, 66f
Opportunist species, 79
Oral rehydration therapy, 242
Organic compounds, **23**
Organic fertilizers, **156**
Organisms, **28.** See also names of specific
    organisms, eg., Animal(s)
  genetically modified, 59, 60f
  limiting factors on, 34
Organization of Petroleum Exporting
    Countries (OPEC), 204–5
Outdoor air pollution, 256–59
  acid deposition and, 257–60
  factors influencing levels of, 257
  industrial smog as, 255
  major pollutants, 255t
  photochemical smog as, 256–57
  potential effects on tropospheric warming,
    271
  preventing and reducing, 262–66
Output pollution control, **11.** See also Cleanup
    approach
Overnutrition, **149**
Overpopulation, human, 92
Overshoot, population, 79
Oxygen, dissolved content of, in water,
    34
Oxygen-demanding wastes as water
    pollutant, 188f, 193f
Oxygen sag curve, 188
Ozone (O₃)
  as air pollutant, 255t
  in stratosphere, 252–53, 278

Ozone depletion in stratosphere, 278–81
  causes of, 278–79
  at earth's poles, 279
  protection against, 281
  reasons for concern about, 279, 280f
  reducing exposure to UV radiation linked
    to, 280f
  threats to levels of, 278
Ozone hole/ozone thinning, 279

Pandemic, 238
Paper products, recycling, 287
Parasitism and parasites, 74
Partial-zero-emission-vehicles (PZEV), 264
Passenger pigeon, extinction of, 129–30
Passive smoking, 238
Passive solar heating system, **224**f, 225f
Pastures, 149
Pathogens, 238–39. See also Disease; Infec-
    tious disease
  antibiotic resistance and immunity of,
    239–40
Pay-as-you-throw (PAUT) system, recycling
    solid wastes and, 287
Pension systems, birth rates and availability
    of, 87
Per capita GDP, **8,** 303
Perception of risk, 249–50
Permafrost, 30
Perpetual resource, 9
Persistent organic pollutants (POPs), 247
  international treaty on control of, 297
Pest(s), **164**
  alternative methods of controlling, 167–69
  chemical control of, 164–67
  natural control of, 164
  species extinction threats in control of, 141
Pesticides, **164**
  advantages of modern synthetic, 164–65
  alternatives to, 167–68
  bioaccumulation and biomagnification of
    DDT, 140f
  R. Carson on environmental effects of, 165b
  circle of poison and, 167b
  disadvantages of, 165–66
  reducing individual exposure to, 167f
  threats of, to wild species, 139, 140f
  U.S. laws on, 166–67
Petrochemicals, **204**
Petroleum, **203.** See also Oil
pH, **42**
  scale, 44f
  of soils, 42
Pheromones, controlling insects using, 168
Phosphorus cycle, 49f
  human impact on, 49
Photochemical smog, **256**–57
  in Mexico City, 102
  preventing and cleaning up, 264f
Photosynthesis, **34**–35, 46
  effects of higher carbon dioxide levels on,
    271
Photovoltaic (PV) cells, **226**f–27
Physical change in matter, **26**
Phytoplankton, 34
Phytoremediation of hazardous waste, 293,
    294f
  advantages and disadvantages of, 294f
Pioneer species, 75
Planetary management worldview, **16, 318,**
    319f
Plant(s). See also Trees
  botanical gardens and arboreta for protec-
    tion of, 144–45
  epiphytes, 75

as food source, 150
lack of, in urban areas, 99
pharmaceutical, 133*f*
Plantation agriculture, **150**
Plastics recycling, 288
Plug-in wind-powered vehicles, 221
Point sources of pollution, **11**
reducing, 195–96, 198
water pollution, 186–87, 195–96, 198
Poisons, **244.** *See also* Toxic chemicals
arsenic, 191–92
lead as, 295–96
pesticides and circle of, 167*b*
toxicity ratings and lethal doses for, 245*t*
Politics, 309–17
environmental action by student groups, 316
environmental groups, networks and, 315
environmental leadership and, 311
environmental policy and laws in U.S., and, 311–12
environmental problems and, in democracies, 309–10
individual's role in environmental policy, 310, 311*f*, 312*f*
managing public lands in U.S. and role of, 312–15
principles guiding environmental policy decisions and, 310
sustainable energy resources and, 233–35
threats to U.S. environmental law and regulations, 316–17
Pollutants. *See* Air pollutants; Water pollutants
Polluter-pays principle, environmental policy and, 310
Pollution, **11**
air (*see* Air pollution)
economic solutions for (see Economics)
light, 100
noise, 99, 100*f*
point, and nonpoint sources of, **11**, 186–87, 195–96
prevention versus cleanup of, **11** (*see* Prevention approach)
sources and effects of, 11
species extinction linked to, 139–40
tradable, 263, 305
in urban areas, 100
water (*see* Water pollution)
Pollution cleanup, **11.** *See also* Cleanup approach
Pollution prevention, 11. *See also* Prevention approach
Polyculture, **152**
Polymer, 23
Polyvarietal cultivation, **152**
Population(s), **30.** *See also* Human population; Human population growth; Population dynamics
biological evolution in, 52
biotic potential of, 77
carrying capacity and size of, 78–79
growth of, 7, 78
limits on growth of, 77–78
range of tolerance in, 33, 34*f*
reproductive patterns in, 79–80
Population change, calculation of, **85,** 88*t*
Population dynamics, 77–80
exceeding carrying capacity, 78–79
exponential and logistic population growth, 78
limits on population growth, 77–78
reproductive patterns and, 79–80

Possibility, risk and, 237
Postconsumer wastes, recycling of, 287
Potential energy, **25**
Poverty, **12,** 306
environmental problems linked to, 12–13
food security, hunger, malnutrition and, 13, 148–49
global distribution of wealth and, 306, 307*f*
as human health risk, 247, 248*f*, 249*f*
improving environmental quality by reducing, 307–8
microlending as solution for, 307*b*
in urban areas, 100–101
Power tower, 225
Prairie potholes, 68
Precautionary principle, 80–**81.** *See also* Prevention approach
chemical hazards and, 247, 298
environmental policy and, 310
reducing flood risks, 177
Precipitation
acid deposition, 48, 257–60
climate and average, 61, 62*f*, 253, 254
infiltration of soils by, 42
Precision in making measurements, 21
Preconsumer (internal) waste, recycling of, 287
Predation, **74**
Predator-prey relationship, **74**
Predators, **74**
human elimination of, 141
sharks as, 72*b*
wolves in national parks, 119*b*
Prevention approach
to acid deposition, 259, 260*f*
to air pollution, 264*f*, 265*f*, 266
to chemical hazards, 247
environmental policy and, 310
flood risk reduction, 177
global warming and, 274*f*, 281
groundwater depletion, 180*f*
integrated pest management as, 168
to lead poisoning, 297*f*
to pollution, 11
to preventing water pollution, 192*f*, 195*f*
to soil salinization, 154*f*
to water pollution, 199–200
Prey species, 74
Primary air pollutants, **255,** 256*f*
Primary consumers, **35**
Primary ecological succession, **75,** 76*f*
Primary (closed-loop) recycling, 287
Primary sewage treatment, **196,** 197*f*
Primm, Stuart, 58, 132
Probability, risk and, 237
Producers (autotrophs), 33, **34**
rate of production by, 39–41
Profit-making organizations, 310
Proteins, 23
Protons (p), **24**
Proverbs, quotation from, 323
Public lands in United States, 312–15
forests, 114–15, 312
map of, 314*f*
national parks, 118–20, 313
threats to, 314–15
wilderness reserves, 123, 313
Public participation principle, environmental policy and, 310
Pyramid of energy, **39***f*

Quality of life in urban areas
improving, 104–7
problems of, 97, 98*f*, 99–102

Radiation, ultraviolet. *See* Ultraviolet (UV) radiation
Radioactive isotopes (radioisotopes), **211**
Radioactive radon, 255*t*, 260–62
Radioactive wastes, high-level, 216, 217*b*
Radon gas, radioactive, 255*t*, 260–62
Rain. *See* Precipitation
Rangelands, 149
Range of tolerance, **33**
for temperature, 34*f*
Rapid rail as transportation, 105*f*
Reconciliation ecology, 145–46
Recreational value of wild species, 133
Recycling, **10**–11, 287–89
advantages and disadvantages of, 288–89
composting as, 287–88
encouraging, 289
of municipal solid wastes, 287
of plastics, 288
two types of, 287
Red-cockaded woodpeckers, 56
Red Lists, 130
Red tides, 193*f*
Refined petroleum (oil), 204
as water pollutant, 194
Refinery, crude oil, 204*f*
Reforestation
in Kenya, 118*b*
neem tree useful for, 117
Regulations. *See* Environmental regulations
Reindeer, excessive population growth of, 79*f*
Reliable runoff, **173**
Religious beliefs, 87
Renewable resources, **9**
as energy source, 201, 224–31
tragedy of the commons and degradation of, 9–10
Replacement-level fertility, **85,** 88*t*
Reproducibility in making measurements, 21
Reproduction
differential, 52
patterns of, 79–80
Reproductive isolation, speciation and, **57***f*
Reproductive time lag, 79
Reserves of nonrenewable resources, **11**
Reservoirs. *See* Dams and reservoirs
Resource(s), **9.** *See also* Natural capital
common-property/free-access, 9
ecological footprint affecting, 10, 80–81
environmental degradation of, 9 (*see also* Natural capital degradation)
environmental problems and excessive consumption of, 13–14
human population, wealth/income, and use/waste of, 8*f*
nonrenewable, 10–11 (*see also* Nonrenewable resources)
perpetual, and renewable, 9 (*see also* Renewable resources)
sustainable yield of, 9
three types of, 300
tragedy of the commons in caring for, 9–10
waste of, 283–84
Resource Conservation and Recovery Act (RCRA), U.S., 292
Resource partitioning, **73***f*
Resource productivity (material efficiency), **25**
Resource-use permits, 305
Respiratory system, effects of air pollutants on human, 262
Response to toxic chemicals, **244**

Reuse, **11**, 285–87
  encouraging, 289
  individual's role in, 286f
  refillable containers as, 286
Reverse osmosis, water and, 182
Reversibility principle, environmental policy
    and, 310
Rhinoceros
  mutualism between birds and black,
    75f
  reduction in range of black, 135f
Risk, **237**–38
  perception of, 249–50
  probability, possibility and, 237
  of technology, 249
Risk analysis, 247–50
  comparative, 247
  estimating risks as, 247–48
  estimating risks of technology, 249
  improving, 250
  perception of risk and, 249–50
Risk assessment, **237**, 247
Risk communication, 247
Risk management, **237**, 247
Rivers and streams, 67–68
  channelization of, 177
  downhill flow of water in, 68f
  ecological services of, 180f
  pollution in, 188–89
  stresses on global systems of, 174, 175f
RNA (ribonucleic acid), 23
Rock, **50**
Rock cycle, 50f
Roosevelt, Theodore, 144
Rosenblatt, Roger, 200
Rowland, Sherwood, 279
Runoff, **67**
  reliable, 173
  surface, 44, 45, 67, 68f, 173

Sachs, Jeffrey D., 308
Salinity, as ecosystem limiting factor, **34**
Salinization of soils, 153, **154f**
  solutions for, 154f
Salt marsh, 65
Saltwater desalination, 182
Saltwater intrusions, 179f
Saltwater life zones, 65. *See also* Marine life
    zones
Sanctuary approach to protecting wild
    species, 144–46
Sanitary landfill, **290**, 291f
  advantages and disadvantages of, 292f
SARS (severe acute respiratory syndrome),
    240
Saudi Arabia, oil reserves in, 204, 205
Science, **20**–22
  accurate measurements in, 21
  consensus science, frontier science, and
    junk science, 21–22
  environmental, 1, 5, 22
  hypotheses in, 20, 21
  process of, 20f, 21
  sound science, 6, 21–22
  theories and laws of, 21
Scientific Certification Systems (SCS), 113
Scientific hypotheses, **20**
  testing of, 21
Scientific (natural) law, **21**
Scientific methods, **20**
Scientific theory, **20**, 21
Scientists
  consensus about climate change and
    global warming among, 268–69
  process followed by, 21f, 22

Sea levels, climate change and, 272f, 273
Seasonal wetlands, 68
Secondary air pollutants, **255**, 256f
Secondary consumers, **35**
Secondary ecological succession, **75**, 125
Secondary recycling, 287
Secondary sewage treatment, **196**, 197f
Second-growth forests, 110
Second law of thermodynamics, **27**–28
  in living systems, 28f
Secure hazardous waste landfills, 295, 296f
Sedimentary rock, **50**
Selective breeding, 58–59
Selective cutting, tree harvesting by, **111**, 112f
Sense of place, 321
Septic tank, 196f
Service-flow economy, shift to, 305–6
Sewage
  ecological treatment of, 197
  prevention of, 196–97
  primary and secondary treatment of, 196,
    197f
  reducing water pollution through treat-
    ment of, 196
Shale oil, **207**, 208f
Sharks, reasons for protecting, 72b
Sheep, logistic population growth of, on Tas-
    mania, 78f
Shellfish as food source, 161–62
Shelterbelts, 155f, **156**
Shiva, Vandana, 156
Sick-building syndrome, 260
Silent Spring (Carson), 165b
Simple carbohydrates, 23
Simple living, 321
Single-variable analysis, 21
Skin cancer, UV radiation and, 280
Smart growth concept and tools, 104–5, 106f
Smog
  industrial, 256
  photochemical, 256–57
Smoking, health hazards of, 237–38
Social problems in urban areas, 100–101
Soil, **41**–43
  erosion of (*see* Soil erosion)
  formation of, 41f
  impact of food production on, 152–54
  layers in mature, 42
  microorganisms in, 29
  pH of, 42
  profiles of, 41f, 42, 43f
  sequestering carbon dioxide in, 275f, 276
Soil conservation, **154**–56
  fertilizers and, 156
  solutions for soil-salinization problem,
    154f
  tillage methods for, 154, 155f
Soil erosion, **152**–54
  global, 153
  salinization, and waterlogging as, 153, 154f
  of topsoil, 152
  in United States, 153f
Soil fertility, erosion and loss of, 153
Soil horizons, 41f, 42
Soil profile, 41f, **42**, 43f
Soil texture, **42**
Solar capital, **6**. *See also* Solar energy
Solar cells, **226f**–27
Solar cookers, 225, 226f, 227
Solar energy, 201
  flow of energy to and from earth, 31f, 32f
  generating high-temperature heat and
    electricity with, 225
  heating buildings and water with, 224–25
  natural greenhouse effect and, 32f

  passive, and active, 224f, 225f
  solar cells and, 226–27
  sustainability principles and reliance on,
    16, 17f
Solar thermal plant, 225
Solid wastes, **283**
  burning, 289, 290f
  burying of, 289–92
  hazardous (*see* Hazardous waste)
  individual's role in reducing, 285f
  industrial, 283
  integrated management of, 284–85
  municipal, 283–84
  producing less, and reducing amount of,
    285
  production of, in U.S., 283–84
  recycling to reduce amount of, 287–89
  reuse as alternative to producing, 285–87
  sources of, 283f
Solutions
  acid deposition, 260f
  air pollution, 264f, 265f, 266f
  alternatives to cars, 103, 104f
  developing environmentally-sustainable
    economies, 81–83
  developing environmentally-sustainable
    societies, 321f
  establishing priorities for protecting biodi-
    versity, 127–28
  for global warming, 274–77
  groundwater protection, 192f
  for hazardous waste management, 293–95
  lead poisoning, 297f
  managing fisheries, 127f
  national parks management, 120
  pollution prevention versus cleanup, 11
  preventing and reducing water pollution,
    195–200
  preventing coastal water pollution, 195f
  principles of sustainability (*see* Sustainabil-
    ity, principles of)
  recycling municipal solid waste, 287
  reducing automobile use, 103
  reducing childhood death from hunger
    and malnutrition, 149
  reducing energy waste, 219f
  reducing incidence of infectious disease,
    242, 243f
  reducing poverty, 307b
  reducing risks of flooding, 177
  reducing threat of nonnative species,
    137–39
  reducing water waste, 183–85
  reuse to reduce solid waste, 286–87
  smart growth tools in urban areas as,
    104–6
  soil salinization, 154f
  sustainability principles and, 18f
  sustainable agriculture, 169f, 170
  sustainable forestry, 112, 113f, 116–17
  sustainable water use, 185–86
  wind turbines and wind farms, 228f
Soot in atmosphere, 271
Sound science, **6**, 21–22
Source separation, municipal solid wastes,
    287
Space heating, net energy ratios for, 203f
Special-interest groups, 310
Specialist species, **56**
  in tropical rain forest, 56f
Speciation, **56**–57
  crisis of, 132
Species, **28**
  commensalism among, 74–75
  competition among, 73–74

competitor, 79–80
ecological niches of, 55–56
endemic, 57, 136
evolution of, 56–57
extinction of, 57–58 (*see also* Extinction of species)
foundation, 72–73
generalist, 55b, 56
indicator, 71
keystone, 71–72
microbial, 28–29
native, 70
nonnative, 70 (*see also* Nonnative species)
opportunist, 79
parasitism and mutualism among, 74, 75f
pioneer, 75
predation among, 74
preservation of (*see* Ecosystem approach to sustaining biodiversity; Species approach to sustaining biodiversity)
reasons for preserving wild, 132–33
specialist, 56
Species approach to sustaining biodiversity, 129–47
causes of premature extinction of wild species, 134–41
ecosystem approach versus, 109f
goals, strategies, and tactics for, 109f
importance of wild species, 132–34
legal and economic approaches to protecting wild species, 141–43, 144b
reconciliation ecology and, 146b
sanctuary approach to protecting wild species, 144–46
species extinction and, 129–32 (*see also* Extinction of species)
Species-area relationship, 130
Species diversity, 36f, 37. *See also* Biodiversity
Squamous cell skin cancer, 280
S-shaped curve of population growth, 78
Stegner, Wallace, 123
Stewardship environmental worldview, **16, 318,** 319f
Stewart, Richard B., 276
Stratosphere, **30, 252**–53
ozone in, 252
ozone depletion in, 278–80
Streams. *See* Rivers and streams
Strip cropping, **155**f
Strip cutting, 112f
Strong, Maurice, 201
Student environmental groups, 316
Subatomic particles, 24
Subsidies. *See* Taxes, fees, and subsidies
Sudan, water resources, 174
Sulfur dioxide (SO_2) as air pollutant, 255t
emissions trading of, 263
Sun. *See* Solar energy
Superfund Act, U.S., 292
Superinsulated house, 223
Surface impoundment of hazardous waste, 295
Surface litter layer (O horizon), soil, 41f, 42
Surface runoff, 44, 45, 67, 68f, **173**
Surface water, **67, 172**
preventing and reducing pollution in, 195–98
zones in downhill flow of, 68f
Suspended particulate matter (SPM) as air pollutant, 255t
Sustainability, **5** 7
ecocity concept and, 105–7
environmental worldviews and, 15–18

four principles of, 16, 17f, 18f (*see also* Sustainability, principles of)
themes of path to, 5f
Sustainability, principles of, 16, 17f, 18f
agriculture and, 170
energy resources and, 219, 235
global warming and ozone depletion, 281
hazards and risks to human health, 250
reducing solid and hazardous waste, 298
transition to eco-economies and, 308, 309f
water resources and, 200
Sustainable agriculture, 154–56, 169–70
Sustainable energy future, 233–35
decentralized micropower system, 233, 234f
economics, politics, education, and, 233–35
energy policy and, 233
individual's role in, 235f
new vision for energy, 233
strategies for, 235f
Sustainable forestry, 112–13, 116–17
Sustainable living, 5–7, 320–23
becoming better environmental citizens as, 321–22
components of sustainability revolution and, 322–23
environmental literacy and, 320
learning from the earth, 320–21
living more simply as, 321
Sustainable water use, 185f, 186
Sustainable yield, **9**
Swamps, 68
Synfuels, 210, 211f
Synthetic natural gas (SNG, syngas), 210, 211f
Systems
characteristics of natural and human-dominated, 80f
unsustainable, 99, 100f

Tar sand, **206**–7
Taxes, fees, and subsidies, 304–5
alternative transportation and shifts in, 103
for food production, 163
green, 304
improving environmental quality using, 302, 304–5
reducing threat of climate change using, 276
Technological optimists, 15
Technology
estimating risks of, 249
transfer of, to reduce threat of climate change, 276
Tectonic plates, 55
Temperature
climate and average, 61, 62f, 253, 254 (*see also* Climate change)
of earth, as favorable for living organisms, 32–33
organism's range of tolerance for, 33, 34f
Temperature inversions, **257**
Teratogens, **243**
Terracing, **155**f
Terrestrial life zones. *See also* Biome(s)
ecosystem approach to preserving biodiversity in, 109–25
ecological succession in, 75–76, 125
effects of acid deposition on, 259
net primary productivity of, 40f
Terrorism
bioterrorism, 238
nuclear technology and potential, 213, 215
Tertiary (third) consumers, **35**
Test group, 246

Theory
of evolution, **52**
scientific, 21
Thermodynamics, laws of, 27–28
Thoreau, Henry David, 18
Threatened (vulnerable) species, **130,** 131f
Three Mile Island nuclear accident, 213
Threshold dose-response model, 246f
Tiger, reduced range of Indian, 135f
Tiger salamanders, 56
Tigris-Euphrates River basin, 175f
Timber
certifying sustainably-grown, 113
harvesting methods, 111, 112f
Todd, John, 171, 197
Tool libraries, 286
Topsoil
A horizon, 41f, 42
erosion of, 152–53
Total fertility rate (TFR), **85,** 88t
in United States, 86f
Toxic chemicals, **243**
as air pollutants, 255t
case reports on, 245
dose and lethal dose of, 244, 245t
effects of, on human body, 243–44
estimating toxicity of, 244–45
in homes, 293f
laboratory experiments on, 245–46
persistent organic pollutants (POPs), 247
prevention and precautionary principle applied to, 247
reasons for lack of knowledge about, 246–47
response to, 243
trace levels of, 244
types of, 243
Toxicity, **244**
dose, response, lethal dose ratings, 244, 245t, 246f
estimating, 244–45
studies and experiments to determine, 245–46
of trace levels of chemicals, 244
Toxicological studies, 245
Toxicology, **244**–47
assessment of trace levels, 244
case reports and studies on, 245
dose, response, and lethal doses, 244, 245t, 246f
estimating toxicity of chemicals, 244–45
laboratory experiments for estimating toxicity, 245–46
prevention and precautionary principle applied in, 247
Toxic wastes, **283**
Trace levels of toxic chemicals, 244
Tradable pollution, 263, 305
Trade, international illegal, of wild species, 140–41
Trade-offs (advantages/disadvantages), 4, 6
of alternative transportation, 104f, 105f
of animal feedlots, 161f
of aquaculture, 162f
of clear-cutting in forests, 112f
of coal as fuel, 209, 210, 211f
of conventional crude oil as energy source, 205, 206f
of dams and reservoirs, 179–80
of economic development, 8f
of ethanol and methanol fuels, 230f
of genetically modified crops and foods, 160f
of geothermal energy, 231f
of groundwater withdrawal, 178, 179f

Trade-offs (cont'd)
of hazardous waste disposal, 295f
of hydrogen as energy source, 232f
of large-scale hydropower, 227f
of motor vehicle use, 102–3
of natural gas as fuel, 208, 209f
of nuclear fuel cycle, 215f, 216
of oil shale and oil sand fuels, 208f
of recycling, 288f, 289
of sanitary landfills, 292f
of solar cells, 226f
of solar energy, 225f
of solid/biomass incineration, 229f, 290f
of synthetic fuels, 210, 211f
of wind power, 229f
Traditional agriculture, subsistence and intensive, **152**
Tragedy of the commons, 9–10
Transgenic organisms, **59**
Transmissible diseases, **238**–39. See also Infectious disease
Transportation
motor vehicles, 221–22
net energy ratios for, 203f
saving energy in, 220–22
in urban areas, 102–3
Treaties
banning of ozone-depleting chemicals, 281
banning of persistent organic pollutions, 298
proposed climate, 276–77
protecting endangered species with international, 141
Tree farms, **110**
Tree plantations, **110**
Trees. See also Forest(s)
climate regulation and role of, 109
harvesting methods, 111, 112f
Trophic levels in ecosystems, 37f, **39**
Tropical forests
deforestation of, 115–16, 135
neem tree for reforesting, 117
pharmaceutical plants from, 133f
reducing deforestation and degradation of, 116, 117f
reforestation in Kenya, 118b
soils of, 43f
Troposphere, **30, 252**
effects of warmer, 271–74
greenhouse gases in, 32f, 254
greenhouse gases and warming of, 266–68, 269f
Tuberculosis, 240
Turkey, water resources in, 175
Turner, Ray, 281b

Ultraviolet (UV) radiation
effects of, on amphibians, 70–71
ozone depletion and human health problems caused by, 279, 280
reducing exposure to, 280f
stratospheric ozone as protection from, 252–53
Uncertainty in making measurements, 21
Undernutrition, chronic, 148–49
Underpricing of water, 182
United Nations Children's Fund (UNICEF), 149
United Nations Conference on the Human Environment, 317
United Nations Development Programme (UNDP), 317
United Nations Environment Programme (UNEP), 108, 126, 166, 273, 292, 317

United States
baby boom generation in, 89, 90f
biodiversity hot spots in, 142f
Bowash megalopolis in, 99f
changes in life quality in years 1900-2000, 86f
Chesapeake Bay, pollution in, 193, 194f
commercial energy flow in, 219f, 233f
commercial energy by source in, 202f
consumption of resources in, 15
death rates in, 87
deaths from air pollution in, 262f
deaths from tobacco use and smoking in, 238f
demographic indicators for Brazil, Nigeria, and, 90f
ecological footprint of, 10f
environmental history of, S2–S7
environmental law (see Environmental law)
environmental policy in, 311–17
environmental quality and, 14
fertility and birth rates in, 86
forests and forest management in, 113–15
freshwater resources in, 173f, 174
grassroots environmental action in, 296–97
Great Lakes pollution and cleanup in, 190
greenhouse gas emissions by, 266
groundwater withdrawal in, 178, 179f
high-level radioactive wastes in, 217b
immigration to, 88
land use in, 108
motor vehicle use in, 102
oil reserves and oil use in, 205
overnutrition in, 149
public lands in (see Public lands in United States)
soil erosion in, 212–13, 215, 217b
solid waste production in, 283–84
student environmental action in, 316
urbanization in, 97–98
water-deficit regions in, 174f, 181f
water waste in, 182–83
wind energy in, 228
U.S. Department of the Interior, 174
U.S. Fish and Wildlife Service (USFWS), 142
U.S. Forest Service (USFS), 312
Unsustainable systems, **99**
in urban areas, 100f
Urban areas
ecocity Curitiba, Brazil, 106–7
environmental and resource problems in, 99–102
food production in, 163
global, 97f
growth of, 96–99
increasing sustainability and quality of life in, 104–7
Mexico City, Mexico, 101–2
poverty in, 100, 101
transportation and development of, 102–3
in United States, 97–98
water pollution caused by, 193f
Urban heat island, 100
Urbanization, 87, 96–99
benefits of, 99
disadvantages of, 99–101
growth of megacities, 96, 97f, 98f
poverty and, 101–2
undesirable effects of urban sprawl, 98f, 99
in United States, 97–98
Urban sprawl, natural capital degradation and problems associated with, 98f, 99, 103, 193f

User-pays approach, 103

Van Doren, Mark, 1
Variables in scientific experiments, 21
Vegetation, flooding caused by removal of, 176. See also Plant(s)
Viral disease, 240–41
emerging, 241
Vitamins, human nutrition and, 149
Volatile organic compounds (VOCs) as air pollutant, 255t
Volcanic eruptions, 55
Voluntary simplicity, 321
Vulnerable species. See Threatened (vulnerable) species

Ward, Barbara, 5
Waste, 283–99
achieving society low in, 296–98
burning of, 289, 290f
burying, 289–92
of energy, 218–21
hazardous, 292–96
integrated management of, 284–85
oxygen-demanding, as water pollutant, 188f, 193f
production of, in United States, 283–84
radioactive, 216, 217b
recycling and reduction of, 287–89
reuse and reduction of, 285–87
solid, 283–84
sources of, 283f
types of, 283
of water resources, 182–86
Waste-to-energy incinerator, 289, 290f
Water. See also Water pollution; Water resources
desalination of seawater, 182
drinking (see Drinking water)
fresh (see Freshwater)
groundwater, 171–73 (see also Groundwater)
hydrologic cycle, 44, 45, 171
in hydrosphere, 30
importance of, 171
infiltration of, in soil, 42
producing electricity from moving, 227
surface, 67, 68f, 171–73
Water conservation, 182–86
Water (hydrologic) cycle, **44,** 45f, 171
effects of human activities on, 45
Water diversion projects, 180, 181f
Water-filled pools (dry casks) in nuclear reactors, 212
Water hot spots, U.S., 174f
Waterlogging of soils, 153–**54**f
Water pollutants
arsenic as, 191–92
degradable and non-degradable wastes as, 190–91
major categories of, 187t
MTBE (methyl tertiary butyl ether) as, 191
oil as, 194–95
oxygen-demanding wastes as, 188f, 193f
Water pollution, **186**–200
drinking water quality and, 198–200
in freshwater streams, lakes, and aquifers, 188–92
individual role in reducing, 200f
in oceans, 192–95
point and nonpoint sources of, 186–87
pollutants (see Water pollutants)
preventing and reducing surface, 195–98
soil erosion as cause of, 153

Water Quality Act of 1987, 195
Water resources, 171–86
    conflict over, in Middle East, 174–75
    drinking water, 198–200
    effects of global warming on, 272f
    excessive amounts of (flooding),
        175–76
    global population and supplies of,
        171f
    increasing supplies of, 178–82
    polluted (see Water pollution)
    problems with, in urban areas, 99–100
    reducing waste of, 182–86
    shortages of, 174–75
Watershed, **67,** 68f, **173**
Water table, **172**
Water vapor, 30
Wavelengths, 26
Wealth, distribution of global, 306, 307f. See
        also Poverty
Weather, **253**
    atmospheric conditions and, 253–54
    global warming and extremes of, 272f
West Nile virus, 240
Wet acid deposition, 257
Wetlands
    coastal, 65
    draining of, 176
    flooding caused by destruction of, 177
    inland, 68
    seasonal, 68

Wet steam geothermal energy, 231
Wiener, Jonathan B., 276
Wilderness, **122**
    protection of, 122–23
    in United States, 123
Wilderness Act of 1964, U.S., 123
Wilderness Society, U.S., 122
Wildlife
    bats as, 133–34
    effects of climate change on, 139–40
    effects of habitat destruction,
        degradation, and fragmentation on,
        134–36
    effects of ozone depletion on, 280f
    effects of pesticides on, 140f
    extinction of, 134–41 (see also Extinction of
        species)
    intrinsic value of, 134
    poaching and smuggling of, 140–41
    protecting, 141–46
    reasons for preserving, 132–33
    reintroduction of gray wolf, 119b
    U.S. refuge system for, 144
Wildlife refuges, 144, 205
Wildlife studies on toxic chemicals, 245
Wilson, Edward O., 58, 72, 130, 132, 146
    priorities for protecting biodiversity by,
        108, 127
Wind, generation of electricity with,
        227–29
Windbreaks, 155f, **156**

Wind energy, 227–28
    advantages/disadvantages of, 229f
    cars powered by, 221
Wind farms, 228f
Wind turbines, 228f
Wolf, reintroduced populations of gray, 119b
Women
    birth rates in, and opportunities for, 87
    reducing birth rates by empowering, 94
World Bank, 317
World Conservation Union (IUCN), 130, 317
World Environmental Organization (WEO),
        proposed, 318
World Health Organization (WHO), 13, 166,
        187, 237, 239f, 274, 317
World Resources Institute (WRI), 110
World Wildlife Fund, 110
Wright, Frank Lloyd, 4

Xeriscaping, 184
Xerox Corporation, 306

Yellowstone National Park, reintroduction of
        gray wolf into, 119b
Yucca Mountain nuclear waste storage facil-
        ity, 217b

Zone of saturation, **172**
Zoning rules, protection of marine biodiver-
        sity using, 126
Zoos, protecting endangered species in, 145